Internet and Computers for Agriculture

Internet and Computers for Agriculture

Editor

Dimitre Dimitrov

MDPI • Basel • Beijing • Wuhan • Barcelona • Belgrade • Manchester • Tokyo • Cluj • Tianjin

Editor
Dimitre Dimitrov
Department of University
Transfer, NorQuest College,
Edmonton, AB, Canada

Editorial Office
MDPI
St. Alban-Anlage 66
4052 Basel, Switzerland

This is a reprint of articles from the Special Issue published online in the open access journal *Agriculture* (ISSN 2077-0472) (available at: https://www.mdpi.com/journal/agriculture/special_issues/Internet_Computers_Agriculture).

For citation purposes, cite each article independently as indicated on the article page online and as indicated below:

LastName, A.A.; LastName, B.B.; LastName, C.C. Article Title. *Journal Name* **Year**, *Volume Number*, Page Range.

ISBN 978-3-0365-6630-6 (Hbk)
ISBN 978-3-0365-6631-3 (PDF)

Contents

About the Editor

Dimitre Dimitrov

My name is Dimitre Dikov Dimitrov, and I currently teach at NorQuest College (Canada). I received my PhD from the University of Alberta (Canada). My broad scientific specialization is in the interdisciplinary area of the computational mathematical modelling of physical, chemical and biological processes in natural and agricultural ecological systems through coupled hydrology–soil–vegetation–atmosphere schemes, driven by meteorological and environmental data, and constrained by site observations conducted by fellow researchers and collaborators, and large datasets obtained from big data collections on the Internet, websites with application programming interface end points, remote sensing, and/or GIS. Within the scope of my research, I developed an innovative state-of-the art mathematical and computational internet platform, DIMONA, for the modelling and study of various ecosystem processes at site- to large- spatial scales, and along various regional ecological gradients. My passion for mathematical and environmental sciences has motivated my dedication to developing and promoting various computational and smart solutions to assist in solving the most pressing issues at the planetary scale today.

agriculture
MDPI

Editorial

Internet and Computers for Agriculture

Dimitre D. Dimitrov

Department of University Transfer, Faculty of Arts & Sciences, NorQuest College, Edmonton, AB T5J 1L6, Canada; dimitre@ualberta.ca

The Special Issue "Internet and Computers for Agriculture" reflects the rapidly growing need for new information and communication technology (ICT) involvement in global change agriculture nowadays. The aim was to cover the recent and current progress in various aspects of ICT applications in precision agriculture, such as web applications and mobile apps, Internet of Things (IoT) platforms and smart devices, cloud technologies, artificial intelligence (AI), machine learning (ML), and deep learning (DL)-based solutions via neural networks (NNs) and convolutional neural networks (CNNs) for detection, classification, computer/machine vision, and language processing purposes, as well as scientific modeling in agriculture and natural ecosystems. This Special Issue brought together twenty peer-reviewed articles, including eighteen original research articles, [1–18] (chronologically presented), one case report article, [19], and one review article, [20], as summarized in Table 1 and described below.

In article [1], a novel CNN-based DL method for grape variety identification was proposed based on the canonical correlation analysis (CCA) applied to fuse selected deep features from various CNNs, i.e., AlexNet, GoogLeNet, ResNet18, ResNet50, and ResNet101, and a multi-class support vector machine (SVM) classifier trained in these fused features. To test the proposed method, grape images from the open-source Embrapa Wine Grape Instance Segmentation Dataset (WGISD) were initially resized to meet the CNN requirements and then used for selected deep feature extraction. In general, the fused deep feature approach outperformed the single deep feature approach, as indicated by the best performance of the former (AlexNet and ResNet50) with an F_1 score of 96.9% compared to the best performance of the latter (ResNet101) with an F_1 score of 88.2%. The proposed method can be applied in developing the computer/machine vision of smart machinery for a more targeted and accurate identification of grape varieties, thus improving grape yield.

In article [2], an adversarial contextual embeddings-based model for agricultural diseases and pests (ACE-ADP) was proposed to be implemented as a web application for named entity recognition in Chinese agricultural diseases and pests domains (CNER-ADP). While the adversarial training enhanced the robustness of identifying rare named entities, the ACE-ADP dealt with the polysemous issues (multiple meanings of the same word) and quality of text representation by fine-tuning the bidirectional encoder representations for transformers (BERT) ML framework, a neural-network-based technique originally developed by Google for natural language processing (NLP). Thus, the multi-vector BERT-based ACE-ADP performed better by 4.23%, reaching an F_1 score of 98.31%, compared to the single-vector baseline word2vec-based BiLSTM-CRF model when both applied to the Chinese named entity recognition dataset for agricultural diseases and pests (AgCNER).

In article [3], a novel, end-to-end, AI-powered, IoT-based platform and agro-weather station were introduced into smart farming. The multi-agent agile and containerized system consisted of low-cost hardware and software components, organized in five layers, for continuous real-time monitoring and AI-based forecasting of various meteorological factors, e.g., air temperatures (minimum, maximum), humidity, pressure, precipitation, wind speed, and dew point. These meteorological factors were surveyed by heterogeneous nodes, located at the base station and at various distances from it, thereby constituting the perception back-end layer of various sensors. The collected data were transmitted to the

Citation: Dimitrov, D.D. Internet and Computers for Agriculture. *Agriculture* **2023**, *13*, 155. https://doi.org/10.3390/agriculture13010155

Received: 21 December 2022
Revised: 2 January 2023
Accepted: 5 January 2023
Published: 7 January 2023

local webserver (Apache) for pre-processing and to the cloud server for backup using the transmission back-end layer, tasked with ensuring the wired and wireless communication of heterogeneous protocols, e.g., MQTT, NRF24L01, HTTP, and WiFi. To spare resources in the local webserver, Cronjob was run to send series of measurements from the database to Google Collaboratory free computing resources (Colab), where the AI-based predictions of meteorological factors were conducted using ML regression methods on the TensorFlow framework, using the Keras neural network library, optimizer Adam, and the mean squared error (MSE) as a loss function. Upon completion of model training using Keras, Colab would generate a TensorFlow Lite (tflite) outcome to be sent back to the base station and be used by the local webserver for AI modeling embedded within the platform website. The middleware back-end layer was designed to orchestrate the performance of all agents, including the virtual private network (VPN) and the database. The two front-end layers with user-friendly web-based graphical user interfaces (GUIs) formed the presentation layer for reporting weather observations and AI-based predictions to different users based on their personal profiles, and the purpose of the management layer was for operation and maintenance. The agro-weather station was successfully tested in Casablanca, Morocco, with mean absolute scale error (MASE), root mean square deviation (RMSE), and Willmott's index of agreement stabilizing after the fifth epoch of AI modeling at 0.0012, 0.034, and 0.987, respectively.

Table 1. Summary of presented research.

Article	Implemented Application Types	Agricultural Activities	Outcomes
[1]	AI application (DL and CNN)	Grape variety identification	Computer/machine vision
[2]	Web application, AI application (NN)	Disease and pest name recognition	Natural language processing system
[3]	Web application (cloud-based), IoT platform, AI application (ML)	Weather measurements and prediction	Meteorological station
[4]	AI application (DL and CNN)	Maize variety seed recognition and classification	Computer/machine vision
[5]	Modeling software	Quantifying uncertainties in modeling hydrology	Decision support system
[6]	Modeling software	Product distribution	Transport optimization system
[7]	Mobile app, modeling software	Farm management	Information system
[8]	AI application (DL and CNN)	Grape disease recognition and classification	Computer/machine vision
[9]	AI application (DL and CNN)	Fishery product price prediction	Forecasting system
[10]	Web application (cloud-based), mobile app, IoT platform	Collecting soil, air, and light properties	Smart monitoring system
[11]	Web application (cloud-based), modeling software	Predicting olive crop yield	Management and decision making system
[12]	AI application (DL and CNN)	Grape detection	Computer/machine vision
[13]	AI application (DL and CNN)	Buffalo breed recognition and classification	Computer/machine vision
[14]	AI application (DL and CNN)	Tea picking	Computer/machine vision

Table 1. *Cont.*

Article	Implemented Application Types	Agricultural Activities	Outcomes
[15]	Modeling software	Agricultural e-commerce	Behavior system
[16]	AI application (DL and CNN)	Monitoring duck flocks	Real-time detection system
[17]	IoT application	Measuring environmental, Plant, and soil water status	Monitoring system
[18]	Modeling software	Planning uncertainties in horticulture market	Decision-making system
[19]	Modeling software	Selecting leisure agricultural parks	Intelligence approach
[20]	Review	Digital technologies in agriculture	Summary

In article [4], the open-source software framework PyTorch was used to build upon the existing ResNet CNN [21] and to create a P-ResNet network with 17,960,232 parameters, optimizer Adam, batch normalization between convolutions, and rectified linear unit (ReLU) during training. P-ResNet models were developed using PyCharm Integrated Development Environment (IDE) and were trained for classification using a server with NVIDIA GeForce GTX 1660 SUPER GPU and 16 GB GDDR4 on-board memory. The proposed DL network was intended to be used in machine/computer vision tasks, particularly in the classification of various seeds. Thus, P-ResNet was applied to the classification of seeds of five main maize varieties in China, i.e., BaoQiu, ShanCu, XinNuo, LiaoGe, and KouXian, by using 6464 RGB images for training and 1616 ones for validation. According to classification accuracy, P-Resnet outperformed well-known DL networks, such as AlexNet, VGGNet, GoogLeNet, MobileNet, and DenseNet, by several percentage points.

In article [5], a multi-model hydrological framework decision support system (DSS) was proposed to deal with water security modeling in the context of environmental sustainability and climate resilience. As water supply is critical to life on Earth and soil water contents are key controls in many biogeochemical processes in natural and modeled ecosystems [22], the DSS was applied to quantify the uncertainty in inputs of various hydrological models in order to improve their climate resilience. The water security modeling was coupled to food security in different model development scenarios. As a result, a four-dimensional dynamic space mapping source, resource availability, infrastructure, and economic options were suggested to capture the climate resilience phenomenon. The outcomes of the DSS can be made available to farmers to help with sustainable food production. The proposed DSS was tested for four catchments in Australia.

In article [6], a model for timely management of the distribution of various agricultural products was introduced to enhance the benefits and satisfaction that both agriculture producers and consumers experience. The proposed model was based on solving the time-dependent, split delivery green vehicle routing problem with multiple time windows (TDSDGVRPMTW) for both economic cost and customer satisfaction purposes, and thus could be seen as a modern version of the transportation math problem in contemporary agriculture. The objective to minimize the sum of the economic cost and maximize average customer satisfaction was achieved by optimizing time-varying vehicle speeds, fuel consumption, and carbon emissions in multiple time windows. Applying the model with real data from China suggested reduced total distribution costs, balanced energy conservation, and improved customer satisfaction.

In article [7], a mobile app on an Android application interface platform was introduced to farm management information system (FMIS) services to assist farmers in managing their farms. To reduce the price of and enhance the access to FMIS services, a new conceptual FMIS model for farm efficiency was proposed based on identifying commodity and research areas, and performing information needs assessments. The new model consisted of five layers for information needs, data quality assessment, data extraction, split, match and merge (SMM) processes, and presentation. The new FMIS model was used to address the needs of smallholder chili farmers in Indonesia and outlined areas for improvement in FMIS services.

In article [8], a novel lightweight CNN GrapeNet was introduced to deal with inherent difficulties in identifying crop diseases at different stages due to their wide gamut of symptoms and various plant tissues and color changes. GrapeNet was designed as a modern deep network of residual blocks and convolutional block attention modules (CBAMs) for extracting rich features and key disease information. Special residual feature fusion blocks (RFFBs) were introduced to achieve feature fusion at different depths, the article are dealing with vanishing gradient issues of ultra-deep networks [23]. Identification accuracy of GrapeNet outperformed other deep networks, such as ResNet34, DenseNet121, MobileNetV2, MobileNetV3_large., by ~1.5 to 4 percentage points, while the training time of GrapeNet decreased due to a reduced number of parameters. GrapeNet was tested for grape leaves of the AI challenger 2018 dataset.

In article [9], a recurrent neural network (RNN)-based long short-term memory (LSTM) model was coupled with a novel adaptive signal decomposition method, called variational modal decomposition (VMD), and a new improved bald eagle search algorithm (IBES) to propose the innovative fishery product price forecasting model VMD-IBES-LSTM, capable of dealing with time series data efficiently. Compared to other ML forecasting models, VMD-IBES-LSTM showed high prediction accuracy and better explained the seasonality and trends of changes in China's aquatic product consumer price index. Thus, VMD-IBES-LSTM was shown to be an effective tool for addressing management and decision-making tasks related to predicting the aquatic product consumer prices in China.

In article [10], a cloud-based web application interconnected to an IoT smart system was introduced to address the needs of rural farmers in Pakistan in an attempt to overcome the illiteracy-related absence of proactive decision making in all phases of crop production. The smart system was connected to accessible devices and sensors for real-time capturing of soil moisture, temperature, pH, light intensity, and air humidity. The system was designed to help farmers understand environmental factors related to soil fertility, suitable crop cultivation, automated irrigation, harvest schedule, pest and weed control, crop diseases, and fertilizer usage. The system was upgraded to a mobile app for bilingual usage, i.e., in 'Urdu' and 'English', and was equipped with visual, audio, and voice components as well as iconic and textual menus designed for farmers of various literary levels.

In article [11], predictive software was proposed for management and decision-making purposes in profitability and the economic balance of agricultural farms. The software consisted of a cloud-based web application with a nested user-friendly model for predicting crop yields based on different ML regression algorithms, such as the generalized linear model (GLM), and the Gaussian and Linear kernel support vector machine (SVM). As part of the training, the model was fed more than 20 spatio-temporal meteorological parameters and data for the yields of eight consecutive years. The proposed software performed well in the early prediction of crop yield with absolute errors being less than 20%. The results were crucial for decision making related to tillage investments and crop marketing. The web application was tested on an olive orchard in Spain.

In article [12], a novel method for grape detection was proposed related to computer/machine vision. The method was based on the lightweight network Uniformer, capable of capturing long-range dependencies while improving feature extraction, and the bi-directional path aggregation network (BiPANet), capable of fusing low- and high-resolution feature maps for optimizing semantic and detailed information. The reposition non-maximum suppression (R-NMS) algorithm improved the localization accuracy, and the novel cross-layer feature enhancement strategy in BiPANet resulted in a significant reduction in the number of parameters and computational complexity. The novel method for grape detection outperformed other CNN-based algorithms for computer vision, such as YOLOx, YOLOv4, YOLOv3, Faster R-CNN, SSD, RetinaNet, and the mAP. The proposed method was tested on grape datasets from China.

In article [13], a self-activated multilayered CNN, consisting of five blocks of convolution, batch normalization, ReLU activation function, and max pooling, was proposed as a computer-vision-based recognition framework to identify different buffalo breeds. Particularly, the Nili-Ravi breed, one of the best worldwide for milk and meat production, was successfully identified from other breeds in an attempt to satisfy the great demand for breed selection and breed production. All seven of the classifiers that were tested and compared for breed identification, i.e., Fine-KNN, Med-KNN, Coarse-KNN, LP-Boost, Total-Boost, Bag-Ensemble, and the support vector machine (SVM), performed well and managed to recognize and classify the Nili-Ravi breed from the Khundi breed and a miscellaneous class of other buffalo breeds. The accuracy of identification reached 93% for the CNN performance with the SVM classifier and exceeded 85% for the other classifiers. The CNN framework was tested in Pakistan.

In article [14], a lightweight CNN, named MC-DM (Multi-Class DeepLabV3+), was proposed as a computer vision approach for tea sprout segmentation and picking point localization, based on improved Mobile Networks Vision 2 (MobileNetV2) with an inverted residual structure [24]. The MC-DM architecture allowed for a reduced number of parameters and calculations. In addition, an image dataset of high-quality tea sprout picking points was built to train and test the MC-DM network. The atrous spatial pyramid pooling module in MC-DM acted to obtain denser pixel sampling for the purpose of enhancing the accuracy of picking point identification, which reached 82.52%, 90.07%, and 84.78% for a single bud, one bud with one leaf, and one bud with two leaves, respectively. The MC-DM has been proposed and tested as an effective method for fast segmentation and visual localization for automated machine picking of tea sprouts in China.

In article [15], e-commerce interest linkage mechanisms were studied using the theory of planned behavior and the evolutionary game model involving the causal relationship between farmers' characteristics, experiences, cognition, behaviors, and willingness, and government policies. The influence of government policies on farmers' cognition, participation, and behaviors surrounding e-commerce interest linkage mechanisms were studied using the structural equation model. The results showed that the basic characteristics and experiences of farmers affected their cognition surrounding e-commerce interest linkage mechanisms, their willingness to participate, and the way in which they behave in e-commerce activities. While government policies had a positive effect on farmers' cognition surrounding e-commerce, it was found that they did not directly stimulate farmers to participate. Despite this, government policies and farmers' basic characteristics interacted and acted together when it came to willingness to participate and the behavior of farmers in e-commerce. The proposed methodology was tested in China.

In article [16], a CNN-based DL algorithm was proposed for real-time monitoring of dense hemp duck flocks as an alternative to manual duck counting in the intelligent farming industry. Particularly, the authors applied a modified YOLOv7 DL algorithm for the recognition and detection of moving objects in real time. The YOLOv7 algorithm has been further improved by implementing a convolutional block attention module (CBAM) for feature extraction, which can perform attention operations in the channel and spatial dimensions. A large-scale image dataset consisting of 1500 hemp ducks was introduced for the purposes of full-body frame labeling and head-only frame labeling. The results showed that CBAM-YOLOv7 had outstanding precision. The comparison between the two labeling methods demonstrated that the head-only labeling method resulted in a loss of feature information, while the full-body frame labeling method appeared to be better suited to detection in real time. The proposed algorithm was tested in China.

In article [17], the LoRaWAN point-to-multipoint networking protocol was used for implementing an IoT application of sensors for inexpensive and continuous monitoring of environmental, plant, and soil water status in a vineyard. Results showed that the IoT system communicated data continuously and without loss. LoRaWAN was already known as an alternative with reduced cost and superior range compared to WiFi and Bluetooth. Its importance in IoT was justified by its applicability to resource management in a time of global change, especially in remote, rural areas where cellular networks have little coverage and 5G networks of prohibitive costs still lack infrastructure. The IoT system was tested in Portugal.

In article [18], MAXQDA software for qualitative and mixed-methods data was used to investigate the planning uncertainties and to provide data-driven support in the decision-making process along the supply chain of horticultural companies for ornamental plants, perennials, and cut flowers. Real-life planning issues were explored by interviewing experts and the management of typical companies operating in the market. The results showed that tactical planning domains of material/product requirement and production and demand planning are especially critical for the market. An outstanding need emerged for practically developing relevant decision support systems, in addition to some existing ones of a limited extent that were not fully compatible with marketing requirements in the horticultural sector. The methodology was tested in Germany.

In article [19], a case report was presented proposing a fuzzy collaborative intelligence (FCI) approach for selecting leisure agricultural parks during times of great restrictions, such as the recent COVID-19 pandemic. The novelty of the proposed approach was in combining the asymmetrically calibrated fuzzy geometric mean (acFGM), fuzzy weighted intersection (FWI), and fuzzy Vise Kriterijumska Optimizacija I Kompromisno Resenje (fuzzy VIKOR) function. The approach was tested for Taiwan and showed that agricultural parks were among the favorite locations for traveling for leisure during the COVID-19 pandemic.

In article [20], a review was conducted on the impact of new information and communication technologies (ICT) on sustainable food systems (SFSs) and their transformation in the context of global food security and nutrition. The main focus was on digital agriculture technologies involving IoT, AI, and ML, such as drones, robots, autonomous vehicles, and advanced materials, as well as various gene technology, such as biofortified crops, genome-wide selection, and genome editing. Eight action initiatives were suggested, which coupled to appropriate incentives, regulations, and permits, and would be expected to critically influence adoption and usage of modern technologies for promoting various SFS types.

All of the above original research demonstrated the potential for worldwide application in corresponding or similar domains. The contributions of this Special Issue may be seen in the background of a rapidly growing human population with needs surrounding sustainable and secure food production, water management, and reduced GHG emissions, which clarify the need for smart agriculture solutions as an imminent priority on a planetary scale.

Conflicts of Interest: The author declares no conflict of interest.

References

1. Peng, Y.; Zhao, S.; Liu, J. Fused Deep Features-Based Grape Varieties Identification Using Support Vector Machine. *Agriculture* **2021**, *11*, 869. [CrossRef]
2. Guo, X.; Hao, X.; Tang, Z.; Diao, L.; Bai, Z.; Lu, S.; Li, L. ACE-ADP: Adversarial Contextual Embeddings Based Named Entity Recognition for Agricultural Diseases and Pests. *Agriculture* **2021**, *11*, 912. [CrossRef]
3. Faid, A.; Sadik, M.; Sabir, E. An Agile AI and IoT-Augmented Smart Farming: A Cost-Effective Cognitive Weather Station. *Agriculture* **2022**, *12*, 35. [CrossRef]
4. Xu, P.; Tan, Q.; Zhang, Y.; Zha, X.; Yang, S.; Yang, R. Research on Maize Seed Classification and Recognition Based on Machine Vision and Deep Learning. *Agriculture* **2022**, *12*, 232. [CrossRef]
5. Abu Shoaib, S.; Rahman, M.M.; Shalabi, F.I.; Alshayeb, A.F.; Shatnawi, Z.N. Climate Resilience and Environmental Sustainability: How to Integrate Dynamic Dimensions of Water Security Modeling. *Agriculture* **2022**, *12*, 303. [CrossRef]
6. Wu, D.; Wu, C. Research on the Time-Dependent Split Delivery Green Vehicle Routing Problem for Fresh Agricultural Products with Multiple Time Windows. *Agriculture* **2022**, *12*, 793. [CrossRef]
7. Henriyadi, H.; Esichaikul, V.; Anutariya, C. A Conceptual Model for Development of Small Farm Management Information System: A Case of Indonesian Smallholder Chili Farmers. *Agriculture* **2022**, *12*, 866. [CrossRef]
8. Lin, J.; Chen, X.; Pan, R.; Cao, T.; Cai, J.; Chen, Y.; Peng, X.; Cernava, T.; Zhang, X. GrapeNet: A Lightweight Convolutional Neural Network Model for Identification of Grape Leaf Diseases. *Agriculture* **2022**, *12*, 887. [CrossRef]

9. Wu, J.; Hu, Y.; Wu, D.; Yang, Z. An Aquatic Product Price Forecast Model Using VMD-IBES-LSTM Hybrid Approach. *Agriculture* **2022**, *12*, 1185. [CrossRef]
10. Cheema, S.M.; Ali, M.; Pires, I.M.; Gonçalves, N.J.; Naqvi, M.H.; Hassan, M. IoAT Enabled Smart Farming: Urdu Language-Based Solution for Low-Literate Farmers. *Agriculture* **2022**, *12*, 1277. [CrossRef]
11. Cubillas, J.J.; Ramos, M.I.; Jurado, J.M.; Feito, F.R. A Machine Learning Model for Early Prediction of Crop Yield, Nested in a Web Application in the Cloud: A Case Study in an Olive Grove in Southern Spain. *Agriculture* **2022**, *12*, 1345. [CrossRef]
12. Su, S.; Chen, R.; Fang, X.; Zhu, Y.; Zhang, T.; Xu, Z. A Novel Lightweight Grape Detection Method. *Agriculture* **2022**, *12*, 1364. [CrossRef]
13. Pan, Y.; Jin, H.; Gao, J.; Rauf, H.T. Identification of Buffalo Breeds Using Self-Activated-Based Improved Convolutional Neural Networks. *Agriculture* **2022**, *12*, 1386. [CrossRef]
14. Yan, C.; Chen, Z.; Li, Z.; Liu, R.; Li, Y.; Xiao, H.; Lu, P.; Xie, B. Tea Sprout Picking Point Identification Based on Improved DeepLabV3+. *Agriculture* **2022**, *12*, 1594. [CrossRef]
15. Wei, X.; Ruan, J. Influences of Government Policies and Farmers' Cognition on Farmers' Participation Willingness and Behaviors in E-Commerce Interest Linkage Mechanisms during Farmer–Enterprise Games. *Agriculture* **2022**, *12*, 1625. [CrossRef]
16. Jiang, K.; Xie, T.; Yan, R.; Wen, X.; Li, D.; Jiang, H.; Jiang, N.; Feng, L.; Duan, X.; Wang, J. An Attention Mechanism-Improved YOLOv7 Object Detection Algorithm for Hemp Duck Count Estimation. *Agriculture* **2022**, *12*, 1659. [CrossRef]
17. Valente, A.; Costa, C.; Pereira, L.; Soares, B.; Lima, J.; Soares, S. A LoRaWAN IoT System for Smart Agriculture for Vine Water Status Determination. *Agriculture* **2022**, *12*, 1695. [CrossRef]
18. Drechsler, M.; Holzapfel, A. Decision Support in Horticultural Supply Chains: A Planning Problem Framework for Small and Medium-Sized Enterprises. *Agriculture* **2022**, *12*, 1922. [CrossRef]
19. Wu, H.-C.; Lin, Y.-C.; Chen, T.-C.T. Leisure Agricultural Park Selection for Traveler Groups Amid the COVID-19 Pandemic. *Agriculture* **2022**, *12*, 111. [CrossRef]
20. Khan, N.; Ray, R.L.; Kassem, H.S.; Hussain, S.; Zhang, S.; Khayyam, M.; Ihtisham, M.; Asongu, S.A. Potential Role of Technology Innovation in Transformation of Sustainable Food Systems: A Review. *Agriculture* **2021**, *11*, 984. [CrossRef]
21. He, K.; Zhang, X.; Ren, S.; Sun, J. Deep residual learning for image recognition. In Proceedings of the IEEE Conference on Computer Vision and Pattern Recognition, Las Vegas, NV, USA, 26 June–1 July 2016; pp. 770–778.
22. Dimitrov, D.D.; Lafleur, P.; Sonnentag, O.; Talbot, J.; Quinton, W.L. Hydrology of peat estimated from near-surface water contents. *Hydrol. Sci. J.* **2022**, *67*, 1702–1721. [CrossRef]
23. Tan, H.H.; Lim, K.H. Vanishing gradient mitigation with deep learning neural network optimization. In Proceedings of the 2019 7th International Conference on Smart Computing & Communications (ICSCC), Miri, Malaysia, 28–30 June 2019; pp. 1–4.
24. Alzubaidi, L.; Zhang, J.; Humadi, A.J.; Al-Dujaili, A.; Duan, Y.; Al-Shamma, O.; Santamaria, J.; Fadhel, M.A.; Al-Amidie, M.; Farhan, L. Review of deep learning: Concepts, CNN architectures, challenges, applications, future directions. *J. Big Data* **2021**, *8*, 53. [CrossRef] [PubMed]

agriculture

MDPI

Article

Fused Deep Features-Based Grape Varieties Identification Using Support Vector Machine

Yun Peng [1,2], Shenyi Zhao [1] and Jizhan Liu [1,*]

1 Key Laboratory of Modern Agricultural Equipment and Technology, Ministry of Education, Jiangsu University, Zhenjiang 212013, China; 2111716004@stmail.ujs.edu.cn (Y.P.); 2111916017@stmail.ujs.edu.cn (S.Z.)
2 School of Electronic Engineering, Changzhou College of Information Technology, Changzhou 213164, China
* Correspondence: 1000002048@ujs.edu.cn; Tel.: +86-511-88797338

Abstract: Proper identification of different grape varieties by smart machinery is of great importance to modern agriculture production. In this paper, a fast and accurate identification method based on Canonical Correlation Analysis (CCA), which can fuse different deep features extracted from Convolutional Neural Network (CNN), plus Support Vector Machine (SVM) is proposed. In this research, based on an open dataset, three types of state-of-the-art CNNs, seven species of deep features, and a multi-class SVM classifier were studied. First, the images were resized to meet the input requirements of a CNN. Then, the deep features of the input images were extracted by a specific deep features layer of the CNN. Next, two kinds of deep features from different networks were fused by CCA to increase the effective classification feature information. Finally, a multi-class SVM classifier was trained with the fused features. When applied to an open dataset, the model outcome shows that the fused deep features with any combination can obtain better identification performance than by using a single type of deep feature. The fusion of fc6 (in AlexNet network) and Fc1000 (in ResNet50 network) deep features obtained the best identification performance. The average F1 Score of 96.9% was 8.7% higher compared to the best performance of a single deep feature, i.e., Fc1000 of ResNet101, which was 88.2%. Furthermore, the F1 Score of the proposed method is 2.7% higher than the best performance obtained by using a CNN directly. The experimental results show that the method proposed in this paper can achieve fast and accurate identification of grape varieties. Based on the proposed algorithm, the smart machinery in agriculture can take more targeted measures based on the different characteristics of different grape varieties for further improvement of the yield and quality of grape production.

Keywords: grape varieties identification; Support Vector Machine (SVM); Convolutional Neural Network (CNN); deep feature fusion; Canonical Correlation Analysis (CCA); smart machinery

Citation: Peng, Y.; Zhao, S.; Liu, J. Fused Deep Features-Based Grape Varieties Identification Using Support Vector Machine. *Agriculture* **2021**, *11*, 869. https://doi.org/10.3390/agriculture11090869

Academic Editor: Dimitre Dimitrov

Received: 10 August 2021
Accepted: 8 September 2021
Published: 10 September 2021

1. Introduction

Grape is one of the most popular fruits which can be used for wine production or fresh food. There are many varieties of grapes in the world, with more than 10,000 varieties and about 3000 cultivars. Through the improvement and screening of grape varieties, there are dozens of wine grapes widely planted at present. Many wine producers need to mix a larger number of different varieties of grapes to produce high-quality wines, such as the "Blend D. Antónia", which has more than 30 grape varieties. This means that more than one grape variety is planted per parcel and even per row [1]. Although the basic management methods of grapes are similar, the different varieties have their own characteristics, and they have different requirements for pruning, spraying, fertilization, and harvest time. Scientific and accurate field management is the key to the production of wine grapes of high quality. With the development of modern agriculture, more smart machines are used for pruning [2], spraying [3,4], and harvesting grapes [5]. Accurate identification of grape varieties is necessary for the smart machinery to make more targeted decisions for different varieties. Therefore, it is important to develop reliable methods that could automatically identify the varieties of grapes.

The traditional methods of grape variety identification mainly include manual and chemical identification [6]. Manual identification requires a high level of experience and expertise of the inspector and has disadvantages such as being time-consuming and subjective. In contrast, chemical identification methods such as mass spectrometry, chromatography, and fluorescence spectrometry are widely used, have demonstrated great efficiency, and are considered to be reliable for identifying grape varieties [7–10]. However, it is not convenient to apply the chemical methods to smart machinery, and these methods cannot produce real-time identification results for the real-time operation of smart machinery. Machine vision, which only needs an image sensor, is very easy and convenient to integrate into smart machinery. Thus, in this paper, the method of machine vision was considered to identify grape varieties. With the development of machine vision, image inspection as a non-destructive technology is widely used in industry [11,12], agriculture [13,14], and fisheries [15,16], showing great application prospects.

In the agriculture field, machine vision-based methods are widely used for the classification or grading of agricultural products. Nasirahmadi et al. used a Bag of Features (BoF) model to solve the classification problem of 20 kinds of almonds. In their research, three classifiers (L-SVM, Chi-SVM, and kNN) based on five keypoint detectors and a SIFT descriptor were investigated and achieved 79–87%, 83–91%, and 67–78% accuracy, respectively [17]. Bhargava et al., aiming at the problem of apple (fresh, rotten), six different varieties of apple (Fuji, York, Golden Crown, Red Crown, Granny Smith, and Jonagold) were selected. Firstly, the fruit region in the image was segmented by grab-cut method and Fuzzy C-Means Clustering, and then six features were extracted from feature space by principal component analysis to train an SVM classifier, and finally, 98.42% classification accuracy was obtained [18]. In [19], Ponce et al. focused on the variety identification of olive fruits. Six different Convolutional Neural Networks were trained by 2800 images with seven kinds of olive, and a top accuracy of 95.91% was obtained by Inception-ResnetV2. In addition, there are also some studies about the identification of grape varieties based on machine learning or deep learning. Bogdan et al. developed a model which is a combination of deep learning ResNet50 classifier model with multi-layer perceptron for grape varieties identification. A well-known benchmark dataset, which provided the instances from five different grape varieties taken from the field, was used for training and testing on the developed model. The test results showed that the classification accuracy of the model for different grape varieties can reach 99% [20]. El-Mashharawi et al. adopted a CNN for the identification of grape varieties. A dataset provided by Kaggle, which contains 4565 images with six species of grapes, was used to train and test the network, and a validation accuracy of 100% was achieved on the test set. The main reason for such high accuracy could be attributed to that each image in the dataset contains only one grape grain, and the background is pure white without any interference [21]. Besides, the AlexNet was trained by Pereira et al. with 10 different generated datasets and the highest accuracy of 77.3% was obtained with the four-corners-in-one preprocessed dataset [1].

Two kinds of methods were mainly adopted in the above studies: (1) SVM-based models for classification and (2) deep learning-based models for classification. SVM-based classifiers have a simple classification idea, i.e., to maximize the interval between samples and decision surfaces. Thus, using kernel functions to map features to high-dimensional space to solve nonlinear classification problems can often achieve excellent results [22,23]. However, the classification effect often depends on the class, number, and robustness of the selected features, which is often specific to the research and limited by their experience, and thus, the performance of classifiers trained by different people often has large performance differences. Compared to traditional machine learning methods, deep learning often could achieve better performance by the application of a Convolutional Neural Network (CNN), which consists of a variety of filters, nonlinearities, and pooling operators. The filters with different sizes are used for learning. A nonlinear operator such as hyperbolic tangents, rectified linear units, or logistics sigmoid are added to improve the nonlinear fitting ability of the model. Convolution and nonlinearities are usually followed by a pooling operator such as subsampling, average pooling, or maximum pooling. The CNN model can automatically discover features, and as the number of CNN layers increases, the feature discovered becomes more advanced. Compared to handcraft features by manual selection, deep learning can automatically learn the hierarchical features hidden into the images, which is not only more effective but also avoids the tedious features selection procedure.

However, when dealing with a small dataset, it is difficult to train a CNN from scratch. Based on the discussion in the previous section, it is natural to carry out a method that can train the SVM model with deep features to make full use of the advantages of the two techniques. The literature reports that this method has been applied in agriculture and achieved good performance. In [24], Sethy et al. evaluated the classification performance of a Support Vector Machine classifier for rice disease identification. The features adopted for SVM classifier training were extracted by 13 pre-trained CNN models (AlexNet, Vgg16, Vgg19, Xception, Resnet18, Resnet50, Resnet101, Inceptionv3, Inceptionresnetv2, GoogLeNet, Densenet201, Mobilenetv2, shufflenet). The use of pre-trained models can not only ensure that the model could extract effective features, but at the same time, it avoids the need for large datasets and computational resources to train a CNN. The results show that the classifier trained by features extract by ResNet50 is superior to other models, and the F1 Score is 0.9838. In the same year, this method was used by the team to detect nitrogen deficiency of rice, and the classifier of ResNet50 + SVM achieved the best classification accuracy of 99.84% [25]. In addition, Jiang et al., adopting the same idea as [24,25], developed an SVM classifier for rice leaf diseases diagnosis, and an average correct recognition rate of 96.8% was obtained. The above studies show that training an SVM classifier with the deep features extracted by a CNN model can achieve classification results that are not inferior to those of applying the corresponding deep model directly. Moreover, the computational resources, as well as the training time, are significantly reduced compared with training a deep network from scratch. However, the currently proposed methods of deep features plus SVM have two problems. (1) The dimensionality of the deep features is usually very large; for example, the dimensionality of fc6, fc7, and fc8 of AlexNet are 4096, 4096, and 1000, respectively. The high-dimensional features may affect the classification performance since the SVM model is more suitable for dealing with low-dimensional features. (2) The research mainly focuses on evaluating the classification performance of an SVM classifier trained by deep features of a specific layer of a single CNN model to find the best deep feature for classification.

With all the above, the objective of this research is to develop a new method for training the SVM classifier with CCA fused deep features for the identification of different varieties of grapes and thereby to overcome the problems with the current SVM+ deep feature methods. Inspired by Haghighat et al. [26] and Sun et al. [27], which demonstrated the outstanding performance by the fused feature with Canonical Correlation Analysis (CCA) fusion techniques, and considering working with small datasets, we hypothesize that a method can be developed that can fuse two kinds of deep features and reduce the dimensionality of fused deep features to improve the classification performance and reduce the training time. Thus, our new method would train the SVM classifier with CCA fused deep features for the identification of different varieties of grapes. By utilizing two CNNs to extract the deep features of the training set, and then CCA to fuse the obtained deep features and train the SVM with the fused features, we expect to deliver a better classifier for grape variety recognition. We also address the following specific hypotheses. (1) Compared with the deep feature extracted by a single CNN model, the fused deep feature from different networks can make the SVM classifier learn more features to improve the classification performance. (2) The CCA algorithm can eliminate redundant and invalid information when fusing different deep features, and significantly reduce the feature dimensionality to alleviate the problem that the model performance is affected when the SVM model learns high-dimensional features, and the reduction of feature dimensionality helps to speed up the model training speed.

The remainder of the article is organized as follows. In Section 2, the studied dataset and models and the proposed method are given. Then, in Section 3, the experiment results of different models are presented. Further, in Section 4, a comprehensive discussion based on the experiment results and the other studies are presented. Finally, a conclusion of the research is given in Section 5.

2. Materials and Methods

2.1. Dataset

One publicly available grapes image dataset named WGISD was adopted to evaluate the performance in this study, which can be downloaded from https://github.com/thsant/wgisd (accessed on 10 June 2021). The WGISD dataset consists of 300 images (2048 × 1365 pixels) with 5 different varieties (Chardonnay, Cabernet Franc, Cabernet Sauvignon, Sauvignon Blanc, and Syrah), and the detailed distribution of the dataset is as shown in Figure 1. In the experiment, the dataset was randomly split into 70:30 for training and testing, respectively.

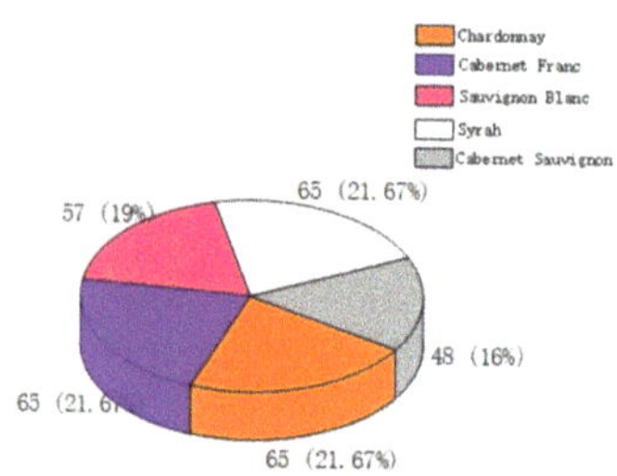

Figure 1. The varieties distribution of WGISD dataset.

2.2. Network Architecture and Deep Features Layers

In this research, the deep feature of three state-of-the-art CNN models, i.e., AlexNet [28], GoogLeNet [29], and ResNet [30], were adopted to evaluate the performance of the proposed method. The deep features are extracted from fully connected layer of a CNN model. Generally, a CNN may include several different fully connected layers (deep feature layers). For example, AlexNet consists of three deep feature layers, namely fc6, fc7, and fc8. Then, in this research, some typical deep feature layers were examined, and the detailed information of the selected layers was listed in Table 1.

Table 1. The deep feature layer and feature vector of the studied CNN.

CNN Model	Feature Layer	Feature Vector
AlexNet	fc6	4096
	fc7	4096
	fc8	1000
GoogLeNet	loss3-classifier	1000
ResNet18	Fc1000	1000
ResNet50	Fc1000	1000
ResNet101	Fc1000	1000

2.2.1. AlexNet

AlexNet was first proposed by Alex Krizhevsky et al. in the ImageNet competition and won first place in 2012. AlexNet is a deepening of the layers of the network based on LeNet, which enables it to learn richer and higher dimensional features. The proposal of AlexNet is the beginning of deep learning. It is a basic, simple, and effective CNN architecture, which is mainly composed of cascade stages, namely a convolutional layer, pooling layer, rectified linear unit (ReLU) layer, and fully connected layer. Specifically, the AlexNet consists of 5 convolutional layers, as shown in Figure 2, with pooling layers behind the first, second, third, and fourth layer, and 3 fully connected layers behind the fifth layer. The success of AlexNet can be attributed to some practical strategies, such as using ReLU nonlinear layers instead of sigmoid function as activation functions, using dropout to suppress overfitting, and using multi-GPU training. ReLU is a half-wave rectification function that can significantly accelerate the training phase and prevent overfitting. In addition, the dropout can be considered as a regularization to reduce the co-adaptation of neurons by setting the number of input neurons or hidden neurons to zero at random, which is usually used in the fully connected layers of AlexNet architecture. In this research, the deep features of fc6, fc7, and fc8 (which could be observed in Figure 2) of AlexNet were adopted to evaluate the performance of the SVM classifier.

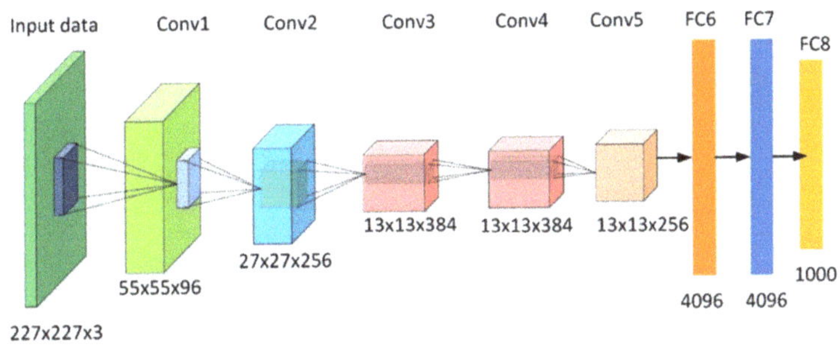

Figure 2. The architecture of AlexNet.

2.2.2. GoogLeNet

GoogLeNet, as shown in Figure 3, is a new deep learning structure proposed by Christian Szegedy in 2014. Before that AlexNet, VGG, and other networks obtained affordable performance by increasing the depth of the network (layers), but the increase in layers brought many negative effects, such as overfitting, gradient disappearance, and gradient explosion. To address this issue, a new module, Inception, was proposed by Szegedy et al. to construct the GoogLeNet network. The architecture of Inception is shown in Figure 4, which puts multiple convolutions or pooling operation together to form a single network unit. The proposal of Inception is to improve the performance from another perspective: it can use computing resources more efficiently and can extract more features with the same amount of calculation.

When designing a network that does not adopt the inception, we tend to use only one operation in a layer, such as convolution or pooling, and the size of the convolution kernel for the convolution operation is also fixed. However, in practical situations, for different sizes of images, different sizes of convolution kernels are needed to make the best performance, or, for the same image, different sizes of convolution kernels behave differently because they have different receptive fields. Therefore, we want to let the network choose by itself, and Inception can meet such needs. An Inception module provides multiple convolutional kernels in parallel, and the network chooses to use them by adjusting the parameters during training. In our research, the last fully connected layer (as shown in Figure 3) was chosen as the feature extraction layer. This layer is named "loss3-classifier" in the Deep Learning Toolbox of MATLAB and is usually chosen as the feature extraction layer of GoogLeNet for different applications [24,31,32].

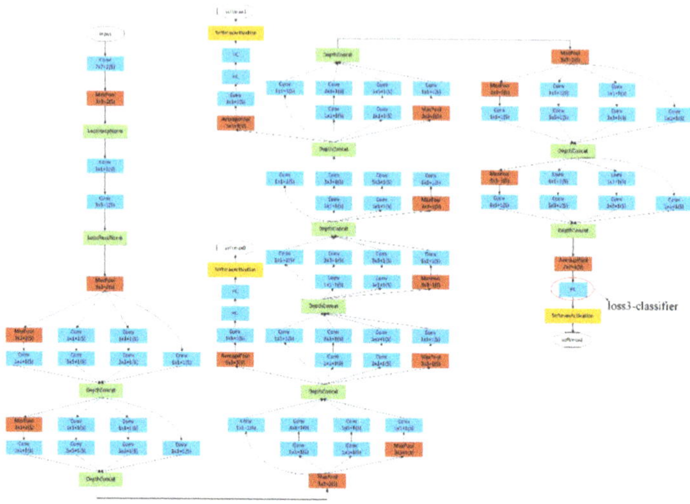

Figure 3. The architecture of GoogLeNet.

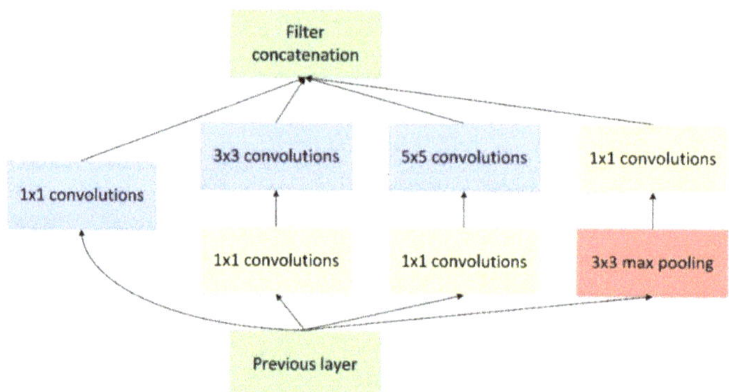

Figure 4. The Inception module proposed by GoogLeNet.

2.2.3. ResNet

As the network deepens, there is a decrease in the accuracy of the training set, and it is certain that this is not caused by overfitting (the accuracy should be high enough in the case of overfitting). To address this problem, He et al. proposed a new network named deep residual network (ResNet), which introduces a residual block structure, as shown in Figure 5, to solve the problem degradation in deep networks while allowing the network to be as deep as possible. The X Identity in Figure 5 is called "shortcut connection" and is the significant difference between the deep residual network and other networks. Two kinds of mapping were proposed by ResNet, the one is identity mapping and the other one is residual mapping. The output of a residual block is y = F(x) + x (do not consider nonlinear activation), and identity mapping refers to the input itself, which is x in the equation, while residual mapping refers to the F(x) part. If the network has reached the optimum, even if the network deepens, the residual mapping will be pushed to 0, leaving only identity mapping, so that the network is always in an optimum state and the performance of the network will not decrease as the layers increase.

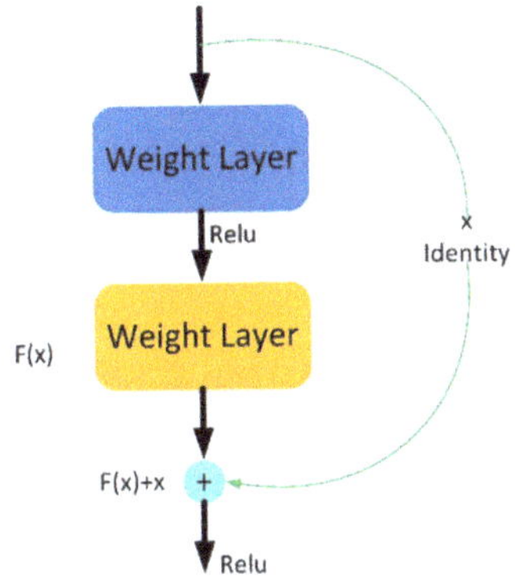

Figure 5. Residual block.

As shown in Equation (1), if the dimensions of x and F(x) are different, a linear projection W should be adopted on x such that the dimensions of x could match the dimensions of the F(x).

$$y = \mathrm{F}(\mathrm{x}, \{\mathrm{W}_i\}) + \mathrm{W}_x\mathrm{x}. \quad (1)$$

In our study, three widely used ResNet architectures, i.e., ResNet18, ResNet50, and ResNet101, were chosen as the deep feature extraction network. Similar to the "loss3-classifier", the fully connected layer before each ResNet was chosen as the deep feature extraction layer. These layers are named "Fc1000" in their respective networks and have also been wildly adopted for deep feature extraction for different applications [24,31,32].

2.3. Fusion of Deep Features by Canonical Correlation Analysis

In this research, two kinds of deep features extracted by the different networks of different deep feature layers were fused into a single feature vector by using a feature fusion technique based on Canonical Correlation Analysis (CCA) [27]. The fused feature is more discriminative than any of the input feature vectors. The Canonical Correlation Analysis (CCA) has been widely adopted to analyze associations between two sets of variables. The detailed mathematical derivation of CCA is shown in Appendix A.

As defined in [27], the deep features extracted by different CNN models could be fused by summation of the transformed features (canonical variates X^* and Y^*), and the fusion equation is shown in Equation (2).

$$\mathrm{Z} = \mathrm{X}^* + \mathrm{Y}^* = \mathrm{W}_x^{\mathrm{T}}\mathrm{X} + \mathrm{W}_y^{\mathrm{T}}\mathrm{Y} = \begin{pmatrix} \mathrm{W}_x \\ \mathrm{W}_y \end{pmatrix}^{\mathrm{T}} \begin{pmatrix} \mathrm{X} \\ \mathrm{Y} \end{pmatrix}, \quad (2)$$

where X^* and Y^* are the transformation of original extracted deep features X and Y by matrices W_x^T and W_y^T, respectively. Through matrix transformation, the matrices X and Y, whose dimensionalities are not necessarily equal, become equal. Further, the fused deep features Z can be obtained through numerical summation of corresponding positions of X^* and Y^*.

2.4. Proposed Methodology

The processing flow of the proposed method is illustrated in Figure 6.

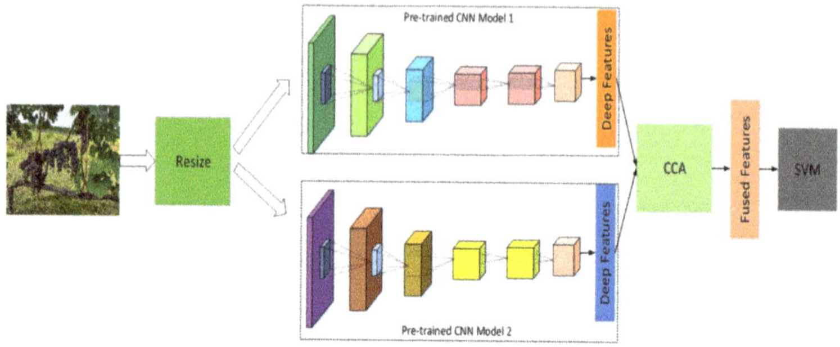

Figure 6. The processing flow of the proposed method.

Firstly, we resize the image to make it fit the input requirement of the CNN models. The CNN models used in this article were AlexNet, ResNet (18, 50, and 101), and GoogLeNet, and their input requirements are $227 \times 227 \times 3$, $224 \times 224 \times 3$, and $224 \times 224 \times 3$, respectively.

Second, we input the image to the pre-trained CNN model to extract the deep features on specific layers of the CNN model. A typical CNN usually consists of two parts: convolutional base and classifier. The convolutional base is mainly used to automatically learn hierarchical features representations of the input image. Models trained with large datasets often have stronger feature learning abilities than trained with a small dataset. This means that a model pre-trained by a large dataset to extract deep features probably captures statistics of natural images much better than could have been learned from the WGISD dataset because of its limited size. Therefore, all the CNNs adopted for deep features extraction are pre-trained by ImageNet, which is a famous public dataset that contains 14 million images with 20 thousand classes.

Third, fuse the deep features extracted from different CNN models by Canonical Correlation Analysis (CCA) algorithm. The Canonical Correlation Discriminant Features is more discriminative than any of the input feature vectors. Compared with directly concatenating the deep features extracted from different CNN models, CCA can effectively eliminate the invalid features and reduce the feature dimensionality, which can avoid overfitting training and reduce the training time and computational resources of the SVM classifier.

Finally, the fused deep features were fed into a well-trained SVM classifier, then the SVM classifier could distinguish the grape variety and output the result. In the training stage, the function of "fit class error correcting output codes" (fitcecoc) was used, which could train a multi-class SVM classifier. The function of "fitcecoc" uses $K(K-1)/2$ binary SVM model with One-Vs-All coding design, which enhances the classification performance of the classifier.

2.5. Experiment Environment

The experiment computer is a Dell-T7920 (Austin, TX, USA) workstation running on Windows 10, and the hardware configuration is 2 Intel Xeon Gold 6248R CPUs, 64GB of RAM, and 2 Nvidia Quadro RTX 5000 graphics cards with 32 GB memory. The software environment is MATLAB 2020b (Netik, MA, USA), which supports the operation of CNN models such as AlexNet, GoogLeNet, and ResNet by installing the DeepLearning toolbox. In addition, the network parameters of AlexNet, GoogLeNet, and ResNet (18, 50, and 101) are pre-trained by ImageNet and have powerful feature extraction capability.

2.6. Performance Evaluation Metrics

To evaluate the performance of the proposed method, four metrics have been applied, i.e., accuracy [33], recall [33], precision [33], and F1 Score [33], which are most widely used to evaluate the performance of classification. The accuracy represents the ratio of the number of correctly classified samples to the total number of samples. The recall represents the ratio of the number of true positive samples to the positive samples. The precision represents the ratio of the number of true positive samples to the number of the samples predicted as positive, and the F1 Score is a harmonic metric, which takes into account both the recall and precision of the classification model. The equation of the adopted metrics is shown in Equations (3)–(6).

$$Accuracy = \frac{TP + TN}{TP + FP + TN + FN} \tag{3}$$

$$\text{Recall} = \frac{TP}{TP + FN} \tag{4}$$

$$\text{Precision} = \frac{TP}{TP + FP} \tag{5}$$

$$\text{F1}Score = 2 \times \frac{recall \times precision}{recall + precision} \tag{6}$$

where TP is true positive, TN is true negative, FP is false positive, and FN is false negative.

3. Results

In the experiment, in order to obtain more reliable experiment data, 10 independent runs of training and validation of each SVM classifier were made on the dataset, and the mean of results and deviation on the test set was adopted to represent its performance. Besides, since the standard deviations of the results of each model are very small, only the mean of the results were discussed.

3.1. Performance Analysis Based on Single Type of Deep Feature

The classification results and training time of the SVM classifier trained with a single type of deep feature are shown in Table 2. It is observed that the Fc1000 layer of ResNet101 obtained the best classification performance among all the examined deep layers, and the average accuracy, precision, recall, and F1 Score are 87.7%, 88.4%, 88%, and 88.2%, respectively. Besides, the fc6 of AlexNet can achieve better performance than fc7 and fc8, which was also indicated by other studies [24,34], reporting that the deep feature of fc6 is more distinguishable than that of fc7 and fc8. Then, in the following sections, the fc6 deep feature of AlexNet is considered only. Figure 7 shows the visualization of same sample with different deep feature. In addition, the training time of an SVM classifier with the fc6 layer of AlexNet is the longest, with an average of 2.5 s, while the average time of all other single type deep features was within 0.5 s. The dimension of the fc6 deep feature is 4096, and all the others are 1000, which may be the reason why the training time of fc6 is longer than that of other deep features.

Table 2. Performance metrics and training time of SVM classifier with a single deep feature (bold font shows the best performance).

Metrics	**AlexNet**			**GoogLeNet**	**ResNet18**	**ResNet50**	**ResNet101**
	fc6	**fc7**	**fc8**	**Loss3-Classifier**	**Fc1000**	**Fc1000**	**Fc1000**
Accuracy (%)	86.3 ± 1.1	81.5 ± 1.8	79.3 ± 0.9	72.5 ± 2.7	79 ± 2.4	83.4 ± 2	**87.7** ± 1.2
Precision (%)	87.8 ± 0.9	83.8 ± 1.4	81.4 ± 0.7	75.2 ± 2.8	80.9 ± 2.7	84.4 ± 2	**88.4** ± 1.4
Recall (%)	86.8 ± 1.1	82.3 ± 2	79.7 ± 1	73.3 ± 2.7	79.7 ± 2.2	84.1 ± 1.9	**88** ± 1.2
F1 Score (%)	87.3 ± 0.9	83.1 ± 1.7	80.6 ± 0.8	74.2 ± 2.7	80.3 ± 2.4	84.3 ± 1.9	**88.2** ± 1.2
Training time (S)	2.5 ± 0.39	0.4 ± 0.32	0.1 ± 0.04	0.09 ± 0.04	0.1 ± 0.02	0.1 ± 0.04	0.2 ± 0.27

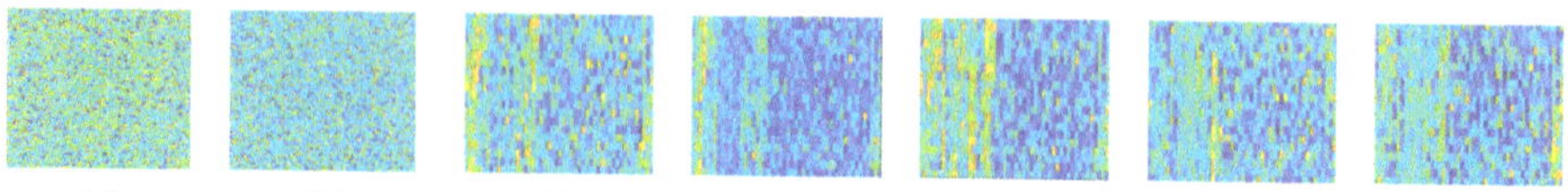

Figure 7. Visualization of different deep feature with the same sample: (**a**) fc6; (**b**) fc7; (**c**) fc8; (**d**) loss3-classifier; (**e**) Fc1000 of ResNet18; (**f**) Fc1000 of ResNet50; (**g**) Fc1000 of ResNet101.

3.2. Performance Analysis Based on CCA Fused Deep Features

Table 3 shows the performance of the SVM classifier trained by CCA fused deep features extracted from different CNNs on the test set. It can be observed that the combination of fc6 and ResNet50 (Fc1000) achieves the best classification performance, and the accuracy, precision, recall, and mean F1 Score are 96.5%, 97.0%, 96.7%, and 96.8%, respectively. The combination of GoogLeNet (loss3-classifier) and ResNet18 (Fc1000) obtained the worst performance, and the mean accuracy, precision, recall, and F1 Score are 90.6%, 91.1%, 91%, and 91.1%, respectively. For all the fused deep features, the dimensionality is 211 and the training times are usually within 0.1 s, which seems to verify the speculation about the relationship between training time and feature dimensionality in the previous section.

Table 3. Performance metrics and training time of SVM classifier with fused deep features (bold font shows the best performance).

Fused Features	Accuracy (%)	Precision (%)	Recall (%)	F1 Score (%)	Train Time (S)	Dimension
fc6 + GoogleNet	94.5 ± 2.4	95.2 ± 2.1	94.7 ± 2.3	95.0 ± 2.2	0.07 ± 0.02	211
fc6 + ResNet18	92 ± 1.6	92.5 ± 1.3	92.2 ± 1.8	92.4 ± 1.5	0.06 ± 0.001	211
fc6 + ResNet50	**96.5** ± 1.4	**97.0** ± 1.2	**96.7** ± 1.5	**96.8** ± 1.3	0.06 ± 0.007	211
fc6 + ResNet101	96.1 ± 1	96.6 ± 0.9	96.2 ± 0.9	96.4 ± 0.9	0.07 ± 0.02	211
GoogleNet + ResNet18	90.6 ± 0.9	91.1 ± 0.7	91 ± 1	91.1 ± 0.8	0.07 ± 0.04	211
GoogleNet + ResNet50	92 ± 0.8	92.8 ± 0.5	92.1 ± 0.9	92.4 ± 0.6	0.09 ± 0.03	211
GoogleNet + ResNet101	95.2 ± 1.2	95.7 ± 1.1	95.4 ± 1	95.5 ± 1.1	0.05 ± 0.006	211
ResNet18 + ResNet50	92.2 ± 1.2	92.9 ± 1.3	92.3 ± 1.5	92.6 ± 1.4	0.07 ± 0.005	211
ResNet18 + ResNet101	92.7 ± 1	93.1 ± 0.8	92.9 ± 0.8	93.0 ± 0.7	0.07 ± 0.02	211
ResNet50 + ResNet101	96.1 ± 1	96.4 ± 1	96.2 ± 0.9	96.3 ± 1	0.18 ± 0.008	211

Table 4 shows the performance comparison between the proposed fused deep features and a single feature. The first row in the table is the best performance obtained by the fused deep features, while the second row is the worst performance, the third row is the best performance obtained by a single deep feature, and the fourth and fifth are the components of the fused deep features in the first row. Since the calculation of F1 Score combines precision and recall, we analyze the performance differences with mean F1 Score in this section.

Table 4. Performance comparison between the fused deep features and a single feature. (Bold font shows the best performance).

Fused Features	Accuracy (%)	Precision (%)	Recall (%)	F1 Score (%)	Fusion Time (S)	Train Time (S)	Dimension
fc6 + ResNet50	**96.5** ± 1.4	**97.0** ± 1.2	**96.7** ± 1.5	**96.8** ± 1.3	0.05 ± 0.008	0.06 ± 0.007	211
GoogleNet + ResNet18	90.6 ± 0.9	91.1 ± 0.7	91 ± 1	91.1 ± 0.8	0.03 ± 0.002	0.07 ± 0.04	211
ResNet101	87.7 ± 1.2	88.4 ± 1.4	88 ± 1.2	88.2 ± 1.2	-	0.2 ± 0.27	1000
fc6	86.3 ± 1.1	87.8 ± 0.9	86.8 ± 1.1	87.3 ± 0.9	-	2.5 ± 0.39	4096
ResNet50	83.4 ± 2	84.4 ± 2	84.1 ± 1.9	84.3 ± 1.9	-	0.1 ± 0.04	1000

The best classification performance is obtained by the fusion of deep feature extracted from fc6 of AlexNet and ResNet50 (Fc1000), and the mean F1 Score is 96.8%, which is 8.6% higher than the best single deep feature. Further, the mean F1 Score of the worst fused deep features reaches 91.1%, which is also 2.9% higher than the best single feature. For fc6 and ResNet50 (Fc1000), the mean F1 Score of the fused deep feature is 9.5% and 12.5% higher than that of the single deep feature before the fusion, which indicates that the method proposed in this paper could improve the classification performance significantly. In addition, the training time plus feature fusion time of the proposed method is not more than the training time of a single feature, indicating that the proposed method is helpful to shorten the model training time.

Although the proposed method achieved satisfactory performance on the entire test set, the identification performance of the classifier on each variety of grape in the dataset is not clear. We further obtained the confusion matrix of the classifier's (fc6 deep feature + Fc100 of ResNet50) 10 independent runs on the test set, as shown in Table 5. The confusion matrix is a tool to evaluate the performance of supervised learning classification algorithm. In our research, each row of the confusion matrix represents the actual variety of grapes while each column represents the predicted variety, and the value on the diagonal represents the correct classification of each variety. It can be intuitively observed that the classifier achieved the best classification effect for Syrah. All 140 samples of Syrah were correctly identified. Following is Cabernet Sauvignon, for which all 164 samples were also correctly identified, but some other varieties were misidentified as Cabernet Sauvignon. However, the classification effect on the other three varieties cannot be directly observed from the confusion matrix, so we calculate the F1 Score of the classifier for each variety. The mean F1 Score of Syrah, Cabernet Sauvignon, Cabernet Franc, Sauvignon Blanc, and Chardonnay are 1, 0.982, 0.9738, 0.9399, and 0.9394, respectively. Therefore, we can obtain the actual ranking of the classification effect on each variety of grape by the classifier: Syrah > Cabernet Sauvignon > Cabernet Franc > Sauvignon Blanc > Chardonnay. Furthermore, through the intuitive observation of the test set, we seem to find that the closer the grape clusters to the camera, the better the classification performance. This gives us some inspiration that the camera should be as close to the grapevine as possible to capture the more clear clusters of grapes to improve the identification performance in practice.

Table 5. The confusion matrix of the 10 independent runs on the test set with fc6 + Fc1000 (ResNet50) fused deep features.

Grape Varieties	Cabernet Franc	Sauvignon Blanc	Chardonnay	Cabernet Sauvignon	Syrah
Cabernet Franc	187	2	0	5	0
Sauvignon Blanc	0	180	12	1	0
Chardonnay	3	8	178	0	0
Cabernet Sauvignon	0	0	0	164	0
Syrah	0	0	0	0	140

3.3. Performance Comparison with Using Deep Network Directly

Further, the performance of the proposed method (fused deep features plus SVM) was compared with the performance obtained by using CNNs directly. Commonly, the number of classification classes of the original CNN is often inconsistent with the number of classification classes we study, which makes it impossible for us to directly apply the original network to our applications. Therefore, the meaning of directly using CNN in this article essentially refers to directly using the classification output of a CNN. In order to make a CNN suitable for our study, the parameter of the final "softmax" (the classifier in CNN) should be adjusted to be 5 (number of grape varieties in the WGISD). Table 6 shows the training parameters for the examined CNN models. In the experiment, "stochastic gradient descent with momentum" (sgdm) was selected as the optimization algorithm (solver), and the parameters of "MiniBatchSize" (the number of samples utilized in one iteration), "InitialLearnRate", and "MaxEpochs" (the number of passes of the entire training dataset) were 20, 1e-3, and 50, respectively. Moreover, because the workstation has two GPU, the parameter of "ExecutionEnvironment" was set as "multi-gpu" to speed up the training.

Table 6. Training parameters of the examined CNN models.

Parameters	Value
solver	sgdm
MiniBatchSize	20
InitialLearnRate	1e-3
MaxEpochs	30
ExecutionEnvironment	multi-gpu

The performance comparison between the method proposed in this article and using the CNN directly is obtained in Table 7. On the one hand, from the perspective of classification performance, the mean F1 Scores of the AlexNet, GoogLeNet, ResNet18, ResNet50, and ResNet101 are 87.0%, 94.2%, 90.7%, 88.5%, and 82.3%, respectively. The mean F1 Scores of the proposed method are 9.8%, 2.6%, 6.1%, 8.3%, and 14.5% higher than AlexNet, GoogLeNet, ResNet18, ResNet50, and ResNet10, respectively, which indicate that, for the identification of grape varieties, the proposed method has better classification performance than using CNN directly. However, the CNN often contains a large number of parameters, which requires a large number of samples to train and fine tune the parameters. Therefore, we believe that the small number of samples of the public dataset studied in this paper is the reason why it is difficult to obtain better classification performance by using the CNN directly. It is not recommended to train a deep network with small datasets to avoid the problem of overfitting. However, the method proposed in this paper can still achieve good classification performance (the mean F1 Score is as high as 96.8%) with the small dataset, showing obvious advantages. On the other hand, the training time also improved with respect to the experimental environment of our study. To sum up, the results indicate that the method proposed in this paper have advantages both in identification performance and training time compared with using CNN directly.

Table 7. Performance comparison between the proposed method and using CNN directly (bold font shows the best performance).

Method	Accuracy	Precision	Recall	F1 Score	Train Time
AlexNet	85.7 ± 1.3	88.4 ± 1.2	85.7 ± 1.1	87.0 ± 0.9	4228 ± 53
GoogLeNet	93.8 ± 0.8	94.6 ± 0.9	94.0 ± 1.1	94.2 ± 1.0	4222 ± 47
ResNet18	89.8 ± 1.3	91.6 ± 1.2	90 ± 1.1	90.7 ± 1.2	4247 ± 38
ResNet50	87.7 ± 1.1	89.1 ± 1.2	88.0 ± 1.4	88.5 ± 1.3	4310 ± 51
ResNet101	81.6 ± 1.4	83.0 ± 1.3	81.7 ± 1.5	82.3 ± 1.3	4377 ± 63
Fused deep features (proposed by this study)	**96.5** ± 1.4	**97.0** ± 1.2	**96.7** ± 1.5	**96.8** ± 1.3	0.06 ± 0.007

4. Discussion

In this study, a new method based on the CCA feature fusion algorithm plus SVM was proposed. Donahue et al. [35] and Razavian et al. [36] demonstrated the outstanding performance of the pre-trained CNN features in various recognition tasks. Although the model trained from scratch may achieve comparative or better performance, it will be more time-consuming. Besides, one potential limitation is that the amount of dataset used in our research is not particularly large, so training the network directly from scratch may cause over-fitting problems, which may cause degrading of the performance. Therefore, the pre-trained CNN models were adopted to extract features. Haghighat et al. [26] and Sun et al. [27] demonstrated an outstanding performance by the fused features with CCA fusion techniques. In our experiment, we also tried the fusion methods of direct concatenation and concatenation based on PCA, but there is a certain performance margin compared with CCA. Therefore, the CCA was selected to fuse the deep feature, which can find the most relevant features based on the two sets of features, so that better features can be used to extract further features from the comparative poorer features to enhance the performance.

The given CCA scheme is different from other fusion techniques (e.g., ensemble learning and concatenation). To be specific, ensemble learning and concatenation does not consider the correlations between the two input items from the perspective of features and may lose some valuable information. The experiment results on the public WGISD show that the proposed method could achieve satisfactory performance. However, there are many fusion algorithms in practice. In the future, more studies should be focused on selecting a more effective feature fusion algorithm to further improve the performance.

Furthermore, we compared our experimental results with some of the existing literature on grape varieties identification. Bogdan et al. cascaded the output of ResNet50 into a multi-layer perceptron, which can improve the classification accuracy, but this method needs to train two models. The evaluated dataset literature [20] is the same as ours, but the background was removed and each cluster of grapes was extracted separately for training and testing and an accuracy of 99% was obtained. However, in [22], when the background was not removed (the dataset is exactly the same as ours), the accuracy dropped by 26%, from 99% to 63%. This phenomenon suggests that (1) compared with their method, our proposed method can achieve higher classification results, and That (2) the preprocessing, the removal of the background, also plays a key role in the classification effect of the model. Pereira et al. used AlexNet to identify grape varieties collected under natural conditions. Although it carries out a variety of preprocessing on the dataset (does not include the removal of background), the accuracy obtained is 77.3%, which is 19.2% lower than ours. However, when this preprocessing was adopted to another popular Flavia leaf dataset, an accuracy of 89.75% was achieved, which indicates that a different dataset also has a large difference in results. This shows that although our approach has great advantages over theirs, the impact of different datasets on the results cannot be ignored. El-Mashharawi et al. also adopted a CNN for grape varieties identification, and an accuracy of 100% was achieved on a public dataset from Kaggle. Each image on the dataset only contains one grape grain and the background is pure white (the background pixels are all in white color). Based on the influence of background processing on classification performance in the above discussion of literature [20], we believe that the clear background is also an important reason for such high accuracy. In summary, the method proposed in this study could obtain satisfactory performance for grape varieties identification with the images collected under natural conditions. However, image preprocessing should be considered to improve the performance of our method. In addition, although the proposed method in this paper has made promising progress with the method of single deep feature + SVM, a satisfactory result was achieved on the WGISD dataset. As mentioned above, datasets also play an important role in classification performance. The method that we propose needs to be verified on different datasets to prove its versatility.

Agriculture is rapidly evolving towards a new paradigm—Agriculture 4.0, and digital technology, artificial intelligence (AI), and automation will play a particularly important role at this stage. Traditional manual machinery is gradually being replaced by smart machinery. To make the machinery smart like a human being as much as possible, an intelligent algorithm is necessary because different varieties of grapes have different harvesting time, nutrient requirements, and susceptibility to diseases and insect pests. With the help of the proposed algorithm, the smart machinery could carry out more scientific and reasonable operations for harvesting, fertilization, and management of different varieties of grapes, reducing input and improving grape quality and yield at the same time. In addition, the algorithm proposed in this paper is not only suitable for grapes but it can also be used for the identification of other fruits or vegetables for a more extensive application. However, there are still a lot of efforts that need to be made to enable our algorithm to be adopted in smart machineries. (1) The algorithm is an independent module, and when it is mounted on the smart machinery, it needs to consider the interconnection with the front-end image sensor and the back-end actuator. This way, only the proposed algorithm will have practical significance. (2) The image sensors are easily affected by the changeable field environment, and further affect the performance of the algorithm. More efforts should be adopted to deal with such problems.

5. Conclusions

In this research, a new method was proposed for the identification of different varieties of grapes based on small datasets. Our results showed that public WGISD datasets can be successfully used for identification of grape varieties and indicated that the fusion of two kinds of deep features by CCA can not only produce more distinguishable features for improving the classification performance, but can also eliminate redundant and invalid information, and thereby speed up the model training. Based on the proposed algorithm, smart machinery will have the potential of taking more targeted measures according to the different characteristics of different varieties of grapes, thus further improving their performance that involves grape varieties recognition. However, to apply more effectively the proposed algorithm to smart machinery, more images should be collected in the future for different varieties from different vineyards with various conditions to make the model more versatile. In addition, a complete software should be implemented, which includes not only the grape varieties identification module but also all the functional modules.

Author Contributions: Conceptualization, Y.P. and J.L.; methodology, Y.P.; software, Y.P. and S.Z.; writing—original draft preparation, Y.P.; writing—review and editing, Y.P.; supervision, J.L.; funding acquisition, J.L. All authors have read and agreed to the published version of the manuscript.

Funding: The research was funded by grants from the National Science Foundation of China (Grant No. 31971795), A Project Funded by the Priority Academic Program Development of Jiangsu Higher Education Institutions (No. PAPD-2018-87) and Project of Faculty of Agricultural Equipment of Jiangsu University (4111680002).

Institutional Review Board Statement: Not applicable.

Data Availability Statement: Not applicable.

Conflicts of Interest: The authors declare no conflict of interest.

Appendix A

Suppose that two different methods used to extract the p dimensional and q dimensional deep features of each sample, and two matrices, $X \in R^{p\times n}$ and $Y \in R^{q\times n}$, are obtained, where n is the number of samples (the number of samples trained in training phase, or the number of test samples in testing phase). In other words, a total of $(p+q)$ dimensional features are extracted for each sample.

Let $S_{xx} = R^{p\times p}$ and $S_{yy} = R^{q\times q}$ denote the within-sets covariance matrices of X and Y, and $S_{xy} = R^{p\times q}$ denote the between-sets covariance matrix between X and Y. The matrix S shown below is the overall $(p+q)\times(p+q)$ covariance matrix, which contains all the information on associations between the pairs of deep features.

$$S = \begin{pmatrix} \mathrm{cov}(x) & \mathrm{cov}(x,y) \\ \mathrm{cov}(y,x) & \mathrm{cov}(y) \end{pmatrix} = \begin{pmatrix} S_{xx} & S_{xy} \\ S_{yx} & S_{yy} \end{pmatrix}. \tag{A1}$$

However, the correlation between these two sets of deep feature vectors may not follow a consistent pattern, and therefore, it is difficult to understand the relationship between these two sets of deep features from this matrix [37]. The aim of CCA is to find a linear transformation,$X^* = W_x^T X$ and $Y^* = W_y^T Y$, and to maximize the pair-wise correlation between the two datasets:

$$\mathrm{corr}(X^*, Y^*) = \frac{\mathrm{cov}(X^*, Y^*)}{\mathrm{var}(X^*)\cdot \mathrm{var}(Y^*)}, \tag{A2}$$

where $\text{cov}(X^*, Y^*) = W_x^T S_{xy} W_y$, $\text{var}(X^*) = W_x^T S_{xx} W_x$ and $\text{var}(Y^*) = W_y^T S_{yy} W_y$. The covariance between X^* and Y^* (X^*, Y^* $\hat{I}R^{d \times n}$ are known as canonical variables) is maximized by Lagrange multiplier method, and the constraint condition is $\text{var}(X^*) = \text{var}(X^*) = 1$. Further, the linear transformation matrix W_x and W_y can be obtained by solving the eigenvalue equation as below [37]:

$$\begin{cases} S_{xx}^{-1} S_{xy} S_{yy}^{-1} S_{yx} \hat{W}_x = \Lambda^2 \hat{W}_x \\ S_{yy}^{-1} S_{yx} S_{xx}^{-1} S_{xy} \hat{W}_y = \Lambda^2 \hat{W}_y \end{cases}, \tag{A3}$$

where $\hat{W}_x$ and $\hat{W}_y$ are the eigenvectors and Λ^2 is a diagonal matrix of eigenvalues or squares of the canonical correlations.

The number of non-zero eigenvalues of each equation is $d = rank(S_{xy}) \leq \min(n, p, q)$, further arranged in descending order, $\lambda_1 \geq \lambda_2 \geq \cdots \geq \lambda_d$. The transformation matrix W_x and W_y are composed of eigenvectors corresponding to sorted non-zero eigenvalues. For the transformed data, the form of the sample covariance matrix defined in Equation (7) is as follows:

$$S^* = \left(\begin{array}{cccc|cccc} 1 & 0 & \cdots & 0 & \lambda_1 & 0 & \cdots & 0 \\ 0 & 1 & \cdots & 0 & 0 & \lambda_2 & \cdots & 0 \\ \vdots & & \ddots & & \vdots & & \ddots & \\ 0 & 0 & \cdots & 1 & 0 & 0 & \cdots & \lambda_d \\ \lambda_1 & 0 & \cdots & 0 & 1 & 0 & \cdots & 0 \\ 0 & \lambda_2 & \cdots & 0 & 0 & 1 & \cdots & 0 \\ \vdots & & \ddots & & \vdots & & \ddots & \\ 0 & 0 & \cdots & \lambda_d & 0 & 0 & \cdots & 1 \end{array} \right). \tag{A4}$$

As shown in the above matrix, the upper left and lower right identity matrices indicate that the canonical variates are uncorrelated within each dataset, and canonical variates have non-zero correlation only on their corresponding indices.

References

1. Pereira, C.S.; Morais, R.; Reis, M.J.C.S. Deep learning techniques for grape plant species identification in natural images. *Sensors* **2019**, *19*, 4850. [CrossRef]
2. Botterill, T.; Paulin, S.; Green, R.; Williams, S.; Lin, J.; Saxton, V.; Mills, S.; Chen, X.; Corbett-Davies, S. A robot system for pruning grape vines. *J. Field Robot.* **2017**, *34*, 1100–1122. [CrossRef]
3. Monta, M.; Kondo, N.; Shibano, Y. Agricultural robot in grape production system. In Proceedings of the 1995 IEEE International Conference on Robotics and Automation, Nagoya, Japan, 21–25 May 1995; pp. 2504–2509.
4. Ogawa, Y.; Kondo, N.; Monta, M.; Shibusawa, S. Spraying robot for grape production. In *Springer Tracts in Advanced Robotics*; Springer: Berlin/Heidelberg, Germany, 2003; pp. 539–548.
5. Kondo, N. Study on grape harvesting robot. *IFAC Proc.* **1991**, *24*, 243–246. [CrossRef]
6. Versari, A.; Laurie, V.F.; Ricci, A.; Laghi, L.; Parpinello, G.P. Progress in authentication, typification and traceability of grapes and wines by chemometric approaches. *Food Res. Int.* **2014**, *60*, 2–18. [CrossRef]
7. Duhamel, N.; Slaghenaufi, D.; Pilkington, L.I.; Herbst-Johnstone, M.; Larcher, R.; Barker, D.; Fedrizzi, B. Facile gas chromatography–tandem mass spectrometry stable isotope dilution method for the quantification of sesquiterpenes in grape. *J. Chromatogr. A* **2018**, *1537*, 91–98. [CrossRef] [PubMed]
8. Karimali, D.; Kosma, I.; Badeka, A. Varietal classification of red wine samples from four native Greek grape varieties based on volatile compound analysis, color parameters and phenolic composition. *Eur. Food Res. Technol.* **2020**, *246*, 41–53. [CrossRef]
9. Pérez-Navarro, J.; Da Ros, A.; Masuero, D.; Izquierdo-Cañas, P.M.; Hermosín-Gutiérrez, I.; Gómez-Alonso, S.; Mattivi, F.; Vrhovsek, U. LC-MS/MS analysis of free fatty acid composition and other lipids in skins and seeds of Vitis vinifera grape cultivars. *Food Res. Int.* **2019**, *125*, 108556. [CrossRef]
10. Kyraleou, M.; Kallithraka, S.; Gkanidi, E.; Koundouras, S.; Mannion, D.T.; Kilcawley, K.N. Discrimination of five Greek red grape varieties according to the anthocyanin and proanthocyanidin profiles of their skins and seeds. *J. Food Compos. Anal.* **2020**, *92*, 103547. [CrossRef]
11. Benbarrad, T.; Salhaoui, M.; Kenitar, S.B.; Arioua, M. Intelligent machine vision model for defective product inspection based on machine learning. *J. Sens. Actuator Netw.* **2021**, *10*, 7. [CrossRef]
12. Penumuru, D.P.; Muthuswamy, S.; Karumbu, P. Identification and classification of materials using machine vision and machine learning in the context of industry 4.0. *J. Intell. Manuf.* **2020**, *31*, 1229–1241. [CrossRef]

13. Mavridou, E.; Vrochidou, E.; Papakostas, G.A.; Pachidis, T.; Kaburlasos, V.G. Machine vision systems in precision agriculture for crop farming. *J. Imaging* **2019**, *5*, 89. [CrossRef]
14. Radcliffe, J.; Cox, J.; Bulanon, D.M. Machine vision for orchard navigation. *Comput. Ind.* **2018**, *98*, 165–171. [CrossRef]
15. Monkman, G.G.; Hyder, K.; Kaiser, M.J.; Vidal, F.P. Using machine vision to estimate fish length from images using regional Convolutional Neural Networks. *Methods Ecol. Evol.* **2019**, *10*, 2045–2056. [CrossRef]
16. Sung, H.-J.; Park, M.-K.; Choi, J.W. Systems. Automatic grader for flatfishes using machine vision. *Int. J. Control. Autom. Syst.* **2020**, *18*, 3073–3082.
17. Nasirahmadi, A.; Ashtiani, S.-H.M. Bag-of-Feature model for sweet and bitter almond classification. *Biosyst. Eng.* **2017**, *156*, 51–60. [CrossRef]
18. Bhargava, A.; Bansal, A. Classification and grading of multiple varieties of apple fruit. *Food Anal. Methods* **2021**, *14*, 1–10. [CrossRef]
19. Ponce, J.M.; Aquino, A.; Andújar, J.M. Olive-fruit variety classification by means of image processing and Convolutional Neural Networks. *IEEE Access* **2019**, *7*, 147629–147641. [CrossRef]
20. Franczyk, B.; Hernes, M.; Kozierkiewicz, A.; Kozina, A.; Pietranik, M.; Roemer, I.; Schieck, M. Deep learning for grape variety recognition. *Procedia Comput. Sci.* **2020**, *176*, 1211–1220. [CrossRef]
21. El-Mashharawi, H.Q.; Abu-Naser, S.S.; Alshawwa, I.A.; Elkahlout, M. Grape type classification using deep learning. *Int. J. Acad. Eng. Res.* **2020**, *3*, 14–45.
22. Acortes, C.; Vapnik, V. Support vector networks. Machine Learning. *Mach. Learn.* **1995**, *20*, 273–297. [CrossRef]
23. Aizerman, M.A. Theoretical foundations of the potential function method in pattern recognition learning. *Autom. Remote. Control.* **1964**, *25*, 821–837.
24. Sethy, P.K.; Barpanda, N.K.; Rath, A.K.; Behera, S.K. Deep feature based rice leaf disease identification using Support Vector Machine. *Comput. Electron. Agric.* **2020**, *175*, 105527. [CrossRef]
25. Sethy, P.K.; Barpanda, N.K.; Rath, A.K.; Behera, S.K. Nitrogen deficiency prediction of rice crop based on Convolutional Neural Network. *J. Ambient. Intell. Humaniz. Comput.* **2020**, *11*, 5703–5711. [CrossRef]
26. Haghighat, M.; Abdel-Mottaleb, M.; Alhalabi, W. Fully automatic face normalization and single sample face recognition in unconstrained environments. *Expert Syst. Appl.* **2016**, *47*, 23–34. [CrossRef]
27. Sun, Q.-S.; Zeng, S.-G.; Liu, Y.; Heng, P.-A.; Xia, D.-S. A new method of feature fusion and its application in image recognition. *Pattern Recognit.* **2005**, *38*, 2437–2448. [CrossRef]
28. Krizhevsky, A.; Sutskever, I.; Hinton, G.E. Imagenet classification with deep Convolutional Neural Networks. *Adv. Neural Inf. Process. Syst.* **2012**, *25*, 1097–1105. [CrossRef]
29. Szegedy, C.; Liu, W.; Jia, Y.; Sermanet, P.; Reed, S.; Anguelov, D.; Erhan, D.; Vanhoucke, V.; Rabinovich, A. Going deeper with convolutions. In Proceedings of the IEEE Conference on Computer Vision and Pattern Recognition, Boston, MA, USA, 7–12 June 2015; pp. 1–9.
30. He, K.; Zhang, X.; Ren, S.; Sun, J. Deep residual learning for image recognition. In Proceedings of the IEEE Conference on Computer Vision and Pattern Recognition, Las Vegas, NV, USA, 27–30 June 2016; pp. 770–778.
31. Sethy, P.K.; Behera, S.K.; Ratha, P.k.; Biswas, P. Detection of coronavirus disease (covid-19) based on deep features and Support Vector Machine. *Int. J. Math. Eng. Manag. Sci.* **2020**, *5*, 643–651. [CrossRef]
32. Kadhim, M.A.; Abed, M.H. Convolutional Neural Network for satellite image classification. In *Studies in Computational Intelligence*; Springer: Cham, Switzerland, 2019; pp. 165–178.
33. Castelli, M.; Vanneschi, L.; Largo, Á.R. Supervised learning: Classification. In *Encyclopedia of Bioinformatics and Computational Biology*; Elsevier: Amsterdam, The Netherlands, 2018; Volume 1, pp. 342–349.
34. Chan, G.C.; Muhammad, A.; Shah, S.A.; Tang, T.B.; Lu, C.-K.; Meriaudeau, F. Transfer learning for diabetic macular edema (DME) detection on optical coherence tomography (OCT) images. In Proceedings of the 2017 IEEE International Conference on Signal and Image Processing Applications (ICSIPA), Kuching, Malaysia, 12–14 September 2017; pp. 493–496.
35. Donahue, J.; Jia, Y.; Vinyals, O.; Hoffman, J.; Zhang, N.; Tzeng, E.; Darrell, T. Decaf: A deep convolutional activation feature for generic visual recognition. In Proceedings of the 31th International Conference on Machine Learning, Beijing, China, 21–26 June 2014; pp. 647–655.
36. Sharif Razavian, A.; Azizpour, H.; Sullivan, J.; Carlsson, S. CNN features off-the-shelf: an astounding baseline for recognition. In Proceedings of the IEEE Conference on Computer Vision and Pattern Recognition Workshops, Columbus, OH, USA, 23–28 June 2014; pp. 806–813.
37. Krzanowski, W. *Principles of Multivariate Analysis*; OUP Oxford: Oxford, UK, 2000; Volume 23.

agriculture

Article

ACE-ADP: Adversarial Contextual Embeddings Based Named Entity Recognition for Agricultural Diseases and Pests

Xuchao Guo [1], Xia Hao [2], Zhan Tang [1], Lei Diao [1], Zhao Bai [1], Shuhan Lu [3] and Lin Li [1,*]

1 College of Information and Electrical Engineering, China Agricultural University, Beijing 100083, China; gxc@cau.edu.cn (X.G.); styx_tang@cau.edu.cn (Z.T.); S20193081368@cau.edu.cn (L.D.); s20193081367@cau.edu.cn (Z.B.)
2 College of Information Science and Engineering, Shandong Agricultural University, Tai'an 271000, China; haoxia@sdau.edu.cn
3 School of Information, University of Michigan, Ann Arbor, MI 48104, USA; shuhanlu@umich.edu
* Correspondence: lilinlsl@cau.edu.cn

Abstract: Entity recognition tasks, which aim to utilize the deep learning-based models to identify the agricultural diseases and pests-related nouns such as the names of diseases, pests, and drugs from the texts collected on the internet or input by users, are a fundamental component for agricultural knowledge graph construction and question-answering, which will be implemented as a web application and provide the general public with solutions for agricultural diseases and pest control. Nonetheless, there are still challenges: (1) the polysemous problem needs to be further solved, (2) the quality of the text representation needs to be further enhanced, (3) the performance for rare entities needs to be further improved. We proposed an adversarial contextual embeddings-based model named ACE-ADP for named entity recognition in Chinese agricultural diseases and pests domain (CNER-ADP). First, we enhanced the text representation and overcame the polysemy problem by using the fine-tuned BERT model to generate the contextual character-level embedded representation with the specific knowledge. Second, adversarial training was also introduced to enhance the generalization and robustness in terms of identifying the rare entities. The experimental results showed that our model achieved an F_1 of 98.31% with 4.23% relative improvement compared to the baseline model (i.e., word2vec-based BiLSTM-CRF) on the self-annotated corpus named Chinese named entity recognition dataset for agricultural diseases and pests (AgCNER). Besides, the ablation study and discussion demonstrated that ACE-ADP could not only effectively extract rare entities but also maintain a powerful ability to predict new entities in new datasets with high accuracy. It could be used as a basis for further research on other domain-specific named entity recognition.

Keywords: digital agriculture; Chinese agricultural diseases and pests; named entity recognition; adversarial training; semantic enhancement

Citation: Guo, X.; Hao, X.; Tang, Z.; Diao, L.; Bai, Z.; Lu, S.; Li, L. ACE-ADP: Adversarial Contextual Embeddings Based Named Entity Recognition for Agricultural Diseases and Pests. *Agriculture* **2021**, *11*, 912. https://doi.org/10.3390/agriculture11100912

Academic Editor: Dimitre Dimitrov

Received: 5 September 2021
Accepted: 22 September 2021
Published: 24 September 2021

1. Introduction

Agricultural diseases and pests (ADPs) are one of the major disasters in the world. According to the statistics from the Food and Agriculture Organization of the United Nations (FAO), the global annual economic loss caused by ADPs exceeds US$290 billion [1]. Therefore, how to realize the early detection and early control of ADPs is very important to reduce the losses. With the rapid development of the Internet, agricultural diseases and pests-related text data have shown explosive growth, but it is difficult to be directly recognized and used by computers because of its irregularities and unstructured. The knowledge graph is essentially a semantic web, which can integrate scattered, irregular, and unstructured text data into the agricultural knowledge base. As the basic component of knowledge graph construction and question answering, the named entity recognition task is applied into digital agriculture by some knowledge graph-based human-computer diagnostic systems (e.g., website-based AI question answering systems and diagnostic systems) to identify the agricultural diseases and pests-related nouns such as "Wheat scab", "Echinocereus squameus", and "Carbendazim" from the texts collected on the Internet or users' inputs on the diagnostic systems so that to extend the agricultural knowledge graph and provide the general public with the solutions for the crop diseases and pests. It has gradually extended from the general field that extracting person and location to specific fields such as geography [2], clinical medicine [3,4], and finance [5]. However, there is still room for improvement to identify the agricultural diseases and pests-related named entities, which has important research value and practical significance for the prevention and control of agricultural diseases and pests and serving modern agriculture.

The purpose of CNER-ADP is to identify the named entities related to agricultural diseases and pests from texts. However, the following limitations in text data and NER models increase the difficulties of recognizing the named entities in agricultural diseases and pests. (1) It is insufficient annotated data in the agricultural domain, and even it is very difficult to collect enough raw text, which also occurs in other domain-specific fields [6]. Taking agriculture as an example, apart from our self-annotated corpus AgCNER [7], there is no publicly available annotated dataset, which directly hinders the research of agricultural named entity recognition. (2) Furthermore, it is impractical to solve the problem of named entity recognition in the field of agricultural diseases and pests with the help of datasets or pre-trained models in other fields, since the texts in different fields usually contain different proper nouns [8,9]. Taking Figure 1 as an example, the agricultural texts contain many domain-specific proper nouns such as "Carbendazim" and "Edifenphos", which are different in semantics from the nouns such as "right hip" and "back hip" in the field of clinical medicine and "wall calendar" and "Ware" in literature [10].

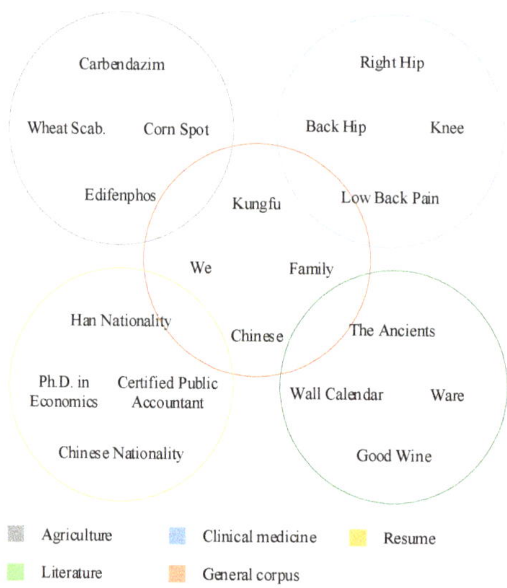

Figure 1. Visualizations of data spaces of the datasets in different fields.

In the case of NER models, as far as we know, the research of named entity recognition in agricultural diseases and pests starts relatively late compared with other domains such as social media and biomedical science [11]. The traditional methods such as rule-based methods, dictionary-based methods, and machine learning-based methods were mainly used to recognize the agricultural diseases and pests named entities [12–14]. The rule- and dictionary-based methods need to pre-design the rules or collect the dictionaries; all of them are less flexible. The machine learning-based methods such as support vector machine (SVM), naive Bayes, and conditional random field (CRF) heavily rely on the manual features, which results in not only the waste of time and effort but also an inability to meet the requirements of massive and complex texts, which have been reported in many previous works [3,15]. In recent years, deep learning, which can realize end-to-end learning without using hand-designed features, has brought breakthroughs for computer vision fields such as image classification [16–18], semantic segmentation [19], and object detection [20], and natural language processing such as text classification, machine translation, and knowledge question answering. However, there only a few works have begun to consider it, recognizing the agricultural named entities [7,15,21]. The common problem of the above models is that all of them utilize the traditional word embedding methods (e.g., word2vec [22]) to generate context-independent embeddings, which cannot effectively solve the polysemy problem, i.e., the same word may have different meanings in different contexts. For example, "Kung Fu" refers to a sport in the general field, while in the field of agricultural diseases and pests, and it refers to the name of a drug. In addition, word2vec can only learn the shallow semantic features, but it is limited in the extraction of syntax, semantics, and other high-level features [23]. In the previous work [24], the pre-trained model, such as the bidirectional encoder representation from transformers (BERT), was used to generate the context-sensitive embeddings, and the CRF was also considered as the decoder to predict the final labels. Due to the difference in data distribution between the domain-specific texts, the original BERT may be limited in the representation of specific knowledge. Recent studies have shown that there is a certain proportion of rare entities in agricultural texts, and the performance of most existing models for such entities needs to be further improved [7].

1.1. Recent Developments Related to NER Models

Researchers try to improve the models' performance mainly from two aspects, i.e., contextual encoders and text representation. Most models try their best to improve the ability to capture the useful text representation by designing efficient neural network architectures, including the commonly used contextual encoders such as Convolutional Neural Networks (CNN), Bi-directional Long Short-Term Memory (BiLSTM) [25], and their variants [26–31]. The former has significant advantages in extracting global context features, and the latter is good at capturing local context features, which are as useful as global context features. Some studies integrated the above two architectures and then proposed hybrid models, e.g., CNN-BiLSTM-CRF, to make full use of the two types of context features [7]. Other works integrated the self-attention mechanism to enhance the ability to captured long-term dependencies [3,32,33]. Moreover, some other typical variants of CNN, such as Gated CNN [27,34], RD_CNN [28], GRN [30], and CAN [31], were also proposed to recognize the entities. Besides, the transformer-based models (e.g., TENER [29] and FLAT [35]) and graph neural network-based models [36,37] have gradually attracted considerable attention in recent years. However, high-quality text representation is the prerequisite and basis for the improvement of the overall performance of the NER models; that is, text representation should contain as much knowledge as possible, such as syntax, semantics, word meaning, and so on. Otherwise, even if the context encoder maintains a strong ability of feature extraction, it may not significantly improve the final recognition accuracy [38]. There is another challenge, i.e., existing models cannot effectively identify the rare words in agricultural texts [8].

Text representation is an effective method that describes the text features by converting discrete text sequences into low-dimensional dense vectors [39]. In the early stage, non-contextual embeddings models were often used to learn shallow semantic features. Some works utilized word2vec or glove to pre-train the lookup table of word embeddings and applied it into named entity recognition [3,40]. Until now, the non-contextual embeddings models are still used to generate the word-level or character-level embeddings [41,42]. Xin Liu et al. [43] introduced a deep neural network, named OMINer, for online medical entity recognition; they also pre-trained the word2vec on a large-scale corpus to produce a lookup table that can be used for Chinese online medicine query text. Besides, some works attempt to use CNN and BiLSTM to further extract and integrate external knowledge such as radical and morphological features [3,44,45]. Although the performance of the NER model is slightly improved, word2vec has obvious limitations, i.e., they fail to distinguish different semantic information of the polysemous words and cannot extract high-level features such as syntactic structure. Recently, the language models have brought a milestone breakthrough for many natural language processing tasks. However, the available BERT model for Chinese was pre-trained on the Chinese Wikipedia corpus, which belongs to the general field. It is undeniable that the pre-trained BERT performs well in the general domain but is not efficient in specific fields [9]. Furthermore, due to the limited corpus in agricultural domains, it is unable to provide enough data for pre-training. Fine-tuning is a commonly used compromise method, which can not only solve the problem of limited data but also help the language model to learn the knowledge of specific fields [46]. Different from English and other Latin characters, Chinese characters are hieroglyphs, and their morphological structure contains rich glyph features, which can be extracted from the perspective of images and are helpful to Chinese named entity recognition. For example, Song and Sehanobish [47] managed to integrate the fine-tuned BERT and extract the glyph features for Chinese NER. Based on the above work, Xuan et al. [48] proposed a fusion glyph network to further explore the interaction between the glyph features and the contextual embeddings. In short, fine-tuning can improve the quality of text representation in the case of lacking data. However, because the BERT is task-agnostic, the limited training dataset may not cover all the semantic features in the field, which will affect the overall robustness and generalization of the NER models.

1.2. Objectives and Hypotheses

To address the abovementioned issues, a general method for agricultural diseases and pests named entity recognition, named ACE-ADP, was proposed in this paper. The objective of ACE-ADP was to use the pre-trained language model (i.e., BERT, which would be fine-tuned on the agricultural training dataset) to learn the domain-specific features. The text representation would be enhanced by the fine-tuned BERT with agricultural knowledge. Besides, adversarial training would also be introduced to enhance robustness and generalization in terms of identifying rare entities. In the course of this study, the following hypotheses were tested:

(1) An adversarial contextual embeddings-based model could be applied for agricultural diseases and pests named entity recognition. As far as we know, it was the first time that combined BERT and adversarial training to recognizing the named entities in the field of agricultural diseases and pests;
(2) The BERT, which was fine-tuned on the agricultural corpus, could generate the high-quality text representation so that to enhance the quality of text representation and solve the polysemous problem;
(3) Adversarial training could also be adopted to solve the rare entity recognition problem. Besides, it could also exert its maximum performance when the text representation was of high quality. As far as we know, the previous research had not explicitly raised this point;
(4) ACE-ADP could significantly improve the F_1 of CNER-ADP with an improvement of 4.31%, especially for rare entities, in which an F_1 was increased by 9.83% on average.

We organized the rest of the paper as follows. The experimental corpora, parameter settings, evaluation metrics, and the proposed method were introduced in Section 2. The experimental results and ablation study are presented in Section 3. The discussions are conducted in Section 4. The conclusion and future directions are described in Section 5.

2. Materials and Methods

We implemented the ACE-ADP with the TensorFlow framework and ran on a single GTX 1080 Ti GPU, Windows 10. The source code will be released at https://github.com/guojson/ACE-ADP.git (accessed date: 15 September 2021).

2.1. Datasets

We assessed the performance of our proposed method on four benchmark datasets, i.e., AgCNER [7], CLUENER [49], CCKS2017, and Resume [50]. Among them, AgCNER is an agricultural dataset that was annotated by ourselves in previous works [7] and includes 11 types of entities related to agricultural diseases and pests. The data distribution for each category in AgCNER is illustrated in Figure 2a. It can be seen from the figure that in addition to a large number of entities such as crop, disease, and pest, there are also some rare entities such as pathogeny, weed, and fertilizer, which undoubtedly increases the difficulty of CNER-ADP task. CLUENER is collected from THUCNews and contains 10 fine-grained entity categories such as Finance and Stock. CCKS2017, released by the 2017 China Conference on Knowledge Graph and Semantic Computing, contains five clinical medicine-related categories and 2231 annotated samples. According to Figure 2b, its data distribution for each category is relatively balanced, but there are also some difficulty-to-identify categories address, scene, and book need to be further considered [49]. The details of all datasets were listed in Table 1. Note that all datasets were labeled by BIO scheme (i.e., Begin, Inside, and Other) and divided into the training set and test set according to the ratio of 8:2. For word2vec-based models, we exploited the character-level embeddings that pre-trained on the Baidu Baike corpus [51].

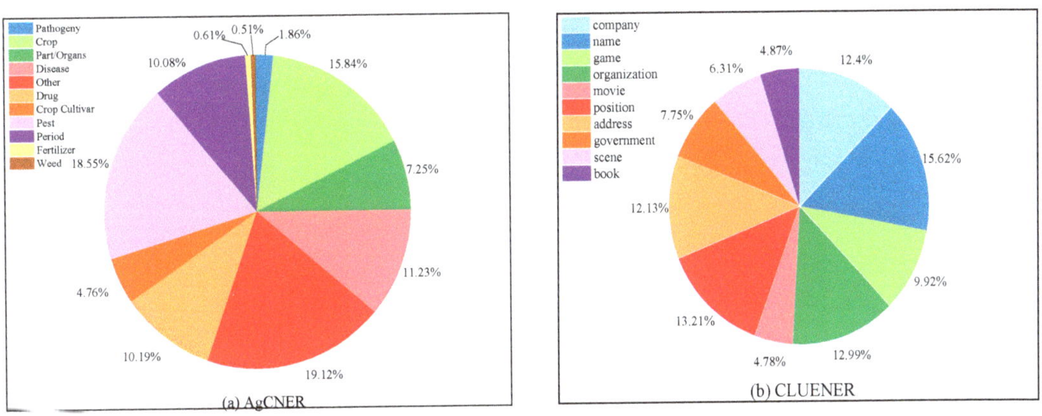

Figure 2. The proportion of each category in the data sets. (**a**) Illustrates the data distribution of each category in AgCNER; (**b**) shows the data distribution of each category in CLUENER.

Table 1. The detailed information of all datasets.

Dataset	Domain	Samples	Entities	Class	Categories
AgCNER	Agriculture	24,696	248,171	11	Crop, Disease, Drug, Fertilizer, Part/Organs, Period, Pest, Pathogeny, Crop Cultivar, Weed, Other
CLUENER	News	12,091	26,320	10	Person, Organization, Position, Company, Address, Game, Government, Scene, Book, Movie
CCKS2017	Clinic	2231	63,063	5	Body, Symptoms, Check, Disease, Treatment
Resume	Resume	4740	16,565	8	Country, Educational institution, Location, Personal name, Organization, Profession, Ethnicity, Background and Job, Title

2.2. Parameter Setting

During the training process, the exponential decay function was used to dynamically control the learning rate and thus to control the speed of parameter updating. In this paper, the decay rate was set to 0.9, and the decay step was 5000. The learning rate for BERT was set to 5×10^{-5} and 0.0001 for the NER model during the fine-tuning process. Moreover, it was set to 0.002 on AgCNER and 0.001 on other data sets during the training process. In this paper, early stopping [52] and a patience of 10 was used to prevent the over-fitting problem. Other hyper-parameters are listed in Table 2.

Table 2. Parameter settings for ACE-ADP model.

Hyper-Parameter		Value
Character embedding		768
Hidden units		256
Dropout		0.25
Optimizer		Adam
Batch_size	fine-tuning	8
	model training	32
Max_epoch	Word2vec	100
	BERT	50

2.3. Evaluation Metrics

In this paper, *Precision* (*P*), Recall (*R*), and F_1-*score* (F_1) were used as the evaluation metrics; only the boundary and type were both correctly identified, the entity could be correctly predicted. Note that their units were %, the below was the same. We ran the experiment three times according to [8], and the average results with standard deviation were listed:

$$Precision = \frac{TP}{TP + FP}, \tag{1}$$

$$Recall = \frac{TP}{TP + FN}, \tag{2}$$

$$F_1\text{-}score = \left(1 + \frac{FP + FN}{2TP}\right)^{-1}. \tag{3}$$

where TP represents the number of labels that are positive and predicted to be positive. FP represents the number of labels that are negative and predicted to be positive. FN represents the number of labels that are negative and predicted to be negative.

2.4. ACE-ADP Method

2.4.1. Problem Definition

In this paper, we regard the named entity recognition task as the sequence labeling problem. Given a sentence $S = (c_1, c_2, \ldots, c_{n},)$ with length n, where c_i represents the i-th Chinese character. Generally speaking, the discrete sentence S will be converted into low-dimensional dense embeddings, i.e., $E = (e_1, e_2, \ldots, e_{n},)$, where $e_i \in R^d$ donates the embedding vector of c_i. Then E will be fed into context encoders (e.g., BiLSTM or CNN) to extract the context features. Next, the decoder (e.g., softmax and CRF) will be exploited to predict the gold label $\hat{y}_i$ (e.g., B-LOC, I-LOC, and O) for each character c_i. Finally, the predicted labels $\hat{Y} = (\hat{y}_1, \hat{y}_2, \ldots, \hat{y}_i)$ for sentence S to be obtained, and the entities maintained in a sentence will be recognized. Formally, the object of the NER is to learn a function $f_\theta : S \rightarrow \hat{Y}$ to predict the labels for all characters.

2.4.2. Fine-Tuned BERT

As described in Section 1, obtaining high-quality embeddings is the first step for the NER model to predict the labels. Different from the early works that utilized word2vec to generate the context-independent embeddings, in this paper, BERT was considered as a generator to produce the context-sensitive embeddings according to the different contexts. BERT is composed of N layers of bidirectional Transformer blocks, which is more efficient to capture the deeper bidirectional relationships by jointly modeling the forward and backward contexts of each word. Formally, we define transformer blocks as $Trans(x)$, then the embedding vector E will be obtained as follows:

$$E_0 = S'W_e + W_p, \tag{4}$$

$$E_l = Trans(E_{l-1}),\ l \in [1,\ N], \tag{5}$$

where S' is the one-hot matrix corresponding to sentence S, W_e represents the embedding matrix pre-trained by BERT, W_p donates the positional embeddings that can be calculated by Equations (6) and (7). E_l represents the contextual embedding at the l-th layer. N is the number of layers of transformer blocks. In this paper, N was set to 12.

$$W_{(p_i,2i)} = \sin\left(p_i/10000^{2i/d}\right), \tag{6}$$

$$W_{(p_i,2i+1)} = \cos\left(p_i/10000^{2i/d}\right), \tag{7}$$

In terms of CNER-ADP, there is a general lack of corpus, which cannot provide sufficient data support for the pre-training of BERT. In this paper, fine-tuning was regarded as a compromise solution to alleviate the insufficient corpus to a certain extent. First of all, BERT parameters were initialized by using the original weights pre-trained on the Chinese Wikipedia corpus (https://storage.googleapis.com/bert_models/2018_11_03/chinese_L-12_H-768_A-12.zip), which belongs to the general domain. Then, a fully connected network was used on the top layer of the BERT to obtained the 768-dimensional context representation. Different from the BERT-CRF architecture proposed in [48], the context encoder, i.e., BiLSTM was integrated between the BERT and CRF to further extract global context features. The fine-tuning architecture for CNER-ADP was shown in Figure 3 without the component of adversarial perturbation. Besides, the fine-tuned weights were saved separately to initialize another BERT used in the models of CNER-ADP, for the reason that the learning rate for fine-tuning is minimal while the training requires a larger one. Moreover, freeze BERT contributes to decreasing the computation and storage load, which is also an important factor to be considered.

2.4.3. Context Encoder and Decoder

In this paper, BiLSTM was used as the encoder to further extract the contextual features from the text representation. It is a variant of RNN and can efficiently solve the problems of gradient vanishing and gradient explode. The formal description for a single LSTM cell is shown in Equations (8)–(10):

$$\begin{bmatrix} f_t \\ i_t \\ o_t \\ \widetilde{C}_t \end{bmatrix} = \begin{bmatrix} \sigma \\ \sigma \\ \sigma \\ tanh \end{bmatrix} \left(\begin{bmatrix} W \\ U \end{bmatrix}^T \begin{bmatrix} e_t \\ h_{t-1} \end{bmatrix} + b \right), \tag{8}$$

$$C_t = f_t * C_{t-1} + i_t * \widetilde{C}_t, \tag{9}$$

$$h_t = o_t * \tan\mathrm{h}(C_t), \tag{10}$$

where f_t, i_t, o_t, $\widetilde{C}_t$, and C_t represent the forget gate, input gate, output gate, candidate cell state, and the memory state at time step t, respectively. The sigmoid is used as the activation function $\sigma(*)$, W, U, and b are trainable parameters. e_t is input at time step t and h_{t-1} is the hidden state at the last timestep. At each time step t, BiLSTM will generate forward and backward hidden vectors $\overrightarrow{h_t}$ and $\overleftarrow{h_t}$, which maintain the forward and backward context information, respectively. The output of BiLSTM at timestep t will be obtained, i.e., $h_t = \left[\overrightarrow{h_t};\overleftarrow{h_t}\right]$ with dimension $2d_c$ and the final output for sentence S is defined as $H = (h_1, h_2, \ldots, h_n)$.

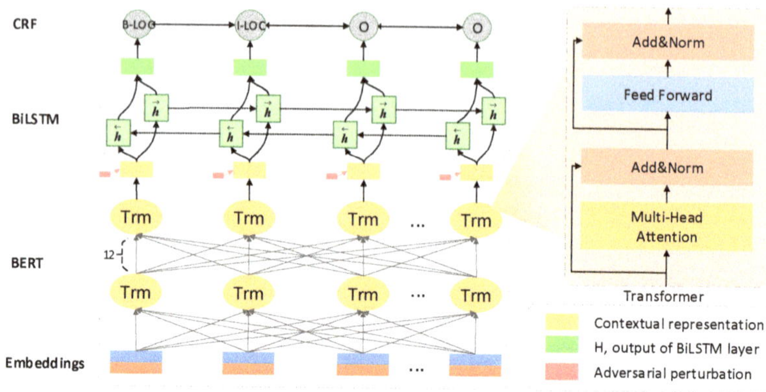

Figure 3. The architecture of the ACE-ADP model for agricultural diseases and pests.

In this paper, the CRF was considered as the decoder because of the strong dependency between adjacent labels in sequence labeling tasks. For example, B-LOC is usually followed by I-LOC but cannot be followed by B-PER or I-PER. Therefore, joint decoding may be more beneficial than independent decoding [53]. As shown in Equation (11), given the predicted tags $\hat{Y}$ of sentence S and its corresponding embedding vector E, its score is calculated by the state score $P \in R^{n \times d_r}$ and state transition matrix $T \in R^{d_r \times d_r}$. Among them, P is mapped from H by a fully connected layer (Equation (12)). Thus, the probability of predicted labels $\hat{Y}$ in all possible tag sequences Y_{all} is calculated by Equation (13):

$$Score(\hat{Y}, S) = \sum_{i=0}^{n} P_{i,\hat{y}_i} + \sum_{i=0}^{n-1} T_{\hat{y}_i,\hat{y}_{i+1}}, \tag{11}$$

$$P = HW_p + b_p, \tag{12}$$

$$p(\hat{Y}|S) = \frac{\exp(Score(S, \hat{Y}))}{\sum_{\hat{Y}' \in Y_{all}} \exp(Score(S, \hat{Y}'))}, \tag{13}$$

where $W_p \in R^{2d_c \times d_r}$ and $b_p \in R^{n \times d_r}$ are trainable parameters, d_r is the number of tags. The gold tag sequence with the highest score is obtained by the Viterbi algorithm.

2.4.4. Adversarial Training

To solve the over-fitting problem and enhance the ability to recognize the rare entities, we treated adversarial training as a data augmentation method, i.e., a new adversarial sample would be generated after adding a small perturbation to the training sample. Assuming that the loss function of the ACE-ADP model without adversarial training was shown in Equation (14). Y represents the ground truth labels. The goal of our model is to minimize the loss by training the weights θ.

$$loss(\hat{Y}, Y) = -\sum logp(\hat{y}|E;\theta), \tag{14}$$

$$loss(\hat{Y}_{adv}, Y) = -\sum logp(\hat{y}|E + \eta_{adv};\theta), \tag{15}$$

As shown in Equation (15), the adversarial training guided loss function would be obtained after adding a worst-case perturbation η_{adv} to the embeddings. In general, η_{adv} can be calculated by the following function:

$$\eta_{adv} = \underset{\eta, ||\eta|| \le \varepsilon}{\text{argmin}} logp(\hat{y}|E + \eta;\hat{\theta}), \tag{16}$$

where η is a perturbation, ε is the bounded norm, which can be calculated by $\varepsilon = \gamma\sqrt{d}$ according to [8], d is the dimension of embeddings, γ is perturbation size that should be reasonably selected for the reason that if γ is too small to play the role of perturbation. Conversely, it will easily introduce noise that can destroy the original semantic information. $\hat{\theta}$ presents the current training weights of the model. Due to the non-differentiability of Equation (16), similar to [54], the approximation is used to replace η_{adv}, as shown in Equation (17).

$$\eta_{adv} = -\frac{\varepsilon g}{\| g \|_2}, \text{where} g = \nabla_E logp(\hat{y}|E;\hat{\theta}), \tag{17}$$

The loss function of the model with adversarial training was defined as follows in Equation (18).

$$loss = loss(\hat{Y}, Y) + loss(\hat{Y}_{adv}, Y), \tag{18}$$

The steps of our proposed model can be found in Appendix A. Based on the above descriptions, the advantages of innovative works for the CNER-ADP task can be summarized as follows:

(1) Contextual-sensitive. BERT can dynamically generate the context-dependent embeddings according to the contexts, which is beneficial for solving the problem of polysemous words that are often caused by context-independent methods such as word2vec and glove;
(2) Domain-aware. In this paper, domain knowledge can be injected into BERT by fine-tuning, which is essential to handle the NER task in specific domains;
(3) Stronger robustness and generalization. The experimental results in Section 4.4 showed that compared with previous models, our proposed model maintains high robustness and generalization.

3. Results

To verify the effectiveness of the ACE-ADP, we conducted comprehensive experiments with several state-of-the-art models on the self-annotated agricultural datasets AgCNER and three other corpora that belong to different fields. The experimental results showed that the proposed model achieved remarkable results and could significantly improve the accuracy of difficult-to-identify entities such as the entities with fuzzy boundaries and the rare ones.

3.1. Main Results Compared with Other Models

The results of all models (i.e., ACE-ADP and several state-of-the-art models proposed in recent years) on four datasets were listed in Table 3. Note that we exploited the fine-tuning BERT for IDCNN, Gated CNN, and AR-CCNER [3] to obtain the best results, and others were set according to their original papers. Our proposed model achieved the highest F_1 of 93.68%, 98.31%, 95.72%, and 96.83% on CLUENER, AgCNER, CCKS2017, and Resume, respectively. For example, ACE-ADP outperformed the IDCNN with improvements of 3.57%, 15.7% in terms of F_1 on AgCNER and CLUENER due to the effectiveness of adversarial training. Moreover, compared with IDCNN, Gated CNN tended to achieve slightly better F_1 on AgCNER and Resume, which benefits from the gated structure that can filter useful features according to their importance. However, due to the blurring boundaries of entities in CLUENER and ccks2017, it performed slightly worse than IDCNN. In contrast, AR-CCNER has achieved a few higher F_1-*scores* than IDCNN and Gated CNN on most datasets, thanks to the fact that radical features may provide rich external knowledge, and the self-attention mechanism helps to enhance the model's ability to capture the long-distance dependencies.

Moreover, we also conducted experiments with other state-of-the-art models, i.e., FGN [48], TENER [29], and Flat-Lattice [35]. FGN not only outperformed IDCNN, Gated CNN as reported by Xuan et al. [48], who integrated the interactive information between the contextual embeddings generated by the fine-tuning BERT and the glyph information extracted by novel CNN structure, but also better than TENER, a transformer-based model, indicating that the fine-tuning BERT may outperform a single transformer-based model in

NER task. Moreover, Flat-Lattice, which benefits from the flat-lattice Transformer and the well-designed position encoding, also presented remarkable results. However, the number of potential words will be increased as the length of the sentence increases, which tends to result in a significant increase in the structural complexity. Unlike FGN and Flat-Lattice, apart from the basic framework BiLSTM-CRF, our model only utilized the fine-tuned BERT and adversarial training to enhance the robustness and generalization, showing lower structural complexity. The results of ACE-ADP went beyond previous reports and slightly lower standard deviations showing its positive effect on domain-specific (e.g., the agricultural diseases and pests) named entity recognition task.

Table 3. Experimental results for all models on four different datasets.

Algorithms	**CLUENER**			**AgCNER**			**CCKS2017**			**Resume**		
	P	R	F_1	P	R	F_1	P	R	F_1	P	R	F_1
BERT-IDCNN-CRF	78.37	77.60	77.98 ± 0.11	94.39	95.08	94.74 ± 0.07	90.55	93.52	92.01 ± 0.13	95.47	96.59	96.03 ± 0.13
BERT-Gated CNN-CRF	75.85	77.98	76.90 ± 0.34	94.32	95.20	94.76 ± 0.08	89.43	92.93	91.15 ± 0.14	95.75	96.81	96.27 ± 0.16
AR-CCNER	78.34	77.74	78.04 ± 0.28	94.60	94.73	94.67 ± 0.06	90.23	93.36	91.77 ± 0.28	95.89	97.22	96.55 ± 0.27
FGN [48]	79.50	79.71	79.60 ± 0.15	94.33	94.56	94.45 ± 0.03	90.44	93.09	91.75 ± 0.16	96.67	97.09	96.88 ± 0.10
TENER	72.94	74.21	73.57 ± 0.17	93.01	95.22	94.10 ± 0.09	91.24	93.08	92.15 ± 0.13	94.91	95.03	94.97 ± 0.21
Flat-Lattice [35]	79.25	80.68	79.96 ± 0.13	93.52	94.31	93.91 ± 0.08	91.55	93.40	92.46 ± 0.16	95.22	95.72	95.47 ± 0.18
ACE-ADP	**93.03**	**94.36**	**93.68 ± 0.18**	**98.30**	**98.32**	**98.31 ± 0.02**	**95.17**	**96.27**	**95.72 ± 0.13**	**96.22**	**97.44**	**96.83 ± 0.17**

3.2. Ablation Study

3.2.1. Macro-Level Analysis

The contextual embeddings based on fine-tuned BERT and adversarial training were the focus of this paper. An ablation study was first conducted to verify their effectiveness and necessity from a macro-level perspective. The experimental results are listed in Table 4. Taking the AgCNER and Resume as an example, according to groups 1 and 4, the model with adversarial training improves the F_1 on AgCNER and Resume by 3.43% and 0.97%, respectively, indicating that adversarial training helps to improve the performance of named entity recognition of the models. Besides, the F_1-*scores* in groups 1, 2, and 5 presented that adversarial training was positively related to the quality of text embeddings, i.e., the higher the quality of text representation, the better the effect of adversarial training. From the results of groups 2, 3, and 1, we could observe that the word2vec-based model (group 2) showed the worst performance on all datasets, the possible reason is that the text embeddings generated by word2vec are context-independent, which cannot provide enough semantic information for the model training. In contrast, the models that integrated the original and fine-tuned BERT delivered significantly better F_1-*scores*, i.e., 96.11% and 98.31% on AgCNER, and 96.38% and 96.83% on Resume respectively, due to the high-quality context embeddings based on BERT in the case of adversarial training. Meanwhile, the fine-tuned BERT-based model (group 1) presented better performance than the original BERT-based model (group 3) for the reason that the contextual embeddings generated by BERT express abundant semantic information, which has a positive effect on improving the performance of the model. Moreover, fine-tuning enables BERT to obtain domain awareness and makes the contextual embeddings contain more domain-specific knowledge, which is crucial for domain-specific NER tasks. In particular, compared with the baseline model listed in group 5 (i.e., word2vec-based BiLSTM-CRF), the F_1-*score* of the proposed model on the AgCNER was increased by 4.23%. Similar results could also be presented on other datasets. In short, the presented findings confirmed the necessity and effectiveness of contextual embeddings and adversarial training.

Table 4. Recognition results of ACE-ADP and its variants on four datasets.

#	Algorithms	CLUENER			AgCNER			CCKS2017			Resume		
		P	R	F_1	P	R	F_1	P	R	F_1	P	R	F_1
1	ACE-ADP	93.03	94.36	93.68 ± 0.18	98.30	98.32	98.31 ± 0.02	95.17	96.27	95.72 ± 0.13	96.22	97.44	96.83 ± 0.17
2	-BERT	68.43	67.15	67.78 ± 0.29	94.01	93.89	93.95 ± 0.06	90.27	91.86	91.05 ± 0.23	91.25	93.15	92.19 ± 0.15
3	-fine-tuning	92.02	93.16	92.58 ± 0.13	95.99	96.23	96.11 ± 0.17	95.01	97.15	96.07 ± 0.16	95.78	96.85	96.38 ± 0.09
4	-AT	78.83	77.39	78.11 ± 0.02	94.59	95.16	94.88 ± 0.04	90.30	92.84	91.56 ± 0.14	95.12	96.60	95.86 ± 0.28
5	-BERT-AT	68.48	66.95	67.70 ± 0.41	94.18	93.99	94.08 ± 0.06	89.16	91.42	90.27 ± 0.13	92.09	93.56	92.82 ± 0.12

'-' means not participating in training.

3.2.2. Effect of BERT

To further verify the effectiveness of BERT in detail, several experiments with word2vec-, original BERT-, and fine-tuned BERT-based models were conducted on four benchmark datasets. Their F_1-*scores* are presented in Table 5. As expected, the word2vec-based models tended toward the lower F_1-*scores* than BERT-based ones on all datasets, as discussed in Section 3.2.1. BERT, which consists of 12-layer of bidirectional Transformers, could dynamically generate high-quality embeddings according to the different contexts. For example, BiLSTM with original BERT achieved significant improvement of F_1-*scores* with +9.07%, +0.11%, +1.35%, and +2.06% on CLUENER, AgCNER, CCKS2017, and Resume, respectively. However, there was still room for improvement because of the task independence of the original BERT. The fine-tuning BERT-based models have presented the best performance in multiple domain-specific datasets for the reason that BERT not only maintains the strong ability of semantic representation but also obtains the domain awareness after fine-tuning, which may encourage BERT to represent the domain-specific features efficiently [9]. Besides, in the case of BERT, the performance of IDCNN, Gated CNN, AR-CNER, and CNN-BiLSTM-CRF were also improved. Therefore, the present findings demonstrated the effectiveness of BERT, and it would be more suitable for the domain-specific NER tasks after fine-tuning.

3.2.3. Effect of Adversarial Training

The F_1 of the adversarial training-based models with word2vec, original BERT, and fine-tuning BERT were presented in Table 6. Combining with the details presented in Table 5, several important conclusions could be summarized: (1) The word2vec-based model with adversarial training tended towards slightly worse results, indicating that in the case of poor text representation, adding perturbation would be counterproductive. (2) The F_1 of original BERT-based models with adversarial training were significantly improved compared with those listed in Table 5, indicating the effectiveness of the adversarial training to enhance the robustness and generalization. Taking BiLSTM as an example, its F_1 increased by +1.92% on AgCNER and +1.5% on Resume. (3) The F_1 of BiLSTM with fine-tuning BERT and adversarial training were further increased by +2.2% on AgCNER and +0.45% on Resume, which indicated that in the case of original BERT and fine-tuned BERT, the recognition performance could be further improved by using adversarial training. (4) There was very little difference in terms of F_1-*scores* between the BiLSTM, AR-CCNER, and CNN-BiLSTM-CRF, which indicated that the complex architectures such as radical features, self-attention, and CNN might be unnecessary. (5) The experimental results in Tables 5 and 6 show that Gated CNN and RD_CNN achieved better performance than BiLSTM. In actual uses, they could replace BiLSTM as feature encoders to extract local and global context features when integrating high-quality text representation and adversarial training. Furthermore, most of the standard deviations listed in Table 6 are lower than those in Table 5, indicating that the adversarial training may contribute to improving the stability of the model. In short, the above experimental results verified the effectiveness of adversarial training and once again demonstrated that adversarial training could enhance the robustness of the NER model.

Table 5. F_1 of models with word2vec, original BERT, and fine-tuning BERT without adversarial training.

Algorithms	CLUENER			AgCNER			CCKS2017			Resume		
	W	O	F	W	O	F	W	O	F	W	O	F
BiLSTM	67.70 ±0.41	76.77 ±0.35	78.11 ±0.18	94.08 ±0.06	94.19 ±0.07	94.88 ±0.02	90.27 ±0.13	91.62 ±0.16	91.56 ±0.13	92.82 ±0.12	94.88 ±0.11	95.86 ±0.17
IDCNN	66.58 ±0.38	76.33 ±0.23	77.98 ±0.11	93.99 ±0.06	93.91 ±0.13	94.74 ±0.07	91.46 ±0.29	91.20 ±0.31	92.01 ±0.13	92.71 ±0.42	94.44 ±0.33	96.03 ±0.13
Gated CNN	66.26 ±0.25	75.23 +0.22	76.90 ±0.34	93.56 ±0.11	93.72 ±0.02	94.76 ±0.08	91.02 ±0.28	89.86 ±0.12	91.15 ±0.14	89.25 ±0.35	93.12 ±0.23	96.27 ±0.16
RD_CNN	66.16 ±0.15	75.73 ±0.21	77.95 ±0.18	93.20 ±0.08	93.89 ±0.04	94.82 ±0.05	89.18 ±0.23	90.03 ±0.19	91.52 ±0.17	89.56 ±0.23	93.39 ±0.17	95.87 ±0.19
AR-CCNER	68.67 ±0.35	77.08 ±0.26	78.04 ±0.28	94.46 ±0.08	94.12 ±0.06	94.67 ±0.06	91.45 ±0.30	91.10 ±0.15	91.77 ±0.28	93.09 ±0.25	95.01 ±0.19	96.55 ±0.27
CNN-BiLSTM-CRF	68.45 ±0.37	76.88 ±0.22	78.18 ±0.12	94.07 ±0.12	94.53 ±0.02	94.78 ±0.05	92.03 ±0.16	91.49 ±0.24	91.28 ±0.25	93.84 ±0.18	95.18 ±0.16	95.26 ±0.24

"W" represents word2vec, "O" means the original BERT, and "F" donates the fine-tuned BERT.

Table 6. F_1 of adversarial training-based models with word2vec, original BERT, and fine-tuning BERT.

Algorithms	CLUENER			AgCNER			CCKS2017			Resume		
	W	O	F	W	O	F	W	O	F	W	O	F
BiLSTM	67.78 ±0.29	92.58 ±0.13	93.68 ±0.18	93.95 ±0.06	96.11 ±0.17	98.31 ±0.02	91.05 ±0.23	96.07 ±0.16	95.72 ±0.13	92.19 ±0.15	96.38 ±0.09	96.83 ±0.17
IDCNN	66.12 ±0.25	94.72 ±0.21	94.45 ±0.17	93.71 ±0.08	96.98 ±0.14	98.23 ±0.05	91.27 ±0.19	96.12 ±0.13	95.25 ±0.17	93.13 ±0.12	96.91 ±0.11	96.16 ±0.12
Gated CNN	66.07 ±0.14	95.03 ±0.16	96.33 ±0.13	93.48 ±0.03	97.48 ±0.15	98.42 ±0.08	90.88 ±0.12	96.27 ±0.11	96.19 ±0.14	91.57 ±0.16	96.73 ±0.11	97.57 ±0.15
RD_CNN	65.88 ±0.16	94.51 ±0.18	96.68 ±0.13	92.86 ±0.07	97.35 ±0.14	98.95 ±0.05	90.18 ±0.16	95.56 ±0.10	95.61 ±0.15	91.20 ±0.17	96.01 ±0.13	97.34 ±0.14
AR-CCNER	62.30 ±0.36	91.64 ±0.24	89.66 ±0.25	92.80 ±0.11	97.50 ±0.12	97.70 ±0.06	90.96 ±0.20	96.08 ±0.16	95.97 ±0.12	90.36 ±0.15	96.74 ±0.12	97.26 ±0.14
CNN-BiLSTM-CRF	67.23 ±0.26	90.81 ±0.19	89.82 ±0.22	93.67 ±0.11	96.57 ±0.12	97.66 ±0.12	91.63 ±0.16	95.33 ±0.19	95.14 ±0.17	93.07 ±0.16	96.77 ±0.13	96.91 ±0.16

"W" represents word2vec, "O" means the original BERT, and "F" donates the fine-tuned BERT.

4. Discussion

4.1. Performance for Rare Entities

In this section, AgCNER and CLUENER were selected as comparable datasets to verify the performance of ACE-ADP in identifying the rare entities, which are challenging for NER models.

The experimental results were illustrated in Figures 4 and 5, which showed that ACE-ADP outperformed other comparable models and significantly improved the F_1-*scores* of all categories both in AgCNER and CLUENER, especially the categories that are difficult-to-identify or have a low percentage of entities. In terms of AgCNER, ACE-ADP still maintained the highest F_1 on easy-to-identify entities such as disease, pest, and crop, which were 98.98%, 98.73%, and 98.88%, respectively. Meanwhile, it significantly improved the F_1 of rare entities such as fertilizer, weed, and pathogeny by +11.99%, +3.95%, and +8.25% (8.06% on average) compared with the word2vec-based BiLSTM. Moreover, according to the results of CLUENER reported in [49], ACE-ADP could effectively improve the Precision, Recall, and F_1 of difficult-to-recognize entities such as address, scene, and book, all of them achieved F_1-*scores* above 90%. The possible reasons are that adversarial training is helpful to improve the robustness of the model, and fine-tuned BERT can generate the character-level embeddings with rich domain-specific semantic information, which also contributes to improving the NER performance.

In addition, we took the Ft-BERT-BiLSTM and ACE-ADP as examples and visualized their confusion matrices on the AgCNER to further illustrate the effectiveness of the proposed model. As shown in Figure 6, ACE-ADP obtained more correctly predicted labels of the rare entities (e.g., fertilizer, weed, and pathogeny) than Ft-BERT-BiLSTM, which means that ACE-ADP can achieve higher TP, while FP and FN are relatively low.

According to Equation (3), ACE-ADP tends to obtain a higher F_1. Thus, the experimental results illustrated that the high-quality text representation and adversarial training could effectively enhance the NER models' robustness and were useful to identify the rare and difficulty-to-identify entities.

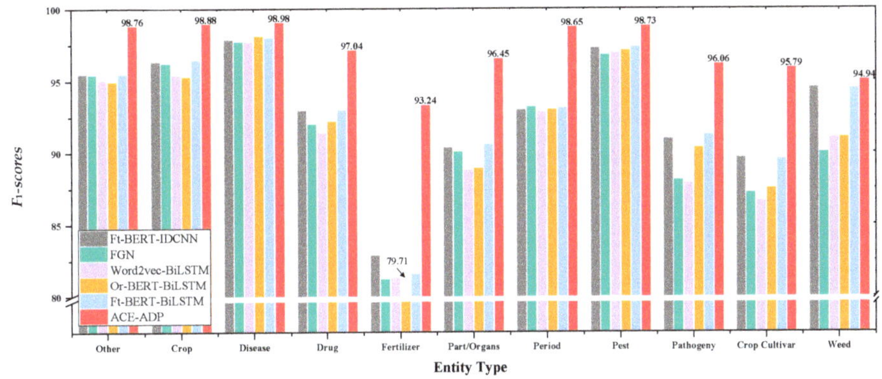

Figure 4. Detailed results of F_1-*scores* for each category on AgCNER. "Ft" means the fine-tuned BERT, "Or" describes the original BERT.

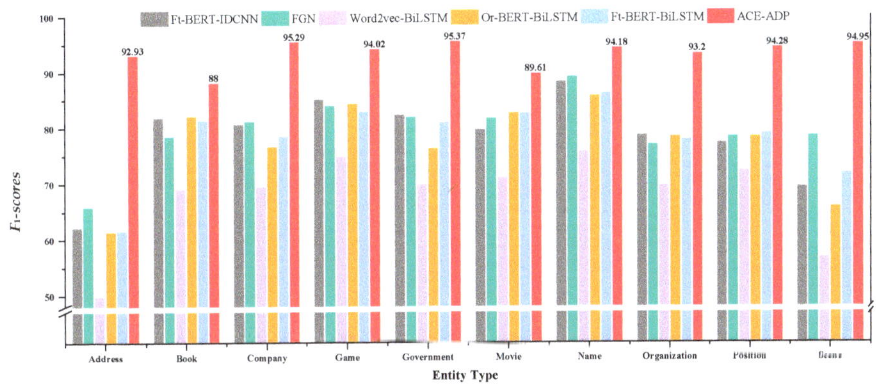

Figure 5. Detailed results of F_1-*scores* for each category on CLUENER.

4.2. Robustness and Generalization

The curves of training and validation loss for Word2vec-BiLSTM-CRF, BERT-BiLSTM-CRF, and ACE-ADP on AgCNER were visualized as Figure 7. As shown in Figure 7a,b, the validation losses of Word2vec-BiLSTM-CRF and BERT-BiLSTM-CRF decrease with the increase in iteration and then gradually increase, showing the obvious over-fitting characteristics, while that of ACE-ADP decreases first and then tends to be flat, indicating that ACE-ADP could effectively alleviate the over-fitting problem.

Apart from the training set and testing set of AgCNER used in this paper, another dataset, which has never been used before and contains 2223 agricultural samples, was considered as the final testing dataset to further verify the model's robustness and generalization. Besides, the experiments were also conducted on the standard Resume, which contains the standard training set, development set, and testing set, and is widely used in Chinese NER tasks. The experimental results on the extended AgCNER and standard Resume are listed in Table 7. Taking AgCNER as an example, the models that integrated fine-tuned BERT and adversarial training outperformed the state-of-the-art models, i.e., FGN, Flat-Lattice, and TENER, and delivered significantly better Precision, Recall, and F_1-*scores* on both development and testing sets. The same conclusion could also be drawn on Resume.

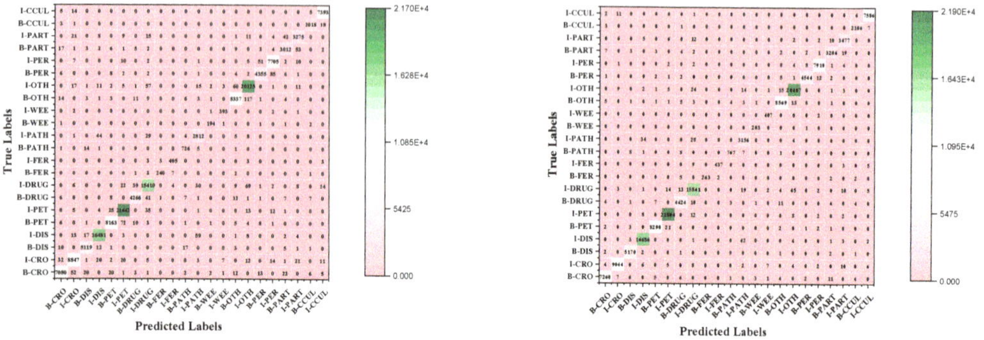

(**a**) Ft-BERT-BiLSTM (**b**) ACE-ADP

Figure 6. Confusion matrixes of (**a**) Ft-BERT-BiLSTM, and (**b**) ACE-ADP on AgCNER dataset. *x*-axis: predicted labels; *y*-axis: true-axis; numbers on the cell where x = y represents the TP values.

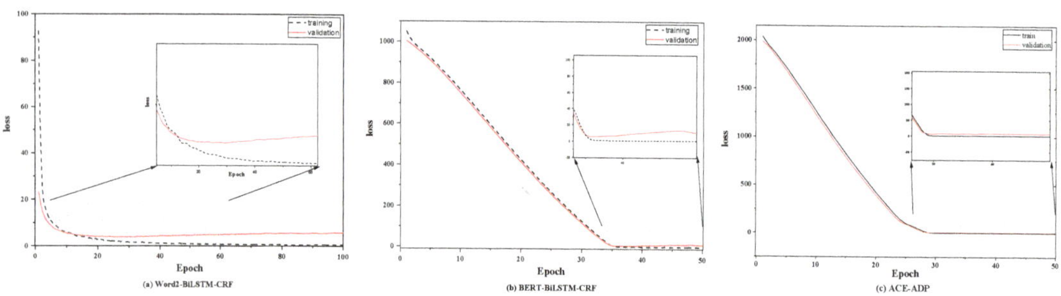

Figure 7. The training and validation losses for (**a**) Word2vec-BiLSTM-CRF, (**b**) BERT-BiLSTM-CRF, and (**c**) ACE-ADP on the AgCNER dataset.

Table 7. Experimental results on extended AgCNER and Resume.

Algorithms	AgCNER						Resume					
	Dev			Test			Dev			Test		
	P	*R*	F_1	*P*	*R*	F_1	*P*	*R*	F_1	*P*	*R*	*F*
ACE-ADP	98.30	98.32	98.31	98.50	98.47	98.49	96.43	97.79	97.11	96.63	97.66	97.14
IDCNN	98.15	98.30	98.23	98.18	98.25	98.21	95.52	96.69	96.10	94.55	96.13	95.33
Gated CNN	98.07	98.76	98.42	98.17	98.75	98.46	96.57	98.47	97.51	96.86	99.00	97.92
RD_CNN	98.71	99.20	98.95	98.69	99.19	98.94	96.87	98.65	97.75	97.11	98.93	98.01
AR-CCNER	97.38	98.03	97.70	97.80	97.88	97.84	95.91	97.79	96.84	97.10	98.33	97.71
CNN-BiLSTM-CRF	97.51	97.81	97.66	97.63	97.58	97.61	95.91	96.38	96.14	95.64	96.79	96.22
FGN	94.33	94.56	94.45	94.26	94.62	94.44	93.13	95.82	94.46	92.12	94.73	93.41
Flat-Lattice	93.52	94.31	93.91	93.71	94.11	93.91	94.74	96.26	95.49	94.90	95.83	95.36
TENER	92.88	95.09	93.97	93.03	95.09	94.05	94.45	95.09	94.77	93.71	94.52	94.11

"dev" represents the development set, "test" donates as the testing set.

Therefore, the above experimental results demonstrated that our works, i.e., integrating the contextual embedding and adversarial training, may contribute to alleviating the over-fitting problem and enhancing the robustness and generalization of the NER models, which would provide us with a feasible solution for the issue of agricultural diseases and pests named entity recognition.

4.3. Convergence

As shown in Figure 8, to evaluate the impact of contextual embeddings and adversarial training on convergence, we took AgCNER as an example and visualized the change curves of F_1 with each iteration by using the BiLSTM that integrated the word2vec, original BERT (Or), fine-tuning BERT (Ft), and adversarial training (AT), respectively. The word2vec-based model showed the slowest convergence speed. Meanwhile, the adversarial training (Word2vec-AT) seems to speed up the convergence of the model a lot at the beginning, but the effect was limited. In contrast, from the change curve of BERT-Or, it could be seen that BERT significantly accelerated the convergence, for the reason that compared to word2vec, the contextual embeddings generated by BERT contain deeper semantic information and would provide better initialization for the model [23]. However, due to the task independence and lack of domain knowledge, the F_1 was slightly lower than that of the word2vec-based model when it tended to be stable. Fine-tuning could equip BERT with domain awareness and provide abundant domain-specific features for the contextual encoders. Therefore, the convergence of BERT-Ft was greatly accelerated; its F_1-*scores* are generally higher than word2vec- and original BERT-based models. In terms of BERT-Or-AT and BERT-Ft-AT, adversarial training could not only accelerate the convergence speed but also significantly improve the recognition performance of the model in the case of high-quality text representation. For example, during the entire training process, the values of F_1 of BERT-Ft-AT were always higher than those of BERT-Ft. Therefore, the above experimental results showed that fine-tuning BERT and adversarial training could improve not only the NER performance but also accelerate the convergence.

4.4. Visualization of Features

To intuitively illustrate the effective effect of BERT on the agricultural and other domain-specific text representation, we visualized the sentence-level embeddings produced by the embeddings-based methods on the training data of the four datasets from From four different perspectives. As shown in Table 8, 100 samples for each dataset were randomly selected, and each sentence-level embedding was projected into a three-dimensional vector by using T-SNE [55]. All the images were obtained by rotating clockwise about the Z-axis by 0°, 90°, 180°, and 270°, respectively. In the first row of Table 8, all data points are mixed indiscriminately in space, illustrating that the text representation generated by word2vec cannot effectively represent the semantic features in different domains. In the second row of Table 8, the same type of data points, especially those belonging to AgCNER, CCKS2017, and Resume, were clustered well. However, the data points belonging to CLUENER were relatively loose, and there were several data points mixed with other types of data points, confirming that the original BERT does have a positive effect on the text representation, but due to the task independence, it may not be enough that only using it to generate the domain-specific embeddings. For the fine-tuned BERT (i.e., the last row of Table 8), the data points were correctly divided into four clusters, and the similar data points were more closely distributed, verifying that fine-tuning makes BERT have domain-awareness. Moreover, compared to Word2vec and Original BERT in Table 8, it was more clear of the boundaries between the different types of data points, which was consistent with [9], indicating that injecting domain-specific knowledge by fine-turning may be helpful to the domain-specific NER task.

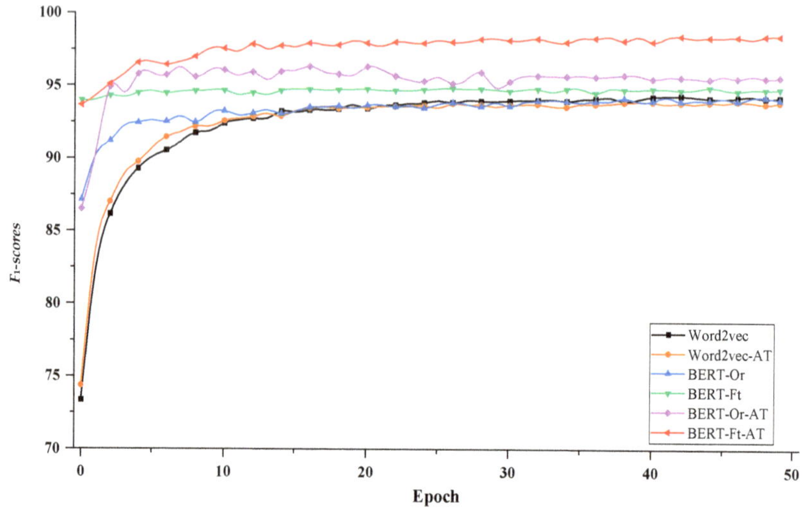

Figure 8. The trend of F_1 for BiLSTM-based models on AgCNER.

Table 8. Visualized results for each dataset.

Rotation Angles	0°	90°	180°	270°
Word2vec	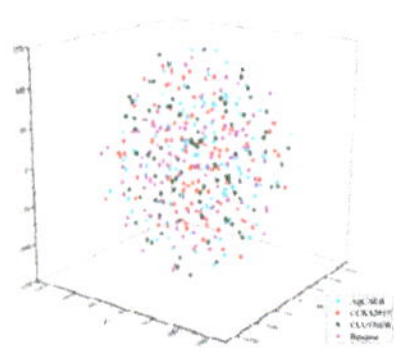	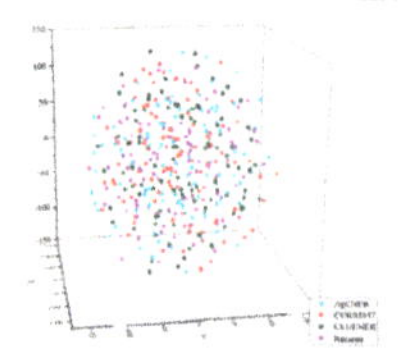	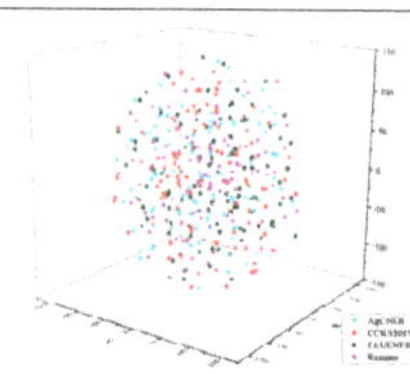	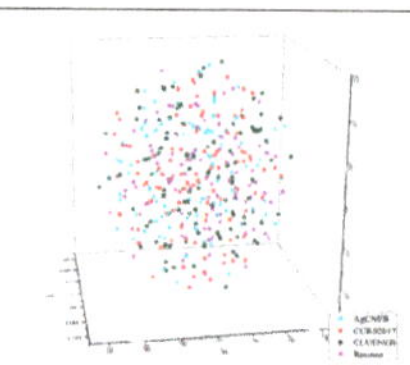
Original BERT	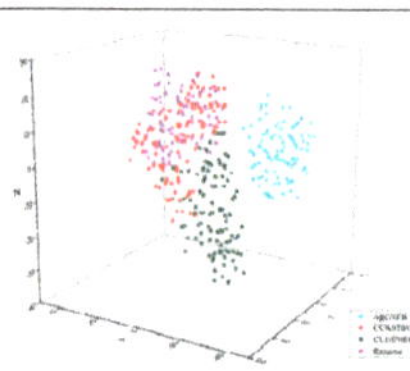	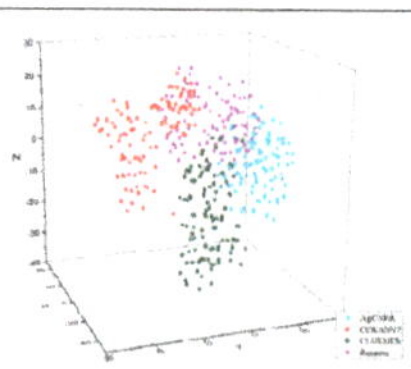	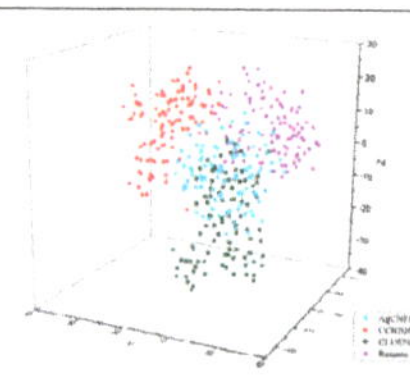	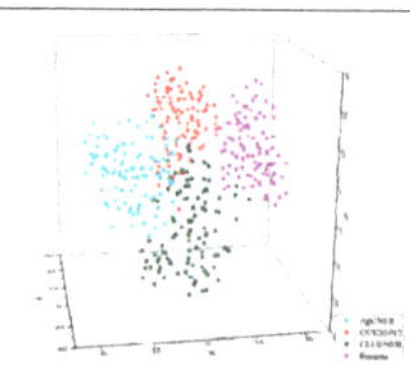
Fine-tuned BERT	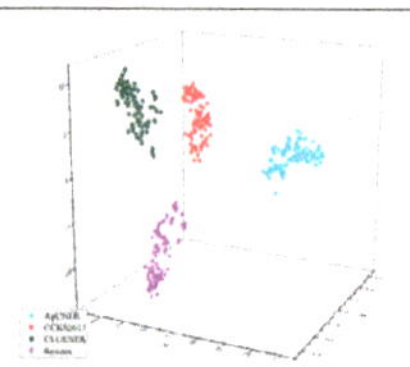	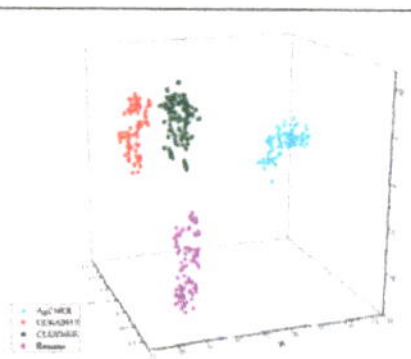	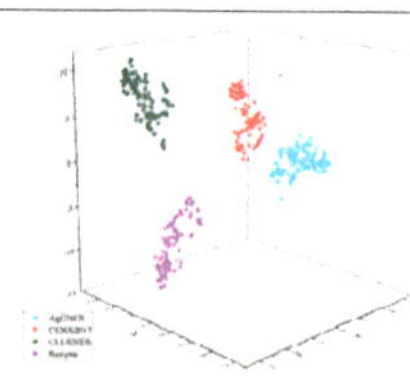	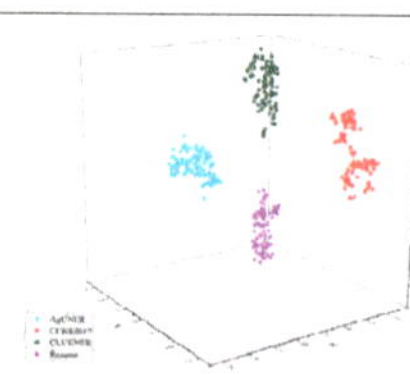

4.5. Parameter Analysis

Perturbation size γ represents the degree of perturbation to embedding representation. It is one of the most important parameters during the adversarial training process. According to [56,57], a group of γ range from 0.001 to 0.1 were selected as the candidate perturbation sizes to find the most suitable γ. Similar to [57], the bigger γ was not considered since the larger perturbation may destroy the semantic information of the text representation. As shown in Figure 9, the model achieved different F_1-*scores* for AgCNER and Resume when the perturbation size γ was set to different values, which proved that different perturbation sizes could affect the performance of the model in different degrees. Besides, the final results were not obvious when γ was varied from 0.001 to 0.01, indicating that when $\gamma < 0.01$, the effect of adversarial training on the NER model was hardly observed. We could also find that when $\gamma = 0.1$, the model obtained the optimal F_1-*scores* of 98.31% and 96.83% on AgCNER and Resume, respectively, indicating that adversarial training could give full play to its performance. Thereby, 0.1 was selected as the perturbation factor during the entire experiment.

In summary, comprehensive experiments and discussions demonstrated the effective performance of ACE-ADP in identifying the agricultural named entities. Besides, an ablation study further demonstrated that it could effectively identify rare entities while accelerating convergence. In the future, we will extend our model to other specific fields to further verify its robustness and generalization. Moreover, we also attempt to improve the model to make it suitable for relation extraction and the joint intent recognition and slot filling task so that to play a role in the construction of agricultural diseases and pests question answering systems based on the knowledge graph.

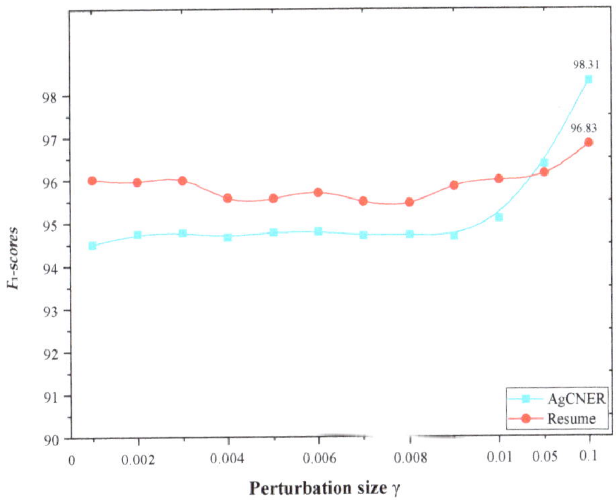

Figure 9. Results by different perturbation sizes γ.

5. Conclusions

To address the problems of rare entity recognition and polysemous words in CNER-ADP tasks, we presented a universal ACE-ADP framework, which effectively enhances the semantic feature representation by introducing and integrating contextual embeddings and adversarial training to recognize the named entities in agricultural diseases and pests. The high-quality context embeddings with agricultural knowledge and high-level features were generated by adopting the BERT that fine-tuned on the agricultural corpus. Furthermore, adversarial training was also introduced to enhance the robustness and generalization of the NER model. Comprehensive experimental results showed that ACE-ADP could significantly improve the F_1-*scores* of the agriculture-related dataset. Moreover, the ablation study and discussion not only verified that ACE-ADP maintained strong robustness and generalization but also showed that it had a strong ability to recognize the rare entities, which is of great benefit to the construction of agricultural diseases and pests question answering systems based on the knowledge graph. To serve digital agriculture, the proposed model will be integrated into the knowledge graph-based question answering systems so that to improve the accuracy in identifying the agricultural diseases and pets-related nouns and make the question answering systems provide more accurate solutions for the agricultural diseases and pests control.

Author Contributions: Conceptualization, methodology, X.G. and X.H.; validation, Z.T., L.D., Z.B. and S.L.; supervision, project administration, L.L. All authors have read and agreed to the published version of the manuscript.

Funding: This research was funded by the National Key Research and Development Program, grant number 2016YFD0300710.

Institutional Review Board Statement: Not applicable.

Informed Consent Statement: Not applicable.

Data Availability Statement: Not applicable.

Conflicts of Interest: The authors declare no conflict of interest.

Appendix A

As shown in Algorithm A1, the steps of our proposed model can be summarized as follows:

Algorithm A1 Pseudocode for domain-specific named entity recognition task with adversarial training and contextual embeddings.

Input: Fine-tuned BERT model for a specific field, global learning rate l_r, perturbation size γ, the number of iterations T, a domain-specific sentence S, and their ground-truth labels Y.
Output: the predicted labels $\hat{Y}$, the training weights of the model $\hat{\theta}$.
1: Converting the sentence S into the contextual embeddings $E = (e_1, e_2, \ldots, e_n)$ by fine-tuned BERT on the texts in the field of agricultural diseases and pests.
2: For t = 1, . . . , T do
3: $H = BiLSTM(E)$, according to Equation (8) to Equation (10).
4: $P = HW + b$, according to Equation (12).
5: Calculating the $loss(\hat{Y}, Y)$ by using the CRF algorithm.
6: $g \leftarrow \nabla_E logp(\hat{y}|E;\hat{\theta})$,
7: $\varepsilon \leftarrow \gamma\sqrt{d}$
8: $\eta_{adv} \leftarrow -\varepsilon \times g/l2\_normalize(g)$
9: $E_{adv} = E + \eta_{adv}$
10: $loss(\hat{Y}_{adv}, Y) \leftarrow$ Repeat lines 3–9
11: $loss = loss(\hat{Y}, Y) + loss(\hat{Y}_{adv}, Y)$
12: F_1-*scores* $\leftarrow$ conlleval(Y, $\hat{Y}$), calculating the overall F_1-*scores* for predicted labels.
13: If $F_{1\text{-}max} > F_1$-*scores* then
14: $F_{1\text{-}max} \leftarrow F_1$-*scores*
15: Save the weights $\hat{\theta}$ of the model
16: end for
17: Output: the best-predicted labels $\hat{Y}$, the best training weights of the model $\hat{\theta}$.

1. The sentence is converted into contextual embeddings by using BERT, which is fine-tuned on the texts of agricultural diseases and pests.
2. The character-level embeddings are used as input of the BiLSTM to extract the global context features. Note that other contextual encoders such as Gated CNN and RD_CNN can also be used to extract the context features according to the experimental results in Section 3.2.3.
3. The possible labels are predicted, and the loss is calculated by the CRF layer.
4. Calculating the perturbation according to Equation (17) and adding it to the original character-level embeddings.
5. Steps (1) to (4) are repeated until a maximum iteration is reached.

References

1. Lu, J.; Tan, L.; Jiang, H. Review on Convolutional Neural Network (CNN) Applied to Plant Leaf Disease Classification. *Agriculture* **2021**, *11*, 707. [CrossRef]
2. Molina-Villegas, A.; Muñiz-Sanchez, V.; Arreola-Trapala, J.; Alcántara, F. Geographic Named Entity Recognition and Disambiguation in Mexican News using Word Embeddings. *Expert Syst. Appl.* **2021**, *176*, 114855. [CrossRef]
3. Yin, M.; Mou, C.; Xiong, K.; Ren, J. Chinese clinical named entity recognition with radical-level feature and self-attention mechanism. *J. Biomed. Inform.* **2019**, *98*, 103289. [CrossRef]
4. Huang, K.; Altosaar, J.; Ranganath, R. ClinicalBert: Modeling clinical notes and predicting hospital readmission. *arXiv* **2019**, arXiv:1904.05342.
5. Francis, S.; Van Landeghem, J.; Moens, M.F. Transfer learning for named entity recognition in financial and biomedical documents. *Information* **2019**, *10*, 248. [CrossRef]
6. Lee, J.; Yoon, W.; Kim, S.; Kim, D.; Kim, S.; So, C.H.; Kang, J. BioBERT: A pre-trained biomedical language representation model for biomedical text mining. *Bioinformatics* **2019**, *36*, 1234–1240. [CrossRef]
7. Guo, X.; Zhou, H.; Su, J.; Hao, X.; Tang, Z.; Diao, L.; Li, L. Chinese agricultural diseases and pests named entity recognition with multi-scale local context features and self-attention mechanism. *Comput. Electron. Agric.* **2020**, *179*, 105830. [CrossRef]
8. Yasunaga, M.; Kasai, J.; Radev, D. Robust multilingual part-of-speech tagging via adversarial training. *arXiv* **2017**, arXiv:1711.04903.
9. Du, C.; Sun, H.; Wang, J.; Qi, Q.; Liao, J. Adversarial and domain-aware bert for cross-domain sentiment analysis. In Proceedings of the 58th Annual Meeting of the Association for Computational Linguistics, Online. 5–10 July 2020; pp. 4019–4028.
10. Xu, J.; Wen, J.; Sun, X.; Su, Q. A discourse-level named entity recognition and relation extraction dataset for chinese literature text. *arXiv* **2017**, arXiv:1711.07010.
11. Malarkodi, C.S.; Lex, E.; Devi, S.L. Named Entity Recognition for the Agricultural Domain. *Res. Comput. Sci.* **2016**, *117*, 121–132.
12. Nadeau, D.; Sekine, S. A survey of named entity recognition and classification. *Lingvisticae Investig.* **2007**, *30*, 3–26. [CrossRef]
13. Liu, W.; Yu, B.; Zhang, C.; Wang, H.; Pan, K. Chinese Named Entity Recognition Based on Rules and Conditional Random Field. In Proceedings of the 2018 2nd International Conference on Computer Science and Artificial Intelligence, ShenZhen, China, 8–10 December 2018; pp. 268–272.
14. WANG Chun-yu, W.F. Study on recognition of chinese agricultural named entity with conditional random fields. *J. Hebei Agric. Univ.* **2014**, *37*, 132–135. [CrossRef]
15. Zhao, P.; Zhao, C.; Wu, H.; Wang, W. Named Entity Recognition of Chinese Agricultural Text Based on Attention Mechanism. *Nongye Jixie Xuebao/Trans. Chin. Soc. Agric. Mach.* **2021**, *52*, 185–192. [CrossRef]
16. Saleem, M.H.; Potgieter, J.; Arif, K.M. Plant Disease Detection and Classification by Deep Learning. *Plants* **2019**, *8*, 468. [CrossRef]
17. Hasan, R.I.; Yusuf, S.M.; Alzubaidi, L. Review of the State of the Art of Deep Learning for Plant Diseases: A Broad Analysis and Discussion. *Plants* **2020**, *9*, 1302. [CrossRef] [PubMed]
18. Zhao, S.; Peng, Y.; Liu, J.; Wu, S. Tomato Leaf Disease Diagnosis Based on Improved Convolution Neural Network by Attention Module. *Agriculture* **2021**, *11*, 651. [CrossRef]
19. Chen, S.; Zhang, K.; Zhao, Y.; Sun, Y.; Ban, W.; Chen, Y.; Zhuang, H.; Zhang, X.; Liu, J.; Yang, T. An Approach for Rice Bacterial Leaf Streak Disease Segmentation and Disease Severity Estimation. *Agriculture* **2021**, *11*, 420. [CrossRef]
20. Hao, X.; Jia, J.; Gao, W.; Guo, X.; Zhang, W.; Zheng, L.; Wang, M. MFC-CNN: An automatic grading scheme for light stress levels of lettuce (Lactuca sativa L.) leaves. *Comput. Electron. Agric.* **2020**, *179*, 105847. [CrossRef]
21. Biswas, P.; Sharan, A. A Noble Approach for Recognition and Classification of Agricultural Named Entities using Word2Vec. *Int. J. Adv. Stud. Comput. Sci. Eng.* **2021**, *9*, 1–8.
22. Mikolov, T.; Chen, K.; Corrado, G.; Dean, J. Efficient estimation of word representations in vector space. *arXiv* **2013**, arXiv:1301.3781.
23. Jawahar, G.; Sagot, B.; Seddah, D. What Does BERT Learn about the Structure of Language? In Proceedings of the 57th Annual Meeting of the Association for Computational Linguistics, Florence, Italy, 6 July 2019; pp. 3651–3657.
24. Zhang, S.; Zhao, M. Chinese agricultural diseases named entity recognition based on BERT-CRF. In Proceedings of the 2020 5th International Conference on Mechanical, Control and Computer Engineering (ICMCCE), Harbin, China, 25–27 December 2020; pp. 1148–1151.
25. Huang, Z.; Xu, W.; Yu, K. Bidirectional LSTM-CRF models for sequence tagging. *arXiv* **2015**, arXiv:1508.01991.
26. Strubell, E.; Verga, P.; Belanger, D.; McCallum, A. Fast and Accurate Entity Recognition with Iterated Dilated Convolutions. In Proceedings of the 2017 Conference on Empirical Methods in Natural Language Processing, Copenhagen, Denmark, 1 September 2017; pp. 2670–2680.
27. Dauphin, Y.N.; Fan, A.; Auli, M.; Grangier, D. Language modeling with gated convolutional networks. In Proceedings of the International Conference on Machine Learning, Sydney, Australia, 6–11 August 2017; pp. 933–941.
28. Qiu, J.; Wang, Q.; Zhou, Y.; Ruan, T.; Gao, J. Fast and accurate recognition of Chinese clinical named entities with residual dilated convolutions. In Proceedings of the 2018 IEEE International Conference on Bioinformatics and Biomedicine (BIBM), Madrid, Spain, 3–6 December 2018; pp. 935–942.
29. Yan, H.; Deng, B.; Li, X.; Qiu, X. Tener: Adapting transformer encoder for named entity recognition. *arXiv* **2019**, arXiv:1911.04474.

30. Chen, H.; Lin, Z.; Ding, G.; Lou, J.; Zhang, Y.; Karlsson, B. GRN: Gated relation network to enhance convolutional neural network for named entity recognition. In Proceedings of the AAAI Conference on Artificial Intelligence, Honolulu, HI, USA, 27 January–1 February 2019; pp. 6236–6243.
31. Zhu, Y.; Wang, G. CAN-NER: Convolutional Attention Network for Chinese Named Entity Recognition. In Proceedings of the 2019 Conference of the North American Chapter of the Association for Computational Linguistics: Human Language Technologies, Volume 1 (Long and Short Papers), Minneapolis, MN, USA, 2–7 June 2019; pp. 3384–3393.
32. Li, L.; Zhao, J.; Hou, L.; Zhai, Y.; Shi, J.; Cui, F. An attention-based deep learning model for clinical named entity recognition of Chinese electronic medical records. *BMC Med. Inform. Decis. Mak.* **2019**, *19*, 1–11. [CrossRef]
33. Cao, P.; Chen, Y.; Liu, K.; Zhao, J.; Liu, S. Adversarial transfer learning for Chinese named entity recognition with self-attention mechanism. In Proceedings of the 2018 Conference on Empirical Methods in Natural Language Processing, Brussels, Belgium, 2–4 November 2018; pp. 182–192.
34. Wang, C.; Chen, W.; Xu, B. Named entity recognition with gated convolutional neural networks. In *Chinese Computational Linguistics and Natural Language Processing Based on Naturally Annotated Big Data*; Springer: Berlin/Heidelberg, Germany, 2017; pp. 110–121.
35. Li, X.; Yan, H.; Qiu, X.; Huang, X.-J. FLAT: Chinese NER Using Flat-Lattice Transformer. In Proceedings of the 58th Annual Meeting of the Association for Computational Linguistics, Seattle, WA, USA, 5–10 July 2020; pp. 6836–6842.
36. Cetoli, A.; Bragaglia, S.; O'Harney, A.; Sloan, M. Graph Convolutional Networks for Named Entity Recognition. In Proceedings of the 16th International Workshop on Treebanks and Linguistic Theories, Prague, Czech Republic, 1 July 2017; pp. 37–45.
37. Gui, T.; Zou, Y.; Zhang, Q.; Peng, M.; Fu, J.; Wei, Z.; Huang, X.-J. A lexicon-based graph neural network for chinese ner. In Proceedings of the 2019 Conference on Empirical Methods in Natural Language Processing and the 9th International Joint Conference on Natural Language Processing (EMNLP-IJCNLP), Hong Kong, China, 3 November 2019; pp. 1039–1049.
38. Li, J.; Sun, A.; Han, J.; Li, C. A survey on deep learning for named entity recognition. *IEEE Trans. Knowl. Data Eng.* **2020**. [CrossRef]
39. Pre-trained models for natural language processing: A survey. *arXiv* **2020**, arXiv:2003.08271.
40. Zhang, R.; Lu, W.; Wang, S.; Peng, X.; Yu, R.; Gao, Y. Chinese clinical named entity recognition based on stacked neural network. *Concurr. Comput. Pract. Exp.* **2020**, e5775. [CrossRef]
41. Suman, C.; Reddy, S.M.; Saha, S.; Bhattacharyya, P. Why pay more? A simple and efficient named entity recognition system for tweets. *Expert Syst. Appl.* **2021**, *167*, 114101. [CrossRef]
42. Yang, Z.; Chen, H.; Zhang, J.; Ma, J.; Chang, Y. Attention-based multi-level feature fusion for named entity recognition. *IJCAI Int. Jt. Conf. Artif. Intell.* **2020**, *2021*, 3594–3600. [CrossRef]
43. Liu, X.; Zhou, Y.; Wang, Z. Deep neural network-based recognition of entities in Chinese online medical inquiry texts. *Futur. Gener. Comput. Syst.* **2021**, *114*, 581–604. [CrossRef]
44. Chiu, J.P.C.; Nichols, E. Named entity recognition with bidirectional LSTM-CNNs. *Trans. Assoc. Comput. Linguist.* **2016**, *4*, 357–370. [CrossRef]
45. Ma, X.; Hovy, E. End-to-end Sequence Labeling via Bi-directional LSTM-CNNs-CRF. In Proceedings of the 54th Annual Meeting of the Association for Computational Linguistics (Volume 1: Long Papers), Berlin, Germany, 3 August 2016; pp. 1064–1074.
46. Peters, M.E.; Ruder, S.; Smith, N.A. To Tune or Not to Tune? Adapting Pretrained Representations to Diverse Tasks. In Proceedings of the 4th Workshop on Representation Learning for NLP (RepL4NLP-2019), Florence, Italy, 5 August 2019; pp. 7–14.
47. Song, C.H.; Sehanobish, A. Using Chinese Glyphs for Named Entity Recognition (Student Abstract). In Proceedings of the AAAI Conference on Artificial Intelligence, New York, NY, USA, 7–12 February 2020; Volume 34, pp. 13921–13922.
48. Xuan, Z.; Bao, R.; Jiang, S. FGN: Fusion glyph network for Chinese named entity recognition. *arXiv* **2020**, arXiv:2001.05272.
49. Xu, L.; Dong, Q.; Liao, Y.; Yu, C.; Tian, Y.; Liu, W.; Li, L.; Liu, C.; Zhang, X. CLUENER2020: Fine-grained named entity recognition dataset and benchmark for chinese. *arXiv* **2020**, arXiv:2001.04351.
50. Zhang, Y.; Yang, J. Chinese NER Using Lattice LSTM. In Proceedings of the 56th Annual Meeting of the Association for Computational Linguistics (Volume 1: Long Papers), Melbourne, Australia, 15–20 July 2018; pp. 1554–1564.
51. Li, S.; Zhao, Z.; Hu, R.; Li, W.; Liu, T.; Du, X. Analogical Reasoning on Chinese Morphological and Semantic Relations. In Proceedings of the 56th Annual Meeting of the Association for Computational Linguistics (Volume 2: Short Papers), Melbourne, Australia, 15–20 July 2018; pp. 138–143.
52. Prechelt, L. Early stopping-but when? In *Neural Networks: Tricks of the Trade*; Springer: Berlin/Heidelberg, Germany, 1998; pp. 55–69.
53. Bekoulis, G.; Deleu, J.; Demeester, T.; Develder, C. Joint entity recognition and relation extraction as a multi-head selection problem. *Expert Syst. Appl.* **2018**, *114*, 34–45. [CrossRef]
54. Miyato, T.; Dai, A.M.; Goodfellow, I. Adversarial training methods for semi-supervised text classification. *arXiv* **2016**, arXiv:1605.07725.
55. der Maaten, L.; Hinton, G. Visualizing data using t-SNE. *J. Mach. Learn. Res.* **2008**, *9*, 2579–2605.
56. Zhao, S.; Cai, Z.; Chen, H.; Wang, Y.; Liu, F.; Liu, A. Adversarial training based lattice LSTM for Chinese clinical named entity recognition. *J. Biomed. Inform.* **2019**, *99*, 103290. [CrossRef] [PubMed]
57. Liu, X.; Cheng, H.; He, P.; Chen, W.; Wang, Y.; Poon, H.; Gao, J. Adversarial training for large neural language models. *arXiv* **2020**, arXiv:2004.08994.

agriculture

Article

An Agile AI and IoT-Augmented Smart Farming: A Cost-Effective Cognitive Weather Station

Amine Faid [1,*], Mohamed Sadik [1] and Essaid Sabir [1,2]

1 NEST Research Group, LRI Lab, ENSEM, Hassan II University of Casablanca, Casablanca 20000, Morocco; m.sadik@ensem.ac.ma (M.S.); e.sabir@ensem.ac.ma (E.S.)
2 Department of Computer Science, University of Quebec at Montreal (UQAM), Montreal, QC H2L 2C4, Canada
* Correspondence: a.faid@ensem.ac.ma; Tel.: +212-6-523-523-06

Abstract: Internet of Things (IoT) can be seen as the electricity of 21st century. It has been reshaping human life daily during the last decade, with various applications in several critical domains such as agriculture. Smart farming is a real-world application in which Internet of Things (IoT) technologies like agro-weather stations can have a direct impact on humans by enhancing crop quality, supporting sustainable agriculture, and eventually generating steady growth. Meanwhile, most agro-weather solutions are neither customized nor affordable for small farmers within developing countries. Furthermore, due to the outdoor challenges, it is often a challenge to develop and deploy low-cost yet robust systems. Robustness, which is determined by several factors, including energy consumption, portability, interoperability, and system's ease of use. In this paper, we present an agile AI-Powered IoT-based low-cost platform for cognitive monitoring for smart farming. The hybrid Multi-Agent and the fully containerized system continuously surveys multiple agriculture parameters such as temperature, humidity, and pressure to provide end-users with real-time environmental data and AI-based forecasts. The surveyed data is ensured through several heterogeneous nodes deployed within the base station and in the open sensing area. The collected data is transmitted to the local server for pre-processing and the cloud server for backup. The system backbone communication is based on heterogeneous protocols such as MQTT, NRF24L01, and WiFi for radio communication. We also set up a user-friendly web-based graphical user interface (GUI) to support different user profiles. The overall platform design follows an agile approach to be easy to deploy, accessible to maintain, and continuously modernized.

Keywords: smart farming; IoT; WSN; containerization; multi-agent; neural network; LSTM

Citation: Faid, A.; Sadik, M.; Sabir, E. An Agile AI and IoT-Augmented Smart Farming: A Cost-Effective Cognitive Weather Station. *Agriculture* **2022**, *12*, 35. https://doi.org/10.3390/agriculture12010035

Academic Editor: Dimitre Dimitrov

Received: 13 October 2021
Accepted: 22 December 2021
Published: 29 December 2021

1. Introduction

1.1. Smart Farming

Nowadays, the traditional crop management practices remain insufficient to follow the persistent global needs for food. This challenge is mainly driven by the exponential population growth, climate change, and bad agriculture practices. The UN estimates that by 2050, the world population will stand between 9.4 and 10.1 billion. The population numbers will continue rising through the years, and it will reach between 9.4 and 12.7 billion by 2100 [1]. Thus, food production is expected to grow by 70% by 2100 to meet the population's expansion [2]. Meanwhile, the impact of climate change, is putting productive lands and production under tremendous stress [3], with yearly damages expected to range between 0.1 and 1.0 percent of the gross world product by 2100 [4]. Furthermore, farmers' mismanagement of crops and the extensive exploitation of resources often lead to soil deterioration through acidification, erosion, and heavy metals pollution [5,6]. Thus, farmers must use different resources such as fertilizers, water, and nutrients in a very optimized way for sustainable food production [7]. Therefore, smart farming presents a tremendous opportunity for farmers to overcome agricultural challenges and protect the environment. Smart farming is described as the application of information and communication technology (ICT) to the agriculture industry in order to improve crop management efficiency. The idea behind this concept consists of combining the most relevant trends in monitoring technologies with the best field practices to assist farmers to overcome the agriculture challenges [8].

Internet of Things (IoT) based systems for smart farming are considered to be the backbone of the fourth industrial revolution. It consists of the implementation of IoT-based systems for continuous monitoring and data analysis. Different heterogeneous sensor nodes are deployed across yields for key environmental processes' monitoring. Sensor nodes that communicate continuously using wireless sensor networks (WSN) technology to overcome various geographical constraints that wired technologies have in topographic features such as deserts, mountains, rivers, valleys, and lakes. The wireless sensor network is the skeleton for an IoT-based system. It is based on a set of wirelessly connected nodes for network establishment as presented in Figure 1. The wireless nodes are the front gate elements for environmental sensing, data collection, and it can even play the role of the actuator through activities automation such as pumps control, irrigation, etc. However, sensor nodes with limited size and resources, such as power supplies, CPUs, and memory, require an additional optimization layer for optimum efficiency [9–11]. Therefore, techniques such as wireless network clustering, as investigated in our work in [12], can improve considerably the network's energetic performance through the creation of sub-networks and multi-hope communication to limit distances between remote IoT nodes and base stations. Thus, avoiding energy dissipation in long distances. Additionally, the usage of light communication protocols dedicated for IoT sensor nodes, and decentralization where operations that consume the system's resources are not performed on the sensor node's level but the BS level side are part of the optimization approaches that enhance different performance.

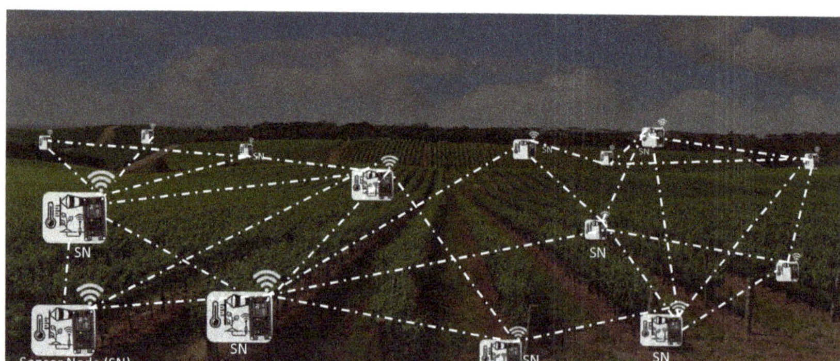

Figure 1. High level presentation of wireless sensor network.

Undoubtedly, IoT-based solutions for smart farming are tremendously requested by large facilities. However, the penetration of such solutions within small farms is considered limited, especially in the developing economies [13]. This limitation is mainly due to the expensive costs, the complications of maintenance, and the overall performance of different solutions and services. Moreover, the lack of technological knowledge presents a barrier for IoT adoption among farmers [14]. Nevertheless, many researchers around the globe are continuously developing low-cost systems for small farmers, different solutions were gathered in detailed reviews such as [15–19]. In [20], the authors present a low-cost-based agro-ecological management system for smart farming in India. The system performance was tested using the OMnet++ simulation tool. OMnet++ which stands for Objective Modular Network Testbed in C++, is a framework and library based on C++

language. The framework is allowing researchers to create and simulate wireless networks' communication. In Egypt, the authors of [21] present an agriculture monitoring system based on IoT and artificial intelligence (AI) for decision making. The implemented expert system enables the proposed framework to mimic the ability of a human expert to make decisions on plant diseases. In [22], the authors present a real-time monitoring system for farmers, the "Expert Advisory System" is developed to improve the crop productivity in Uzbekistan. This is can be achieved based not only on the provided environment's data that supports farmers for their expert judgment (weather status, plant's diseases, etc.), but also to evaluate precisely the soil needs for healthy plants and therefore good crop and better natural resources' preservation. The simulation was conducted using Contiki Simulator. In [23], the authors proposed an IoT-based low-cost system for smart irrigation. The system is based on MQTT, HTTP, and Neural Network (NN) for intelligent decision-making. Another low-cost prototype has been presented in [24] in which authors have created a user-friendly system for smart farming. In South Korea, the authors of [25] proposed an adaptive network mechanism for a reliable smart farming system. The technique feature is based on the ability to choose suitable transmission protocols based on the network condition. The system implements Long Range Wide Area Network (LoRa-WAN) and IEEE protocols for transmission. As a smart farming system, the framework allows having a salable transmission system to scale out the platform performance in large farms. Using such as transmission approach will guarantee continuous communication between sensor nodes and the base station. Therefore, it will provide a continuous hands on the farm and its environmental parameters such as humidity, temperature, etc. Additionally the paper investigates some metrics such as latency and data reliability for the system's validation. The authors of [26] presented an experimental analysis of energy harvesting for IoT devices, as well as a comparison of various wireless technologies for agriculture systems in Canada. IEEE 801.11 g, IEEE 802.15.4, and LoRaWAN protocols are also being investigated. In [27], an IoT system has been proposed for precise ecological monitoring in agriculture domains. The proposed platform is based on different views for different end-users. According to the authors, the architecture is open for an extra layer of protocols and it can be implemented on different servers and cloud-based architectures. In Turkey, an architectural approach based on Farm Management Information System (FMIS) has been introduced in [28]. In [29], the authors proposed an energy-efficient weather station, the system focused on the algorithm optimization to deal with the energy consumption dilemma such as devices high power's consumption in data sending, data receiving, wireless communication, energy dissipation in different bands within different ranges, etc. Thus, the algorithm focused on the optimization in data transmission through the deployment of asynchronous algorithms that is based on measurement threshold and sleep mode approach. The station considers measurement data is ready when the anemometer's rotations are reaching a certain value. Thus, if no rotation is detected the BS puts the system in sleep mode for 500 ms. When data is ready, the transmission module doesn't send the measurement of wind speed unless if it is greater than a predefined threshold. According to the authors, the optimized algorithm had the potential to optimize the energy consumption of the station by more than 60%. In Tunisia, the authors of [30] proposed a low-cost monitoring system for smart farming based on IoT and Unmanned Aerial Vehicles (UAVs). The proposed platform relies on a set of under and above-ground environmental sensors.

To the best of our knowledge, current monitoring systems focus on providing systems to the researchers' community, rather than providing customized solutions to farmers for their crops management on a real-time basis. The proposed solutions do not address the energy consumption dilemma since most papers focus on technologies comparisons rather than hardware and software enhancement. Current platforms principally rely on light communication protocols to control the energy consumption issue and improve network lifetime. Additionally, the studied solutions in this article rely on very basic approaches when handling the operation and maintenance (OAM) of different components and services. In addition, current systems don't answer the security dilemma. Some of the most common challenges in IoT are presented in Table S1 (see Supplementary Materials). Additionally, different authors propose low-cost solutions without breaking down the proposed systems' costs into the cost of investments and the cost of operations. Thus, to overcome some of the challenging issues that the IoT-based systems have, this paper focuses on designing and implementing a sophisticated yet low-cost meteorological system for smart farming. The system is mainly designed based on open-sources software and platforms. The system design takes into consideration the algorithmic optimization in data acquisition, system security, and data presentation. The algorithm design is implemented to overcome the energy consumption dilemma, and it is oriented to improve the end-user experience.

1.2. *Rationale of the SW and HW Architecture of the Proposed Method*

When targeting deployment of agile systems, we should always put in the heart of our approach the continuous integration and continuous deployment CI/CD philosophy. Thus, in this work, we use the latest technological advancement with new conceptual ideas to deliver an added value. Unlike virtualization which consumes resources, is slow to run, and doesn't offer isolation between virtual machines (VMs), the containerization technology takes the software design to a new level thanks to the resources' optimization, high level of isolation, and high speed. Therefore, we build a fully containerized base station based on the open-source containerization platform (Docker). All services within the base station are presented as containers to facilitate the operation and maintenance of the base station's software components such as database, MQTT broker, etc. Deploying software on top of Docker will allow us to have features such as interoperability, ease of use, scalability, and high performance. Meanwhile, performance can't be achieved only by using the good software infrastructure (Docker) but also using a highly adequate technology adapted for the exact use case (i.e., weather monitoring). Consequently, classical databases cannot grant high performance. Hence, the time series databases (TSDB) are considered to be the most powerful databases to deal with massive monitoring data. In our work, we use InfluxDB as one of the best current TSDBs. Meantime, when using low-cost hardware, we should avoid pushing the usage of the BS' available resources to the maximum when we can outsource some of the heavy activities, therefore we use Google Colaboratory (Colab) free computing resources to train our models and offload our base station for other tasks. Once a set of predefined amounts of measurement is recorded, the Debian distribution which is a Linux-based Operating System (OS) within the BS initiates a cronjob to export data into Google Drive, at that point Colab recover the data and start training the prediction's model. When the model is ready, Tensorflow generates "tflite" and store it back into Google Drive. The model is then sent back to the BS for internal usage within the local webserver. We chose to deploy the open-source LAMP (Linux Apache MySQL PHP) server within our BS. Security is also a challenge in modern systems due to the high exposure to the internet. Thus, in our system, we use different techniques to ensure the system's security and data privacy. Adoption of a separated cloud server that can only receive data from the base station is very important. The idea is to allow third-party users to consult the dashboard on a cloud server without being able to connect to the local server. Connection to a local server is only dedicated to researchers and engineers through VPN tunnels based on a pre-generated license.

1.3. *Objectives and Hypotheses*

In the course of pursuing our main objective to build an agile, low-cost, and AI-powered meteorological station, we address the following hypotheses by implementing the corresponding item as described against the bullet points below.

- The different trained models, within the proposed platform, support the generation of AI-based information to assist different end-users roles such as farmers, researchers, and engineers.
- The proposed architecture is mainly based on open-source technologies to serve the low-cost philosophy adopted in this paper. The open-source environments chosen to

allow us to customize source codes and adapt them for our exact needs with minimal cost of use. Therefore, usage of Debian OS rather than Microsoft OS will allow us to modify easily the OS Kernel to optimize the performance of our station in terms of the boot, background processes, usage of command lines, and enhancing the CPU usage and speed which lead eventually to enhance energetic performance. Furthermore, the long-term goal of the system is to propose a flexible platform based on the layering approach to enable the system's flexibility and continuous upgrade.

- The system design follows a hybrid architecture that combines centralized and distributed techniques for devices' connection which will lead to higher performance. The design enhances the system's portability, scalability, interoperability, and compatibility with different technologies, protocols, and equipment's vendors. The Hardware (HW) design is based on deploying heterogeneous nodes that are not dependable on the vendor of the components, the type of the sensors, or the end node performance. The proposed distributed agro-weather station plays a Base Station (BS) role that embeds different agents. The BS is connected to several heterogeneous wireless nodes. The nodes, including the BS, are equipped with different heterogeneous sensors to measure the field's temperature and humidity, wind speed and direction, and atmospheric pressure. The network continuously monitors the field and provides insights and predictions through a local server implemented within the BS. Since energy efficiency is a crucial metric in IoT systems, our network ensures good energy efficiency management without impacting system performance.
- The wireless nodes design which is based on standardized transmission protocols such as MQTT allows us to have good interoperability between nodes and different BS. This design approach is also done in order to fulfill the plug-and-play feature in this platform. Therefore, any wireless node within our system can be redeployed within other platforms as long as they follow the same protocol.
- We design the Graphical User Interface (GUI) layer to ensure visibility, accessibility, ease of use, efficiency, and attractiveness. The software design ensures real-time and near-real-time data communication and processing between network sensors and the base station. The agro-weather station is AI-powered to support farmers and researchers with environment's data trends and eventually allow them to make fact-based decisions such as when planting, spraying, and even harvesting. Within the BS, the AI is an agent that is connected to an external Google Collaboratory environment (Colab) which is a standalone hosted Jupyter notebook that provides free computing resources for background data analysis and modeling. Even though it is possible to bypass the usage of Colab and rely only on a local Jupyter netbook running on the local server, however, we should avoid consuming local BS resources on the training tasks and allocate these resources for other services such as sensing, wireless transmission, cronjob services, local web-server, etc. The Graphical User Interface (GUI) ensures utility and warranty and avoids complications whenever end users are connected, and data is consulted.
- The system's services such as remote access, web services, database updates, and telegram chat-bot are built on top as standalone agents. The multi-agent system (MAS) approach helps enhance the BS performance by creating scalable, reliable, efficient, and maintainable modular services. In addition, the MAS approach enables the stateless configuration of services. Therefore, services can be updated through a Start, Upgrade, and Stop (SUS) approach. The SUS approach allows instant execution of predefined scripts for each operation. The scripts are hosted in accessible repository under/usr/local/bin. The file system/usr/local/bin contains different scripts that normal users can access and use, this repository protects scripts from being modified when the system's updates are planned or executed. Consequently, facilitating different modules' operation, maintenance, and troubleshooting.
- System security approach and design have been implemented in our platform. Access to the BS is ensured locally with user credentials or remotely through a secured

VPN tunnel. We deploy OpenVPN within the BS. Meanwhile, because OpenVPN needs to follow an assigned IP address, while our system is connected to a public address, we can overcome this challenge by deploying a NoIP client within our BS that plays the role of a dynamic DNS pointing continuously to a static hostname such as "www.ensem-aws.tk" accessed on 13 April 2021. Different user roles are configured with different roles and privileges. Additionally, offline data snapshots of the OAM & Data dashboards are stored in the cloud and consulted separately.

- The system building costs are essential in our platform's design. Capital expenditures (Capex) and operating expenses (Opex) are the leading Key Performance Indicators (KPIs) taken into consideration for solution design and prototyping. Thus, the cost analysis is ensured to support presenting the system's short-term and long-term benefits for different users.

2. Materials and Methods

Many contributions have been proposed, and several works are continuously developing IoT-based systems for smart farming. Thus, we propose an agro-weather station (AWS) designed to serve farmers and researchers at once. The system is developed to provide real-time data and AI-based insights for different end-users. The logical architecture illustrated in Figure 2 is based on a multi-layer approach. The system can be seen as a natural evolution and implementation of our previous work in [31]. The proposed system's high-level design (HLD) is decomposed into four main layers: the perception layer, the transmission layer, the presentation layer, and the management layer. The main role of the perception layer is sensing and collecting data through the network's deployed heterogeneous sensors. In addition, the perception layer contains different sensors to collect environmental parameters such as temperature, humidity, wind direction, and wind speed. Meanwhile, the transmission layer can be considered as the skeleton of our system. This layer connects the network's dispersed sensors to the BS through different deployed protocols such as WIFI, NRF24L01, MQTT, etc. The presentation layer's role is to ensure data collection, processing, and data transformation to trends and insight reports. The presentation layer is based on a graphical user-friendly GUI accessible through different devices such as laptops, mobiles, and tablets. Finally, the management layer is responsible for surveying the system and providing real-time status and alarms of different deployed nodes for reliable system management. In this section, we discuss the research methods in our system's design. Moreover, we review the several opportunities and challenges that IoT-based systems present for smart farming use case.

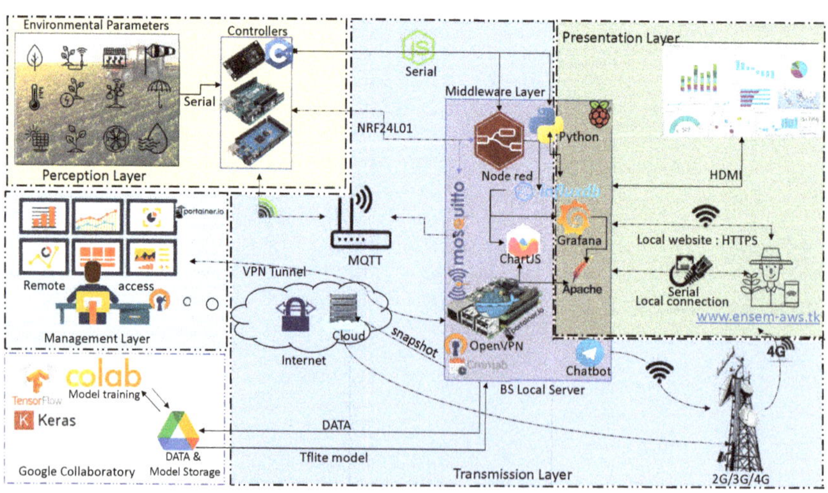

Figure 2. The End-to-End high-level decomposition of the proposed agro-weather station.

2.1. System Design

To effectively implement our system, we adopt an agile methodology approach to develop and deliver the system throughout its different phases as presented in Supplementary Part II (see Supplementary Materials). The agile approach enables a collaborative environment in which different stakeholders such as farmers, researchers, and developers continuously improve the system through its different releases and iterations. The approach starts with the requirement collection phase. In this step, we gather the system requirements that should be implemented according to the stakeholders, such as data visualization, system maintenance, and intelligent-based insights. The requirements are then prioritized according to their relevance and timeline within the analysis phase. The design phase focuses on the hardware and software design approach in complete alignment with the system baseline, such as the low-cost strategy, the portability, the scalability, the interoperability, etc. The development phase consists of the system implementation, where the system is transformed from a design into a prototype. The release phase is the extensive operation and exploitation of the current prototype version for performance analysis. Finally, the monitor phase measures and tracks the system performance and identifies any potential problem or improvement for corrective actions. This last phase is conducted in a proactive way to allow the system's continuous enhancement.

The design of the agro-weather station follows the multi-layer approach. Thus, in following, we discuss the various system's layers:

2.1.1. Perception Layer [Back-End]

The perception or access layer is the lowest level of the proposed system model. It consists of a set of interconnected sensors within different nodes. The Perception layer's primary role is to ensure data collection of different environmental parameters through physical sensors. The deployed sensors collect different agriculture data such as temperature, humidity, pressure, and solar radiation as depicted in Table 1. The standalone nodes are designed in hybrid mode. Nodes can either collect parameters locally and communicate with the BS through serial protocols such as I2C, UART or transmit the data to the BS wirelessly. The BS uses the data to build and continuously enhance the AI-based model for weather forecasting. A model that is fully based on the LSTM approach. The hybrid energy design of the nodes supports a hybrid power supply through solar energy or AC power. Meanwhile, to overcome the IoT devices' power limitation, the integrated firmware supports different techniques to enhance the energy efficiency of nodes. The technique mainly involves deploying a change point detection algorithm with adaptive communication frequency with the BS [32]. Multiple environmental characteristics are consistent over time in agriculture, allowing the measurement frequency to be adjusted in response to parameter variation. In time-series data analysis, the change point is a critical component. It's the point in time when a signal's property (variance, mean, etc.) rapidly changes. The problem of detecting sudden changes in a time series is known as the change point detection [33]. In addition, different agents are deployed to follow the perception layer's performance. Alerts are programmed to track different nodes' energy performance and status. The nodes design takes into consideration several baselines such as:

- Nodes portability: nodes are portable across the network without any physical or logical constraints.
- Nodes scalability: the nodes can handle additional functions by deploying different sensors in any future nodes' expansion or upgrade.
- Nodes interoperability: the nodes can communicate with different BS based on different vendors without critical constraints.

Table 1. Sensors' technical specifications.

Sensor	Range	Accuracy	Resolution
Air Temperature	0° to 50°	±1°	1°
Air Humidity	0–100%	±%	0.5%
Soil Temperature	0°–60°	±2°	1°
Relative Permittivity	1–81	±3%	<0.02
Wind Speed	5–100 km/h	±1 km/h	1 km/h
Wind Direction	0°–360°	±2°	22.5°
Solar Radiation	360 to 1120 nm	±5%	1 W/m^2

2.1.2. Transmission Layer [Back-End]

The transmission layer is the skeleton of the network that ensures proper communication between its different elements. It focuses on the establishment of transparent and reliable end-to-end data transportation links. The layer transport either wirelessly or locally the nodes' data from the perception layer to the presentation layer for eventual data formatting. Although wired communication is less energy-consuming than wireless communication, not limited by dedicated radio bandwidth, and it is more secured to remote network attacks. However, wireless communication is highly flexible and easy to deploy in vast areas. It also allows considerable agility in terms of deployment and maintenance with remote access from anywhere and anytime. Several standards and protocols are deployed in our proposed system in a full-duplex mode, such as WIFI (IEEE802.11), NRF24L01, and Bluetooth. Furthermore, the system is open-source, enabling its smooth upgrade and supporting new technologies such as the extension to wireless technologies 2G/3G/4G, Lora, LPWAN, Zigbee (IEEE802.15.4), and Sigfox. Comparison between these protocols is shown in Table 2.

The proposed system supports a connection-oriented transmission mode for reliable transmission or connection-less mode to enhance network connectivity. Wire communication is also deployed in our system architecture to present the different use cases and scenarios the system can handle regardless of data transmission medium.

Table 2. Comparison between wireless transmission protocols.

Parameters	Standards	Band	Data Rate	Range	Energy Effeciency
WIFI	IEEE 802.11	2.4–60 GHz	1 Mbps–7 Gbps	20–100 m	Low
ZigBee	IEEE 802.15.4	2.4 GHz	20–250 Kbps	10–20 m	High
Bluetooth	IEEE 802.15.1	2.4 GHz	24 Mbps	8–10 m	High
MQTT	OASIS	2.4 GHz	259 Kbps	-	High
Cellular	2G/3G/4G	9000 MHz, 18,000 MHz, 21,000 MHz, 2700 MHz	-	-	Medium
LoRaWAN	LoRa R1	868/900 MHz	0.3–50 Kbps	30 km	Very High
SigFox	SigFox	200 KHz	100–600 bps	30–50 km	Very High

2.1.3. Presentation Layer [Front-End]

The presentation layer is responsible for reports presentation through formatted data. The designed user-friendly GUI is compatible with different platforms such as tablets, phones, and laptops. The collected data support the creation and enhancement of the AI-based model for weather forecasting. The design of the presentation layer takes into consideration several vital metrics such as:

- Data visibility: to ensure a high level of clarity that allows farmers and researchers to monitor, analyze and make smart in-depth decisions.
- Data accessibility: to provide an open and accessible data format that different system's agents can explore for different use cases such as modeling, training, evaluation performance, etc.

- Ease of use: to ensure that different user roles can use our GUI comfortably. Thus, we make a baseline that different users should explore report overview in the 30s without compromising data quality.
- System attractiveness: to create a dynamic and appealing GUI with an attractive and balanced design.
- System alerts: to monitor the evolution of the critical parameters closely and provide timely alerts through different channels such as emails and telegram chat-bot.

Due to the energy limitations in different IoT nodes, we have chosen to rely on lightweight protocols within the presentation layer, such as the Hypertext Transfer Protocol (HTTP) and The Message-Queue Telemetry Transport (MQTT). HTTP is one of the most popular internet protocols for web messaging that runs over TCP protocol. It is based on request and response architecture [34]. MQTT is a lightweight bandwidth-efficient protocol designed specifically for IoT use cases [35].

2.1.4. Management Layer [Front-End]

The management layer is the central part of the agro-weather station. It provides real-time centralized data monitoring of the BS and different network elements. The layer continuously collects several critical parameters of the base station for an optimized system operation. The main parameters provided by this layer are the processor load, RAM usage, traffic behaviors, network performance, and overall station health checks. The collection of these parameters is done in the background. In addition, the layer allows the remote configuration and monitoring of different thresholds for alerts. The maintenance users can access the management layer through the developed GUI. The access to the GUI is done via an installed end-to-end VPN tunnel from external networks, or locally through the BS LAN port or any device within the same local network.

2.1.5. Middleware Layer [Back-End]

The middleware layer plays the role of the orchestrator entity within the base station, which allows it to create interfaces between:

- The deployed agents that are part of the perception layer and its different heterogeneous sensors and components.
- The presentation layer and its graphical presentation and reporting.
- The transmission layer and its protocols.
- The management layer and its operation and maintenance functionalities.

The middleware layer allows fast deployment of different perception layers' elements. In addition, this layer contains the database for different parameters management, cloud computing for the data and eventual models training, and the decision-making entity as illustrated in Figure 3.

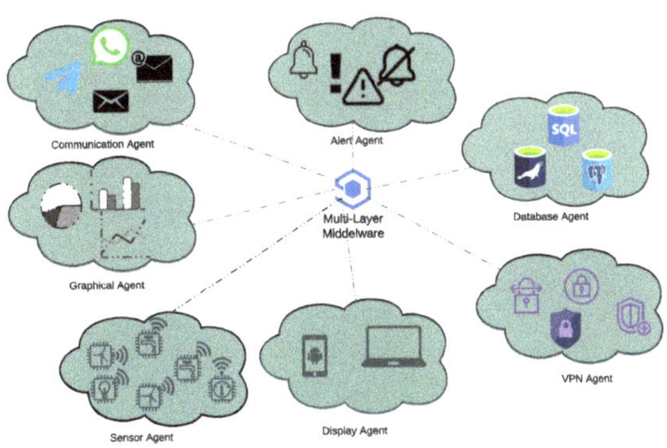

Figure 3. The middleware's Multi-Agents.

Since we have described the layers within the platform, we can present the mapping between these layers and main HW and SW functionalities. A holistic presentation is depicted in Table 3.

Table 3. The HW and SW mapping towards layers.

Layer	Raspberry Pi	Docker	Node-RED	InfluxDB	MQTT	Web Server (Apache)	Cloud	Colab	VPN	Wireless Nodes
Perception Layer	YES		YES		YES					YES
Transmission Layer	YES	YES	YES		YES				YES	
Presentation Layer	YES	YES		YES		YES	YES	YES		
Management Layer	YES	YES		YES		YES				
Middleware Layer	YES									

2.2. *System Implementation*

In the following, we focus on the end-to-end system design and implementation. For the agro-weather station design, the system is broken down into modules for a better-controlled approach.

2.2.1. Hardware Design

In this section, we detail the different elements that constitute our agro-weather station as depicted in Figure 4. We address the different components such as the base station, the remote sensors, the microcontroller, the power supply, the radio frequency, and the cloud.

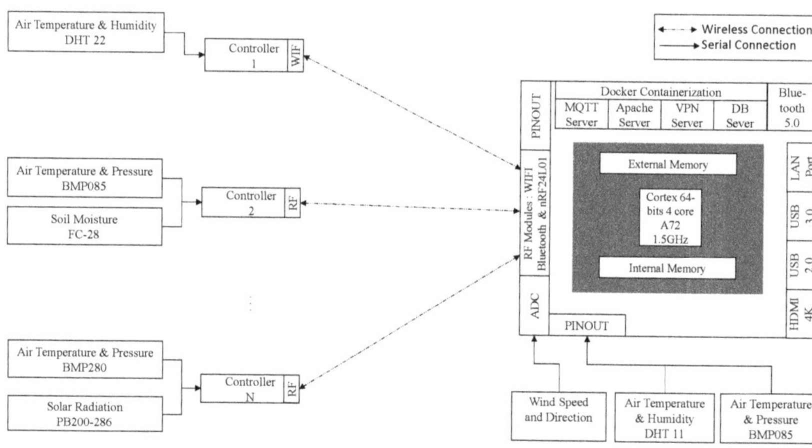

Figure 4. The high level block diagram of the agro-weather station.

The base station: We use a raspberry pi 4 model B based on a Broadcom BCM2835 system-on-a-chip (SoC). The Soc is based on a quad-core 64-bit Advanced RISC Machines (ARM) Cortex-A72 processor running at 1.5 GHz and a 4 Gb LPDDR4 RAM. The Cortex-A72 processor is one of the high-performance and low-power processors that implement ARMv8 architectures. The BS supports 802.11 Wireless LAN, Bluetooth 5.0, 2 micro-HDMI displays up to 4K resolution, 1 Gigabit Ethernet port, and 28 GPIOs supporting UART, I2C, SPI protocols. The designed BS supports radio frequency communication through an external NRF24L01 chip configured as a master.

The remote sensors: We chose to have wireless deployed nodes that are based on the ATmega32U4 microcontroller unit (MCU). The nodes operate on a 5 V to 9 V power range. The high-performance Microchip microcontroller is an 8-bit AVR RSC-based that operates on a voltage range of 1.8 V to 5.5 V. The MCU combines a read-while-write 32 KB ISP Flash memory, a 1024 B EEPROM, 2 KB SRAM and 23 I/O. The MCU supports UART,

SPI, and I2C. The MCU can operate under extreme weather condition that varies between −40 °C to 84 °C [36].

The designed node supports WiFi through an ESP8266 chip and radio frequency communication through the NRF 24L01. The MCU supports up to 126 RF channels with GFSK modulation, which allows a data rate up to 2 Mbps. When a high data rate is configured, the NRF24L01 allows the configuration of different saving techniques such as sleep mode and standby mode [37].

Sensors: To continuously survey meteorological data, keep the low architecture cost and align with the initial system specifications. We have chosen to deploy the sensors that respect the technical specification illustrated in Figure 4.

Capital expenses: To make the agro-weather station affordable for the end-users, the strategy of the design consists of relying on scalable open source components and migrating HW-based services to SW-based services whenever it is possible. The idea behind this migration is to lower our cost while keeping the same functionalities that HW offers. The migration can be seen in:

- The deployment of Telegram Bot messaging services instead of cellular messaging for alert generation.
- The use of remote accessible web-based services for data visualization instead of local screen. Tools such as Grafana and Chart.JS library.

However, even though the migration philosophy is prioritized, the end-to-end hardware architecture is fully open and supports any future upgrade or modernization. The entire system cost is given in Table S2 (see Supplementary Materials).

Operating expenses: While capital expenses are invested in the project startup to accommodate the solution, the Operating Expenses (Opex) represents the day-to-day expenses to run the platform. These expenses can be seen as the fixed costs that need to be ensured to operate the system. Since the system deployment will be a win-win relationship between farmers and researchers, who will be playing the role of system maintainers. The farmers will benefit from customized data for their crops and fields, while researchers will collect in-depth geographic-related environmental data for their uses. Data that can be explored to understand the study area. The OPEX related to the power consumption is negligible due to the minimal power usage and the low price of energy. In a country like Morocco, the KWh cost is less than $0.12. In [38], authors present an in depth analysis on different board models under different scenarios, study shows that the consumed power of an Arduino Uno board in single-byte transmission over Xbee S2B module is around 1.04 W using microwatt-meter. The power consumption of the Raspberry PI 4 B board in full load is around 7.6 W, while it is 2.25 W in idle mode. An extensive study has been done in [39] for a similar Raspberry board (PI 3 B+). The OPEX breakdown is provided in Table S3 (see Supplementary Materials).

2.2.2. Software Design

The system performance, agility, and security are among many metrics chosen to assess the scalability of our agro-weather station. Therefore, in this session, we address the system's software design approach and methodology as they are critical components in the system's operation.

System architecture: Since one of our goals is system interoperability, two architecture scenarios can be deployed—virtualization for hardware abstraction and containerization for system abstraction Figure 5. Virtualization consists of deploying virtual machines (VM) with a dedicated operating system (OS) within the base station. It is possible based on a physical allocation of hardware resources such as memory, storage, and network. However, the approach has many drawbacks on system scalability and performance due to the boot-up process, system speed, cost of implementation, system dependencies, software updates, networking overheads, and virtual machine size. Thus, to deliver the system in an agile methodology and minimize HW and SW resources, we chose the containerization scenario that performs the virtualization to the OS side and allows the ease of modules maintenance without compromising the continuous integration of services and modules. The authors in [40], provides a performance analysis of the two technologies based on the CPU performance, Disk I/O, and other metrics.

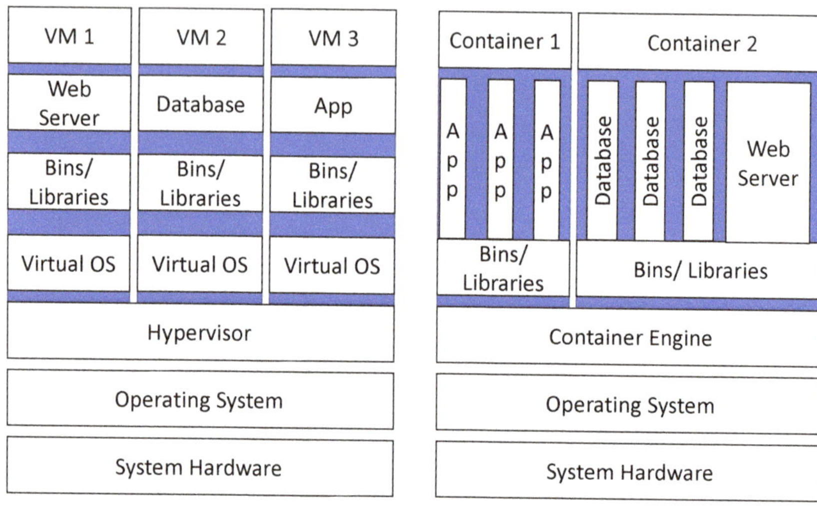

Figure 5. Virtualization versus containerization.

We chose to implement a containerization approach to separate the deployed software architecture from the hardware infrastructure. By adopting this approach, we were able to deliver the system quickly. Even though several open-source operating systems are customized to run on Raspberry Pi, such as kali Linux, Windows 10 IoT Core, Ubunto core, and Pidora. We chose to run over an optimized and stable Debian image called Raspbian. The Raspbian open-source operating system (OS) has multiple advantages, such as being very optimized to be run on Advanced RISC Machines (ARM), a family of Reduced Instruction Set Computing (RISC) architectures for computer processors that is extremely suitable for System on Chips (SoCs) such as Raspberry Pi. The OS is subject to continuous development, and it has a very active community for excellent support. All implemented services act as standalone services. Thus, the development and integration of different services can be done independently from technological bounds. The product development life cycle approach allows continuous improvement of the OS' products and services through 4 main stages: OS' introduction, OS' growth, OS' maturity, and OS' decline.

Even though it is possible to create containers without a specific toolkit, however in our work we use Docker since it is a fully open-source containerization platform that allows mounting container's image efficiently. Docker allows us to build, deploy, upgrade, and manage containers in a simple and straightforward commands' approach command-line interface (CLI). Another important component deployed is the Portainer tool that allows the same tasks through a lightweight management Graphical User Interface. Portainer plays the role of self-service container manager, it allows managing all containers hosted within Docker and grants the privilege of registration of new containers such as OpenVPN, influxDB, etc. It also allows the quick setup of different containers functionalities such as making a container open or secure, public or private within a network which allows the interoperability, scalability, and ease of use in the Operation and Management of the local base station.

Data base: When developing scalable systems, the choice of the type of databases is very crucial for the end-to-end system design, especially in the back-end design, where real-time data storage and data retrieval are important. Several types of databases are used in data management, such as relational and time series. While traditional relational databases are fully supported by SQL and perform incredibly in displaying large datasets, it performs poorly regarding the history of the stored data. Thus, the time series databases (TSDB) such as the open-source InfluxDB are considered very scalable and efficient when handling time-series data [41]. In our system, we use a container-based image of influx DB for different time-series parameters. Dedicated agents ensure data storage of different wireless nodes in the database.

Local Web server: To manage local web-based content such as images, scripts, and HTML files and allow a local web-based interaction between BS and different web users, we have deployed a local web server based on the open-source Apache. The server serves different technologies such as HTML, CSS, HTTP, and PHP. The apache server hosts the complete version of the website and local applications and databases. When an automated script needs to be executed according to planning it is not the local webserver that does that but rather it is the OS that executes it. This can be needed when measurements need to be sent to the Colab server for model training, the Cronjob sends all measurements that are on InfluxDB to Google drive (updating an existing excel). Colab then gets the link from Google drive and recovers the data and runs the modeling. Once the model is obtained Colab is then converting it to "tflite" and storing it on Drive. The model then is sent back to BS to be used in the local server within the BS. The local server runs the "tflite" model from TensorFlow and embeds it within the local website as depicted in Figure 6.

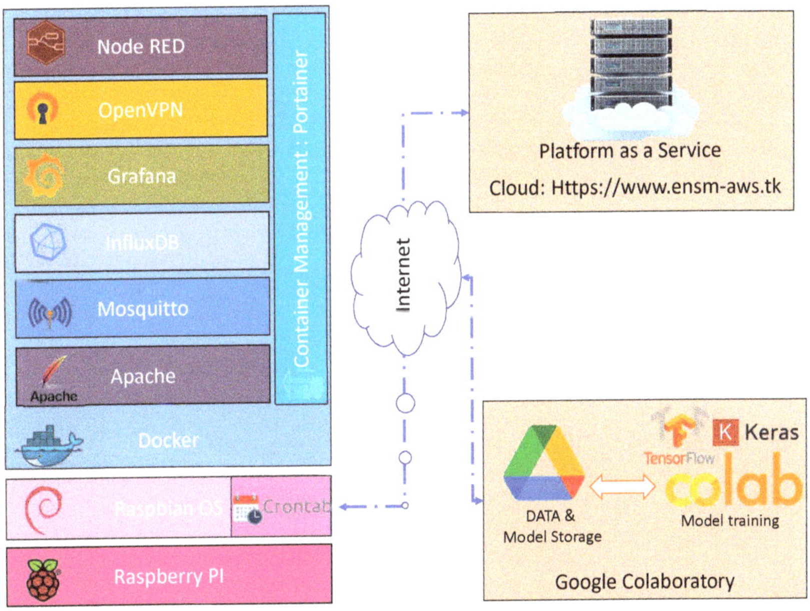

Figure 6. Overview on Docker's containers and interaction with cloud and Colab.

MQTT Brocker: MQTT is an OASIS-recognized Internet of Things communications protocol. It's built as a super-lightweight publish/subscribe messaging transport that is perfect for linking different IoT devices with minimal network resources. MQTT is now used in many industries, including automotive, manufacturing, telecommunications, etc. Within our base station, we deploy the open-source container-based broker named Mosquitto that runs MQTT 3.1.1 version. Once the sensor nodes (clients) are configured and pointing to the server broker, all clients can broadcast messages (publisher mode) or receive messages (subscriber mode). The communication is done without a direct point-to-point connection between publisher and subscriber. The configured client in our platform is an

ESP32-based wireless sensor node. The broker doesn't store the received data and sends it directly to the time-series database.

Configuration: Since the base station is based on Debian distribution, we can run our script using, literally, any programming language. However, to be consistent, we have chosen to configure components using either shell (SH), C/C++ language, Python, or Node-RED. All cronjob scripts for automation are created based on SH. Meanwhile, all remote nodes such as ESP32, Arduino micro, Arduino Uno, and ESP8266 are configured using C++ languages. Additionally, the scripting within the Bs such as remote connection with Google drive, Chatbot, local sensing is done based on Python. Finally, the interaction between the base station and Mosquitto, influxDB, external wireless nodes are done through Node-RED. Node-RED is a visual programming tool that was originally developed by IBM for connecting hardware devices, APIs, and web services as part of the Internet of Things. Node-RED includes a flow editor that is used to construct JavaScript functions in a web browser. Node-RED run-time is built on top of NODE.js.

Cloud: The data storage model in our architecture is based on a hybrid approach where local and cloud-based storage is implemented. Locally, the BS station stores data on the Hard Disk Drive (HDD) and SDCard. The process is also ensured through simultaneous use of the storage as a Service (STaaS) cloud computing model to transfer and backup the critical measurement data, such as Dropbox, Google Drive, and One Drive. The data backup process is ensured based on a pre-configured agent in a cronjob on the BS. The service is deployed to serve as backup to ensure any data restoration in case it is planned. The multiple data duplication strategies are designed to protect the system against data loss or corruption and ensure that the BS services' availability is guaranteed.

We use the cloud also as a "Platform as a Service" (PaaS) to run our web-based server on which the website and different BS services are hosted. The could is being accessed remotely for a read-only mode. For security reasons we don't allow real-time data consultation except through the VPN which requires pre-generated certificates. Meanwhile, other users may need to consult historic data, but still, they aren't granted VPN access and they aren't onsite to connect locally, for this reason, we have created a separate entity which is the cloud. The cloud hosts our website www.ensem-aws.tk which is an offline version of the local server deployed on the base station. The base station has unidirectional communication with the cloud which means only BS can push snapshots of the data (measurement and dashboards) but reversed communication is not allowed. Different service providers propose cloud services as public, private and hybrid solutions, providers such as Amazon Web service, Microsoft Azure, and Google cloud are the market's leaders. In our work, we use Namecheap cloud services for the hosting, the choice is driven by the cost-effectiveness strategy. However, the created services are not related to any vendor, and migration between cloud services is easy with the right set-up (DNS configuration).

VPN access: One of the crucial metrics is the platform security and its immunity to external access and attacks such as Denial of Service attacks DOS or Distributed DOS (DDOS). Since the BS is a lightweight server that needs to be optimized for resource deployment and usage, we configure a virtual private network (VPN) for secure connection establishment. The VPN allows the creation of end-to-end private tunnels between clients and the server, allowing external users to access the BS through open yet dedicated ports. In our BS, we implemented OpenVPN, one of the robust open-source VPN servers that support Secure Socket Layer (SSL) and Transport Layer Security (TLS) protocols. The secured channels are established by creating a Full-tunnel where all clients' traffic is directed through the VPN tunnel or the split tunnel where the only specified type of traffic is redirected. The tool supports IPv6 for the virtual private networks and can be executed over User Datagram Protocol (UDP) or Transmission Control Protocol (TCP). OpenVPN supports up to 500 VPN certificates generation and 100 tunnels connection at the same time. The certificates are deployed on the end user's clients' accounts.

GUI: The Graphical User Interface is deployed for a human-to-machine graphical environment. It is designed in a User-Centered Design (UCD) approach, where the end-users needs are prioritized, and restless re-adapting is applied. The GUI can be seen as a set of web services and applications hosted on an Apache webserver. The web application is organized into different sections with different dynamic pages such as the home page, login page, weather dashboard, BS maintenance dashboard, and contact page. The dashboards are customized to display different parameters within different time ranges. Data presentation is dynamic, and users can specify the time range and frequency of data updates such as 1 s, 5 s, and 10 min. GUI is mainly built on top of Grafana and Chart.JS library for the embedded part within the local website.

2.2.3. Neural Network Model

Weather forecasting or prediction can be seen as the application of various techniques to predict meteorological parameters. Many techniques are used in the literature by different researchers, such as Machine Learning and Deep Learning. Machine learning's popularity comes from its ability to identify the most relevant features within an appropriate model. Various approaches are used, like Support Vector Machine (SVM), Artificial Neural Network (ANN), or Recurrent Neural Network (RNN). Since meteorological data is considered as a non-linear multidimensional time series problem [42] the adopted network model has to reflect the temporal features within the dataset [43]. Thus, the long-term and short-term model (LSTM), introduced by [44] is one of the best approaches to deal with weather forecasting. We use a Recurrent Neural Network System (RNN) that supports time series as inputs in our system.

The usage of LSTM in models building and training allows us to have greater accuracy. In our work, the trained LSTM model enables training and forecasting future data based on historical multi-variate time series (sequential data). A brief description of LSTM structure and workflow is given in Supplementary Part IV (see Supplementary Materials).

2.2.4. Dataset for Model Building

We collect multivariate weather data from the weather station of Mohammad V International Airport in the city of Casablanca, Morocco as presented in Figure 7.

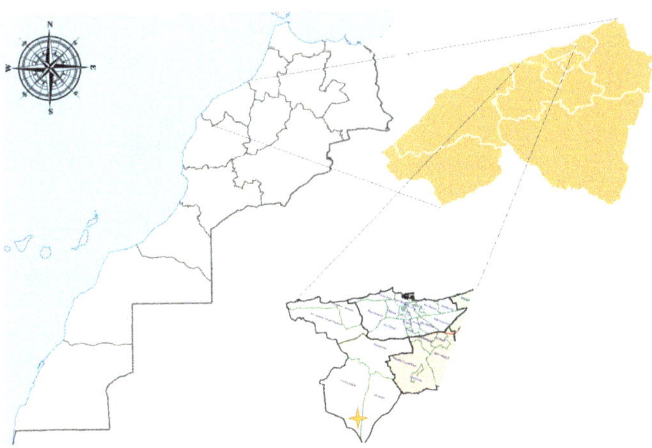

Figure 7. Geographic location of the study area (Casablanca, Morocco).

The collected dataset includes different features such as minimum and maximum temperature, pressure, wind speed, and dew point. The dew point presents the temperature at which the air becomes saturated with moisture [45]. The summary of data features is presented in Table 4. However, to extract the most relevant features useful for model building, we use feature selection based on heatmap as one of the crucial concepts in machine learning. The feature selection will allow our model to reduce the overfitting, reduce the training time, and improve accuracy.

Table 4. Data feature of the our model.

Feature	Unit	Description
Temperature range	Celsius (°C)	Temperature range at a height of 2 m above the earth's surface
Temperature dew	Celsius (°C)	dew/Frost point at a height of 2 m above the earth's surface
Temperature Max	Celsius (°C)	Maximum temperature at a height of 2 m above the earth's surface
Temperature Min	Celsius (°C)	Minimum temperature at a height of 2 m above the earth's surface
Temperature	Celsius (°C)	Temperature at a height of 2 m above the earth's surface
Earth skin temperature	Celsius (°C)	Earth skin temperature at a height of 2 m above the earth's surface
Precipitation	mm	Precipitation
Humidity	g/kg	Specific humidity at a height of 2 m above the earth's surface
Relative humidity	%	Relative humidity at a height of 2 m above the earth's surface
Pressure	kPa	Surface pressure
Wind speed range	m/s	Wind speed range at a height of 10 m above the earth's surface
Wind speed Min	m/s	Minimum wind speed at a height of 10 m above the earth's surface
Wind speed Max	m/s	Maximum wind speed at a height of 10 m above the earth's surface
Wind speed	m/s	Minimum wind speed at a height of 10 m above the earth's surface

The collected data covers a daily measurement from the 1st of January 1981 till the 29th of January 2021 Figure 8. The dataset contains 14 features of 14.638 measurements.

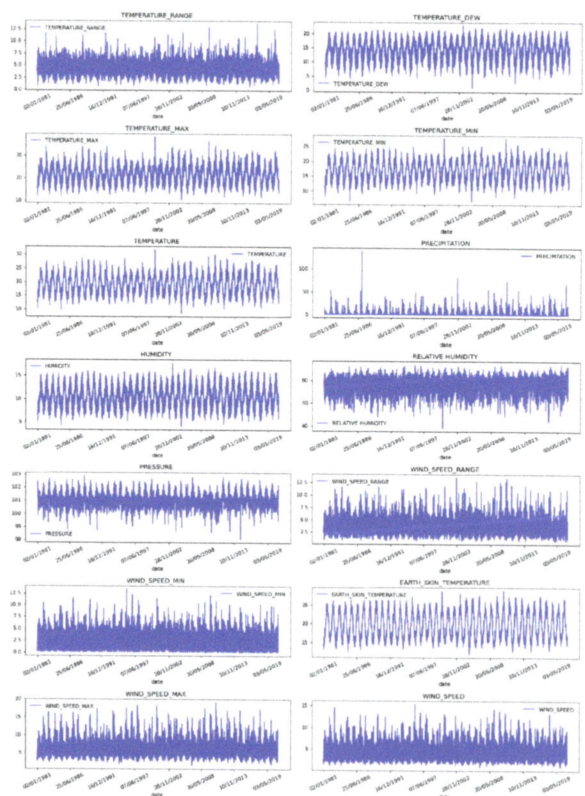

Figure 8. Exemplary plots of Temperature range, Temperature dew point, Temperature Max, Temperature Min, Temperature, Precipitation, Humidity, Relative humidity, Pressure, Wind speed range, Wind speed Max, Wind speed Min, Wind speed.

After analyzing the multi-variate historical data, it was clear that the correlation between temperature (mean), Temperature range, Temperature dew, temperature max, and Temperature min is very high. Thus we only keep the temperature (mean) and delete other features to decrease the overfitting. Further details on the features' correlation heatmap are illustrated in Supplementary Part V (see Supplementary Materials).

2.2.5. Model Accuracy

The model's performance is assessed through the computation of different traditional statistical metrics such as the Mean Absolute Scaled Error (MASE) and the Root Mean Square Deviation (RMSE). Additionally, we use the index of agreement [46] to separate the RMSE into unsystematic and systematic components [47,48]:

- **RMSE, MASE:** The calculation of the Root Mean Square Deviation (RMSE) represents the square root of the average of squared errors. It is mainly computed to measure the deviation between the actual values and the prediction values to define the accuracy of different models. The MASE was proposed in [49] as an assessment technique to define the accuracy of forecasts in regression models. As a mean absolute approach, MASE uses the ratio of errors which is allowing it to be independent of the scale of the forecaster. The RMSE and the MASE can be calculated following the equations:

$$\text{RMSE} = \sqrt{\frac{1}{n}\sum_{i=1}^{n}\left(\hat{X}_i - X_i\right)^2} \quad (1)$$

$$\text{MASE} = \frac{\sum_{i=1}^{n}\left|\hat{X}_i - X_i\right|}{\sum_{i=1}^{n}\left|X_{i-s} - X_i\right|} \quad (2)$$

  where $\hat{X}$ is the predicted value associated with the actual value X, and n is the size of the dataset. The larger the MASE and the RMSE mean, the enormous difference between the predicted and actual values, while the smaller the MASE and the RMSE mean, the closer the prediction values to the actual values.
- **Index of Agreement:** The Willmott's index of agreement [46], d_{index}, measures the model's relative accuracy in a range that varies from 0 to 1, with 0 indicating no agreement between the model predicted values and real observations and 1 indicating a perfect fit [50]. The index can be computed following the equation:

$$d_{index} = 1 - \frac{\sum_{i=1}^{n}\left(X_i - \hat{X}\right)^2}{\sum_{i=1}^{n}\left(\left|\hat{X} - \bar{X}\right| + \left|X_i - \bar{\hat{X}}\right|\right)^2}, \quad 0 \le d_{index} \le 1 \quad (3)$$

  where X is the real measured values associated to the predicted values $\hat{X}$, $\bar{X}$ is the average value of the measurements, and $\bar{\hat{X}}$ is the average value of the predicted measurements.

3. Results

3.1. System Workflow

In agreement with our expectations, the overall system design acted to minimize the human need for agro-weather station programming or maintenance. Thus, platform allows remote supervision and efficient automation to address this need. The deployed node follows a plug-and-play approach, in which the BS listens continuously to old and detects any new nodes under MQTT or NRF24L01 networks. Once a new node has its pre-loaded firmware, it starts sensing and transmitting data to the BS. The BS receives the traffic and ensures data is stored locally to the InfluxDB database for eventual analysis. The aggregated traffic from NRF24L01 and MQTT networks is transferred to the cloud service based on a cronjob program's automated script. The cronjob under the Raspbian operating system allows the execution of specific scripts in specific time and frequency ranges. The Local server ensures data presentation from the linked database and presents this data under dynamic web pages that are locally hosted. The data is also synchronized with the cloud-based server for open access. Alerts are configured for cognitive and context-aware monitoring. Several agents continuously monitor specific thresholds in a parallel way. Once a threshold is reached, system alerts are generated to the BS' admin user in emails and telegram messages. Different user roles are created to allow different levels of data consulting from the dynamic website.

Furthermore, for Operation And Maintenance (OAM) purposes, remote access to the different BS functionalities and programs is ensured throughout the deployed VPN server and the installed clients on the end-users devices such as phones, tablets, and laptops. The pre-generated VPN profiles are mandatory to allow the VPN tunnels establishment. Additional certificates could be generated locally or remotely based on the GUI VPN manager. Additionally, the aggregated data is framed and added to the historically collected data from the international Mohammed V airport weather station. The concatenated file is hosted on a google drive service that is continuously linked to the Colab environment for model training and predictions. The end-to-end workflow is depicted in Figure 9.

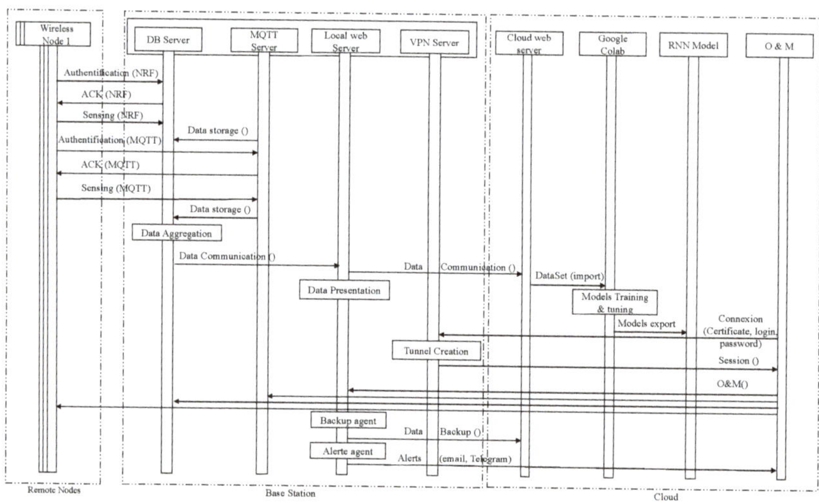

Figure 9. The network's communication workflow.

3.2. Experimental Results

3.2.1. System's Performance

For experimental purposes, we have tested our system in the laboratory and in the field. Different scenarios were applied to monitor the agro-weather station performance and behavior. We have also deployed the end-to-end portable system in an outdoor open wheat field for a real use case study Figure 10. The prototype has been operating for approximately 8 h in a 5-ha farm in Casablanca, Morocco for environment monitoring and continuous data collecting. Several scenarios were applied to test the BS security level, such as remote access using VPNs, on-site access through LAN network, and SSH access through WiFi. The BS monitored its different parameters for 8 h and provided through the designed GUI several clear Key Performance Indicators (KPIs) and data for end-users in an appealing dynamic interface as presented in Figure 11. The web-based GUI, based on Grafana, provides different dynamic figures and charts to reflect the BS' behaviors in terms of different metrics. Each figure is included within a re-sizable block, giving specific users the right to change the specific time frame, resize the block, or even change the position of the entire element. The GUI allows the end-users to filter through a specific time range or configure dashboard updates' frequency. It also allows the configuration of alerts based on predefined thresholds. The re-configurable dashboard was divided into multi sections for different purposes such as quick information, detailed health check, and network performance. The quick KPIs embed the critical data such as CPU and GPU temperature, CPU usage, RAM usage, and overview on threads and processes were accessible within the fixed 30 s as predefined in the initial baseline. A detailed health check section was created to support the operation and maintenance needs by presenting a detailed evolution of different parameters through time, metrics such as CPU load, Memory load, Processes, network usage, and network packets, furthermore network performance metrics such as load average, network errors, and network drops were introduced to assist maintenance users during troubleshooting or root cause analysis if any specified behavior. Other parameters such as Disk read/write load, time, and count was introduced to track the BS storage behavior. Since the dashboard is embedded within the local and the cloud-based website, it is compatible with different end-user devices such as phones (IOS, android) and laptops (Mac, Windows, Linux). The dashboard's dynamicity and responsiveness make it adaptable with different screen sizes. It allows it to behave and perform like a desktop application.

During the experimentation phase, different agents were performing different activities and monitoring multiple metrics. Most of the station's KPIs were below thresholds. The CPU temperature, GPU temperature was below 58 °C, while the CPU load was continuously under 75% (average of 45%) which allows the protection of the station resources. However, due to the operational multi-agents, the number of created processed was typically high reaching 300. The memory load was always under 2.8 Gb. Throughout the simulation, remote access using VPNs was tested several times which explains the multi-spikes in the network usage graph. Data were continuously transferred from the BS to local and external HDD which is tracked by the disk I/O requests and volume.

Figure 10. The agro-weather station in a field in Casablanca, Morocco.

Figure 11. The agro-weather station dashboard.

An offline version of the dashboard can also be consulted through the secured online portal https://ensem-aws.tk/ using the limited access credential login="test" and password="Azerty1+", current snapshots are captured on 13th April 2021. The online portal contains an offline stored version of different parameters in a specific period. Synchronization can be done periodically through a synchronization agent in form of a cronjob.

3.2.2. System's Security

The access to local nodes is limited by using credentials. However, it is known that remote accessibility is a crucial feature in any modern system especially with the emerging Covid-19 pandemic. In our system, remote access doesn't only allow different users to avoid on-site presence to consult data, but it provides also access anytime, from anywhere using any device to connect remotely to the base station and its components. As a result, allowing enormous cost savings. Meanwhile, the remote access functionality comes always with security challenges and threats such as:

- DDOS attacks.
- Phishing attacks
- Password sharing.
- Vulnerable backups.
- Leakage of information.

Therefore, a secured end-to-end connection is always required before any data consultation. To address this challenge an open-source low-cost VPN solution is implemented. Different certifications are generated per user profile. Once a user wants to connect from an external network, the client application connects to the server application to establish the end-to-end tunnel. In Figure 12 we can connect to our local private network from a public external network through the established tunnels. Once the VPN is activated, we can access any node within the network with IP address 192.168.x.x using only basic credentials.

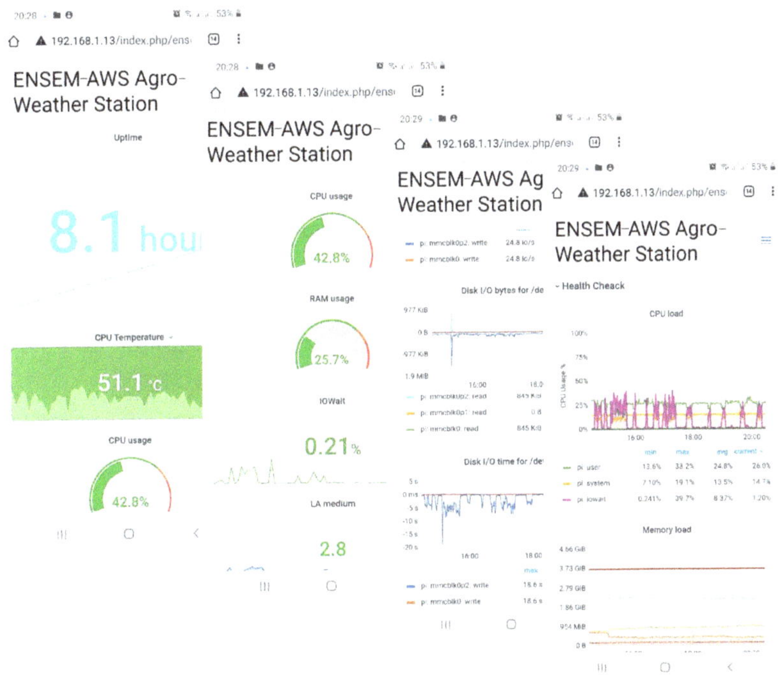

Figure 12. Remote access to the agro-weather station using VPN.

3.2.3. Weather Monitoring

The sensor nodes implemented within the farm collect continuous environmental data through the locally deployed sensors. The aggregated data is then forwarded in the form of JSON block to the base station. The Json block represents a series of JSON files in a specific time range. The data is transmitted via different transmission mediums (wired, wireless) and based on different protocols (NRF24L01, WIFI). The Figure 13 illustrates the plot of the temperature and the humidity recorded within 8h of environmental monitoring. From the readings, the temperature records show a continuous upward trend. The values continuously rise in a range between 15 °C and 25 °C. However, the humidity reading shows a continual downward trend. The humidity values decrease between 100% in the morning down to 40% in the evening. Even though the measurement reflects only 8 h of records, the extensive sensing through the year can be extremely useful for the farmers and researchers through the following:

- Assist in understanding the various effects of temperature and humidity on plants and crop productivity.
- Adapt the crop type (wheat, oats, potatoes, etc.) based on the period.
- Select the seed quality based on the season.
- Select and trigger automation action (e.g., irrigation) based on a set of parameters (e.g., temperature threshold).
- Select the best time for proper soil preparation.
- Prevent plants damaging by choosing the best timing (high humidity and temperature) to apply pesticides, since treatments should be applied in early to allow foliage to dry before reaching 29–32 °C [51]

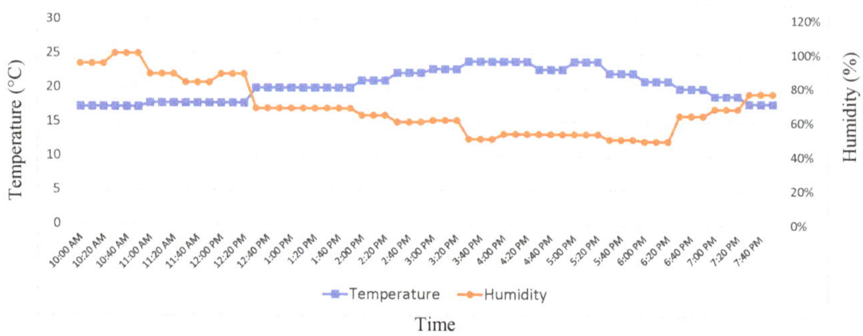

Figure 13. Temperature and humidity records of the agro-weather station.

Due to their significance as ecological controls, the wind speed and direction are very important metrics that should be considered by smart farming. The analyzed collected data can help farmers and researchers through:

- Lower production costs through the usage of the right wind turbines for electricity generation.
- Increase crops profits.
- Understand the impact of winds on plants and crop production (plants seeding, damaging, etc.)
- Secure reliable data for implementation of customized ML algorithms for dedicated farming fields.

In Figure 14 we present the wind rose that illustrates the distribution of wind speed and wind direction in the monitoring period. Detailed wind speed readings are plotted in Figure 15. The plot shows that the dominant wind direction is between North-northwest and the North with 17% of 4 to 8 km/h and 10% of wind with speed between 8 and 12 km/h and 5% of wind in a range of 12 to 16 km/h.

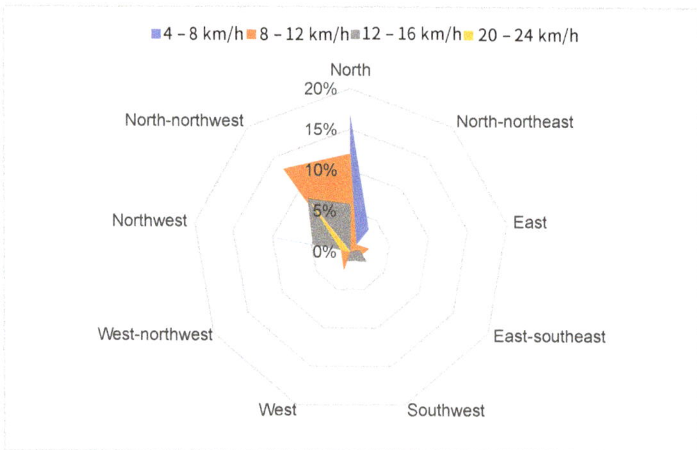

Figure 14. Wind speed, direction and distribution of the agro-weather station.

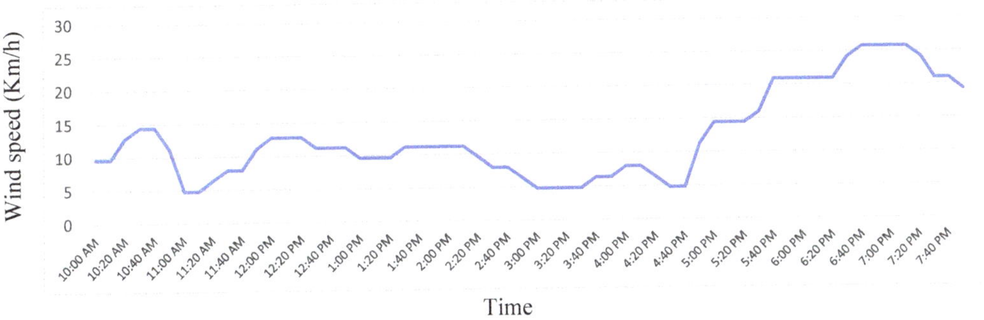

Figure 15. Wind speed records of the agro-weather station.

To summarize, smart farming cannot be achieved if it is not supported by reliable data and automation of recurrent activities such as monitoring and control. Thus, to select and apply the best crop management practices, we should be able to understand the evolution and impact of the meteorological parameters on the plants and on the productivity of any studied field. The proposed agro-weather station in this paper doesn't only allow different users to monitor their crops on a real-time basis but also decreases human activities and improves the adoption of new technologies in the agriculture field. Our tailored system allows real-time data that can support real-time decisions.

3.2.4. LSTM Model Implementation and Validation

To train our LSTM model we used the popular Python package named "Keras" which is embedded within TensorFlow. Applying the right hyperparameters is crucial for an optimized model before any learning process. Thus, an initial parameterization has been implemented and adjusted based on the model outcomes. The key optimized parameters were the number of hidden layers, the number of nodes within each layer, the number of epochs, and the learning rate.

The initial setup relies on the Mohammed V airport base station's data to build a reliable LSTM based model for weather forecasting. However, after a predefined period, data from the local base station is used and concatenated with the historical data to keep the model updated with the latest measurement. Meanwhile, since our initial dataset analysis shows some missing and abnormal observations at different time slots, we proceed with data pre-processing through removing the entire daily record for missing data. However, we keep the abnormal measurements (outliers) to measure the robustness and performance of the model through its ability to provide a good prediction. For model building, the data is split into three subsets, 70% of data is reserved for the training, while 20% is dedicated for validation, and finally, 10% is for the testing.

In the implementation of our model, we used 2 hidden layers which have been enough to avoid unnecessary model complications while being able to detect complex features. The layers were powered up by 50 neurons in each. The learning batch size was tested using 5, 10, and 100 days, while the learning rate was initially set to 0.1. Meanwhile, for the loss function, we have chosen the mean squared error, while the used model's optimizer is Adam. Additionally, we used sigmoid and tanh as the activation functions.

The simulations are performed on a cloud using the Colab platform from Google. The implemented algorithms used Tensorflow platform and Keras as main library to train and test the performance of our models.

After each round the RMSE, the MASE, and Willmott's index are computed. The learning curve in Figure 16 presents the plot of the computed standard deviations against the number of the epoch. The model has been trained during 30 epochs, Even though the network was training itself in 30th epochs, it is clear that in our training process, the MASE and RMSE were minimal in the 5th epochs with a value of 0.0012 and 0.034 respectively, however Willmott's index was around 0.987 with space of improving. The optimum values were in Epoch 10 with minimal MASE, RMSE, and Willmott's with values of 0.0012, 0.0034, and 0.988 respectively. The MASE and RMSE minimum remain stable throughout the epochs, and Willmott's index stabilizes at 0.987.

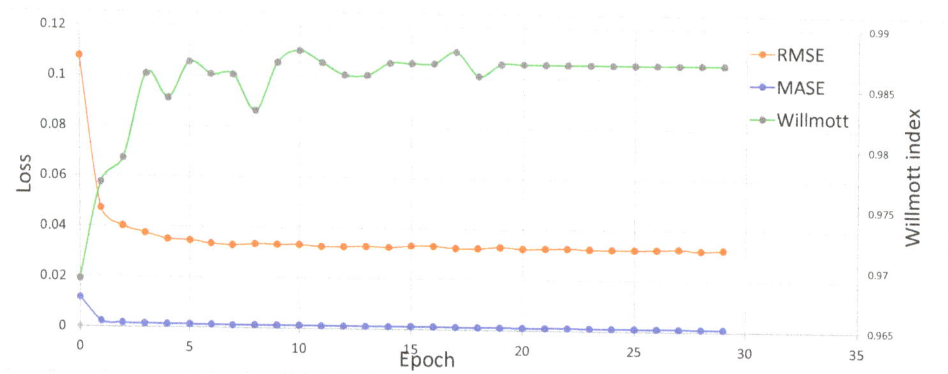

Figure 16. Presentation of MASE-based and RMSE-based and Willmott's-based training curve.

The Records' deviation in Figure 17, and the real temperature versus predicted are plotted in Figure 18. The prediction of our LSTM model is plotted for 1 month, 1 year, and 8 years from April-2013 till April-2021. The prediction values are based on a trained model with 32 years of historical data, which possibly captures the ongoing effects of climate change. The visual comparison in Figure 18 shows that the LSTM model is performing well. The prediction trends follow almost the training trends in the overall plot with a mean of residual values over the 8 years around 0.59 °C. Further details on model validation are presented in Supplementary Part VI (see Supplementary Materials).

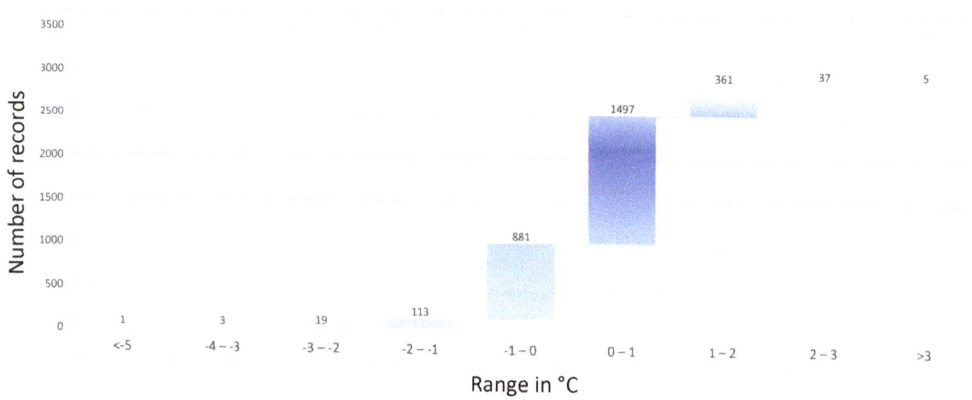

Figure 17. Prediction deviation vs. number of records.

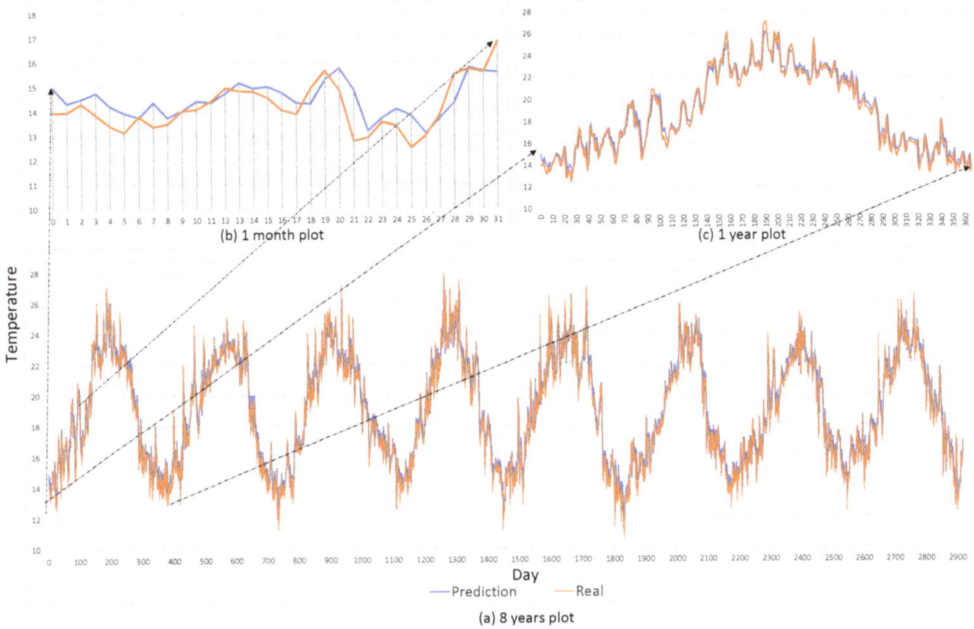

Figure 18. Model validation—Muli-timeframe temperature prediction. (**a**) daily predictions vs. real plot for 8 years (entire dataset). (**b**) daily predictions vs. real plot for 1 month. (**c**) daily predictions vs. real plot for 1 year.

4. Discussion

4.1. Design Issues

Compared to any other field, IoT systems have different bottlenecks that aren't only slowing down the adoption of such solutions but they are dragging more attention by scientific communities through several contributions. In [14,15] the authors describe the most common challenges in IoT solutions for agriculture. Challenges such as resources optimization, cost analysis, lack of knowledge of technology, quality of service, security, and networking. Below we address most of these challenges.

4.2. Comparison with Other Existing Systems

Even though various solutions have been presented under different research outputs. However, to the best of our knowledge, the current solutions suffer from different bottlenecks that impact not only the adoption of such platforms but also the massive penetration within small farms. Challenges that are linked to the systems' cost, ease of use, automation capabilities, operation & maintenance, security, and even remote access and usage of artificial intelligence. We believe that there is no perfect system, and we are convinced that there is always a space for the system's enhancement and adaptation to different use cases. In the following, we compare the results from similar works based on the most common challenges such as resources' optimization, cost analysis, quality of services, networking challenges, artificial intelligence, operation and maintenance, remote accessibility, and security:

- **Resources' optimization**: Our proposed work relies on a fully containerized architecture, where services are not only lightweight but scalable, agile, and portable. This makes the agro-weather station's full software architecture manageable and reproducible within a very short amount of time. It allows us to have very reliable and optimized micro-services such as local web server, time-series database, VPN server, etc that consume wisely the BS's resources such as CPU, memory, and network as presented in Figure 11. Meanwhile, all the studied works in this article [20–30,52–59] are based on classical low-performance system on chips (SoCs) in designing and implementing station. The used SoCs such as Arduino microcontroller doesn't support multitasking such as Raspberry Pi micro-computer, which limit the performance of the stations to basic monitoring tasks.

- **Cost analysis**: In our work, we provide an Opex and Capex deep analysis to support the cost-effectiveness approach targeted in this work. Thus, we estimated the cost of investment to build the agro-weather station at 176 $ as depicted in Section 2.2.1. While, works presented in [22,25,26,30,57–59] have claimed development of a low-cost-based system for smart farming, meanwhile no one of these contributions has provided a detailed operation based analysis and capital based costing.
- **Quality of services**: When studying the literature, we can see that there is a huge limitation when it comes to the deployment of a fully multi-agent-based architecture in systems' design. In [28] the author presents a logical tailored approach for multi-user architecture design. In [21–23], the authors proposed the deployments of advisory systems either for early disease detection or crop productivity management. However, the proposed works focus on the design of a generic prototype rather than a customized approach relevant to each user type. Meanwhile, our system tries to provide different features adapted to different users based on real use-cases. Remote access to engineers and researchers, temperature forecasting for farmers and researchers, data security for all users are all standalone agents within the BS. Each agent ensures the utility and warranty of its functions and services.
- **Networking**: We address the networking challenge based on fully agile architecture embedding different heterogeneous nodes. Our system supports multiple protocols for backbone transmission protocols such as (MQTT, NRF24L01, LAN). The platform is also extended for other protocols. Additionally, we create a system that supports plug-and-play nodes as described in Section 3.1. In our platform, we experiment with the proposed transmission protocols through the end-to-end monitoring ecosystem. Meanwhile, only [25] presents an adaptive mechanism for reliable smart farming. However, the technique is only dealing with an isolated scope which is transmission without validating the system by an end-to-end smart farming platform. While, other contributions rely on singe transmission protocol such as in [20,22,23,52,55–59] or don't even support wireless transmission such as in [21,24].
- **Artificial Intelligence**: AI becomes a necessity in modern applications and services. In our system, we deploy a high-performing LSTM model part of RNN for temperature forecasting. The deployed model as depicted in Section 3.2.3 shows a high-level performance in temperature prediction which allows the possibility to train and deploy similar models for other meteorological data such as humidity, pressure, wind speed, etc. In our system, we propose the hybrid data collection where historical data is continuously enhanced by the BS itself for a continuous model's improvement. The BS supports tasks automation in form of cronjob and through a standalone deployed agent for alerts notification. The proposed work in [54] proposes a temperature foresting model based on ANN (Artificial Natural Network), meanwhile, it is known that the ANN is less powerful than the RNN (adopted in this paper). ANN doesn't support the recurrent connections and is considered to be powerful with tabular data and text data rather than the sequence data which we have in meteorological parameters. Additionally, in [21–23] the authors create advisory systems for decision making. The proposed systems require a minimum level of knowledge, while in the developing countries the lack of technological awareness among farmers will be challenging for the usage of such systems. The same technological bottleneck is impacting the usage of AI within drones framework in crop management such as in [30,56]. In [53] authors only present a conceptual model that lacks testing and validation. Meanwhile, no AI implementation is considered in the works in [20,24–29,52,55,57–59].

- **Operation and Maintenance**: The studied works in this effort focus on the system's utility through the functionalities and services offered by the proposed platforms, rather than focusing on the warranty through the assurance that the developed platforms and services will deliver the needed requirements. Therefore, no system surveyed in this work offers the possibility to have a holistic dashboard for operation and maintenance. Thus, we have created a dedicated dashboard part of the management layer as presented in Section 3.1. The dashboard will allow users such as engineers to consult different performance metrics such as network usage and packet drops, CPU load, operating system's threads, and processes, etc. continuously for management purposes.
- **Remote access**: None of the surveyed works in this study allows the possibility to have ultimate remote access to the deployed platform and its components. Only conceptual architecture with remote accessibility feature was proposed in [28] and partial data consultation was proposed in [54]. Meanwhile, as part of the system's customization adopted in this work, we think that full remote accessibility allows tremendous ease of use and cost-saving when it comes to consulting the platform and the locally collected data. Thus, the proposed platform allows different users to connect remotely to the BS and consult all components through the GUI or the SSH sessions.
- **Security**: Even though smart farming isn't a critical domain when it comes to data sensitivity. However, privacy in today's world is a huge concern for different users. Remote access comes with privacy challenges. As mentioned in Sections 2.2.2 and 3.2.2 our platform supports the establishment of end-to-end VPN tunnels that add an extra layer of privacy and security when using remote access. The deployed VPN server allows peers to connect using the pre-generated secret keys and certificates that are based on strong 256-bit encryption. Even with the high-security concerns especially in the Covid-19 period, none of the others experimental platforms address or implement any feature or approach for system's and services' security and reliability.

A holistic comparison between our proposed system and the existing system is presented in Supplementary Part VII (see Supplementary Materials). In the table, we present the major advantages of the system and the major challenges that we detected during the review. We also categorize the platforms based on their validation. Additionally, we assess their cost-effectiveness based on deployed hardware and software. Finally, we categorized the usage of some features such as operation and maintenance, remote accessibility, and AI implementation.

5. Conclusions

In this paper, we propose an end-to-end AI-powered IoT-based low-cost platform for smart farming. The main goal of the system's creation is to support different users such as farmers and researchers to monitor, understand, and act for better crop management. The followed approach in system design enables the smooth system's development and enhancement through its different iterations, releases, and versions. The platform design and development purpose are to propose a real-time context-aware system for continuous cognitive monitoring. Ease of use, portability, low cost, and robustness are among many metrics considered in the system's design and implementation. The proposed HW and SW prototype was validated on the field and presented a high performance in different activities and operations such as real-time monitoring, temperature forecasting, scalable wireless connections, reliable dashboards, etc. Therefore, we think that our low-cost platform can help farmers and researchers to co-create value and to have an impact on crop management. Additionally, we believe that developed station stress on vital concepts that were missed in similar works such as remote accessibility, security, operation, and maintenance. However, we believe that there is a large space for development and system enhancement. Thus, to enhance the platform performance, our subsequent work will focalize on:

- Add more advanced functions to enhance the system performance, such as AI-based power management and fault detection agents.
- Design and create predefined test cases and scenarios that enable extreme ease of use and service for the enhancement of automation.
- Design AI-based mobility management that can be implemented to BS to reconfigure the platform without the need for human interaction.
- Migrate the solution to a pure native cloud and support more communication technologies like Lora, Sigfox, and NBIoT.

Supplementary Materials: The following supporting information can be downloaded at: https://www.mdpi.com/article/10.3390/agriculture12010035/s1, Table S1: Some of the most common challenges in IoT, Figure S1: the key principales of Agile Methodology, Table S2: Capex investment, Table S3: Opex investment, Figure S2: Structure of single cell within LSM, Figure S3: Model work flow of a single cell, Figure S4: Feature selection using heat map method, Figure S5: Scatter plot of real values vs. prediction values, Figure S6: Residual plot of different time frame, Table S4: Range and percentage of deviation records, Table S5: Comparison between the proposed platform and other existing platforms.

Author Contributions: Conceptualization, methodology, software, validation, formal analysis, investigation, resources, writing, review, editing, visualization, supervision, project administration, funding acquisition, A.F., M.S. and E.S. All authors have read and agreed to the published version of the manuscript.

Funding: This work was carried out within the framework "Agrometeorological Stations Platform" project funded by the Moroccan Ministry of Higher Education and Scientific Research-National Centre for Scientific and Technical Research (CNRST) (PPR2 project).

Institutional Review Board Statement: Not applicable.

Informed Consent Statement: Not applicable.

Data Availability Statement: The data that support the findings of this study are available on request from the corresponding author.

Acknowledgments: We are grateful to the guest editor and the anonymous reviewers for their valuable comments that substantially improved the presentation of this research.

Conflicts of Interest: The authors declare no conflict of interest.

Abbreviations

The following abbreviations are used in this manuscript:

AC	Alternative Current
AI	Artificial Intelligence
ANN	Artificial Neural Network
ARM	Advanced RISC Machines
CLI	Command Line Interface
Covid-19	Coronavirus disease-2019
CPU	Central Processing Unit
DB	Data Base
DC	Direct Current
DOS	Denial of Service Attacks
FTP	File Transfer Protocol
HDD	Hard Disk Drive
HTTP	Hyper Text Transfer Protocol
HW	Hardware
IoT	Internet of Things
GUI	Graphical User Interface
KPI	Key Performance Indicators
LSTM	Long Short Term Memory

M2M	Machine-to-Machine
MQTT	Message Queue Telemetry Transport
MASE	Mean Absolute Scaled Error
NN	Neural network
OASIS	Organization for the Advancement of Structured Information Standards
O&M	Operation and Maintenance
OS	Operating System
PA	Precision Agriculture
PaaS	Platform as a Service
RISC	Reduced Instruction Set Computing
RMSE	Root Mean Square Error
RNN	Recurrent Neural Network
SF	Smart Farming
SSL	Secure Socket Layer
SUS	Start Upgrade Stop
SW	Software
STaaS	Storage as a Service
TLS	Transport Layer Security
UD	User Datagram Protocol
VM	Virtual Machine
VPN	Virtual Private Network
SVM	Support Vector Machine
SSH	Secure Shell
WSN	Wireless Sensor Network

References

1. United Nations; Department of Economic and Social Affairs; Population Division. *World Population Prospects Highlights, 2019 Revision Highlights, 2019 Revision*; United Nations: New York, NY, USA, 2019.
2. O'Grady, M.J.; O'Hare, G.M. Modelling the smart farm. *Inf. Process. Agric.* **2017**, *4*, 179–187. [CrossRef]
3. Julian, Q.; Nat, D. *Climate Change and Land Tenure*; IIED (International Institute for Environment and Development) and Natural Resources Institute, University of Greenwich: London, UK, 2008; p. 68.
4. Ingram, G.K.; Hong, Y.H. (Eds.) *Climate Change and Land Policies*; Lincoln Institute of Land Policy: Cambridge, MA, USA, 2011.
5. Adomako, T.; Ampadu, B. The Impact Agricultural Practices on Environmental Sustainability in Ghana: A Review. *J. Sustain. Dev.* **2015**, *8*, 70. [CrossRef]
6. Mohanavelu, A.; Naganna, S.R.; Al-Ansari, N. Irrigation Induced Salinity and Sodicity Hazards on Soil and Groundwater: An Overview of Its Causes, Impacts and Mitigation Strategies. *Agriculture* **2021**, *11*, 983. [CrossRef]
7. Wheeler, T.; von Braun, J. Climate Change Impacts on Global Food Security. *Science* **2013**, *341*, 508–513. [CrossRef]
8. Walter, A.; Finger, R.; Huber, R.; Buchmann, N. Opinion: Smart farming is key to developing sustainable agriculture. *Proc. Natl. Acad. Sci. USA* **2017**, *114*, 6148–6150. [CrossRef]
9. Akyildiz, I.; Su, W.; Sankarasubramaniam, Y.; Cayirci, E. A survey on sensor networks. *IEEE Commun. Mag.* **2002**, *40*, 102–114. [CrossRef]
10. Yick, J.; Mukherjee, B.; Ghosal, D. Wireless sensor network survey. *Comput. Netw.* **2008**, *52*, 2292–2330. [CrossRef]
11. Dâmaso, A.; Freitas, D.; Rosa, N.; Silva, B.; Maciel, P. Evaluating the Power Consumption of Wireless Sensor Network Applications Using Models. *Sensors* **2013**, *13*, 3473–3500. [CrossRef]
12. Faid, A.; Sadik, M.; Sabir, E. IHEE: An Improved Hybrid Energy Efficient Algorithm for WSN. In *Advances in Information and Communication*; Arai, K., Ed.; Springer International Publishing: Cham, Switzerland, 2021; pp. 283–298.
13. Trendov, N.M.; Varas, S.; Zeng, M. *Digital Technologies in Agriculture and Rural Areas—Status Paper*; FAO: Rome, Italy, 2019; p. 157.
14. Farooq, M.S.; Riaz, S.; Abid, A.; Umer, T.; Zikria, Y.B. Role of IoT Technology in Agriculture: A Systematic Literature Review. *Electronics* **2020**, *9*, 319. [CrossRef]
15. Villa-Henriksen, A.; Edwards, G.T.; Pesonen, L.A.; Green, O.; Sørensen, C.A.G. Internet of Things in arable farming: Implementation, applications, challenges and potential. *Biosyst. Eng.* **2020**, *191*, 60–84. [CrossRef]
16. Jawad, H.; Nordin, R.; Gharghan, S.; Jawad, A.; Ismail, M. Energy-Efficient Wireless Sensor Networks for Precision Agriculture: A Review. *Sensors* **2017**, *17*, 1781. [CrossRef]
17. Khanna, A.; Kaur, S. Evolution of Internet of Things (IoT) and its significant impact in the field of Precision Agriculture. *Comput. Electron. Agric.* **2019**, *157*, 218–231. [CrossRef]
18. Talaviya, T.; Shah, D.; Patel, N.; Yagnik, H.; Shah, M. Implementation of artificial intelligence in agriculture for optimisation of irrigation and application of pesticides and herbicides. *Artif. Intell. Agric.* **2020**, *4*, 58–73. [CrossRef]
19. Navarro, E.; Costa, N.; Pereira, A. A Systematic Review of IoT Solutions for Smart Farming. *Sensors* **2020**, *20*, 4231. [CrossRef]

20. Gangwar, D.S.; Tyagi, S.; Soni, S.K. A conceptual framework of agroecological resource management system for climate-smart agriculture. *Int. J. Environ. Sci. Technol.* **2019**, *16*, 4123–4132. [CrossRef]
21. Khattab, A.; Habib, S.E.; Ismail, H.; Zayan, S.; Fahmy, Y.; Khairy, M.M. An IoT-based cognitive monitoring system for early plant disease forecast. *Comput. Electron. Agric.* **2019**, *166*, 105028. [CrossRef]
22. Muzafarov, F.; Eshmuradov, A. Wireless sensor network based monitoring system for precision agriculture in Uzbekistan. *TELKOMNIKA Telecommun. Comput. Electron. Control* **2019**, *17*, 10. [CrossRef]
23. Nawandar, N.K.; Satpute, V.R. IoT based low cost and intelligent module for smart irrigation system. *Comput. Electron. Agric.* **2019**, *162*, 979–990. [CrossRef]
24. Doshi, J.; Patel, T.; Bharti, S.k. Smart Farming using IoT, a solution for optimally monitoring farming conditions. *Procedia Comput. Sci.* **2019**, *160*, 746–751. [CrossRef]
25. Ramli, M.R.; Daely, P.T.; Kim, D.S.; Lee, J.M. IoT-based adaptive network mechanism for reliable smart farm system. *Comput. Electron. Agric.* **2020**, *170*, 105287. [CrossRef]
26. Sadowski, S.; Spachos, P. Wireless technologies for smart agricultural monitoring using internet of things devices with energy harvesting capabilities. *Comput. Electron. Agric.* **2020**, *172*, 105338. [CrossRef]
27. Popović, T.; Latinović, N.; Pešić, A.; Žarko, Z.; Krstajić, B.; Djukanović, S. Architecting an IoT-enabled platform for precision agriculture and ecological monitoring: A case study. *Comput. Electron. Agric.* **2017**, *140*, 255–265. [CrossRef]
28. Köksal, Ö.; Tekinerdogan, B. Architecture design approach for IoT-based farm management information systems. *Precis. Agric.* **2019**, *20*, 926–958. [CrossRef]
29. Leelavinodhan, P.B.; Vecchio, M.; Antonelli, F.; Maestrini, A.; Brunelli, D. Design and Implementation of an Energy-Efficient Weather Station for Wind Data Collection. *Sensors* **2021**, *21*, 3831. [CrossRef]
30. Almalki, F.A.; Soufiene, B.O.; Alsamhi, S.H.; Sakli, H. A Low-Cost Platform for Environmental Smart Farming Monitoring System Based on IoT and UAVs. *Sustainability* **2021**, *13*, 5908. [CrossRef]
31. Faid, A.; Sadik, M.; Sabir, E. IoT-based Low Cost Architecture for Smart Farming. In Proceedings of the 2020 International Wireless Communications and Mobile Computing (IWCMC), Limassol, Cyprus, 15–19 June 2020; pp. 1296–1302. [CrossRef]
32. Faid, A.; Sadik, M.; Sabir, E. EACA: An Energy Aware Clustering Algorithm for Wireless IoT Sensors. In Proceedings of the 2021 28th International Conference on Telecommunications (ICT), Vancouver, BC, Canada, 29–30 April 2021; pp. 1–6. [CrossRef]
33. Aminikhanghahi, S.; Cook, D.J. A survey of methods for time series change point detection. *Knowl. Inf. Syst.* **2017**, *51*, 339–367. [CrossRef]
34. Grigorik, I. Making the Web Faster with HTTP 2.0. *Commun. ACM* **2013**, *56*, 42–49. [CrossRef]
35. Deschambault, O.; Gherbi, A.; Légaré, C. Efficient Implementation of the MQTT Protocol for Embedded Systems. *J. Inf. Process. Syst.* **2017**, *13*, 26–39.
36. Microchip Technology. ATmega328P Datasheet, 2021. Available online: https://www.microchip.com/wwwproducts/en/ATmega328P (accessed on 13 April 2021).
37. Nordic Semiconductor. nRF24L01 Datasheet, 2021. Available online: https://www.sparkfun.com/datasheets/Components/SMD/nRF24L01Pluss_Preliminary_Product_Specification_v1_0.pdf (accessed on 13 April 2021).
38. Al-Shorman, M.Y.; Al-Kofahi, M.M.; Al-Kofahi, O.M. A practical microwatt-meter for electrical energy measurement in programmable devices. *Meas. Control* **2018**, *51*, 383–395. [CrossRef]
39. Paunski, Y.K.; Angelov, G.T. Performance and power consumption analysis of low-cost single board computers in educational robotics. *IFAC-PapersOnLine* **2019**, *52*, 424–428.
40. Potdar, A.M.; D G, N.; Kengond, S.; Mulla, M.M. Performance Evaluation of Docker Container and Virtual Machine. *Procedia Comput. Sci.* **2020**, *171*, 1419–1428.
41. Struckov, A.; Yufa, S.; Visheratin, A.A.; Nasonov, D. Evaluation of modern tools and techniques for storing time-series data. *Procedia Comput. Sci.* **2019**, *156*, 19–28. [CrossRef]
42. Maqsood, I.; Khan, M.R.; Abraham, A. An ensemble of neural networks for weather forecasting. *Neural Comput. Appl.* **2004**, *13*, 1433–3058. [CrossRef]
43. Kreuzer, D.; Munz, M.; Schlüter, S. Short-term temperature forecasts using a convolutional neural network—An application to different weather stations in Germany. *Mach. Learn. Appl.* **2020**, *2*, 100007. [CrossRef]
44. Hochreiter, S.; Schmidhuber, J. Long Short-Term Memory. *Neural Comput.* **1997**, *9*, 1735–1780. [CrossRef] [PubMed]
45. Kent, R. Chapter 4—Services. In *Energy Management in Plastics Processing*, 3rd ed.; Kent, R., Ed.; Elsevier: Amsterdam, The Netherlands, 2018; pp. 105–210. [CrossRef]
46. Willmott, C.J. On the Validation of Models. *Phys. Geogr.* **1981**, *2*, 184–194. [CrossRef]
47. Delle Monache, L.; Nipen, T.; Deng, X.; Zhou, Y.; Stull, R. Ozone ensemble forecasts: 2. A Kalman filter predictor bias correction. *J. Geophys. Res. Atmos.* **2006**, *111*. [CrossRef]
48. Kang, D.; Mathur, R.; Rao, S.T.; Yu, S. Bias adjustment techniques for improving ozone air quality forecasts. *J. Geophys. Res. Atmos.* **2008**, *113*. [CrossRef]
49. Hyndman, R.J.; Koehler, A.B. Another look at measures of forecast accuracy. *Int. J. Forecast.* **2006**, *22*, 679–688. [CrossRef]
50. Willmott, C.J. On the Evaluation of Model Performance in Physical Geography. In *Spatial Statistics and Models*; Gaile, G.L., Willmott, C.J., Eds.; Springer: Dordrecht, The Netherlands, 1984; pp. 443–460. [CrossRef]

51. Stack, L.; Dill, J.; Pundt, L.; Raudales, R.; Smith, C.; Smith, T. New England Greenhouse Floriculture Guide; A Management Guide for Insects, Diseases, Weeds and Growth Regulators. *Northeast Greenhouse Conference and Expo.* 2017–2018. Available online: https://www.plantgrower.org/uploads/6/5/5/4/65545169/17section_b_2017-18_floriculture_guide.pdf (accessed on 13 April 2021).
52. Codeluppi, G.; Cilfone, A.; Davoli, L.; Ferrari, G. LoRaFarM: A LoRaWAN-Based Smart Farming Modular IoT Architecture. *Sensors* **2020**, *20*, 2028. [CrossRef]
53. Ratnakumari, K.; Koteswari, S. Design & implementation of innovative IoT based smart agriculture management system for efficient crop growth. *J. Eng. Sci.* **2020**, *11*, 607–616.
54. Castañeda-Miranda, A.; Castaño-Meneses, V.M. Internet of things for smart farming and frost intelligent control in greenhouses. *Comput. Electron. Agric.* **2020**, *176*, 105614. [CrossRef]
55. Tiglao, N.M.; Alipio, M.; Balanay, J.V.; Saldivar, E.; Tiston, J.L. Agrinex: A low-cost wireless mesh-based smart irrigation system. *Measurement* **2020**, *161*, 107874. [CrossRef]
56. Roy, S.K.; De, D. Genetic Algorithm based Internet of Precision Agricultural Things (IopaT) for Agriculture 4.0. *Internet Things* **2020**, *1*, 100201. [CrossRef]
57. Cicioğlu, M.; Çalhan, A. Smart agriculture with internet of things in cornfields. *Comput. Electr. Eng.* **2021**, *90*, 106982. [CrossRef]
58. Podder, A.K.; Bukhari, A.A.; Islam, S.; Mia, S.; Mohammed, M.A.; Kumar, N.M.; Cengiz, K.; Abdulkareem, K.H. IoT based smart agrotech system for verification of Urban farming parameters. *Microprocess. Microsyst.* **2021**, *82*, 104025. [CrossRef]
59. Gao, D.; Sun, Q.; Hu, B.; Zhang, S. A Framework for Agricultural Pest and Disease Monitoring Based on Internet-of-Things and Unmanned Aerial Vehicles. *Sensors* **2020**, *20*, 1487. [CrossRef] [PubMed]

agriculture

Article

Research on Maize Seed Classification and Recognition Based on Machine Vision and Deep Learning

Peng Xu [1], Qian Tan [2], Yunpeng Zhang [2], Xiantao Zha [2], Songmei Yang [2] and Ranbing Yang [2,*]

1 College of Information and Communication Engineering, Hainan University, Haikou 570228, China; xupeng@hainanu.edu.cn
2 College of Mechanical and Electrical Engineering, Hainan University, Haikou 570228, China; tanqian@hainanu.edu.cn (Q.T.); zhangyunpeng@hainanu.edu.cn (Y.Z.); zhaxt@hainanu.edu.cn (X.Z.); 995093@hainanu.edu.cn (S.Y.)
* Correspondence: yangranbing@hainanu.edu.cn

Abstract: Maize is one of the essential crops for food supply. Accurate sorting of seeds is critical for cultivation and marketing purposes, while the traditional methods of variety identification are time-consuming, inefficient, and easily damaged. This study proposes a rapid classification method for maize seeds using a combination of machine vision and deep learning. 8080 maize seeds of five varieties were collected, and then the sample images were classified into training and validation sets in the proportion of 8:2, and the data were enhanced. The proposed improved network architecture, namely P-ResNet, was fine-tuned for transfer learning to recognize and categorize maize seeds, and then it compares the performance of the models. The results show that the overall classification accuracy was determined as 97.91, 96.44, 99.70, 97.84, 98.58, 97.13, 96.59, and 98.28% for AlexNet, VGGNet, P-ResNet, GoogLeNet, MobileNet, DenseNet, ShuffleNet, and EfficientNet, respectively. The highest classification accuracy result was obtained with P-ResNet, and the model loss remained at around 0.01. This model obtained the accuracy of classifications for BaoQiu, ShanCu, XinNuo, LiaoGe, and KouXian varieties, which reached 99.74, 99.68, 99.68, 99.61, and 99.80%, respectively. The experimental results demonstrated that the convolutional neural network model proposed enables the effective classification of maize seeds. It can provide a reference for identifying seeds of other crops and be applied to consumer use and the food industry.

Keywords: machine vision; maize seeds; classification; deep learning; convolutional neural network

Citation: Xu, P.; Tan, Q.; Zhang, Y.; Zha, X.; Yang, S.; Yang, R. Research on Maize Seed Classification and Recognition Based on Machine Vision and Deep Learning. *Agriculture* **2022**, *12*, 232. https://doi.org/10.3390/agriculture12020232

Academic Editor: Dimitre Dimitrov

Received: 9 December 2021
Accepted: 3 February 2022
Published: 6 February 2022

1. Introduction

Maize (*Zea mays* L.) is a significant fundamental agricultural product for the economies and markets of countries. With the development of society, the widespread use of biotechnology has improved maize breeding technologies and accelerated the renewal and iteration of varieties. However, the increasing number of varieties of maize seed and their color characteristics overlap to make it more challenging to classify seeds after harvest [1]. In addition, the phenomenon of seeds being mixed may occur during production activities such as planting, harvesting, transportation, and storage [2]. Therefore, variety identification plays a crucial role in the production, processing, and marketing of seeds. It will provide markets and consumers with pure seeds that will ensure yields and stabilize their market value.

Traditionally, there are many methods for variety identification [3]. Morphological identification is limited by the range of morphological characteristics, the interference of human and environmental factors, and the impact of testing period or cost, which will decrease the accuracy of identification. Biochemical identification enables the recognition of seeds with different genetic characteristics, but it is difficult to identify closely related varieties. Molecular identification through DNA markers has the advantage of genetic stability and is independent of environmental conditions. But the cost of primer design is high, and the identification process can damage the sample [4]. In summary, these detection methods are difficult to adapt to be online detection in the seed processing industry [5] and cannot complete the sorting of samples during processing. Therefore, it is necessary to develop non-destructive, rapid, and efficient methods for the variety identification and classification of maize seeds.

Machine vision is the method of image processing adapted to multi-classification, which has been successfully applied in several fields. As for seed classification, the non-destructive nature hereof is undoubtedly a better choice than traditional detection methods. This method extracts color, texture, and shape features from seed images for classification. In [6], 12 color features were extracted to distinguish between the different types of damage in maize, with an accuracy of 74.76% for classifying normal and six damaged maize. In [7], 16 morphological features were extracted to classify dry beans, and the overall correct classification rate of SVM was 93.13%. In [8], developed a machine that automatically extracts shape, color, and texture feature data of cabbage seeds and uses them to classify the quality of seeds. The research of maize seeds has focused on bioactivity screening and quality inspection. However, an additional issue that demands consideration is the classification of maize seeds of different varieties [9].

Deep learning techniques have developed rapidly. The convolutional neural network (CNN) is a part of them, which has strong self-learning ability, adaptability, and generalization [2,10,11]. It has achieved considerable success in image classification, object detection, and face recognition [12]. CNN is a deep feedforward network inspired by the receptive field mechanism, which has the properties of local-connectivity, weight sharing, and aggregation in structure [13,14]. The network was composed of an input layer, convolution layers, pooling layers, fully connected layers, and an output layer [15]. CNN models have emerged since 2012, such as AlexNet [16], GoogLeNet [17], VGGNet [18], ResNet [19], DenseNet [20], MobileNet [21], ShuffleNet [22], EfficientNet [23], and more.

Machine vision has also been combined with deep learning to classify seeds [2,24]. In [15], used a CNN to automatically identify haploid and diploid maize seeds through a transfer learning approach. The experiment showed that the CNN model achieved good results, significantly outperforming machine learning-based methods and traditional manual selection. In [25], a wheat recognition system was developed based on VGG16, and the classification accuracy was 98.19%, which could adequately distinguish between different types of wheat grains. In [26], they used their self-designed CNN and ResNet models to identify seven cotton seed varieties, and it achieved good results, with 80% accuracy of the model identification. Reference [27] determined HSI images of 10 representative high-quality rice varieties in China and established a rice variety determination model using the PCANet, with a classification accuracy of 98.66%. In [24], used the CNN-ANN model to classify maize seeds, completing a test of 2250 instances in 26.8 s, with a classification accuracy of 98.1%.

Many studies have combined deep learning with machine vision because of its high accuracy, speed, and reliability. However, the increasing number of seed varieties and consumption are placing new demands on these studies, and the applicability of previous research methods has diminished. Therefore, inspired by the successful classification of agricultural products by deep CNNs, this paper studied the classification of maize seeds of different varieties. It is an in-depth exploration from another perspective based on the reference [9]. Specifically, this research used maize seed images from [9] and increased the number of samples by data augmentation. The proposed CNN network and transfer learning were used to study this classification task to obtain the best classification performance. This study not only extends [9], but its distinction lies in the attempt to automatically obtain deeper features from the data to achieve end-to-end problem-solving.

In summary, the objective of this study was to propose a non-destructive method for the automatic identification and classification of different varieties of maize seeds from images, thus overcoming the time-consuming and inefficient problems of traditional identification methods. We would pursue this study objective by: (1) implementing machine vision combined with deep learning by applying a CNN network with P-ResNet architecture for varietal detection; (2) establishing a seed dataset and dividing it into training and validation sets in the ratio of 8:2 for experiments; as well as (3) evaluating and comparing the classification performance of the models and using visualization to validate the results. In addition, we address the following specific hypotheses: (1) transfer learning can acquire knowledge learned in other Settings and be used to complete similar tasks in deep learning, thus helping to save the training time of the model; and (2) compared with manual feature extraction methods, the CNN model can be used to automatically extract more depth features from images, thus improving the classification performance.

2. Materials and Methods

2.1. Sample Preparation

In this study, 8080 maize seeds of five common varieties in China were used as a dataset to train the deep learning model for image classification. These maize seeds were provided by the National Seed Breeding Base in Hainan (Longitude 109.17° E, Latitude 18.35° N). These seeds were selected and certified by experts and have manually been cleaned for impurities and dust [9]. The selected for the experiment were of excellent quality, without noticeable defects or damage. The image in the dataset included 1710 BaoQiu, 1800 KouXian, 570 LiaoGe, 2000 ShanCu, and 2000 XinNuo. Figure 1 shows RGB images of five varieties of maize seeds.

(**a**) (**b**) (**c**) (**d**) (**e**)

Figure 1. RGB images of maize seed grains. (**a**) BaoQiu. (**b**) KouXian. (**c**) LiaoGe. (**d**) ShanCu. (**e**) XinNuo.

These seeds were placed individually on a black background for image acquisition. The influence of the seed storage situation in the National Seed Breeding Base in Hainan results in different quantities of seeds for each variety. However, the situation does not complicate the assessment of the accuracy of the different varieties, as there was no reuse of maize seeds. In addition, there were different shapes and sizes of the five maize seeds in the images, which provided some assistance in the classification of this study. In this work, all of the seeds were randomly divided into a training set (80%) and a validation set (20%), then stored in their respective subdirectories. Finally, the training and validation sets contained 6464 and 1616 maize seeds, respectively. To establish the classification model, BaoQiu, KouXian, LiaoGe, ShanCu, and XinNuo samples were collected in 2020, as shown in Table 1.

Table 1. Data for training and validation.

No.	Cultivar Name	Seeds		
		Training Set	Validation Set	Number
1	BaoQiu	1368	342	1710
2	KouXian	1440	360	1800
3	LiaoGe	456	114	570
4	ShanCu	1600	400	2000
5	XinNuo	1600	400	2000
	Total	6464	1616	8080

2.2. Image Acquisition and Segmentation

In the machine vision part, an image acquisition system was built for capturing maize seeds. The system has cameras mounted on top and light sources on either side to provide illumination. It is impractical to capture a single seed in a stretch during image acquisition. Therefore, hundreds of seeds were photographed in an area of 12 cm × 12 cm, with them not touching each other. All of the photos were taken in the same environment, with a camera distance of 16 cm. The resolution of the acquired image was 3384 × 2708 pixels, which contained multiple single seeds that cannot be directly included as input in the CNN model. Therefore, the image was segmented into 350 × 350 pixels size and saved in PNG format for use.

2.3. Image Preprocessing and Data Augmentation

Large amounts of training data can avoid over-fitting and improve the accuracy of CNN, so data augmentation operation is often used to extend the dataset [28]. The training images were randomly rotated [29], flipped horizontally and vertically [30], and normalized, considering the uncertainty of the state of the detected seeds in the actual situations. The enhanced image was trained together with the sample image to improve the classification precision and robustness of the model and further improve its applicability. Detailed information is shown in Figure 2.

(a)

(b)

(c)

(d)

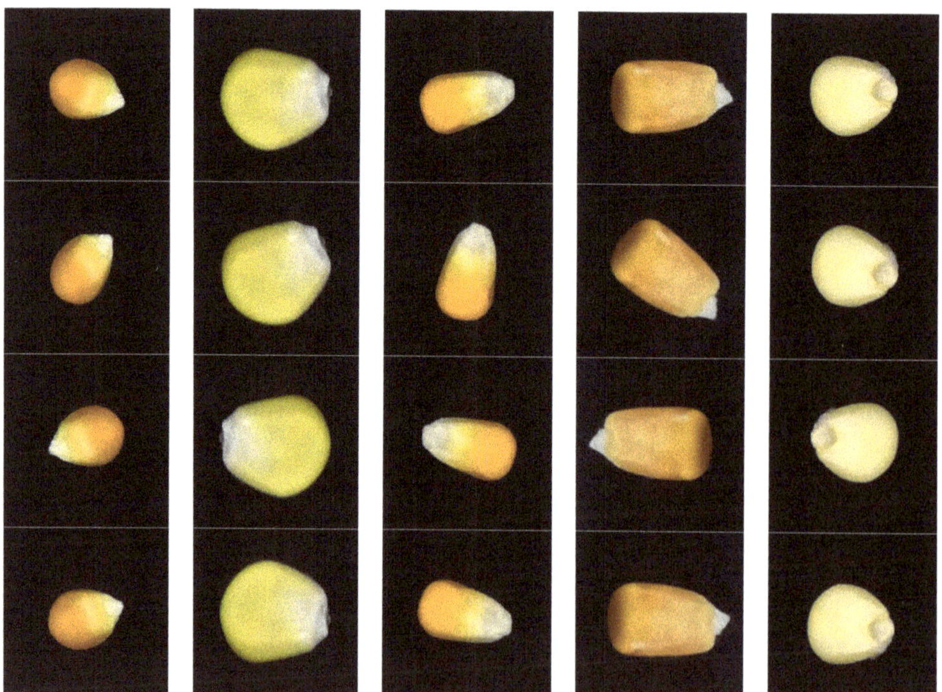

Figure 2. Data enhancement: (**a**) Original images; (**b**) Randomly rotated; (**c**) Flipped horizontally; (**d**) Flipped vertically.

2.4. Convolutional Neural Network

Deep learning is an emerging algorithm in machine learning, which has attracted extensive attention from researchers because of its remarkable effect on learning image features. Deep learning extracts higher-dimensional and abstract features by autonomic learning from training samples through neural networks [10]. This research proposed a new model (P-ResNet) based on an improvement of ResNet, which provides a method to classify maize seeds. The network architecture of P-ResNet consists of six parts, five of which are the convolution layer, and the last one is a fully connected layer. The convolution operation is followed by batch normalization, and then ReLU is applied as the activation function to complete the output of the convolution layer. In addition, to avoid over-fitting and reduce the number of parameters and computation in the network, which adopted a strategy of max pooling and average pooling. The input image was resized to 224 × 224 × 3. According to the prepared dataset, the output of the fully connected layer was fed into softmax to generate a probability distribution to predict the varieties of 5 maize seeds. Table 2 provides a detailed description of the P-ResNet network.

Table 2. Network architecture for P-ResNet.

Layer Name	Output Shape	The Network Layer	Stride
	Input (224 × 224 RGB image)		
Convolution layer 1	112 × 112	7 × 7, 64	2
Max pooling		3 × 3, 64	2
Convolution layer 2	56 × 56	$\begin{bmatrix} 3 \times 3, 64 \\ 3 \times 3, 64 \end{bmatrix} \times 3$	1, 1, 1
Convolution layer 3	28 × 28	$\begin{bmatrix} 3 \times 3, 128 \\ 3 \times 3, 128 \end{bmatrix} \times 3$	2, 1, 1
Convolution layer 4	14 × 14	$\begin{bmatrix} 3 \times 3, 256 \\ 3 \times 3, 256 \end{bmatrix} \times 3$	2, 1, 1
Convolution layer 5	7 × 7	$\begin{bmatrix} 3 \times 3, 512 \\ 3 \times 3, 512 \end{bmatrix} \times 3$	2, 1, 1
Classification	1 × 1	average pooling, 5-d fully-connected, softmax	1

As can be seen from Table 2, the convolutional layer 1 of the P-ResNet network goes through a 7 × 7 convolution. The receptive field is large enough to be used for the feature extraction of images in this database. In order to classify maize seeds more accurately, more subtle features need to be extracted. Furthermore, a suitable network depth was required to be designed and to reduce the size of the presented model. Therefore, the convolution layer of layers 2–5 was improved in the architecture of the network to make it more suitable for the model classification task. The design of this study used twenty-four 3 × 3 stacked convolution layers for learning, with more nonlinear activation functions to make the decision function more accurate; on the other hand, it can effectively decrease the number of parameters in calculation. Furthermore, in online inspection in the seed processing industry, the objective region occupies a small area of the whole image, and the proportion of information obtained is weak. In order to avoid redundant and useless information, this study adds a pooling layer to integrate spatial information before the convolution kernel of the residual module does down-sampling.

2.5. Transfer Learning

The RGB images with labeled data were input into the improved network P-ResNet. In the experiment, 6464 images were utilized for training and 1616 images for validation. Transfer learning [31] was performed for 5 varieties: BaoQiu, ShanCu, XinNuo, LiaoGe, and KouXian. After training the model, its performance was evaluated and compared through the training and validation sets. The models were developed using the open-source software framework of PyTorch 1.9.0, the programming language of Python 3.8.10, and the Integrated Development Environment of PyCharm 1.3. The classification model was trained on a server equipped with one NVIDIA GeForce GTX 1660 SUPER GPU and 16 GB GDDR4 on-board memory.

In this study, as shown in Table 3, some classical CNN models have been used to compare with P-ResNet. The acquired data were fed into a pre-trained network, storing the activation values of each layer as features. The cross-dataset fine-tuning method was used for training. According to the new task, the weights of the presented model were updated and back-propagated through the network. This approach can transfer weights from the pre-trained model to the one we want to train. Details of the hyper-parameters applied during the fine-tuning procedure are listed in Table 4. Using these enables transferring the knowledge gained from the large dataset to the classification problem of maize seeds. For the purpose of this study, the convolution layer was used as a fixed feature extractor. Then a fully connected layer with merely five neurons was constructed. Finally, the categorization results were obtained with the prediction layer.

Table 3. Compare the properties of CNN models.

Network Name	Depth	Image Input Size	Parameters (Millions)	FLOPs (G)	Total Memory (MB)
AlexNet	8	224-by-224	16.63	0.31	2.77
VGGNet	16	224-by-224	138.36	15.50	109.29
P-ResNet	26	224-by-224	17.96	2.75	32.83
GoogLeNet	22	224-by-224	6.99	1.59	30.03
MobileNet	19	224-by-224	3.50	0.32	74.26
DenseNet	121	224-by-224	7.98	2.88	147.10
ShuffleNet	19	224-by-224	1.37	0.04	11.24
EfficientNet	18	224-by-224	21.46	2.87	144.98

Table 4. Hyper-parameters were applied to the fine-tuning procedure.

Parameter	Value
Epochs	30
Batch size	32
Learn rate	0.001
Momentum	0.9
Learn rate weight coefficient	15
Learn rate bias coefficient	15
Learn rate schedule	Exponential
Weight decay	0.005
Decay period	10

The whole network has 17,960,232 parameters. The proposed CNN model uses Adam [32] as the optimizer to train with an initial learning rate value of 0.001, and the loss function of the network was declined by updating the weight parameters. Batch Normalization was used between each convolution and ReLU layer during network training, instead of the traditional dropout to improve training and reduce over-fitting. Epoch is the complete training cycle of the entire dataset with maize seeds, and its maximum value corresponds to the limit value of the minimum loss function. The maximum training epoch was set to 30, and the minimum batch size was set to 32. These parameters achieved better results in the optimizer. The process of transfer learning and classification of maize seeds in the network involved in the experiment was given in Figure 3.

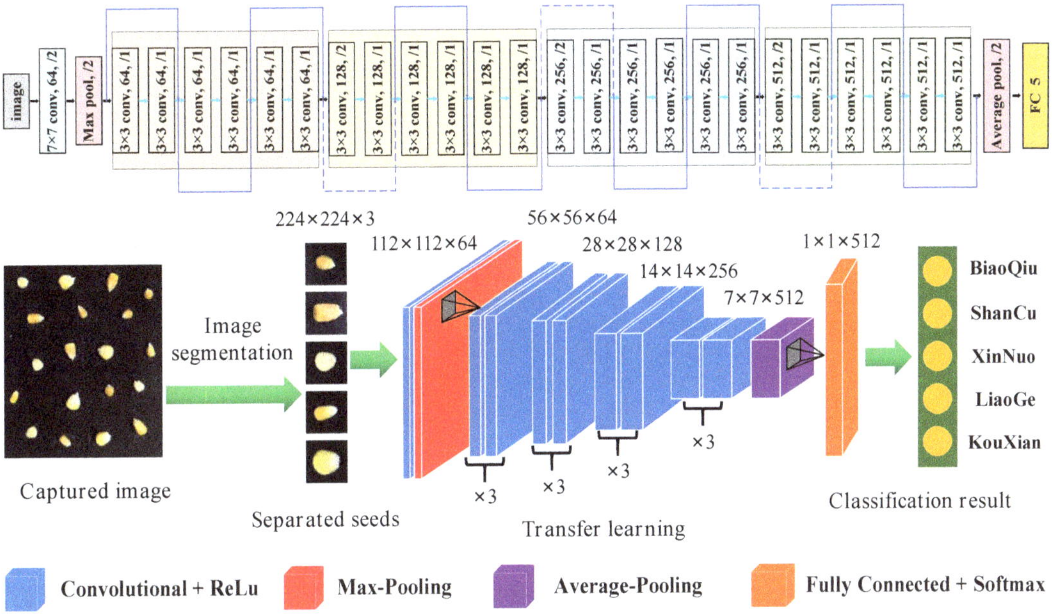

Figure 3. Process of transfer learning and classification of maize seeds.

2.6. Performance Evaluation

In this paper, the confusion matrix was used to visualize the performance of the CNN model. This data on the confusion matrix represents the actual class in the samples and the class predicted by the CNN classifier. The four metrics typically included true positives (TP), true negatives (TN), false positives (FP), and false negatives (FN) [24]. In this work, TP and TN correspond to the correct identification of maize seeds, while FP and FN correspond to false identification of it. The performance of models was evaluated based on some statistical parameters of the confusion matrix, such as accuracy, sensitivity, specificity, precision, and F1-score, which can be obtained from them [33]. The performance evaluation was performed using images from the validation set and their respective labels, which were not used for training. Table 5 represents the formulae for performance evaluation and their evaluation focus.

Table 5. Performance evaluation to measure the performance of the CNN models.

Metrics	Formula	Evaluation Focus
Accuracy	$\frac{TP + TN}{TP + FP + FN + TN}$	It is the sum of correct predictions divided by all the predictions.
Specificity	$\frac{TN}{TN + FP}$	It reflects the ability of the classifier to exclude misclassification images.
Sensitivity	$\frac{TP}{TP + FN}$	It reflects the ability of the model to detect instances of certain classes.
Precision	$\frac{TP}{TP + FP}$	Its high value indicates the low number of false positives hence better classification.
F1-score	$\frac{2 * TP}{2 * TP + FP + FN}$	Its high value means the model classifies well.

3. Results

The parameters given in Table 4 were selected for transfer learning. The prepared dataset was trained using AlexNet, VGGNet, P-ResNet, GoogLeNet, MobileNet, DenseNet, ShuffleNet, and EfficientNet. The optimal parameters in Table 4 were used to prevent over-fitting during training and avoid spending more time. All networks have been trained for 30 epochs. The accuracy and loss of the training and validation data for each epoch are shown in Figure 4. In the initial phase (1–10 epochs), the loss values declined sharply, but the accuracy improved dramatically. Finally, the CNN models reached an accuracy of over 92% in the training phase, and the loss of models was steady below 0.15, indicating that these are very robust and dependable. Also, the model achieved the convergence procedure in approximately 15 epochs. As can be depicted in Figure 4, after this period, the validation accuracy and loss curves smoothed out, and the difference between the accuracy and loss values of the validation and training data decreased. There, for the fact, was some fluctuation with accuracy and loss for GoogLeNet. This condition suggests that the model is not stable until the 25 epochs, possibly because some of the varieties were easily confounded. There are gaps in GoogLeNet's handling of the dataset for this study compared to other models. Nevertheless, even in the worst case, the metrics were above 90% or below 0.2. This result indicates that the classifier's performance is satisfactory and did not prevent it from achieving its final classification purpose.

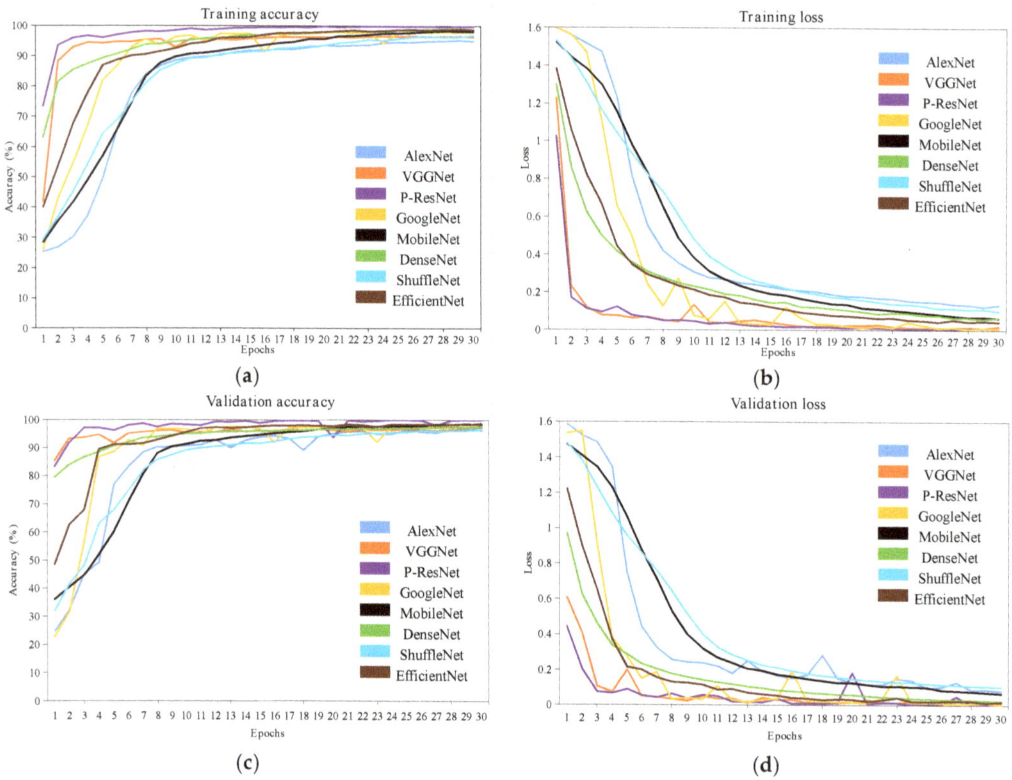

Figure 4. Each epoch of the CNN model: (**a**) Training accuracy, (**b**) Training loss, (**c**) Validation accuracy, (**d**) Validation loss.

After the training, a confusion matrix was created for each classification algorithm, and performance evaluation was visualized using the values on the confusion matrix (TP, TN, FP, FN). The confusion matrices for the validation set of the CNN model are depicted in Figure 5. In addition, the performance metrics depicted from the confusion matrix are presented in Table 6, including their mean values for precision, specificity, sensitivity, accuracy, and f1 scoring. In this experiment, all CNN models can identify five classes of maize seeds, and all had an accuracy rate of over 92%. The highest accuracy was obtained by P-ResNet (99.70%), followed by AlexNet (97.91%), VGGNet (96.44%), GoogLeNet (97.84%), MobileNet (98.58%), DenseNet (97.13%), ShuffleNet (96.59%), and EfficientNet (98.28%). Even though these models were trained with images from self-made datasets, fine-tuning these models can achieve similar results to using the end-to-end models in datasets with limited samples. This situation will make image acquisition more convenient and fast, will save effort and time, and will thus improve efficiency. Besides, the confusion matrix and classification results also prove that P-ResNet has excellent performance. This result also illustrates that the presented network can catch the detailed information of the samples. These can provide relatively high accuracy classification under complex datasets, which is beneficial for transferring it to similar classification tasks. The experimental results also demonstrate that enhancing the data used for training has a positive impact on the performance of the presented model on datasets with a small number of samples. In particular, these include datasets with low sample sizes. At the same time, the deep learning-based feature extraction method can effectively preserve information about the maize seeds, reduce the loss of information due to manual feature extraction.

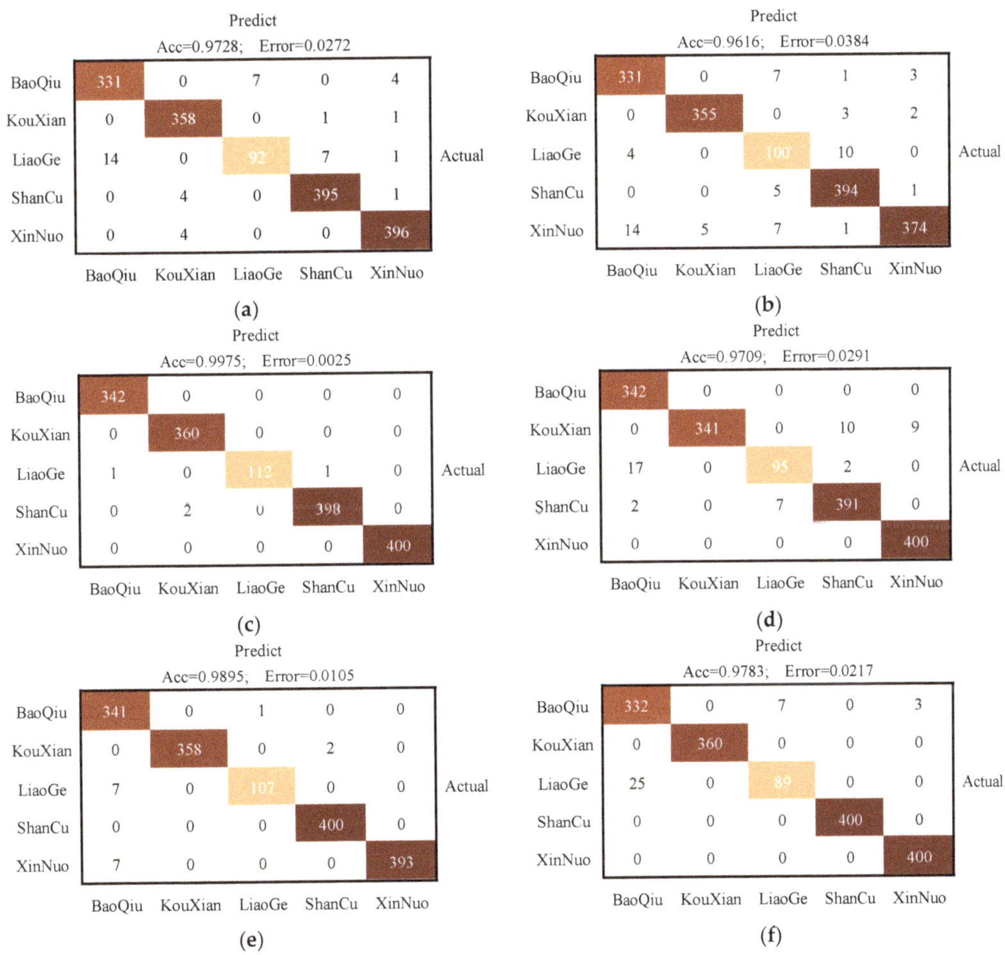

Figure 5. *Cont.*

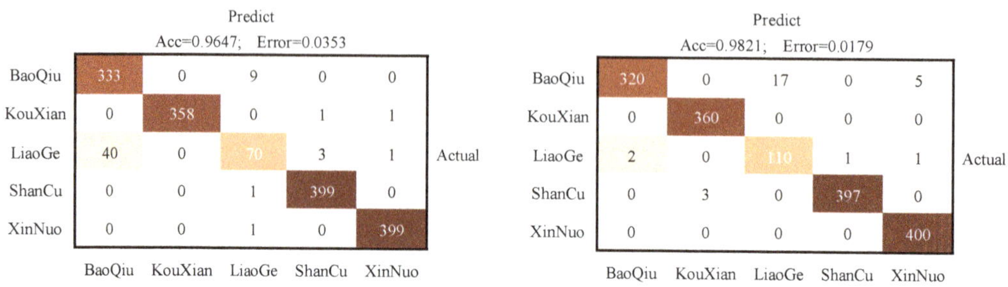

(g) (h)

Figure 5. Confusion Matrix: (**a**) AlexNet; (**b**) VGGNet; (**c**) P-ResNet; (**d**) GoogLeNet; (**e**) MobileNet; (**f**) DenseNet; (**g**) ShuffleNet; and (**h**) EfficientNet.

The experimental analysis showed that the deep learning architecture with updated weights and fine-tuning had good generalization capability in the maize seed dataset. Compared with the networks in the literature, the proposed P-ResNet has relatively better performance and higher accuracy. It also found that the value of the maximum difference in classification accuracy between all models was no more than 3%. Although there were differences between them, they performed similarly for multi-classification. Therefore, an improved ResNet-based network has been used for transfer learning in the study. Due to its better classification results, it confirms that the idea of balancing its depth and width when designing the network is feasible. It would also increase the complexity of the model and consume more computation time. As can be observed in Table 3, the P-ResNet proposed generates a relatively small number of parameters (17.96 Million) and memory (32.83 MB). VGGNet has 7.6 times as many parameters as it does, while the memory footprint is close to GoogLeNet. The FLOPs value (2.75 G) is approximately the same as that of DenseNet and EfficientNet, indicating the low complexity of the model. It also demonstrates the potential for the network to be lightweight and mobile. With the continuous improvement of the model, a version more suitable for mobile and embedded devices can be achieved.

Table 6. Classification results of CNN models.

Name	Accuracy (%)	Specificity (%)	Sensitivity (%)	Precision (%)	F1-Score (%)
AlexNet	97.91	98.31	93.94	95.59	94.80
VGGNet	96.44	97.04	92.98	92.45	92.83
P-ResNet	99.70	99.94	99.55	99.78	99.71
GoogLeNet	97.84	98.26	94.16	95.54	94.99
MobileNet	98.58	98.73	97.25	97.93	97.57
DenseNet	97.13	97.46	93.03	94.99	93.88
ShuffleNet	96.59	97.13	91.54	94.84	92.76
EfficientNet	98.28	99.58	97.86	96.69	97.28

Figure 6 shows a comparison of the performance of the CNN models tested, from which it can be seen that the training time of the proposed network is comparable to that of the lightweight networks (MobileNet and ShuffleNet). AlexNet had the shortest training time (14 min) and VGGNet the longest (62 min). It is important to note that the time required for training the network depends on the hardware resources. The use of advanced GPU can reduce the training time of CNN. However, when time and model complexity were considered, P-ResNet's training time was only 4 min slower than AlexNet, and the values for Parameters, FLOPs, and total memory were relatively better. This situation is because of the use of Adam optimizer in the proposed model, which minimizes error loss, and transforms the training and validation data for each epoch. In this work, when the region of interest of seeds was extracted from the image and applied to the deep learning model, the processing time of training can be reduced.

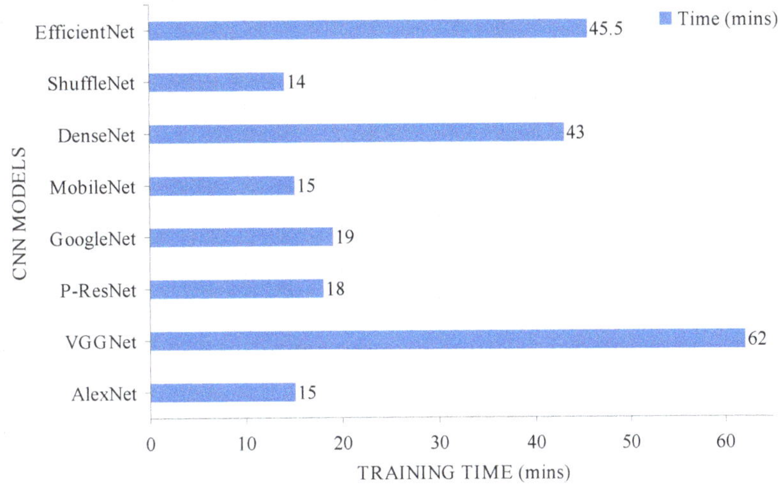

Figure 6. Consumed time for training.

The classification results for all varieties of maize seeds in the different models were shown in Table 7, and this statistic clearly shows how the model performance stays in general and as a whole. As can be seen from Figure 7, the classification accuracy of 8 different models for five maize seeds was over 90%. The P-ResNet network had the best classification performance, with 99.74, 99.68, 99.68, 99.61, and 99.80% accuracy for the BaoQiu, KouXian, LiaoGe, ShanCu, and XinNuo, respectively. However, BaoQiu and LiaoGe had lesser classification performance among all models, and the lowest values were 94.97 and 94.60%, respectively. These results indicated that VGGNet, DenseNet, and ShuffleNet models were not the best adapted for these two varieties. It also revealed that there probably is overlap in features between BaoQiu and LiaoGe and the other three varieties, resulting in poor distinction. In addition, the low number of LiaoGe in the dataset may also have contributed to this situation. Through, their classification results were still very encouraging.

Table 7. Statistics of classification results for all maize varieties.

Model	Classification Accuracy (%)				
	BaoQiu	**KouXian**	**LiaoGe**	**ShanCu**	**XinNuo**
AlexNet	97.45	98.38	97.21	98.20	98.32
VGGNet	96.21	97.38	95.96	96.70	95.96
P-ResNet	99.74	99.68	99.68	99.61	99.80
GoogleNet	97.82	97.82	97.39	97.70	98.44
MobileNet	98.07	98.88	98.50	98.88	98.57
DenseNet	95.83	98.00	96.02	98.00	97.81
ShuffleNet	94.97	97.88	94.60	97.69	97.81
EfficientNet	97.51	98.81	97.70	98.75	98.63

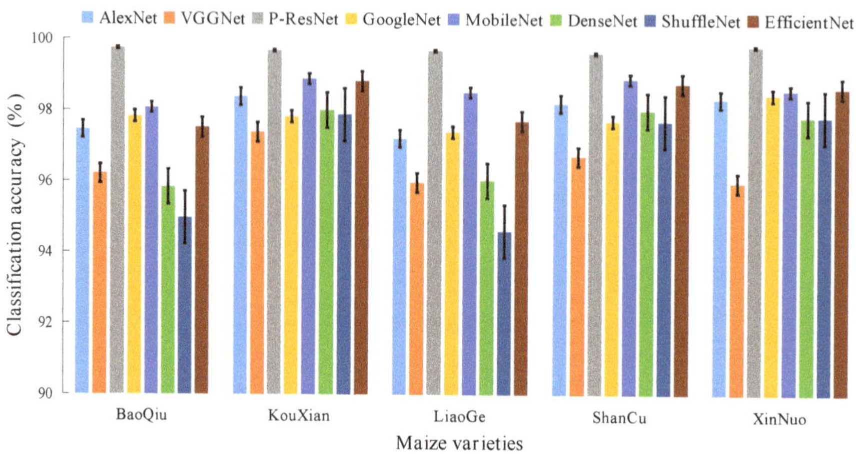

Figure 7. Classification results for maize seeds. The error bars are standard deviations of means.

In this experiment, given the different methods, datasets, and classification criteria employed, relevant studies cannot be compared in detail. Nevertheless, it compared some applications in agricultural classification tasks, and the results are shown in Table 8. These comparisons considered several criteria, such as dataset size, application, the method used, and accuracy. The results showed that the accuracy for the different classification tasks was above 95%, which indicated that the CNN model proposed in this paper and the pre-training method using transfer learning were feasible. It can provide a reference for the classification of agricultural products. The five types of maize seeds utilized in the research are relatively common in China, with a wide distribution of planting areas. In this situation, the credibility of this study has been enhanced. Although the P-ResNet model achieved good results, maize seeds may vary depending on storage time and cultivation conditions (soil or climate). These conditions lead to changes in the dataset, which may influence its accuracy in distinguishing the target varieties. Therefore, it will be necessary to update the algorithm in the future, and the aim is to retain the classification precision and robustness of the model.

Table 8. Comparison of the proposed model and related studies (maize seeds).

Imaging Method	Dataset Size	Application	Approach	Result	References
Hyperspectral imaging	1632	Variety identification	LDA	99.13%	[1]
Digital camera	700	Quality detection	Maximum likelihood	96.67%	[6]
Near-infrared spectroscopy	760	Variety identification	PLS-DA	99.19%	[3]
Digital camera	5400	Variety identification	ANN	98.10%	[24]
Near-infrared spectroscopy	2250	Variety identification	LSTM	95.22%	[34]
Digital camera	8080	Variety identification	SVM	96.46%	[9]
Digital camera	1600	Quality detection	VGG16	98.00%	[2]
Digital camera	8080	Variety classification	P-ResNet	99.70%	Our work

These results were obtained using the Gradient-weighted Class Activation Mapping (Grad-CAM) technique to visualize the used regions of a random input image to extract features for image classification prediction [35]. The gradient of any target feature through the last convolution layer produces a roughly local feature map, highlighting the regions of the image that are important for it. Figure 8 shows the achieved results of the implementation of this method on maize seed images. It can be shown that the image locations for seeds were accurately calculated, with the class activation heat map indicating the importance rank and similarity of the location relative to the particular variety.

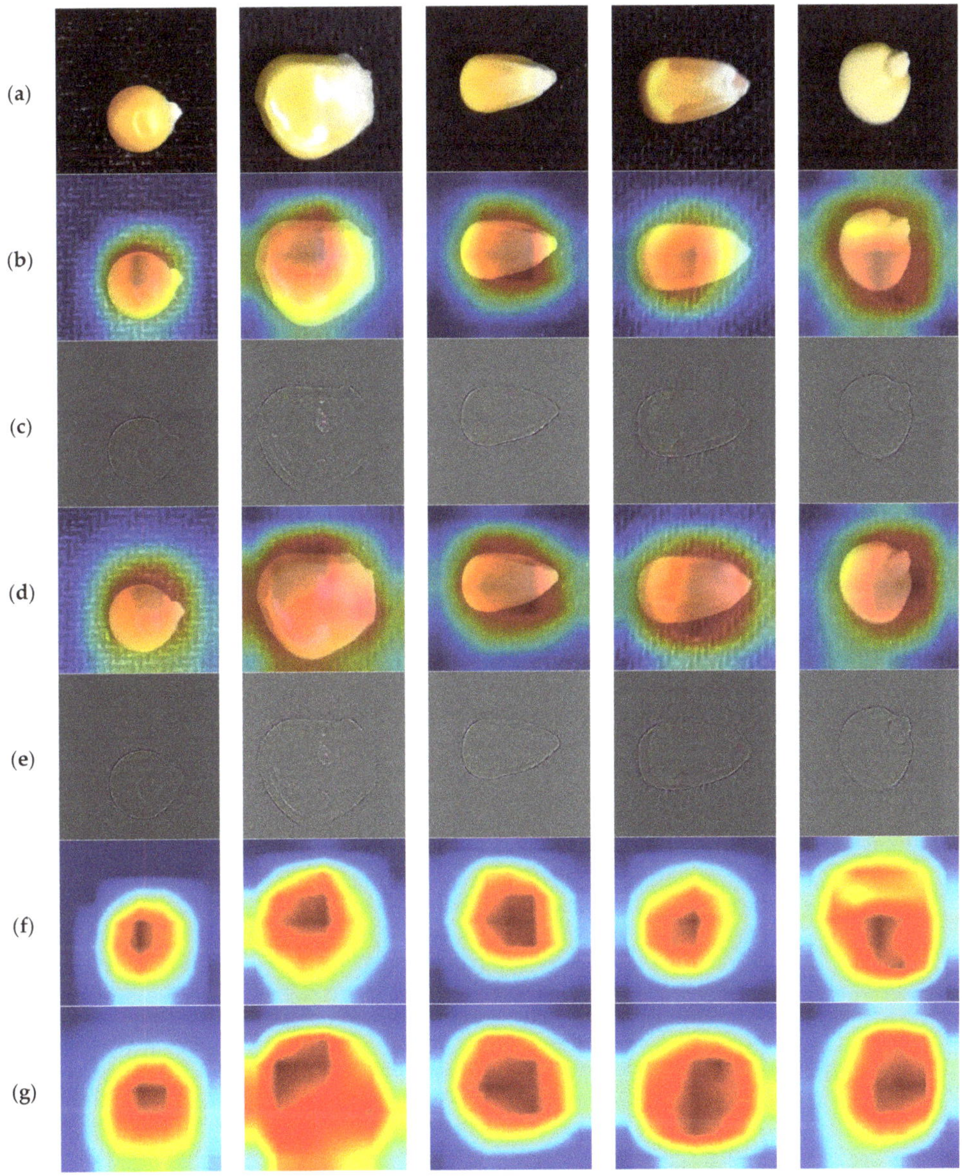

Figure 8. (**a**) Original images; (**b**) Grad-CAM visualization; (**c**) Guided Grad-CAM visualization; (**d**) Grad-CAM++ visualization; (**e**) Guided backpropagation visualization; (**f**) HeatMap visualization; (**g**) HeatMap++ visualization.

4. Discussion

There are two primary methods for training CNN models using sample data: (1) starting from zero; and (2) transfer learning. In practice, while training a CNN model from the ground up gives us the best active control concerning the network, it may not have enough data and time to train in some cases, or the data to create the markers may be difficult to obtain. Moreover, over-fitting and convergence states are also potential problems. In such cases, transfer learning can be applied to gain knowledge gained in other settings. It is a convenient and effective method of knowledge adaptation [31], which is usually more efficient than training a new neural network since all parameter values are not required to start from zero. In higher-layers networks, some features are more applicable to a specific task. However, there are many similar features like color and texture for the lower layers of the network. These can be transferred to other tasks and are very helpful for performing similar tasks in deep learning.

P-ResNet was designed based on the principle of balancing the width and depth of the network according to the specific task, which has a better architecture than GoogLeNet, DenseNet, and EfficientNet. It can reduce parameters for computation and avoid gradient disappearance and gradient explosion during training. Meanwhile, it does not need to crop or scale the input image like AlexNet and VGGNet, which can maximally protect the information integrity. In addition, the underlying implementation of this network was simplified to make it more lightweight as possible as MobileNet and ShuffleNet. Since only images are required, which can be produced by low-cost digital cameras, this approach can be widely deployed and disseminated in intelligent agriculture. Machine vision can only obtain phenotypic information of seeds, while spectral information can reflect the internal quality of seeds. The combination of CNN-based machine vision and spectroscopic techniques for seed classification and detection was considered in the follow-up work.

The proposed method has been compared with related work based on the unification of the research objectives (classification or identification) and the object of study (maize seeds). In Table 7, it can be seen that automatic extraction of image features for recognition using CNN is better than manual extraction, and these results illustrate that deep learning is more effective than traditional machine learning methods in cultivar classification. However, the variety and number of samples collected in this study are limited and cannot represent all maize seeds within China. Therefore, the number of samples should be increased to improve its applicability to the model. Moreover, this experiment only considered the classification effects of seed samples from the same year, so the impact of different planting years, growing regions, and climatic conditions on the classification of seeds of the same variety can be compared in subsequent studies.

5. Conclusions

In this work, a combination of deep learning algorithms and machine vision has been used to automatically classify five varieties of maize seeds using a CNN model. In terms of classification, the model architecture developed can be applied to different regions and types of seeds to ensure the provision of high-quality seeds for agricultural production. Also, the method has application potential in identifying varieties of seeds, and the developed variety classification model can be applied to seed sorting machinery to provide an idea and reference for real-time industrial detection. This study proposes an improved model, P-ResNet, and compares it with AlexNet, VGGNet, GoogLeNet, MobileNet, DenseNet, ShuffleNet, and EfficientNet models. The results showed that the P-ResNet model achieved the best accuracy to classify maize seeds in a non-destructive, fast, and efficient manner. These results highlight the advantages of transfer learning and its potential to work with deep learning using a few quantities of training samples. In addition, the Grad-CAM has been used to visualize the regions of use of the input seed images, making this work more efficient and productive. This machine vision technology based on CNN with high accuracy and reliability can also be applied in other intelligent agricultural equipment to facilitate the analysis of seeds or other crops to save cost, labor, and time.

Based on the work presented in this paper, further studies on more varieties of maize seeds and the environment in which they are grown would be appropriate, and it is in order to optimize the stability of the proposed model. Considering that each seed in maize has its genetic characteristics, grouping seeds of selected varieties might avoid natural variability in seedlings. Given the real-time nature of the data, that helps to develop an integrated and intelligent automated seed sorting system for the food industry and smartphone-based applications used by consumers.

Author Contributions: Conceptualization, P.X., Q.T., Y.Z. and R.Y.; methodology, P.X.; software, P.X.; validation, P.X. and R.Y.; formal analysis, P.X., Q.T., Y.Z., X.Z., S.Y. and R.Y.; investigation, P.X.; resources, R.Y.; data curation, P.X.; writing—original draft preparation, P.X., Q.T. and Y.Z.; writing—review and editing, P.X., X.Z., S.Y. and R.Y.; visualization, P.X.; supervision, X.Z., S.Y. and R.Y.; project administration, P.X.; funding acquisition, R.Y. All authors have read and agreed to the published version of the manuscript.

Funding: This research was funded by the National Talent Foundation Project of China, grant number T2019136.

Institutional Review Board Statement: Not applicable.

Informed Consent Statement: Informed consent was obtained from all subjects involved in the study.

Data Availability Statement: All data are presented in this article in the form of figures and tables.

Conflicts of Interest: The authors declare no conflict of interest.

References

1. Xia, C.; Yang, S.; Huang, M.; Zhu, Q.; Guo, Y.; Qin, J. Maize Seed Classification Using Hyperspectral Image Coupled with Multi-Linear Discriminant Analysis. *Infrared Phys. Technol.* **2019**, *103*, 103077. [CrossRef]
2. Tu, K.; Wen, S.; Cheng, Y.; Zhang, T.; Pan, T.; Wang, J.; Wang, J.; Sun, Q. A Non-Destructive and Highly Efficient Model for Detecting the Genuineness of Maize Variety 'JINGKE 968' Using Machine Vision Combined with Deep Learning. *Comput. Electron. Agric.* **2021**, *182*, 106002. [CrossRef]
3. Qiu, G.; Lü, E.; Wang, N.; Lu, H.; Wang, F.; Zeng, F. Cultivar Classification of Single Sweet Corn Seed Using Fourier Transform Near-Infrared Spectroscopy Combined with Discriminant Analysis. *Appl. Sci.* **2019**, *9*, 1530. [CrossRef]
4. Cui, Y.; Xu, L.; An, D.; Liu, Z.; Gu, J.; Li, S.; Zhang, X.; Zhu, D. Identification of Maize Seed Varieties Based on near Infrared Reflectance Spectroscopy and Chemometrics. *Int. J. Agric. Biol. Eng.* **2018**, *11*, 177–183. [CrossRef]
5. Xie, C.; He, Y. Modeling for Mung Bean Variety Classification Using Visible and Near-Infrared Hyperspectral Imaging. *Int. J. Agric. Biol. Eng.* **2018**, *11*, 187–191. [CrossRef]
6. Li, X.; Dai, B.; Sun, H.; Li, W. Corn Classification System Based on Computer Vision. *Symmetry* **2019**, *11*, 591. [CrossRef]
7. Koklu, M.; Ozkan, I.A. Multiclass Classification of Dry Beans Using Computer Vision and Machine Learning Techniques. *Comput. Electron. Agric.* **2020**, *174*, 105507. [CrossRef]
8. Huang, K.Y.; Cheng, J.F. A Novel Auto-Sorting System for Chinese Cabbage Seeds. *Sensors* **2017**, *17*, 886. [CrossRef]
9. Xu, P.; Yang, R.; Zeng, T.; Zhang, J.; Zhang, Y.; Tan, Q. Varietal Classification of Maize Seeds Using Computer Vision and Machine Learning Techniques. *J. Food Process Eng.* **2021**, *44*, e13846. [CrossRef]
10. Traore, B.B.; Kamsu-Foguem, B.; Tangara, F. Deep Convolution Neural Network for Image Recognition. *Ecol. Inform.* **2018**, *48*, 257–268. [CrossRef]
11. Zhao, G.; Quan, L.; Li, H.; Feng, H.; Li, S.; Zhang, S.; Liu, R. Real-Time Recognition System of Soybean Seed Full-Surface Defects Based on Deep Learning. *Comput. Electron. Agric.* **2021**, *187*, 106230. [CrossRef]
12. Nie, P.; Zhang, J.; Feng, X.; Yu, C.; He, Y. Classification of Hybrid Seeds Using Near-Infrared Hyperspectral Imaging Technology Combined with Deep Learning. *Sens. Actuators B Chem.* **2019**, *296*, 126630. [CrossRef]
13. Qiu, Z.; Chen, J.; Zhao, Y.; Zhu, S.; He, Y.; Zhang, C. Variety Identification of Single Rice Seed Using Hyperspectral Imaging Combined with Convolutional Neural Network. *Appl. Sci.* **2018**, *8*, 212. [CrossRef]
14. De Medeiros, A.D.; Bernardes, R.C.; da Silva, L.J.; de Freitas, B.A.L.; dos Dias, D.C.F.S.; da Silva, C.B. Deep Learning-Based Approach Using X-Ray Images for Classifying Crambe Abyssinica Seed Quality. *Ind. Crops Prod.* **2021**, *164*, 113378. [CrossRef]
15. Altuntaş, Y.; Cömert, Z.; Kocamaz, A.F. Identification of Haploid and Diploid Maize Seeds Using Convolutional Neural Networks and a Transfer Learning Approach. *Comput. Electron. Agric.* **2019**, *163*, 104874. [CrossRef]
16. Krizhevsky, A.; Sutskever, I.; Hinton, G.E. ImageNet Classification with Deep Convolutional Neural Networks. *Commun. ACM* **2017**, *60*, 84–90. [CrossRef]
17. Szegedy, C.; Liu, W.; Jia, Y.; Sermanet, P.; Reed, S.; Anguelov, D.; Erhan, D.; Vanhoucke, V.; Rabinovich, A. Going Deeper with Convolutions. In Proceedings of the IEEE Computer Society Conference on Computer Vision and Pattern Recognition, Boston, MA, USA, 7–12 June 2015.
18. Simonyan, K.; Zisserman, A. Very Deep Convolutional Networks for Large-Scale Image Recognition. In Proceedings of the 3rd International Conference on Learning Representations, ICLR 2015—Conference Track Proceedings, San Diego, CA, USA, 7–9 May 2015.
19. He, K.; Zhang, X.; Ren, S.; Sun, J. Deep Residual Learning for Image Recognition. In Proceedings of the IEEE Computer Society Conference on Computer Vision and Pattern Recognition, Las Vegas, NV, USA, 27–30 June 2016.

20. Huang, G.; Liu, Z.; van der Maaten, L.; Weinberger, K.Q. Densely Connected Convolutional Networks. In Proceedings of the 30th IEEE Conference on Computer Vision and Pattern Recognition, Honolulu, HI, USA, 21–26 July 2017; pp. 2261–2269. [CrossRef]
21. Howard, A.G.; Zhu, M.; Chen, B.; Kalenichenko, D.; Wang, W.; Weyand, T.; Andreetto, M.; Adam, H. MobileNets: Efficient Convolutional Neural Networks for Mobile Vision Applications. *arXiv* **2017**, arXiv:1704.04861.
22. Ma, N.; Zhang, X.; Zheng, H.T.; Sun, J. Shufflenet V2: Practical Guidelines for Efficient Cnn Architecture Design. In Proceedings of the Lecture Notes in Computer Science (Including Subseries Lecture Notes in Artificial Intelligence and Lecture Notes in Bioinformatics), 15th European Conference, Munich, Germany, 8–14 September 2018; Part XIV.
23. Tan, M.; Le, Q.v. EfficientNet: Rethinking Model Scaling for Convolutional Neural Networks. In Proceedings of the 36th International Conference on Machine Learning, ICML 2019, Long Beach, CA, USA, 9–15 June 2019; Volume 2019.
24. Javanmardi, S.; Miraei Ashtiani, S.H.; Verbeek, F.J.; Martynenko, A. Computer-Vision Classification of Corn Seed Varieties Using Deep Convolutional Neural Network. *J. Stored Prod. Res.* **2021**, *92*, 101800. [CrossRef]
25. Özkan, K.; Işık, Ş.; Yavuz, B.T. Identification of Wheat Kernels by Fusion of RGB, SWIR, and VNIR Samples. *J. Sci. Food Agric.* **2019**, *99*, 4977–4984. [CrossRef]
26. Zhu, S.; Zhou, L.; Gao, P.; Bao, Y.; He, Y.; Feng, L. Near-Infrared Hyperspectral Imaging Combined with Deep Learning to Identify Cotton Seed Varieties. *Molecules* **2019**, *24*, 3268. [CrossRef]
27. Weng, S.; Tang, P.; Yuan, H.; Guo, B.; Yu, S.; Huang, L.; Xu, C. Hyperspectral Imaging for Accurate Determination of Rice Variety Using a Deep Learning Network with Multi-Feature Fusion. *Spectrochim. Acta Part A Mol. Biomol. Spectrosc.* **2020**, *234*, 118237. [CrossRef]
28. Shorten, C.; Khoshgoftaar, T.M. A Survey on Image Data Augmentation for Deep Learning. *J. Big Data* **2019**, *6*, 60. [CrossRef]
29. Xie, S.; Tu, Z. Holistically-Nested Edge Detection. *Int. J. Comput. Vis.* **2017**, *125*. [CrossRef]
30. Kok, K.Y.; Rajendran, P. Validation of Harris Detector and Eigen Features Detector. In Proceedings of the IOP Conference Series: Materials Science and Engineering, International Conference on Aerospace and Mechanical Engineering (AeroMech17), Batu Ferringhi, Penang, Malaysia, 21–22 November 2017; Volume 370.
31. Salaken, S.M.; Khosravi, A.; Nguyen, T.; Nahavandi, S. Seeded Transfer Learning for Regression Problems with Deep Learning. *Expert Syst. Appl.* **2019**, *115*, 565–577. [CrossRef]
32. Kingma, D.P.; Ba, J.L. Adam: A Method for Stochastic Optimization. In Proceedings of the 3rd International Conference on Learning Representations, ICLR 2015—Conference Track Proceedings, San Diego, CA, USA, 7–9 May 2015.
33. Ishengoma, F.S.; Rai, I.A.; Said, R.N. Identification of Maize Leaves Infected by Fall Armyworms Using UAV-Based Imagery and Convolutional Neural Networks. *Comput. Electron. Agric.* **2021**, *184*, 106124. [CrossRef]
34. Zhang, C.; Zhao, Y.; Yan, T.; Bai, X.; Xiao, Q.; Gao, P.; Li, M.; Huang, W.; Bao, Y.; He, Y.; et al. Application of Near-Infrared Hyperspectral Imaging for Variety Identification of Coated Maize Kernels with Deep Learning. *Infrared Phys. Technol.* **2020**, *111*, 103550. [CrossRef]
35. Selvaraju, R.R.; Cogswell, M.; Das, A.; Vedantam, R.; Parikh, D.; Batra, D. Grad-CAM: Visual Explanations from Deep Networks via Gradient-Based Localization. *Int. J. Comput. Vis.* **2020**, *128*, 336–359. [CrossRef]

agriculture

Article

Climate Resilience and Environmental Sustainability: How to Integrate Dynamic Dimensions of Water Security Modeling

Syed Abu Shoaib *, Muhammad Muhitur Rahman , Faisal I. Shalabi, Ammar Fayez Alshayeb and Ziad Nayef Shatnawi

Department of Civil and Environmental Engineering, College of Engineering, King Faisal University, Al-Hofuf 31982, Saudi Arabia; mrahman@kfu.edu.sa (M.M.R.); fshalabi@kfu.edu.sa (F.I.S.); afshayeb@kfu.edu.sa (A.F.A.); zshatnwi@kfu.edu.sa (Z.N.S.)
* Correspondence: sabushoaib@kfu.edu.sa; Tel.: +966-056-343-0815

Abstract: Considering hydro-climatic diversity, integrating dynamic dimensions of water security modeling is vital for ensuring environmental sustainability and its associated full range of climate resilience. Improving climate resiliency depends on the attributing uncertainty mechanism. In this study, a conceptual resilience model is presented with the consideration of input uncertainty. The impact of input uncertainty is analyzed through a multi-model hydrological framework. A multi-model hydrological framework is attributed to a possible scenario to help apply it in a decision-making process. This study attributes water security modeling with the considerations of sustainability and climate resilience using a high-speed computer and Internet system. Then, a subsequent key point of this investigation is accounting for water security modeling to ensure food security and model development scenarios. In this context, a four-dimensional dynamic space that maps sources, resource availability, infrastructure, and vibrant economic options is essential in ensuring a climate-resilient sustainable domain. This information can be disseminated to farmers using a central decision support system to ensure sustainable food production with the application of a digital system.

Keywords: decision support systems; agricultural water management; water security; data-driven modeling; conceptual resilience model; input uncertainty; climate extreme; process-based modeling

Citation: Abu Shoaib, S.; Rahman, M.M.; Shalabi, F.I.; Alshayeb, A.F.; Shatnawi, Z.N. Climate Resilience and Environmental Sustainability: How to Integrate Dynamic Dimensions of Water Security Modeling. *Agriculture* **2022**, *12*, 303. https://doi.org/10.3390/agriculture12020303

Academic Editor: Dimitre Dimitrov

Received: 5 December 2021
Accepted: 16 February 2022
Published: 21 February 2022

1. Introduction

The application of computers and the Internet brings new dimensions to agricultural water management and sustainable use of resources. Adequate robustness in recording and transmitting sensing data during farming provides farmers with automatic decision-making processes [1]. Trade-offs in land-use competition and sustainable land development are also possible with an appropriate technological knowledge base [2]. As human action is inherent to the water cycle, the dynamic dimension of water security can be attributed to a dimensional space that maps economic options, physical resource availability, and appropriate infrastructure. Crucial areas for environmental sustainability remain within climate change, biodiversity, air quality, and water quality. Water security, as a result of coupled human-natural system models, accounts for human compliance in the face of external drivers [3]. On the other hand, environmental sustainability increases the resilience of communities. A competent climate-resilient community structures itself around significant urgencies and can adapt to a "new normal" [4–8]. Generally, building climate resiliency is about improving development outcomes rather than implementing development activities in a new dimension [4,5]. In actuality, it helps minimize costs and maximize progress toward sustainable development goals [9,10].

However, resilience is a multi-sectoral, incremental process [11]. It is the capacity to recover quickly from strain/difficulties. In another way, resilience is the ability of a substance or object to spring back into its original shape. A region's resilience is calculated by the magnitude and severity of shocks (e.g., natural disasters such as flood and drought with diverse frequency and magnitude) and the region's capacity to contend with them [11,12]. If the shock overcomes the capacity, then resilience is measured by the region's ability to recover to a pre-shock level of functioning. The ultimate outcome represents the region's capacity to cope with the next shock [9]. Patterson and Kelleher describe this recovered resilience as a new, strengthened resilience level [13]. A conceptual linkage of resilience to water security and environmental sustainability is shown in Figure 1.

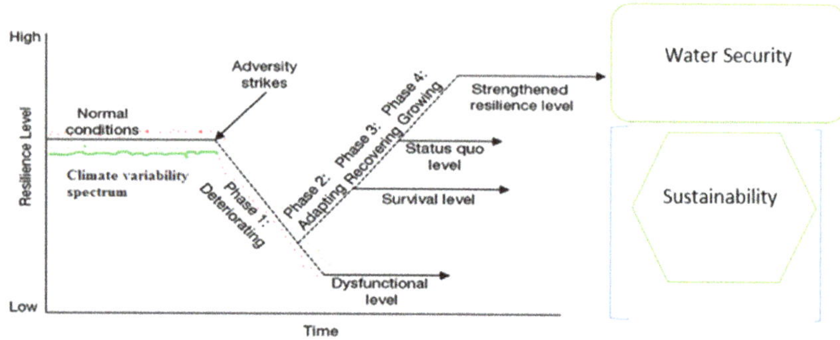

Figure 1. Integration of climate resilience with water security and environmental sustainability.

It is virtually certain that increases (decreases) in the frequency and magnitude of warm (cold) daily temperature extremes will occur in the 21st century at the global scale [14,15]. As sea levels, which have increased globally by 0.19 m over the past century [10], continue to rise, coastal flooding is expected to increase as well [16].

Climate change is already conveying more hydrological inconsistency. Increases in the frequency of floods and the magnitude of droughts are visible. At the same time, climate change uncertainties create more scenarios to explore. Global climate change is anticipated to have a huge impact on our agricultural water resources. Research studies suggest there will be amplification of the global hydrologic cycle with the appearance of more-recurrent and larger-magnitude extremes, such as floods and droughts [14,17–19]. The recent increased frequency of unusual floods and droughts worldwide seem to have only strengthened such findings. As a result, the study of how climate change impacts water security is at the forefront of scientific research today [20,21]. Standard steps are followed to assess the impacts of climate change on water resources: (1) future climate projection [20]); (2) using global climate models (GCMs) to generate data (e.g., precipitation, temperature); (3) downscaling of course-scale GCM outputs to fine-scale data appropriate for hydrological modeling and water resource studies; and (4) estimation of streamflow and groundwater levels [20,22].

In climate change predictions, there are four different sources of uncertainty to consider: input uncertainty, model uncertainty, scenario uncertainty, and internal variability [18,19,22]. Because of an incomplete understanding of physical processes, model uncertainty is prominent. Scenario uncertainty arises because of incomplete information about future emissions. Internal variability is the natural, unforced fluctuation of the climate system [22]. Internal variability is aleatoric and cannot be reduced by the improvement of scientific knowledge. Input uncertainty dominates over the other three sources, leading to the desired output. This paper centered on climate resilience and water security modeling based on environmental sustainability.

The first major objective of this paper is a multi-model hydrological framework decision support system (DSS) attributed to water security modeling in consideration of sustainability and climate resilience. A subsequent vital objective of this paper is the use of water security modeling in various scenarios to ensure water security in light of food security.

2. Materials and Methods

Resilience planning should be based on comprehensive region-specific tools to capture vulnerability in its varied dimensions (for example, biophysical, social, and technological). Dimensions of resilience can vary with a specific purpose. The most significant dimensions in climate resilience are: (i) expectation—a resilient community is competent at anticipating multiple hazards or threats to people and their values. These hazards could be non-routine, discontinuous, or collective events, such as coastal erosion, drought, or economic disinvestment; (ii) cutback—a resilient community takes mitigating action to reduce impacts; (iii) reaction—ability to mobilize resources and coordinate relief efforts [23]; (iv) revival—capacity to re-establish throughout the phases of emergency, restoration, renovation, and community betterment [9,15,16]. To build and strengthen climate resilience, input uncertainty needs to be quantified. For example, flooding is caused by excess rainfall, and drought is due to a shortage of rainfall. At the same time, temperature variation has a significant impact on both of these phenomena. How input uncertainty affects the overall climate resilience system, considering environmental sustainability, is illustrated in this research.

2.1. Data Source and Availability

A high-emission scenario is frequently referred to as "business as usual", suggesting that it is a likely outcome if society does not make concerted efforts to cut greenhouse gas emissions. Projected data are available for 14 general circulation models (GCMs) under each emission scenario. The GCMs are bcc_csm1_1, ccsm4, cesm1_cam5, csiro_mk3_6_0, fio_esm, gfdl_cm3, gfdl_esm2m, giss_e2_h, giss_e2_r, ipsl_cm5a_mr, miroc_esm, miroc5, mri_cgcm3, and noresm1_m. The impacts of three CO2 emission scenarios, the representative concentration pathway (RCP) 4.5, RCP 6.0, and RCP 8.5. The historical data period is 1961–2016, and the projected periods are 2020–2039, 2040–2059, 2060–2079, and 2080–2099 [18,24]. Data from different periods can be used for scenario analysis or mapping future variability.

Additionally, the daily rainfall and evapotranspiration data set was used in this study. The data set was derived from the Australian Water Availability Project (AWAP) [18,24], which is gridded to 0.050° × 0.050° (5 × 5 km) and is extracted for the common 1980–2015 period. The accuracy of this data set is typically low where gauge density is low, as is the case in central-west Australia, for instance [24]. The original meteorological data used in the AWAP product were supplied by the Bureau of Meteorology Australia (BoM). Daily rainfall data are available from 1900 to present, temperatures from 1911 to present, and solar irradiance from 1990 to present. Selected catchments are shown in Figure 2.

In this section, the conceptual flow of the resilience model is shown (Figure 3), where climate extremes were identified with diverse input sources (precipitation, temperature, wind speed). Resilience functions were analyzed with climate extremes. The solution options were output, and a sustainable solution was made as per the resilience index for each resilience function. There are various possible options likely for different climate extremes at a specified location [18,19,21,24–26]. Quantification of input uncertainty was evaluated through the methodical process (Figure 3).

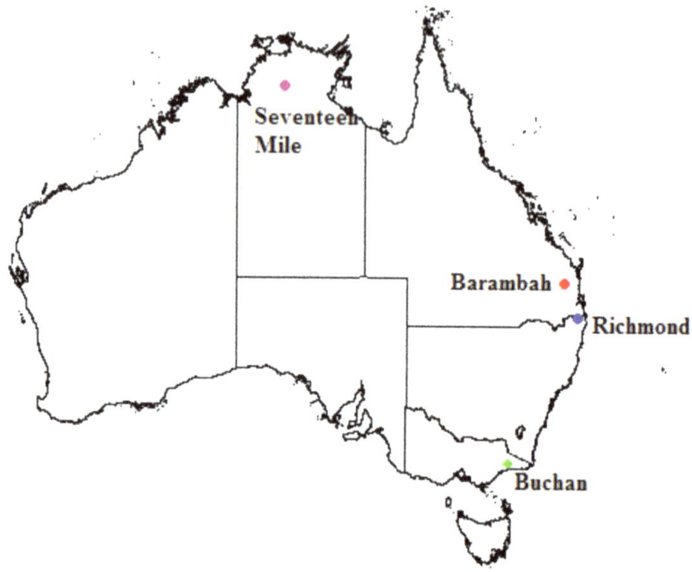

Figure 2. Study area of selected Australian catchment. Four catchments were considered in the study: (i) Richmond, New South Wales, (ii) Buchan River Victoria, (iii) Seventeen Mile Creek, a waterfall view, Northern Territory, (iv) Barambah Creek at Litzows, Queensland.

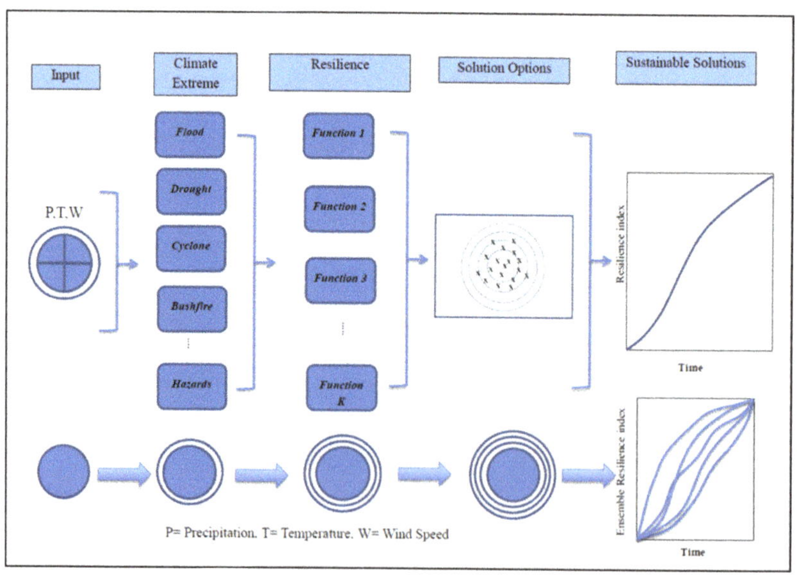

Figure 3. Schematic overview of the experimental design undertaken to find a sustainable solution and quantify the resilience index for input variability. Climate extreme ensembles are analyzed as a function of resilience functions and optimized with different solutions. All the solutions are characterized as the resilience index with time. The circle in the bottom panel of the figure shows the increase in complexity from left to right. Ensembles of solutions are possible with multiple climate extreme possibilities for a specific catchment or community.

The quantification of the extent of climate extremes, resilience functions, and identifiability of the solution options are shown below [18,19,21,24].

Equations (1)–(3) were used to calculate the resilience flow deviation (RsFD) for the selected parameters for climate extremes, resilience functions, and identifiability of the solution options at each percentile (p), denoted as $RsFD_p^M$, $RsFD_p^O$, and $RsFD_p^I$, respectively:

$$(RsFD_p^M)^2 = \underset{I,O}{E}\left[var\left(Q_p^m \middle| Q_p^i, Q_p^o\right)\right] = \left[\frac{1}{IO(M-1)} \sum_{o=1}^{O} \sum_{i=1}^{I} \sum_{m=1}^{M} \left(Q_p^{mio} - \overline{Q}_p^{io}\right)^2\right] \quad (1)$$

$$(RsFD_p^O)^2 = \underset{M,I}{E}\left[var\left(Q_p^o \middle| Q_p^m, Q_p^i\right)\right] = \left[\frac{1}{MI(O-1)} \sum_{i=1}^{I} \sum_{m=1}^{M} \sum_{o=1}^{O} \left(Q_p^{omi} - \overline{Q}_p^{mi}\right)^2\right] \quad (2)$$

$$(RsFD_p^I)^2 = \underset{O,M}{E}\left[var\left(Q_p^i \middle| Q_p^o, Q_p^m\right)\right] = \left[\frac{1}{OM(I-1)} \sum_{m=1}^{M} \sum_{o=1}^{O} \sum_{i=1}^{I} \left(Q_p^{iom} - \overline{Q}_p^{om}\right)^2\right] \quad (3)$$

Additionally, I, M, and O denote the total number of parameter sets representing identifiability of the solution options, the number of the selected climate extremes, and the resilience functions [18,19,21,24].

The prime endeavor of the model is to ensure the sustainable development of cities, towns, and other human settlements, incorporating the uncertainty due to climate extremes, resilience functions, and types of optimization algorithms. One key stake of this plan is ensuring that cities can withstand and recover quickly from catastrophic events. The other objectives are as follows: (i) tools for measuring and increasing resilience to multi-hazard impacts, (iii) building climate change resilience for poor and vulnerable people in cities by creating robust models and methodologies for assessing and addressing risk, (iv) developing statistical methods for the description of the uncertainties related to climate change and economic constraints, (v) developing regionalization methods to improve urban resiliency with a conceptual model, and (vi) developing empirical equations with quantified parameters to relate urban resilience with sustainable development.

Resilience is presented here as a function of the following:

$$\text{Resilience} = f(\text{R}, \text{En}, \text{S}, \text{If}, \text{L}, \text{In}, \text{Ec}, \text{N}, \text{C}, \text{Et}) \quad (4)$$

- R—resources (GDP, population density);
- En—energy consumption and emissions (oil intensity, CO_2 per capita/per USD $GDP, SO_2 per capita/per USD $GDP);
- S—soil type and vegetation (agriculture subsidies, use of pesticides);
- If—infrastructure;
- L—livelihood and land-use pattern (indoor air pollution, child mortality, forest loss);
- In—integration (political risk, among local to national planning and implementing body);
- Ec—ecosystem vitality (biodiversity and habitat, water resources, forest, fisheries);
- N—neighborhood network (governance, regional partnership, quality of local supplies);
- C—coordination (control of corruption, centralization, and decentralization);
- Et—entropy (measure of disorder or randomness of the system).

To consider climate resiliency risk, three drivers were analyzed:

1. Exposure to natural hazards;
2. Quality of natural hazard risk management;
3. Quality of fire risk management.

The functions described above can be grouped in three parts: A = (R, S, If, L); B = (En, Ec, Et); C = (In, N, C). These indexes are developed based on the resilience indicator generated for a specific location or defined catchment.

$$\text{Resilience Index} = \text{A} * \text{B} * \text{C} \quad (5)$$

2.2. Conceptual Hydrological Model and Quantifying Input Uncertainty

Quantifying input uncertainty is vital in hydrologic models; therefore, several conceptual options were applied. Four parent hydrological models (TOPMODEL, ARNOXVIC, PRMS, SACRAMENTO) within the FUSE framework to represent the full spectrum of potential model [19] variability in the absence of specific information on the catchment hydrologic processes. Twenty-two (22) model parameters represent the hydrological system within FUSE [19]. Some of these parameters are inactive depending on the model configuration of interest. Figure 4 outlines the model structure that was considered in the simulation results presented. Flexibility in selecting model structure remains, as it is possible to use the framework for understanding structural errors [18,19,21,24]. The input of the models, such as precipitation (rainfall)/temperature/windspeed, was checked with the variability of different grids in a specified catchment or area compared to the true rainfall of the total catchment grids (Figure 4). At the same time, point gauge rainfall for that specific catchment was considered for the input variation [18,24]. Details of the four parent models could be found in [19].

Architecture of the Upper Soil Layer			
SS[a,b]	ST[d]	CB[c]	
Architecture of the Lower Soil Layer			
BRFS[b]	TRTPT[d]	$BRUS_{fr}$[c]	$BRUS_{pr}$[a]
Surface Runoff			
UZP[b]	UZL[c,d]	SZT[a]	
Percolation: Vertical Drainage			
DAFC[a,c]	GD[d]	SZC[b]	
Evaporation			
Sequential[b]	Root Weighting[a,c,d]		
Interflows			
Denied[a,b]	Allowed[c,d]		
Routing			
Allowed Using Gamma Distribution[a,b,c,d]	Denied		

Figure 4. Multi-model structure built to model catchment processes. Details of the different component multi-model structures are (i) single state (SS), (ii) separate tension storage (ST), (iii) cascading buckets (CB), (iv) baseflow reservoir of fixed size (BRFS), (vi) tension reservoir plus two parallel tanks (TRTPT), (vii) baseflow reservoir of unlimited size (frac rate) (BRUSfr), (viii) baseflow reservoir of unlimited size (power recession) (BRUSpr), (ix) drainage above field capacity (DAFC), (x) gravity drainage (GD), (xi) saturated zone control (SZC), (xii) unsaturated zone Pareto (UZP), (xiii) unsaturated zone linear (UZL), (xiii) saturated zone topographic (SZT). Model structural components used for the 4 models considered in this study are depicted as TOPMODEL = a; ARNOXVIC = b; PRMS = c; SACRAMENTO = d (after [18]).

The chosen experimental design is a combination of climate resilience, environmental sustainability, and water security modeling. The equation was based on the basics of the theoretical context of quantifying the metrics of the methodological framework. More details on [18,19,24].

For a given model (Figure 5), analyses were conducted for different likelihood or objective functions, including the Nash–Sutcliffe efficiency coefficient (NSE) [21], logarithmic Nash–Sutcliffe efficiency (LogNSE), and square root Nash–Sutcliffe efficiency (SqrtNSE). The optimization algorithm (dynamically dimension search (DDS)) was used to find the optimum parameter sets. From each simulation, the quantile variation of streamflow was determined and, thus, compared with different input variabilities of the defined catchment [18,24]. It is said that over-parameterization, dependency on input data bias, and lack of a systematic link between parameter precision and model efficiency are the three main factors [24,27] that complicate the regionalization of conceptual rainfall-runoff models. With the best possible model structure, input uncertainty was quantified to improve climate resilience.

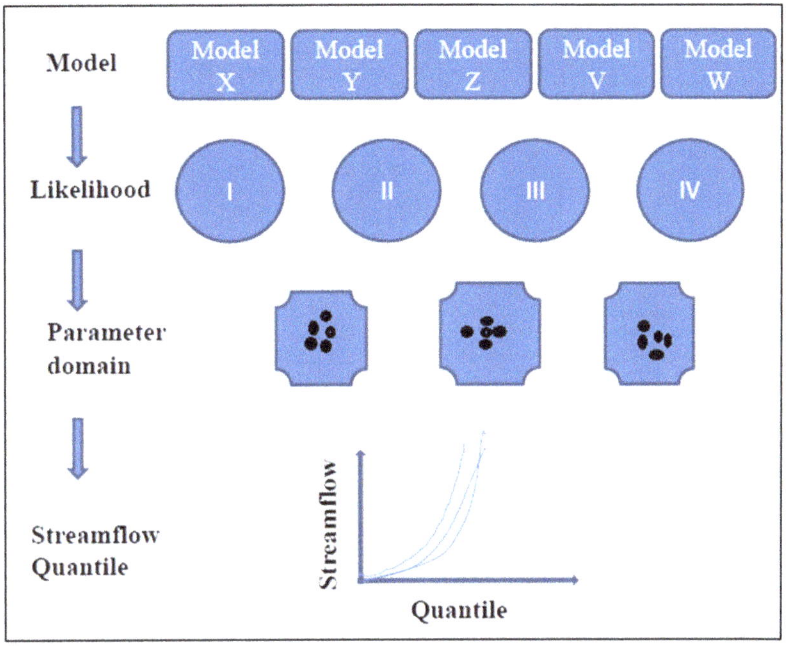

Figure 5. Possible model structure options in hydrological models with alternate descriptions for upper and lower soil layers, types of surface runoff, vertical drainage, evaporation, interflows, and possible routing options. For different conceptual model structures, different likelihood/objective functions can be made. For each likelihood, options of parameter domain can be generated. For each simulation, flow quantiles can be produced to compare with different inputs.

As part of the demonstration of the approach, three objective functions were selected for analysis, including the Nash–Sutcliffe efficiency coefficient (NSE) [28], a logarithmic Nash–Sutcliffe efficiency (LogNSE), and a square root Nash–Sutcliffe efficiency (SqrtNSE). The Nash–Sutcliffe efficiency measures the relative magnitude of the model residual variance in comparison to the observed data variance. This efficiency can be greatly influenced by the peak flow values, which tend to have higher residual error. In contrast, the LogNSE and SqrtNSE are more likely to be influenced by low flows as the log and square root transformation reduces the importance of errors on larger flow magnitudes. These three objective functions are selected to reflect variability in the model optimization process and the desire to provide suitable models that fit different aspects of the hydrograph and the catchment response depending on the purpose of the model.

3. Results and Analysis

In this section, the key outcome is analyzed based on the method and data presented in the previous section. The more-generalized scenario was considered in this analysis. The possible scenarios were: (i) increasing precipitation, constant temperature; (ii) decreasing precipitation, constant temperature; (iii) increasing temperature, constant precipitation;

(iv) decreasing temperature, constant precipitation; and (v) real-time variability with changing conditions.

Figure 6 contains three sub-plots. The left one shows the variability of streamflow/discharge among different developed models (X, Y, Z, W, and V) generated from four parent models. The right plot shows the regional variability of streamflow/discharge due to changing grid with respect to selected parent model ARNOXVIC and the middle one is similar to the right, only testing with different parent model name TOP model.

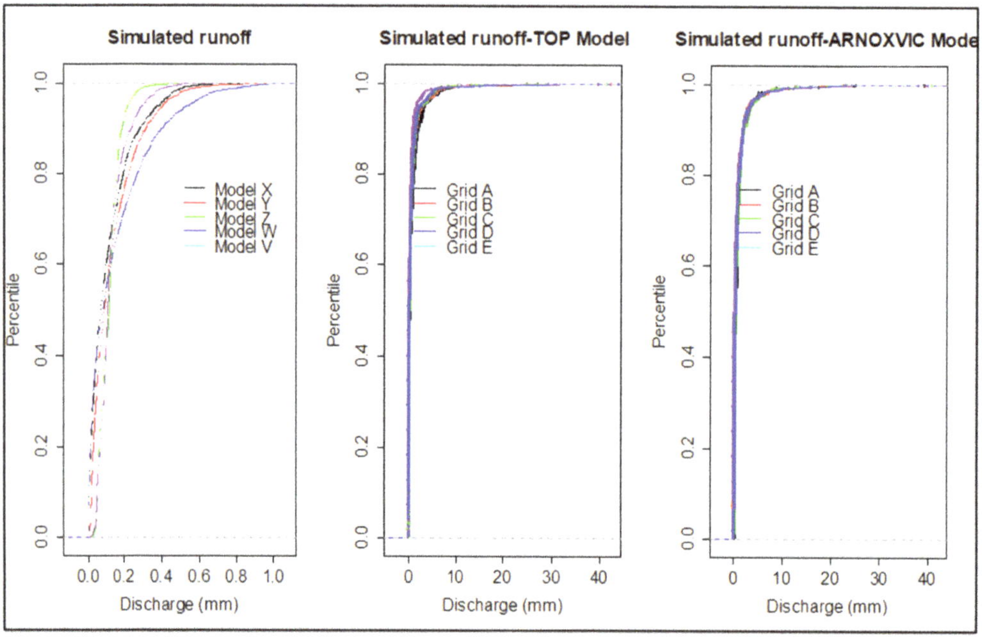

Figure 6. Input variability in different model choices and grid-based rainfall variability with the model input of rainfall estimated with the variability of different grids in a specified catchment or area compared to true rainfall of the total grids of the catchment. At the same time, point gauge rainfall for that specific catchment is considered for the variation in input.

For a specific catchment (e.g., Richmond, NSW), different model structures responded differently to the variability (Figure 6, left) in predicting streamflow variability output. This represents higher prediction input uncertainty and the higher percentile and leads to more uncertainty in climate resiliency, with significant effects on environmental sustainability. By changing the different grid's precipitation levels, variability was reduced (Figure 6, middle and right).

This Figure 7 shows the regional rainfall considering different scenarios to observe the variability of streamflow/discharge due to changing grid with respect to selected parent model PRMS.

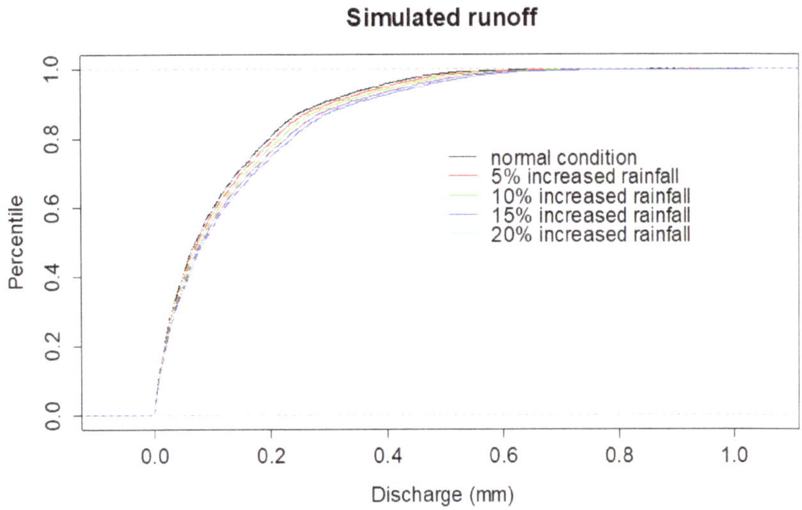

Figure 7. Input rainfall variability to see the changeability of discharge in a defined model.

Input variability in different model choices and grid-based rainfall variability with model input of rainfall was quantified with the variability of different grids in a specified catchment or area compared to true rainfall of total grids of the catchment. At the same time, point gauge rainfall was considered for the variation in input for each specific catchment (Figures 6–8).

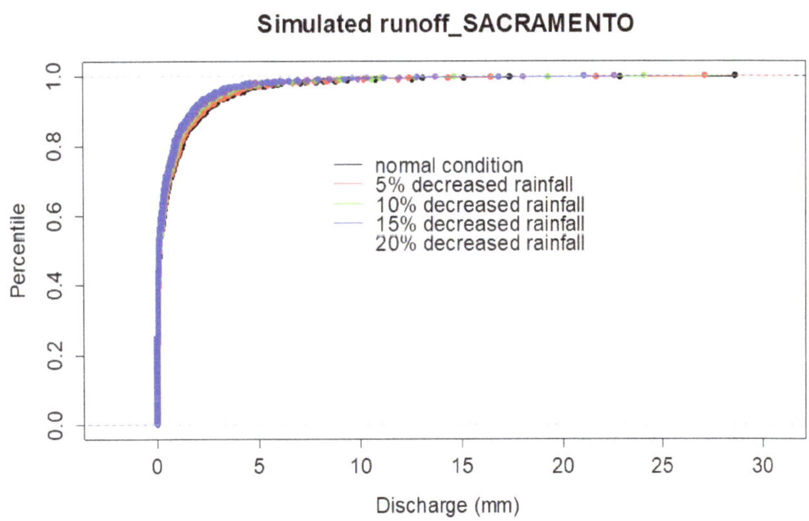

Figure 8. Input rainfall variability (decreasing trend) to observe the changeability of discharge in a defined SACRAMENTO model. Changing rainfall variability is also observed to obtain the overall picture.

Catchment-to-catchment variability was observed due to variable catchment properties. For example, uncertainty in the Richmond catchment was high compared to the Seventeen Mile catchment due to climate variability; one catchment was very wet considering the metrics of quantile flow deviation (QFD) metric [18,19], and the other was very dry. Meanwhile, the Buchan River catchment in Melbourne, Victoria, was more similar to the Barambah River catchment of Queensland. Extreme input uncertainty was quantified by applying a dynamically dimensioned search optimization algorithm [18,19,24]. The analysis of the input uncertainty of four different catchments in Australia (Figure 2) showed the variability in uncertainty due to streamflow uncertainty [19,21,25]. The quantile flow deviation (QFD) metric [18,24,26] was used to estimate the input uncertainty compared to the model structure and parameter uncertainty [19]. This uncertainty was, directly and indirectly, linked with climate resilience and environmental sustainability and was quantified using Equations (1)–(3). Based on the extent of interaction, it is possible to integrate dynamic dimensions of water security modeling. Nash–Sutcliffe efficiency (NSE) has been presented for model simulations of discharge and is widely used to assess the predictive power of hydrological models. Additional analysis considering alternate objective functions to assess how sensitive our findings are to the objective function used is undertaken. NSE values vary in a range of 0.69 to 0.94 considering different catchments and selected model structures.

The three objective functions (NSE, LogNSE, SqrtNSE) are selected to reflect variability in the model optimization process and the desire to provide suitable models that fit different aspects of the hydrograph and the catchment response depending on the purpose of the model (Figure 9). As Figure 9 illustrates, the change in the proportion of uncertainty by considering additional objective functions is not significant.

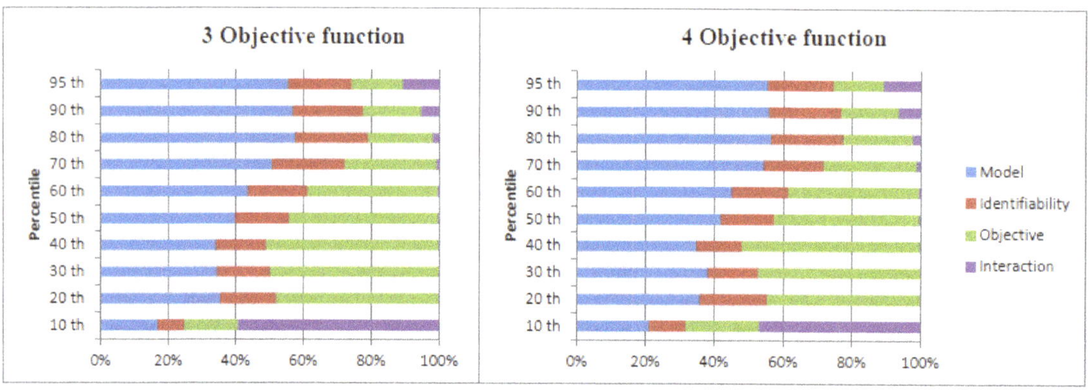

Figure 9. Impact of objective function variability considering model structure, identifiability/parameter, and their interactions.

However, an aggregate sustainability score could be computed based on multiple indicators organized into four dimensions: environment, social, water security and nutrition, and economic. In the multi-dimensional nature of the sustainability score, there was no "natural" or "theoretical" threshold above which a country could be said to be sustainable. In other words, the creation of the metric also highlights the urgent need for governments and other key stakeholders (donors, international development agencies, etc.) to invest in more comprehensive monitoring of the existing water-energy-food systems. Increased monitoring is particularly relevant concerning transformation, transport, retail, and distribution, for which data are still missing, including in some high-income-level countries. The optimal combination and technical validation are crucial for the final computation of the sustainability score [4,29,30].

4. Discussion—Scenarios of Model Development

As this study analyzed the model input and output uncertainty, this approach is useful for precision agriculture. How 1 mm of rainfall can influence cropping patterns and food production can be easily visualized for decision makers. Moreover, streamflow uncertainty has a direct linkage with climate resiliency. In this section, different aspects of the possible integration of the dynamic dimensions of water security were attributed to environmental sustainability and climate resilience. As water security is a multi-faceted dilemma, it goes further than the mere balancing of supply and demand [3,31,32]. On the contrary, static index-based approaches to quantifying water security are impotent to acknowledge the human action inherent to the water cycle. A more flexible and dynamic view of water security is urgent considering human adaptation to environmental change and increasing spatial specialization [3,31]. A four-dimensional dynamic space that maps sources, resource availability, infrastructure, and vibrant economic options is important in a climate-resilient sustainable agriculture domain. These dimensions are based on scenarios considering carbon emissions and possible trends in the world economy.

4.1. Water Security, Water Cycle Modifications, and Climate Resilience

Water security, climate resilience, and environmental sustainability must be closely interrelated for effective agricultural practice, identical to our body parts. As it is complex to quantify, we applied the function with its parameters then applied the regionalization approach to obtain the process extents (Figure 10). As global climate change intensifies, many countries have been hit hard by an unprecedented wave of droughts and water shortages. Water security can be defined as a function of different parameters as follows:

$$\text{Water Security, W} = \text{f (R, C, E, T, G, C2, U)} \tag{6}$$

where R is resource, C is culture, E is economy, T is technology, G is governance, C2 is climate change, and U is uncertainty.

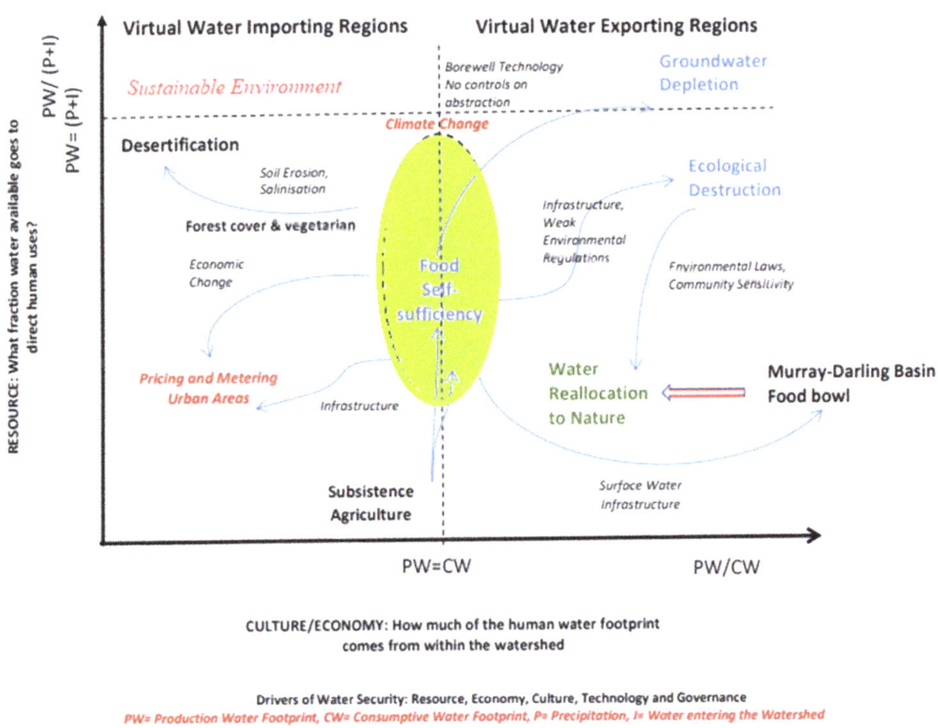

Figure 10. Interaction among drivers of water security that lead to food security for Australia: resource, economy, culture, technology (advancement of computer and Internet use), and governance. After [3].

Establishing safe limits to water cycle modifications and identifying possible spatially explicit methods for quantification is an enormous challenge [33]. Many current and widely relied-upon hydrologic prediction approaches are founded on the assumption of stationarity [34], which permits extrapolation to the future using historical data. In a changing world, however, neither the structure (e.g., patterns of land use and land cover, connectivity between channels, and riparian or wetland environments, or the extent of human-made structures) nor the external drivers (e.g., temperature and precipitation forcing) of the hydrologic response can be treated as fixed [18–22,24–26,35]. Instead, changes in structure and drivers create the potential for new dynamics [36], for example, by hydrologic systems crossing unknown thresholds [19,26,31]. The possibility for the emergence of such new dynamics poses key challenges to predictability, especially on decadal or longer time scales.

Understanding the bloodstream interdependency, availability, and accessibility of surface water and groundwater plays a crucial role in recognizing and managing water security. The Falkenmark water stress index, defined as the per capita annual renewable freshwater available, was an early effort to recognize the relationship between human needs and environmental constraints; regions with less than 1000 m^3 per capita per year were defined as "water-scarce" [35,37]. As populations grow, humanity faces the prospect of uncertain future water supplies due to climate change and the increasing demands on water [35,38,39]. In the virtual water importing and exporting region, some factors dominate the water balance, affecting water security. For example, Murray Darling Basin is considered the food bowl for Australia, where ecological destruction, environmental regulations, laws, community sensitivity, infrastructure, controls on abstraction, and climate change impacts play a key role in the water security domain (Figure 10). This phenomenon instigates the internal integration of the dynamic dimensions of water security modeling with the variability of climate resilience and environmental sustainability.

4.2. *Climate–Water Quality Relationships in Changing Streamflow*

Temperature and precipitation elasticities of water quality parameters, highlighted by N and P nutrients, can be analyzed in large rivers due to the change of streamflow. The spatial and temporal assessments show that precipitation elasticity is more variable in space than temperature elasticity and that seasonal variation is more evident for precipitation elasticity than temperature elasticity [1,32,34,36]. Even small changes to streamflow may have significant agricultural and ecological implications [38]. Therefore, it is vital to integrate the dynamic dimension of water security modeling with climate resiliency and environmental sustainability. It is potentially helpful for investigating the effects of climate change on water quality in large rivers, such as the long-term shift in nutrient concentrations.

4.3. *Drought Management*

Underestimation of solar radiation usually overestimates soil moisture. The concept of reliability-resilience-vulnerability (RRV) is used here in the context of agricultural drought through the analysis of temporal variations in soil moisture. The failure, or unsatisfactory stage, is considered the depletion of soil moisture below the permanent wilting point (PWP), an indicator of agricultural drought (Figure 10). The natural integration of water security modeling with climate resilience and ensuring environmental sustainability integrates the plant, soil, and atmospheric conditions at a particular location with the long-term, spatiotemporal variation of drought susceptibility [3,8,9,26].

5. Conclusions

The development of a decision support system (DSS) using advanced computer data processing tools and Internet systems to analyze large data sets has made the agriculture industry more sustainable. Increased land-use change with industrial development has made sustainable development critical. High-level policy intervention is required, considering climate change. Based on development trends, variability in climate extremes, and water security, a coordinated policy matrix is crucial. Integrating the available resources with the required products could be more effective with the application of a central database using DSS. Considering the diversity in an uncertain world, integrating dynamic dimensions of water security modeling is vital for ensuring environmental sustainability and the associated full range of climate resilience. This is also possible with the use of high-tech computer and Internet systems. This study presents a framework for integrating the dynamic dimensions of water security modeling with climate resilience and environmental sustainability variability. Therefore, a vital contribution of this paper is in accounting for the water security modeling and scenarios of model development. A more flexible and dynamic view of water security is urgent considering human adaptation to environmental change and increasing spatial specialization. With the adoption of high-tech computer data processing, including Internet systems, nature-based solutions can be more effective to model the dynamic indicators as outcomes of coupled human-water systems. Ensuring water security ease, the way of environmental sustainability paradigm. Quantifying the uncertainty in water availability or use projection gives the overall picture of the available water. Thereby, decision makers got the idea of available water for the effective and efficient use of water through enforcing water metering by adopting an integrated water pricing policy. As the amount of water demand is increasing in all sectors, including agriculture, new technology could be used for maximizing water use. This framework will surely help in the policymaking process to allocate the right share of water as per demand to save water, ensuring integrated water management. We have the options of multi-model structures considering four parent models, which gives the policymaker to adopt new model structures according to the purpose of the users. One of this study's shortcomings was finding the real-time data of social structure and economic diversity to interlink the hydrological phenomena of climate and land-use change. Future research will concentrate on the nonlinear dimensions of food security, integrating climate resilience and water security in a sustainable domain.

Author Contributions: Conceptualization, S.A.S.; methodology, S.A.S., M.M.R., F.I.S., A.F.A. and Z.N.S.; software, S.A.S., M.M.R., F.I.S., A.F.A. and Z.N.S.; validation, S.A.S., M.M.R., F.I.S., A.F.A. and Z.N.S.; formal analysis, S.A.S., M.M.R. and F.I.S.; investigation, S.A.S., F.I.S., M.M.R. and F.I.S.; resources, S.A.S., M.M.R. and F.I.S.; data curation, S.A.S., M.M.R., F.I.S. and Z.N.S.; writing—original draft preparation, S.A.S., M.M.R., F.I.S. and Z.N.S.; writing—review and editing, S.A.S., M.M.R., A.F.A. and Z.N.S.; visualization, S.A.S. and Z.N.S.; supervision, S.A.S. and M.M.R.; project administration, S.A.S.; funding acquisition, S.A.S. All authors have read and agreed to the published version of the manuscript and to the work reported.

Funding: The authors acknowledge the Deanship of Scientific Research at King Faisal University for their kind assistance and for funding this research work through Nasher, project number 206076.

Data Availability Statement: Detailed access to the data: In situ streamflow data from Australia Hydrologic Reference Stations: http://www.bom.gov.au/water/hrs/ (accessed on 15 February 2022) and http://www.bom.gov.au/waterdata/ (accessed on 15 February 2022); potential evapotranspiration (PET) and Rainfall data from Australian Water Availability Project (AWAP): http://www.bom.gov.au/jsp/awap/rain/index.jsp and http://www.bom.gov.au/water/landscape/; the C/M ratio data from the GFDS (http://www.gdacs.org/flooddetection/ (accessed on 15 February 2022)); and the DEM data from Shuttle Radar Topographic Mission (SRTM; http://www.cgiar-csi.org/ (accessed on 15 February 2022)). https://climateknowledgeportal.worldbank.org/download-data (accessed on 15 February 2022).

Acknowledgments: The authors acknowledge the Deanship of Scientific Research at King Faisal University for their kind assistance and for funding this research work through the Nasher project, number 206076. We ran all computations using the statistical environment R (https://www.r-project.org/ (accessed on 15 February 2022)) and its graphical interface RStudio (https://www.rstudio.com/ (accessed on 15 February 2022)). The authors extend their appreciation to Nahid Sultana, UNSW, for her sincere effort and the Deputyship for Research and Innovations, Ministry of Education, in Saudi Arabia.

Conflicts of Interest: The authors declare no conflict of interest. The funders had no role in the design of the study, in the collection, analyses, or interpretation of data, in the writing of the manuscript, or in the decision to publish the results.

References

1. Wang, L.; Zhang, M.; Li, Y.; Xia, J.; Ma, R. Wearable multi-sensor enabled decision support system for environmental comfort evaluation of mutton sheep farming. *Comput. Electron. Agric.* **2021**, *187*, 106302. [CrossRef]
2. Jin, G.; Chen, K.; Wang, P.; Guo, B.; Dong, Y.; Yang, J. Trade-offs in land-use competition and sustainable land development in the North China Plain. *Technol. Forecast. Soc. Chang.* **2019**, *141*, 36–46. [CrossRef]
3. Srinivasan, V.; Konar, M.; Sivapalan, M. A dynamic framework for water security. *Water Secur.* **2017**, *1*, 12–20. [CrossRef]
4. Zhou, L.; Tokos, H.; Krajnc, D.; Yang, Y. Sustainability Performance Evaluation in Industry by Composite Sustainability Index. *Clean Technol. Environ. Pol.* **2012**, *14*, 789–803. [CrossRef]
5. Resilience for Sustainability. *Nat. Plants* **2021**, *7*, 101. [CrossRef]
6. Intergovernmental Panel on Climate Change (IPCC). Summary for Policymakers. In *Global Warming of 1.5 °C*; Masson-Delmotte, V., Ed.; World Meteorological Organization: Geneva, Switzerland, 2018.
7. Chaudhary, A.; Gustafson, D.; Mathys, A. Multi-indicator sustainability assessment of global food systems. *Nat. Commun.* **2018**, *9*, 848. [CrossRef] [PubMed]
8. Gleeson, T.; Wang-Erlandsson, L.; Porkka, M.; Zipper, S.C.; Jaramillo, F.; Gerten, D. Illuminating water cycle modifications and Earth system resilience in the Anthropocene. *Water Resour. Res.* **2020**, *56*, e2019WR024957. [CrossRef]
9. Dabson, B.; Heflin, M.C.; Miller, K.K. Regional Resilience: Research and Policy Brief. 2012. Available online: http://nado.org/wp-content/uploads/2012/04/RUPRI-Regional-Resilience-Research-Policy-Brief.pdf (accessed on 15 February 2022).
10. Church, J.A. Sea Level Change. In *Climate Change 2013: The Physical Science Basis*; Stocker, T.F., Ed.; Cambridge University Press: Cambridge, UK, 2013; p. 1535. Available online: http://www.climatechange2013.org/images/report/WG1AR5_ALL_FINAL.pdf (accessed on 15 February 2022).
11. Finka, M.; Tóth, A. Regional Resiliences Improvement by Innovative Approaches in Management of External Shocks. 2014. Available online: http://www.spa-ce.net/pdf/2014/Conference_%202014/Toth_Spa-ce.net-2014.pdf (accessed on 15 February 2022).
12. Martin, R. Regional economic resilience, hysteresis and recessionary shocks. *J. Econ. Geogr.* **2012**, *12*, 1–32. [CrossRef]
13. Patterson, J.L.; Kelleher, P.A. Deeper Meaning of Resilience. In *Resilient School Leaders: Strategies for Turning Adversity into Achievement*; E-Book; Association for Supervision and Curriculum Development: Alexandria, VA, USA, 2005. Available online: http://www.ascd.org/publications/books/104003/chapters/A-Deeper-Meaning-of-Resilience.aspx (accessed on 15 February 2022).
14. Intergovernmental Panel on Climate Change (IPCC). *Climate Change 2007: The Physical Science Basis*; Contribution of Working Group I to the Fourth Assessment Report of the Intergovernmental Panel on Climate Change; Cambridge University Press: New York, NY, USA, 2007.
15. Field, C.B.; Barros, V.; Stocker, T.F.; Dahe, Q. *Managing the Risks of Extreme Events and Disasters to Advance Climate Change Adaptation*; Cambridge University Press: Cambridge, UK, 2012; p. 582.
16. Wong, P.P.; Losada, I.J.; Gattuso, J.-P.; Hinkel, J.; Khattabi, A.; McInnes, K.L.; Saito, Y.; Sallenger, A. Coastal Systems and Low-Lying Areas. In *Climate Change 2014: Impacts, Adaptation, and Vulnerability*; Field, C.B., Barros, V.R., Dokken, D.J., Mach, K.J., Mastrandrea, M.D., Bilir, T.E., Chatterjee, M., Ebi, K.L., Estrada, Y.O., Genova, R.C., et al., Eds.; Cambridge University Press: Cambridge, UK; New York, NY, USA, 2014; pp. 361–409. Available online: https://www.ipcc.ch/site/assets/uploads/2018/02/WGIIAR5-Chap5_FINAL.pdf (accessed on 15 February 2022).
17. Kundzewicz, Z.W.; Mata, L.J.; Arnell, N.W.; DÖLl, P.; Jimenez, B.; Miller, K.; Oki, T.; ŞEn, Z.; Shiklomanov, I. The implications of projected climate change for freshwater resources and their management. *Hydrol. Sci. J.* **2008**, *53*, 3–10. [CrossRef]
18. Shoaib, S.A.; Marshall, L.; Sharma, A. A metric for attributing variability in modelled streamflows. *J. Hydrol.* **2016**, *541*, 1475–1487. [CrossRef]
19. Clark, M.P.; Slater, A.G.; Rupp, D.E.; Woods, R.A.; Vrugt, J.A.; Gupta, H.V.; Wagener, T.; Hay, L.E. Framework for Understanding Structural Errors (FUSE): A modular framework to diagnose differences between hydrological models. *Water Resour. Res.* **2008**, *44*, W00B02. [CrossRef]
20. Sivakumar, B. Global climate change and its impacts on water resources planning and management: Assessment and challenges. *Stoch Env. Res. Risk Assess.* **2011**, *25*, 583–600. [CrossRef]
21. Clark, M.P.; McMillan, H.K.; Collins, D.B.G.; Kavetski, D.; Woods, R.A. Hydrological field data from a modeller's perspective: Part 2: Process-based evaluation of model hypotheses. *Hydrol. Processes* **2011**, *25*, 523–543. [CrossRef]
22. Woldemeskel, F.M.; Sharma, A.; Sivakumar, B.; Mehrotra, R. An error estimation method for precipitation and temperature projections for future climates. *J. Geophys. Res. Atmos.* **2012**, *117*, D22104. [CrossRef]
23. The 2015 FM Resilience Index. Annual Report, Oxford Metrica. 2015. Available online: https://www.fmglobal.com/assets/pdf/Resilience_Methodology.pdf (accessed on 15 February 2022).
24. Shoaib, S.A.; Marshall, L.; Sharma, A. Attributing input uncertainty in streamflow simulations via the Quantile Flow Deviation metric. *Adv. Water Res.* **2018**, *116*, 40–55.
25. Clark, M.P.; Kavetski, D.; Fenicia, F. Pursuing the method of multiple working hypotheses for hydrological modeling. *Water Resour. Res.* **2011**, *47*, W09301. [CrossRef]

26. Shoaib, S.A.; Khan, M.Z.K.; Sultana, N.; Mahmood, T.H. Quantifying Uncertainty in Food Security Modeling. *Agriculture* **2021**, *11*, 33. [CrossRef]
27. Andréassian, V.; Perrin, C.; Oudin, L. From catchment similarity to hydrological similarity: A review of the difficulties hindering the regionalization of hydrological models. *Geophys. Res. Abstr.* **2003**, *13*. Available online: https://meetingorganizer.copernicus.org/EGU2011/EGU2011-13991.pdf (accessed on 15 February 2022).
28. Nash, J.E.; Sutcliffe, J.V. River flow forecasting through conceptual models part I—A discussion of principles. *J. Hydrol.* **1970**, *10*, 282–290. [CrossRef]
29. Xiaoyu, G. When to Use What: Methods for Weighting and Aggregating Sustainability Indicators. *Ecol. Indic.* **2017**, *81*, 491–502.
30. Sirone, S.; Seppala, J.; Leskinen, P. Towards More Non-Compensatory Sustainable Society Index. *Environ. Dev. Sustain.* **2015**, *17*, 587–621. [CrossRef]
31. Wagener, T.; Sivapalan, M.; Troch, P.A.; McGlynn, B.L.; Harman, C.J.; Gupta, H.V.; Kumar, P.; Rao, P.S.C.; Basu, N.B.; Wilson, J.S. The future of hydrology: An evolving science for a changing world. *Water Resour. Res.* **2010**, *46*, W05301. [CrossRef]
32. Morrison, R.R.; Stone, M.C. Spatially implemented Bayesian network model to assess environmental impacts of water management. *Water Resour. Res.* **2014**, *50*, 8107–8124. [CrossRef]
33. Savenije, H.H.; Hoekstra, A.Y.; van der Zaag, P. Evolving water science in the Anthropocene. *Hydrol. Earth Syst. Sci.* **2014**, *18*, 319–332. [CrossRef]
34. Sivapalan, M.; Savenije, H.H.; Blöschl, G. Socio-hydrology: A new science of people and water. *Hydrol. Processes* **2012**, *26*, 1270–1276. [CrossRef]
35. Falkenmark, M.; Lundqvist, J.; Widstrand, C. Macro-scale water scarcity requires micro-scale approaches: Aspects of vulnerability in semi-arid development. *Nat. Resour. Forum* **1989**, *13*, 258–267. [CrossRef]
36. Sivapalan, M.; Konar, M.; Srinivasan, V.; Chhatre, A.; Wutich, A.; Scott, C.A.; Wescoat, J.L.; Rodríguez-Iturbe, I. Socio-hydrology: Use-inspired water sustainability science for the Anthropocene. *Earths Future* **2014**, *2*, 225–230. [CrossRef]
37. Falkenmark, M.; Rockström, J. The new blue and green water paradigm: Breaking new ground for water resources planning and management. *J. Water Resour. Plan. Manag.* **2006**, *132*, 129–132. [CrossRef]
38. Manzoni, S.; Katul, G.; Porporato, A. A dynamical system perspective on plant hydraulic failure. *Water Resour. Res.* **2014**, *50*, 5170–5183. [CrossRef]
39. Michel-Kerjan, E. How resilient is your country? *Nature* **2012**, *491*, 497. Available online: http://www.nature.com/news/how-resilient-is-your-country-1.11861 (accessed on 15 February 2022). [CrossRef]

agriculture

MDPI

Article

Research on the Time-Dependent Split Delivery Green Vehicle Routing Problem for Fresh Agricultural Products with Multiple Time Windows

Daqing Wu [1,2] and Chenxiang Wu [1,*]

1 College of Economics and Management, Shanghai Ocean University, Shanghai 201306, China; dqwu@shou.edu.cn
2 Department of Nanchang Technology, 901 Yingxiong Dadao, Economic and Technological Development Zone, Nanchang 330044, China
* Correspondence: m190501114@st.shou.edu.cn; Tel.: +86-021-61900856

Abstract: Due to the diversity and the different distribution conditions of agricultural products, split delivery plays an important role in the last mile distribution of agricultural products distribution. The time-dependent split delivery green vehicle routing problem with multiple time windows (TDSDGVRPMTW) is studied by considering both economic cost and customer satisfaction. A calculation method for road travel time across time periods was designed. A satisfaction measure function based on a time window and a measure function of the economic cost was employed by considering time-varying vehicle speeds, fuel consumption, carbon emissions and customers' time windows. The object of the TDSDGVRPMTW model is to minimize the sum of the economic cost and maximize average customer satisfaction. According to the characteristics of the model, a variable neighborhood search combined with a non-dominated sorting genetic algorithm II (VNS-NSGA-II) was designed. Finally, the experimental data show that the proposed approaches effectively reduce total distribution costs and promote energy conservation and customer satisfaction.

Keywords: vehicle routing problem; fresh agricultural products; split delivery; NSGA-II algorithm

Citation: Wu, D.; Wu, C. Research on the Time-Dependent Split Delivery Green Vehicle Routing Problem for Fresh Agricultural Products with Multiple Time Windows. *Agriculture* **2022**, *12*, 793. https://doi.org/10.3390/agriculture12060793

Academic Editor: Dimitre Dimitrov

Received: 19 April 2022
Accepted: 28 May 2022
Published: 30 May 2022

1. Introduction

Fresh e-commerce has become one of the main channels for fresh agricultural products, however, logistics distribution has always been the bottleneck of the development of fresh e-commerce. As a major consumer of fresh agricultural products, China has a huge and growing e-commerce business market in fresh agricultural products. According to the market report on e-commerce businesses of fresh agricultural products in China by iResearch, a consulting company, the trade volume of China's e-commerce market of fresh agricultural products will reach 458.5 billion yuan in 2020, an increase of 64% over 2019. The overall trade volume of the fresh food business in China is estimated to reach 11.1971 billion yuan by 2023 [1]. The e-commerce business of fresh agricultural products has developed rapidly, but it is difficult to achieve decent profits due to the low implementation rate of cold-chain facilities, redundant circulation links and considerable cargo damage. Moreover, the "last mile" distribution problem has become the most prominent one, impeding the growth of e-commerce businesses in fresh agricultural products. According to statistics [1], the cargo loss rate in the distribution of fresh agricultural products is around 10%, and the cost of the distribution accounts for around 40% of the overall transportation costs. Furthermore, due to delayed delivery and high cargo loss rate, the complaint rate also remains high, and 90% of the complaints are about poor delivery service. Therefore, management decisions in the distribution process have become the key to the success of delivery service provided by e-commerce platforms of fresh agricultural products, as well as a fascinating subject of academic research.

The e-commerce operation mode of fresh agricultural products allows customers to select the agricultural products they need and submit orders through online platforms. These companies dispatch vehicles to deliver goods for each customer along the pre-planned route from the distribution center after receiving orders from customers. Due to the perishability of fresh agricultural products, maintaining a vehicle's temperature and transport efficiency during delivery is crucial for successful delivery. Customers buying fresh agricultural products think highly of delivery efficiency, which means that the time it takes to deliver agricultural products to customers has a significant impact on customer satisfaction. For the distribution process of small batches and various categories, a variety of products can be delivered by a vehicle with multiple compartments of different temperature settings. According to Hsu and Chen's research [2], the distribution mode of multiple compartments is more appropriate to be applied to handle small but diverse customer demands for fresh agricultural products. Reed et al. [3] and Chen et al. [4] also studied the multi-compartment vehicle routing problem. In former models, the customer's demand is limited so that it does not exceed the capacity of a certain compartment. However, according to the real cases in many studies [5–10], it is possible for a customer's demand to exceed the capacity of a compartment. Therefore, this study focuses on exploring and determining whether the delivery with multi-compartment refrigerated vehicles serves the best option even when the customer's demand exceeds the capacity of a compartment. A model under the condition of split delivery is also developed to research the cold-chain distribution of agricultural products with large and diverse customer demands.

The split delivery vehicle routing problem (SDVRP), as a subtype of vehicle routing problems, has drawn considerable interest from researchers in recent years. The split delivery vehicle routing problem depicts a scenario in which many vehicles of the same weight depart from the same warehouse to deliver items to multiple customers, with each customer accepting the service of multiple vehicles [11]. The benefits of split delivery in vehicle routing problems have been demonstrated [12]. The SDVRP has been applied in many practical cases [13,14]. This paper will study the problem of single compartment vehicle routing optimization based on split delivery, with an aim to provide a more rational distribution strategy and route optimization method for fresh food e-commerce companies, and to reduce the economic costs while improving customer satisfaction. Considering the urban traffic conditions and the need to deliver various products, a split delivery vehicle routing problem model with time-varying road network characteristics is developed, and a multi-objective optimization algorithm is designed to obtain the optimal solution set and screen out the suitable solution from the solution set.

The characteristics of diverse demand for fresh agricultural product delivery explored in this paper are quite similar to the SDVRP model. As such, this paper will expand on earlier research by applying the SDVRP to the problem of fresh agricultural product delivery. Since customers will be serviced more than once after an order is split, it is more practical to have multiple time windows for each customer. To make the model better simulate the actual conditions of reality, the road network congestion constraints, carbon emission and fresh agricultural product loss are considered in the calculation of distribution costs. Therefore, this paper studies a time-dependent split delivery green vehicle routing problem with multiple time windows (TDSDGVRPMTW) based on the aforementioned criteria and takes the total cost and customer satisfaction as optimization objectives. The NSGA-II algorithm framework [15] is improved in this study by combining it with other heuristic algorithms [4,15] to improve the model in this paper. However, the algorithm can only find the Pareto-optimal front but cannot provide a specific solution. In reality, decision-makers must follow a single, well-defined solution when making decisions. In order to solve this issue, the technique for order preference by similarity to the ideal solution (TOPSIS) approach is used to screen the Pareto-optimal front for the best solution of clear distribution that fulfills the actual needs.

Based on the study of the traditional vehicle routing problem, the split delivery vehicle routing problem (SDVRP) was first proposed by Dror and Trudeau [12]. Generally, SDVRP research has been less comprehensive than studies on other VRPs. As with VRP, several SDVRP formulations have been presented while taking into account various restrictions [16–18]. SDVRP with time windows (SDVRPTW) [19–26] is one of those that has gained increasing attention. Notably, this study investigated SDVRPTW in the distribution of fresh agricultural products and presents a heuristic algorithm for solving it. The following parts are a literature review of literature relevant to this paper.

(1) Split delivery vehicle routing problem with multiple time windows

Split delivery vehicle routing problem with multiple time windows (SDVRPTW) is a variant of SDVRP that imposes restrictions on each customer by defining an interval at which a customer must begin to be served. Due to the added time window constraints, the model of SDVRPTW is more complex and more difficult to solve than SDVRP. There is not much former research on SDVRPTW. On the basis of SDVRP, Frizzell and Giffin [16] considered the time window constraint and first proposed SDVRPTW, a path construction method and two ways to improve the path. Ho and Haugland [19] proposed a tabu search-based solution to the SDVRPTW model and analyzed the experiment results from 100 customers. Desaulniers [20] proposed a new exact branch-and-price-and-cut method to solve the SDVRPTW, and the calculation results show that the method is effective. Based on the research of Desaulniers, Archetti et al. [21] proposed an improved strategy of the branch-and-price-and-cut algorithm, which further improved the performance of the algorithm to solve SDVRPTW. Salani and Vacca [22] presented a mixed-integer program for the SDVRPTW based on arc flow formulation, proposed a branch-and-price algorithm and applied an improved approach to the pricing and master problem. Based on a new relaxed compact model, Bianchessi and Irnich [23] proposed a new and tailored branch-and-cut algorithm to solve SDVRPTW. Computational experiments show that the new method can prove the optimality of several previously unsolved examples in the literature. Li et al. [26] assumed that service time is proportionate to demand, that customers have many time windows to choose from and can only be delivered in one of them. They came up with a three-indicator traffic flow model and a combined coverage model to solve this challenge. To overcome this problem, a branch-and-price-and-cut method is presented.

It is important to note, however, that the previous research on SDVRPTW focuses on the solution method of the abstract model and rarely involves the application of the model to specific fields. For example, there is no research that applies the SDVRPTW model to the distribution of fresh agricultural products. Customers often need a variety of fresh agricultural products that require different degrees of refrigeration. Thus, a split delivery of fresh agricultural products is reasonable in such a case. Fresh agricultural products are perishable and require cold-chain transportation, which will increase the complexity of the model and require more efficient solutions. As for the service time, most research adopted constant service time, and only Salani [22] and Li et al. [26] considered the variable service times that depend on the delivery quantity. Considering the service time depending on the delivery quantity represents a more pragmatic way of thinking. Previous research overlooks traffic congestion in road networks, which makes it difficult to apply the research results to reality. This study takes into account variable service time and adds time-varying road network constraints to the SDVRPTW model, so as to make the model fully represent reality.

The VNS-NSGA-II algorithm proposed in this paper belongs to the multi-point heuristic algorithm. Compared with many works of literature [20–24] that use precise algorithms, the method in this paper can solve large-scale problems. Compared with the research from McNabb et al. [25], the method in this study includes an adaptive genetic evolution process and is easier to jump out of the local optimum in the process of searching for a solution. Both the model of this paper and the model of Li et al. [26] set multiple time windows, however, their model only allows customers to select one window to receive products, and the model of this paper allows customers to receive products at each time window. Most research on SDVRP focused on the optimization of a single objective, and only a few of them focused on multi-objective optimization [27–29]. This paper focuses on the application of fresh agricultural products distribution, which requires a high level of delivery efficiency. In addition to transportation costs, customer satisfaction is also a critical aspect. Therefore, establishing a multi-objective optimization model with two distinct objectives of distribution cost and customer satisfaction as optimization objectives is a reasonable choice.

(2) The Distribution of Fresh Agricultural Products

According to previous research [8–10], customers' demand for the distribution of fresh agricultural products mostly occurs in morning rush hours, during which the speed of vehicles is very different from other time periods. Therefore, it is a reasonable choice to regard the vehicle speed as a time-dependent function. At the same time, the amount of carbon emissions produced would change in proportion to the speed of the vehicle [30], hence considering carbon emissions in the model can make the final delivery solution more environmentally friendly. Due to the requirement of high delivery efficiency in the distribution of fresh agricultural products, whether the goods can be delivered in time will have a huge impact on customer satisfaction, which is also an important indicator to measure the service level of enterprises. Therefore, we should take customer satisfaction as an optimization objective and strive to improve it. In order to make the model better, we simulated actual conditions of fresh agricultural product distribution.

(3) Time-Dependent Vehicle Routing Problem

The time-dependent vehicle routing problem is another derivative of VRP. To simulate road traffic congestion, Malandraki and Daskin [31] proposed TDVRP. They treat the vehicle travel time at any two points as a step function of the vehicle departure time. Malandraki and Dial [32] proposed a restricted dynamic programming heuristic for solving the time-dependent traveling salesman problem. However, in the above research, there are situations where vehicles that depart later arrive first, which do not meet the First-In-First-Out (FIFO) criterion. Ichoua et al. [33] proposed a new time-varying model in which the vehicle's travel speed is a step function, and the corresponding travel time is a piecewise linear function. Another time-varying model is proposed by Fleishmann et al. [34], which is based on a smooth step function of travel time. The above two methods confirm the principle of FIFO, which better simulates the actual conditions of the real world.

What is more, based on the research in the previous work and the innovations in this study are summarized as follows:

1. The SDVRPTW model has been applied rarely in the research of fresh agricultural product distribution. The basic SDVRPTW model normally takes into account load constraints, split delivery and the precondition that the consumer has just one time window. In order to make the model better simulate the actual conditions of fresh agricultural product distribution, this study will consider and evaluate a time-varying road network, carbon emissions and customer satisfaction on the basis of SDVRPTW to develop the TDSDGVRPMTW model.
2. While multi-objective optimization problems are frequently aimed at obtaining a Pareto-optimal front, decision-makers expect a complete and feasible solution. Furthermore, the TOPSIS method is employed to select the solution that satisfies the requirements from the Pareto-optimal front.
3. It is verified by a real-world case that as a delivery strategy, the TDSDGVRPMTW model proposed in this paper not only can effectively reduce the total cost of fresh agricultural products distribution, but also improve customer satisfaction.

The description of the problem and the hypotheses are as follows.

The TDSDVRPMTW proposed in this paper can be described as follows: a group of cold-chain vehicles depart from the distribution center, visit customer nodes in the order indicated in the distribution scheme, and then return to the distribution center. $K = \{1, 2, \cdots, k\}$ is the set of vehicles. $N = \{0, 1, 2, \cdots, n\}$ is the set of all nodes in the distribution network where $\{0\}$ represents the distribution center, $N' = N \backslash \{0\}$ denotes the customers. The distance traveled by vehicle k between node i and node j is D_{ij}, and the travel of the vehicle incurs a travel cost TC. Fixed cost FC is incurred when the vehicle is used. Service costs SC are incurred when the vehicle serves the customer. The speed of vehicles is related to time. Vehicles travel at varying speeds during various time periods. $H = \{1, 2, \cdots, h\}$ is the set of all time periods. Carbon dioxide CC is emitted into the atmosphere when a vehicle is driven. Fresh agricultural products are divided into several types based on their respective temperature requirements for transportation. $W = \{1, 2, \cdots, w\}$ is the set of all types of fresh agricultural products. The refrigeration cost RC of fresh product w is e_w. Vehicle k is only allowed to deliver one type of product w. Each customer requires a variety of fresh agricultural products, and each customer has a number of service acceptance time windows equal to the number of fresh product varieties required by the customer. Additionally, it is allowed that a single customer can be served by multiple vehicles. The service time s_{ik} is proportional to the quantity of products delivered. The customer satisfaction ACS_i of customer i is determined by the remaining delivery time after the customer received goods. Total costs include travel costs TC, fixed costs FC, service costs SC, refrigeration costs RC and carbon emissions costs CC. The purpose of this study is to produce a delivery solution that has a lower total cost and a higher level of customer satisfaction. Table 1 lists all notations used in the proposed model. In addition, the following basic assumptions are made in this paper:

1. The customer's demand for fresh agricultural products is split and distributed according to the temperature required for distribution, and each customer has multiple time windows for receiving services.
2. For each service, every customer will complete a satisfaction rating, the value of satisfaction evaluation depends on the deviation degree between the remaining time when the vehicle leaves the customer and the time window.
3. Only one type of produce can be delivered by a single vehicle. A vehicle can only serve a customer once.
4. The day is divided into several time periods, and vehicles travel at various speeds at different periods.

The rest part of this paper is organized as follows. In Section 2, the mathematical formulation of the time-dependent green vehicle routing problem with multiple time windows (TDSDGVRPMTW) is given. Then, the algorithm for obtaining the Pareto-optimal front based on variable neighborhood search combined with non-dominated sorting genetic algorithm II (VNS-NSGA-II) is presented. Besides, the technique for order preference by similarity to the ideal solution method (TOPSIS) is introduced, and the process of experiments is shown in the final part of the section. In Section 3, the results of the numerical experiments are analyzed and compared with the relevant literature, and also the reasons for the similarities and differences are discussed. Finally, conclusions and future research directions are given in Section 4.

Table 1. Symbol definitions in the TDSDGVRPMTW optimization.

Variable	Definition	Parameter	Definition	Set	Definition
TC	The cost of the vehicle's travel	n	The number of customers that need to be delivered	N	Set of all nodes in the distribution network, including the distribution center and all customer points
FC	The cost of vehicle's fixed use	φ	The vehicle's travel cost per unit distance	N'	Set of all customer points $N' = N \backslash \{0\}$
SC	The cost of customer service	δ	The fixed cost per vehicle	K	Set of all vehicles
RC	The cost of refrigeration	e_w	The cost per unit time of transportation temperature of fresh agricultural product w	H	Set of all time periods
CC	The cost of carbon emissions	Q_L	The maximum capacity of the vehicle	W	Set of all types of fresh agricultural products
ACS_i	The average level of satisfaction for customer i	Q_K	Maximum number of vehicles that can be used		
D_{ij}	Distance between nodes i and j in the logistics distribution network	Q_i	The total demand of customer i for various fresh agricultural products		
L_{ik}	The time when vehicle k completes its service and leaves customer i.	q_{iw}	The demand of customer $i$for fresh agricultural product w		
T_{ik}	The time when the vehicle k arrives at the customer i	ET_i	The earliest time for node i to receive service		
T'_{ik}	The time when the vehicle k starts to serve customer i and also when the product is received by customer i	LT_i	The latest time for node i to receive service		
s_{ik}	The time that vehicle k serves customer i				
t_{ijk}	The driving time of vehicle k on section (i, j)				
t^h_{ijk}	The driving time of vehicle k on road section (i, j) in time period h				
d^h_{ijk}	The distance of vehicle k on road section (i, j) in time period h				
st_i	The total number of services received by customer i.				
X_{ijk}	Equal to 1 if the vehicle k runs on the road section (i, j), and 0 otherwise				
x^h_{ijk}	Equal to 1 if the vehicle k runs on the road section (i, j) in time period h, and 0 otherwise				
y_{ik}	Equal to 1 if vehicle k visits client i, and 0 otherwise				
z_k	Equal to 1 if vehicle k is used, and 0 otherwise				

2. Materials and Methods

2.1. TDSDGVRPMTW Model

Based on the needs of building the model, this paper uses the corresponding symbols which are listed in Table 1.

Through the comprehensive analysis above, the multi-objective optimization model of TDSDVRPMTW is given by the following, this model objectives include minimize total cost F_1 and maxmize average customer satisfaction F_2.

$$\text{Minimize } F_1 = TC + FC + SC + RC + CC \tag{1}$$

$$\text{Maximize } F_2 = \frac{1}{n}\left[\sum_{i \in N} ACS_i(L_{ik})\right] \tag{2}$$

Subject to:

$$\sum_{i \in N'} q_{iw} y_{ik} \le Q_L, \forall k \in K, w \in W \tag{3}$$

$$Q_i = \sum_{w \in W} q_{iw}, \forall i \in N' \tag{4}$$

$$\sum_{j \in N'} X_{0jk} \le 1, \forall k \in K \tag{5}$$

$$1 \le \sum_{k \in K} y_{ik} \le Q_K, \forall i \in N' \tag{6}$$

$$\sum_{i \in N} X_{ijk} = \sum_{j \in N} X_{ijk}, \forall k \in K \tag{7}$$

$$t_{ijk} = \sum_{h \in H} X_{ijk} t^h_{ijk}, \forall i \in N, j \in N, k \in K \tag{8}$$

$$D_{ij} = \sum_{h \in H} d^h_{ijk} X_{ijk}, \forall i \in N, j \in N, k \in K \tag{9}$$

$$ET_0 \le X_{i0k}\left(T'_{ik} + s_{ik} + t_{i0k}\right) \le LT_0, \forall i \in N', k \in K \tag{10}$$

$$T'_{ik} \ge T_{ik}, \forall i \in N', k \in K \tag{11}$$

$$L_{ik} = T'_{ik} + s_{ik}, \forall i \in N, j \in N, k \in K \tag{12}$$

$$t_{0i} = T'_{ik}, \forall i \in N', k \in K \tag{13}$$

$$\sum_{k \in K} y_{ik} = st_i, \forall i \in N' \tag{14}$$

$$x^h_{ijk} \in \{0,1\}, X_{ijk} \in \{0,1\}, y_{ik} \in \{0,1\}, z_k \in \{0,1\} \tag{15}$$

Equation (1) is a function to minimize the total cost. It includes the vehicle travel cost (*TC*), fixed cost (*FC*), customer service cost (*SC*), refrigeration cost (*RC*) and carbon emission cost (*CC*). Equation (2) measures the average customer satisfaction (*ACS*). *ACS* is the proportion of fully satisfied customers to the total number of customers. Constraint (3) expresses the carrying weight limit of each vehicle. The combined demand of all customers on each vehicle's delivery route cannot exceed the maximum vehicle load limit. Constraint (4) ensures that the demand of each customer is equal to the sum of the customer's demand for each type of fresh agricultural product. Constraint (5) ensures that each customer can only be served once by each vehicle. Constraint (6) ensures that each customer must be served and the number of times served does not exceed the total number of vehicles. Constraint (7) ensures that the number of vehicles starting and arriving at each node should be balanced. That is to say, when a vehicle arrives at a customer, it will inevitably leave the customer. Constraint (8) expresses the relationship between road section driving time and vehicle driving time within a period. In other words, the total time spent on a road section is equal to the sum of the time spent on that road section in each time period.

Constraint (9) expresses the relationship between road section distance and vehicle driving distance in each time period. That is to say, the distance traveled by the vehicle on the road segment is equal to the sum of the distance traveled by the vehicle on the road section in each time period. Constraint (10) indicates that the vehicles returning to the distribution center should be restricted to the constraint of the working time window of the distribution center. Constraint (11) indicates the relationship between service start time and arrival time. Constraint (12) indicates the relationship between service start time, service duration and departure time. Constraint (13) ensures the time the customer receives the fresh products is equal to the time the service starts. Constraint (14) indicates that the number of times service received by each customer is equal to the total number of vehicles serving this customer. Constraint (15) states the binary decision variable.

- The travel cost

The travel cost (*TC*) refers to the variable cost incurred by each vehicle for delivery activities, mainly consisting of fuel consumption, repairs and driver pay rate. For simplicity, only the effect of distance on travel cost is considered. The cost of the vehicle's travel is generally proportional to the distance and can be calculated as follows:

$$TC = \sum_{i \in N} \sum_{j \in N} \sum_{k \in K} X_{ijk} D_{ij} \varphi \tag{16}$$

where X_{ijk} is a binary variable, $X_{ijk} = 1$ when vehicle k runs through the road section (i, j), otherwise $X_{ijk} = 0$. D_{ij} is the distance between node i and j, φ is the vehicle's travel cost per unit distance.

- The fixed cost

Vehicles' fixed costs are typically constant. It is unrelated to customer demand or delivery distance. It mostly consists of rent or use loss, and other labor costs. Calculate the cost of the vehicle's fixed (*FC*) use by:

$$FC = \sum_{k \in K} z_k \delta \tag{17}$$

where z_k is a binary variable, when vehicle k is used, $z_k = 1$; otherwise, $z_k = 0$. δ is the fixed cost per vehicle.

- The service cost

Since each customer can be serviced by multiple vehicles, the cost impact of the number of times each customer is serviced needs to be considered. The service cost (*SC*) is proportional to the frequency with which each customer is served. The service cost can be calculated as follows:

$$SC = \sum_{i \in N'} st_i \kappa \tag{18}$$

where st_i is the number of times the consumer i receives services and κ is the cost of each service.

- The refrigeration cost

Refrigeration cost (*RC*) refers to the energy cost of refrigeration. According to the study of Wang et al. [35], this paper treats refrigeration costs as a time-dependent function, assigns a fixed unit time cost to each temperature, and considers the refrigeration cost of the unloading service. The cost of refrigeration can be calculated as follows:

$$RC = \sum_{i \in N} \sum_{j \in N} \sum_{w \in W} \sum_{k \in K} X_{ijk} e_w \left(t_{ijk} + s_{ik} \right) \tag{19}$$

where X_{ijk} is a binary variable, $X_{ijk} = 1$ when vehicle k runs through the road section (i, j), otherwise $X_{ijk} = 0$. e_w is the refrigeration cost per unit time of fresh agricultural product

w. t_{ijk} is the driving time of vehicle k on section (i,j). s_{ik} is the time that vehicle k serves customer i.

- The carbon cost

This paper uses the MEET model in reference [30] to explore the calculation of carbon emissions. The carbon emission estimation function for freight vehicles weighing between 3.5 and 40 tonnes is $\varepsilon^h(v) = a_0 + a_1 v + a_2 v^2 + a_3 v^3 + \frac{a_4}{v} + \frac{a_5}{v^2} + \frac{a_6}{v^3}$, while the load correction function is $\lambda^h(v,\varphi) = b_0 + b_1\varphi + b_2\varphi^2 + b_3\varphi^3 + b_4 v + b_5 v^2 + b_6 v^3 + \frac{b_7}{v}$. φ is the ratio of the vehicle's actual load to its capacity, v is the vehicle speed (km/h), and parameters a and b are the load correction coefficient, whose value is related to the vehicle's weight range. If vehicle k's driving distance in time period h is $d^h{}_{ijk}$ (km), the carbon emissions $E^h{}_{ijk}$ (kg) can be computed as follows:

$$E^h_{ijk} = \varepsilon^h \cdot \lambda^h \cdot \frac{d^h_{ijk}}{1000} \tag{20}$$

The cost of carbon emissions (CC) is mainly generated by the energy consumed by vehicles while on the road. The carbon emission calculation function adopted in this paper is related to the vehicle speed, travel time and load capacity. The cost of carbon emissions can be calculated as follows:

$$CC = \sum_{h\in H}\sum_{i\in N}\sum_{j\in N}\sum_{k\in K} x^h_{ijk} E^h_{ijk} \mu \tag{21}$$

where x^h_{ijk} is a binary variable, $x^h_{ijk} = 1$ when the vehicle k runs on the road section (i,j) in time period h, otherwise $x^h_{ijk} = 0$. $E^h{}_{ijk}$ is the carbon emission generated by vehicle k driving on the road section (i,j) in time period h. μ is the unit price of carbon emissions.

2.2. Customer Satisfaction Measurement Method

Customer satisfaction is an important indicator used to gauge the service quality of firms in the study on the vehicle routing problem for fresh agricultural products. The measurement methods of customer satisfaction can be divided into three categories: measure satisfaction by the freshness of products, by product delivery time, and by both freshness and delivery time. Wang et al. [36] used the freshness of perishable products to measure customer satisfaction, and proposed a multi-objective VRP optimization model with mixed time windows and perishability assessment to minimize transportation costs and maximize the freshness of perishable products. Wang et al. [37] established a customer satisfaction evaluation model, in which both the timeliness of the distribution of fresh agricultural products and the loss of freshness of agricultural products are considered. We adopts the satisfaction function based on the time window. Because this paper considers that each customer has many time windows and may receive vehicle services throughout each window, the satisfaction of each customer is averaged by the number of vehicle visits. Customer satisfaction is determined by the time the vehicle completes its servicing. The satisfaction function curve for customer i in the gth time window is depicted in Figure 1. In Figure 1, the horizontal coordinate represents time, and the vertical coordinate represents customer satisfaction. When the vehicle completes the service in $[ET_{ig}, LT_{ig}]$, and customer satisfaction is 100, this is the best time for service. The service can begin when the vehicle arrives at the customer within $[EET_{ig}, ET_{ig}]$ or $[LT_{ig}, ELT_{ig}]$. On the other hand, customer satisfaction is low when the service is completed within this time limit, and satisfaction decreases as the deviation from the optimal service time period increases. If the vehicle provides service to the customer prior to EET_{ig} or after ELT_{ig}, the customer's satisfaction level is 0. $EET_{ig} = ET_{ig} - \theta s_i$, $ELT_{ig} = LT_{ig} + \theta s_i$ are the boundaries of the customer's tolerable time points. Where θ is the customer tolerance coefficient, s_i is the customer i's service time, and L_{ik} is the time when vehicle k completes its service and departs for customer i. G denotes the set of all of customer i's time windows, and st_i denotes the number of times customer i gets served. Then, the average level of satisfaction for each customer (ACS) is as follows:

$$ACS_i(L_{ik}) = \sum_{g\in G} \frac{CS_{ig}(L_{ik})}{st_i} \tag{22}$$

where the satisfaction function is:

$$CS_{ig}(L_{ik}) = \begin{cases} 0, & L_{ik} \le EET_{ig} \\ \left(\frac{L_{ik}-EET_{ig}}{ET_{ig}-EET_{ig}}\right) \times 100, & EET_{ig} < L_{ik} \le ET_{ig} \\ 100, & ET_{ig} < L_{ik} \le LT_{ig} \\ \left(\frac{ELT_{ig}-L_{ik}}{ELT_{ig}-LT_{ig}}\right) \times 100, & LT_{ig} < L_{ik} \le ELT_{ig} \\ 0 & ELT_{ig} < L_{ik} \end{cases} \tag{23}$$

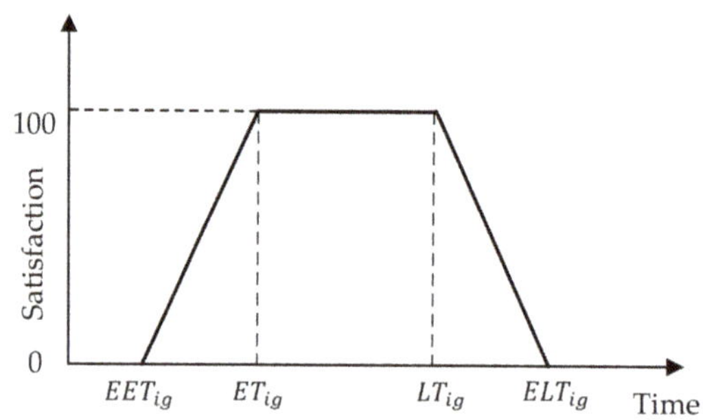

Figure 1. Customer satisfaction curve.

This paper adopts average customer satisfaction to measure the overall satisfaction level of the distribution schemes. The average customer satisfaction function can be expressed as Equation (2).

2.3. *Time-Dependent Vehicle Speed Calculation Method*

Based on existing approaches [38,39], this paper proposed a method of time division for calculating travel time. The distribution center's opening hours are divided into many time periods, and the vehicle speed varies according to the time period. Let F be the length of the time period; $H = \{0, 1, 2, \cdots, h\}$ is a set of all time periods, $[h, h+1]$ said the number of h time period. d_{ijk}^h, t_{ijk}^h, g_{ijk}^h denote the distance, time and speed of vehicle k on the road section (i, j) in time period h, respectively, and D_{ij} denotes the road section's distance (i, j). D_{ij}^h is the distance traveled by vehicle k to complete the remaining distance (i, j) after time h; L_{ik} is the time when vehicle k departs from customer i; and h_k is the remaining drivable time of vehicle k in time period h. As a result, the following steps are used to calculate the driving time t_{ijk} of vehicle k on road section (i, j):

Step 1: Determine how long the initial phase will last. $d_{ijk}^h = g_{ijk}^h h_k$. If $d_{ijk}^h \ge D_{ij}$, then $t_{ijk}^h = \frac{D_{ij}}{g_{ijk}^h}$, $t_{ijk} = t_{ijk}^h$, end of calculation; if $d_{ijk}^h < D_{ij}$, $D_{ij}^h = D_{ij} - d_{ijk}^h$, $t_{ijk}^h = h_k$, continue to step 2.

Step 2: $\zeta = 1$; $d_{ijk}^{h+\zeta} = g_{ijk}^{h+\zeta} F$, If $d_{ijk}^{h+\zeta} < D_{ij}^{h+\zeta-1}$, then $t_{ijk}^{h+\zeta} = F$, $D_{ij}^{h+\zeta} = D_{ij}^{h+\zeta-1} - d_{ijk}^{h+\zeta}$, Step 2 should be repeated; otherwise, $t_{ijk}^{h+\zeta} = \frac{D_{ij}^{h+\zeta-1}}{g_{ijk}^{h+\zeta}}$, $t_{ijk} = \sum_{h\in H} t_{ijk}^h$. The section (i, j) driving time computation is finished.

2.4. VNS-NSGA-II Algorithm

A variable neighborhood search combined with the non-dominated sorting genetic algorithm II (VNS-NSGA-II) was designed. NSGA-II has been extensively applied in research on VRP-related problems as a multi-objective combinatorial optimization algorithm [38–41]. The NSGA-II algorithm decreases the complexity of the non-dominated sorting genetic algorithm and has the advantages of rapid execution and good solution set convergence. A number of heuristic strategies [4,15] are introduced into the NSGA-II algorithm in this study to boost search efficiency and avoid local optimum. Adaptive functions [42] are introduced to the crossover and mutation processes to dynamically alter the likelihood of crossover and mutation, and the variable neighborhood search algorithm (VNS) [4] is added to conduct a variable neighborhood search for good individuals in the population. It has the potential to improve the algorithm's local search capacity. Based on the advantages of the preceding techniques, the VNS-NSGA-II algorithm is constructed to solve the TDSDGVRPMTW model in this study. The VNS algorithm is divided into three stages: initialization, genetic evolution and variable neighborhood search. The initialization step generates the initial population in a random manner. In the process of genetic evolution, there exist crossover and mutation operators whose likelihood of execution is dynamically governed by adaptive functions. In the process of the variable neighborhood search, there are three types of neighborhood search operators. Non-dominated sorting and crowding distance calculation are at the end of the variable neighborhood search process. Figure 2 shows the detailed process of the algorithm.

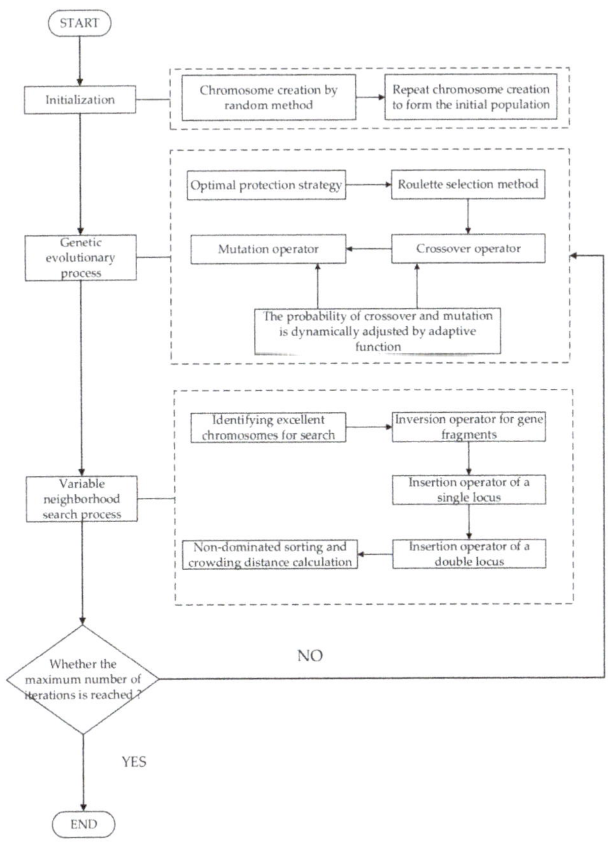

Figure 2. Algorithm flow chart.

2.4.1. Population Initialization

The chromosome in this study uses mixed natural number and letter coding, with natural numbers ranging from 1 to n representing the customer node and capital letters A, B, C, and D representing various types of fresh agricultural products. Each vehicle delivers the same type of fresh agricultural product, and each chromosome holds the delivery schedules of all vehicles and therefore contains each customer's fresh produce needs. A chromosome is depicted in Figure 3 as an example. Where "$A1$" means that customer 1 needs type A fresh agricultural products. The chromosome can be divided into three segments according to the type of fresh products. Each segment is delivered by a different vehicle. When decoding chromosomes into vehicle routes, the chromosomes are first divided into segments based on the type of fresh agricultural product, and then the chromosome segments with the same type of fresh agricultural product are divided into route segments based on the vehicle load capacity and the distribution center's operating time. This study employs a random method to generate the initial population. In other words, the customer numbers are randomly arranged according to the type of fresh agricultural products that suit customers' demands.

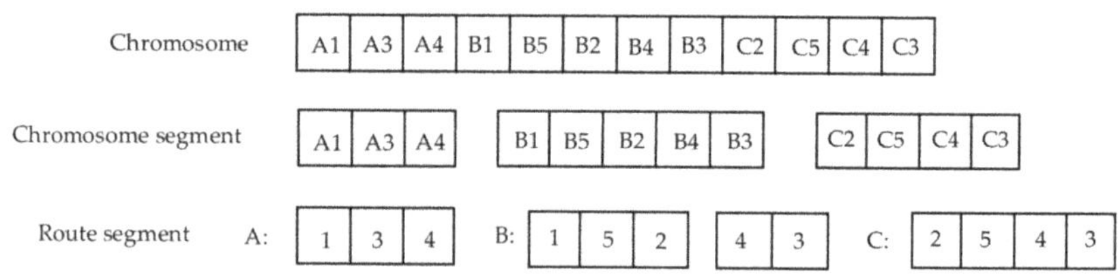

Figure 3. One example of a chromosome.

2.4.2. Genetic Operator

- Selection Operator

This paper combines the optimal protection strategy and roulette selection method to select chromosomes. The specific steps of the optimal protection strategy are to find the two chromosomes with the highest fitness and the lowest fitness in the current population; the fitness value of the chromosome with the highest fitness is compared with the highest fitness value of each generation in history. If the current value is higher, it will be regarded as the chromosome with the best protection; otherwise, the best protection object remains unchanged and remains the best chromosome in history. The chromosome with the worst fitness in the current population is replaced with the one with the best protection. The probability of each chromosome being selected in the roulette method is $p_n = f_n / \left(\sum_n f_n \right)$, and the greater the fitness of the chromosome, the greater the probability of being selected for the cross-mutation operation.

- Crossover Operator

Because chromosomes contain vehicle routes for delivering various types of agricultural products, all vehicle route segments in chromosomes are classified prior to the crossover to ensure that the crossover occurs only between vehicle route segments of the same type. By randomly generating two crossover points on parent chromosomes X and Y and separating the two parent chromosomes into three pieces, we improved the crossover approach in this paper. The center sections of chromosomes X and Y were excised and inserted into the front and back sections of offspring chromosomes Y_1 and X_1, respectively. The remaining front and back sections of parent chromosomes X and Y were placed in the same order into chromosome Y_1's back section and chromosome X_1's front section. In the two offspring chromosomes, leave the chromosome segment in the two crossover sections alone and delete the duplicated chromosome segment in the remaining places. The advantage of this crossover approach is that two identical parent chromosomes produce two different offspring. Figure 4a depicts the unique crossover procedure. In order to make the crossover operator work more efficiently, this study uses the improved adaptive adjustment approach presented in reference [42], which takes into account the number of iterations, the fitness values of chromosomes and populations, and the number of unmodified chromosomes in each generation population, as indicated in Equation (24).

$$P_c = \begin{cases} P_{c1} - \dfrac{(P_{c1}-P_{c2})(f_l - f_{avg})}{(f_{\max}-f_{avg})\left[1+\exp\left(\frac{-gen\times U}{M\times popsize}\right)\right]}, & f_l \geq f_{avg} \\ {P_{c1}}^{[1+\exp(\frac{-gen\times U}{M\times S})]}, & f_l < f_{avg} \end{cases} \tag{24}$$

where p_c stands for adaptive crossover probability, p_{c1} and p_{c2} stand for adaptive adjustment parameters, and $p_{c1} > p_{c2}$. f_l stands for the fitness value of individuals with high fitness in the chromosomes to be crossed; f_{avg} stands for the average fitness value of each generation population, and f_{max} stands for the maximum fitness value of each generation population. The *popsize* denotes the population size, whereas *gen* represents the current iteration number, M represents the maximum iteration number, and U represents the number of individuals with unchanged chromosomes.

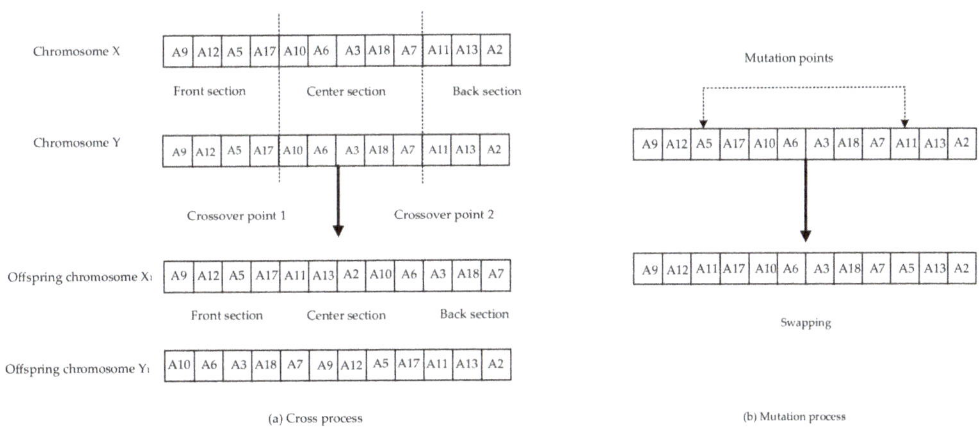

Figure 4. Examples of crossover and mutation.

- Mutation Operator

Before mutation, route segments in chromosomes should also be classified to ensure that mutation occurs between vehicle routes of the same type. The mutation method of the random chromosome point exchange is used in this study. The specific processes are as follows: initially, the chromosomes to be mutated are chosen, and then two random mutation points on the chromosomes are chosen in a random manner. Swap two points to form a new chromosome. Figure 4b depicts the mutation process. Similar to the crossover process, this study adopts an improved adaptive adjustment function to determine the probability of the mutation operator's operation and to improve the efficiency of the mutation operator [42]. Equation (25) is the adaptive function of mutation probability:

$$P_m = \begin{cases} P_{m1} - \dfrac{(P_{m1}-P_{m2})(f_{\max}-f)}{(f_{\max}-f_{avg})\left[1+\exp\left(\frac{gen\times U}{M\times popsize}\right)\right]}, & f \geq f_{avg} \\ {P_{m1}}^{[1+\exp(\frac{gen\times U}{M\times S})]}, & f < f_{avg} \end{cases} \tag{25}$$

where p_m represents the adaptive mutation probability, p_{m1} and p_{m2} are adaptive adjustment parameters and $p_{m1} > p_{m2}$, f is the fitness value of chromosomes to be mutated.

2.4.3. Variable Neighborhood Search Operators

In order to further improve the quality of chromosomes in the population, we added a neighborhood search operator to the algorithm to perform a deep search for some excellent solutions. The excellent chromosomes in the population are the operation object of variable neighborhood search, and the chromosomes in the population are sorted according to fitness from high to low, with the chromosomes in the top half as the excellent chromosomes. In each neighborhood search operator, a node is chosen in a random manner, the distance between it and all other nodes is calculated, and the remaining nodes are organized in ascending order to generate a list of distance values. The variable neighborhood search operator used in this paper is similar to the neighborhood structure proposed by Sánchez et al. [43]. The following are the variable neighborhood search operators proposed in this paper.

- 2-opt operator

Node *i* is randomly selected from the chromosome segment. Select the first node from the list of distance values for node *i* as node *j*. If the first node in the distance value list does not exist in the current chromosome segment, select the next node in turn. The 2-opt operator disconnects node *i* from the node behind it and node *j* from the node behind it and reconnects node *i* with node *j*. This operation will be kept if the fitness improves, otherwise, the next node in the list of distance values is tried, and the process repeats until a better chromosome is found or the maximum number of searches is reached. The process of the 2-opt operator is shown in Figure 5a.

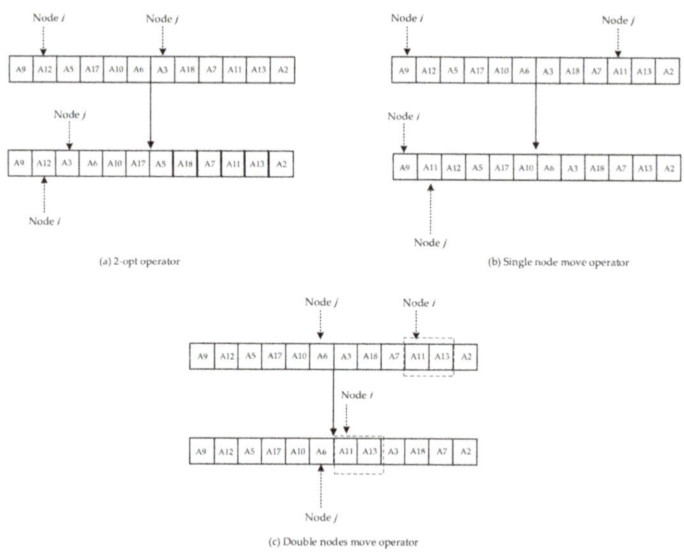

Figure 5. Examples of variable neighborhood search.

- Single node move operator

Node *i* is randomly selected from the chromosome segment. Select the first node in the list of distance values for node *i* as node *j*. If the first node does not exist in the current chromosome segment, the next node in the list is selected as node *j*. The single node move operator removes node *j* from its original position and inserts it behind node *i*, making node *i* adjacent to node *j*. After insertion, the chromosomal fitness value is calculated. The operation is kept if the fitness was increased, otherwise, the next node in the list of distance values is tried, and the process repeats until a better chromosome is found or the maximum number of searches is reached. The process of the single node move operator is shown in Figure 5b.

- Double nodes move operator

Two adjacent chromosome nodes are randomly selected. Node i is the first of the two nodes. Select the first node in the list of distance values for node i as node j. If the first node does not exist in the current chromosome segment, the next node in the list is selected as node j. Insert two adjacent nodes containing node i after node j so that node i and node j are adjacent. The operation was kept if the fitness was increased, otherwise, the next node in the list of distance values is tried, and the process repeats until a better chromosome is found or the maximum number of searches is reached. The process of the double nodes move operator is shown in Figure 5c.

2.4.4. Non-Dominated Sorting and Crowding Distance Calculation

The objective function values of all chromosomes in the population are calculated, and the crowding distance is calculated. Finally, the Pareto-optimal front is obtained. Let F_m^{max} and F_m^{min} be the maximum and minimum values of the mth objective function, respectively, while F_m^{i-1} and F_m^{i+1} are the mth objective function values of the two solutions adjacent to the ith solution. Then, the crowding distance CD_i of the ith solution can be calculated by Equation (26).

$$CD_i = \sum_{m=1}^{M} \left(\frac{F_m^{i+1} - F_m^{i-1}}{F_m^{\max} - F_m^{\min}} \right) \tag{26}$$

The new population is selected according to the Pareto-optimal front and crowding distance. Then, it is judged whether the maximum number of iterations is reached. If the maximum number of iterations is not reached, return to the genetic evolution process and continue to iterate. When the maximum number of iterations is reached, the algorithm is terminated and the Pareto-optimal front is printed. The Pareto-optimal front can be regarded and thought of as a solution set, including numerous potential delivery schemes. To determine which of these delivery schemes best fits the needs, the TOPSIS approach is employed in this study.

2.5. Select the Optimal Solution Strategy

When the Pareto-optimal front is created using the VNS-NSGA-II algorithm, we used the TOPSIS method to analyze all solutions and select the optimal strategy. The following are the specific steps:

Step 1: Complete the index homogenization process. The objective function F_2 is adjusted to $F_2' = 100 - F_2$ in this study so that it can be minimized alongside F_1.

Step 2: Construct the original data matrix. Assume that there are n solutions and m objective functions in the solution set, then matrix B is shown in Equation (27).

$$B = \begin{Bmatrix} b_{11} & b_{12} & \cdots & b_{1m} \\ b_{21} & b_{22} & \cdots & b_{2m} \\ \vdots & & & \vdots \\ b_{n1} & b_{n2} & \cdots & b_{nm} \end{Bmatrix} \tag{27}$$

Step 3: Vector normalization of indicators:

$$z_{ij} = \frac{b_{ij}}{\sqrt{\sum_i^n b_{ij}^2}} \tag{28}$$

Obtain the normalized matrix Z.

$$Z = \begin{Bmatrix} z_{11} & z_{12} & \cdots & z_{1m} \\ z_{21} & z_{22} & \cdots & z_{2m} \\ \vdots & & & \vdots \\ z_{n1} & z_{n2} & \cdots & z_{nm} \end{Bmatrix} \tag{29}$$

Step 4: Determine the optimal scheme Z^+ and the worst scheme Z^- for each indicator:

$$Z^+ = \left(Z_1^+, Z_2^+, \cdots, Z_m^+\right) \tag{30}$$

$$Z^- = \left(Z_1^-, Z_2^-, \cdots Z_m^-\right) \tag{31}$$

Step 5: Calculate the proximity of each solution to the optimal scheme and the worst scheme:

$$S_i^+ = \sqrt{\sum_{j=1}^{m} \iota_j \left(Z_j^+ - z_{ij}\right)^2} \tag{32}$$

$$S_i^- = \sqrt{\sum_{j=1}^{m} \iota_j \left(Z_j^- - z_{ij}\right)^2} \tag{33}$$

where ι_j is the weight of the jth indicator, which is determined based on actual need.

Step 6: Calculate the closeness of each solution to the optimal solution.

$$C_i = \frac{S_i^-}{S_i^+ + S_i^-} \tag{34}$$

where $0 \leq C_i \leq 1$, the closer C_i is to 0, the better the evaluated solution. After selecting the solution that best fits the requirements using the TOPSIS technique, the decision-making process is complete.

2.6. Validation of the Simulation Model

In order to verify the validity of the method proposed in this study, we conducted three experiments, respectively is algorithm comparison experiment, solution selection experiment and real case experiment. The algorithm comparison experiment and solution selection experiment used R201 dataset from Solomon [44] benchmark. The customer demands in the dataset are randomly divided into several parts, whose numbers are limited up to four. The time window is also randomly split into several periods. Python 3.8 programming is used to carry out the experiment. According to the work of Fan et al. [42], the algorithm's parameter settings are connected to the size of the dataset utilized in the experiment as follows: $p_{c1} = 0.7$, $p_{c2} = 0.5$, $p_{m1} = 0.01$, $p_{m2} = 0.008$, $maxit = 100 \sim 300$ iterations, population size $popsize = 100 \sim 200$, and maximum field search times $S_t = 15 \sim 30$. The related parameters for calculating the vehicle travel time and carbon emissions are identical to those in the work of Liu et al. [45], which are shown as follows: the distribution center's time 0 is 7:00 a.m., the traffic congestion periods are 8:00~9:00, 18:00~19:00, and the vehicle speed is 20 km per hour. For the time period h, according to the remainder function $\eta = h \bmod 3$, η is $(1, 2, 0)$ corresponding to $(54, 72, 42)$ km/h, respectively, with three time-varying velocities. The correlation coefficients for the carbon emission model are as follows: $a_0 = 110$, $a_1 = 0, a_2 = 0, a_3 = 0.000375, a_4 = 8702, a_5 = 0, a_6 = 0, b_0 = 1.27, b_1 = 0.0614, b_2 = 0$, $b_3 = -0.0011, b_4 = -0.00235, b_5 = 0, b_6 = 0, b_7 = -1.33$. The carbon emission price $\mu = 0.0528$ yuan per kilogram, the time window correlation tolerance coefficient is $\theta = 0.5$, the driving cost per kilometer is $\varphi = 1.3$, the fixed usage fee is set at $\delta = 20$, and the single service cost is set at $\kappa = 3$. All the experiments were carried out ten times, with the best result being chosen. The results of all experiments will be analyzed in detail in the Section 3.

3. Results and Discussion

3.1. Comparison with Other Efficient Algorithms

This paper compares the VNS-NSGA-II algorithm with other algorithms designed to solve multi-objective VRP problems. The VNS-NSGA-II algorithm is also compared to other algorithms designed to solve multi-objective VRP problems. As the VNS-NSGA-II algorithm is improved on the basis of the NSGA-II algorithm, the NSGA-II algorithm [38] is selected for comparison in order to verify the improvement. Three neighborhood search operators are added to the VNS-NSGA-II algorithm, while the many-objective gradient evolution (MOGE) algorithm [46] also has three search operators. In order to test the search ability of the variable neighborhood operators, the MOGE algorithm is selected for comparison.

Both the MOGE algorithm and NSGA-II algorithm are designed to solve multi-objective VRP. These two algorithms can be applied to the model in this paper. There are some differences between the two algorithms. The MOGE algorithm is designed on the basis of the gradient approximation, while the NSGA-II algorithm is designed based on the laws of biological evolution. The MOGE algorithm uses three operators to explore search space, improve the quality of the solution, avoid local optima, and promote population diversity. The NSGA-II algorithm uses two operators to explore search space, which are used for global search and local search, respectively.

This experiment used datasets of various sizes to verify the efficiency of the VNS-NSGA-II algorithm provided in this paper. The Pareto-optimal front and convergence of the VNS-NSGA-II, NSGA-II [38] and MOGE algorithms [46] were compared. For the experiment, 30, 50, 70, and 100 customers were chosen from the datasets. According to Wang et al. [35], the fresh agricultural products in this experiment are classified into four types based on their required temperatures. For simplicity, we use *A*, *B*, *C* and *D* for different types of products. For each type of agricultural product, a time-sensitive regulatory factor *r* was assigned [26]. Table 2 shows the features of each scale dataset, including the number of customers, the number of vehicles, the total demand, the type of agricultural products, the value of time-sensitive adjustment factors, and the unit price of agricultural products. Figure 6 shows the Pareto-optimal front of each algorithm obtained in the experiment. Table 3 shows the optimal value of each objective function in the Pareto-optimal front of each algorithm. It should be noticed, however, that TTC in Table 3 represents the total cost. The optimal total cost and satisfaction values, as well as the number of iterations, is shown in Figure 7. Table 3 and Figure 7 show the total costs and satisfaction for two different solutions. Next, we will analyze these figures and tables in detail.

Table 2. Characteristics of datasets of different sizes.

Number of Customers	Number of Vehicles	Total Demand	Type of Agricultural Products	Unit Price of Agricultural Products
30	8	520	*A*	10
			B	12
50	15	860	*A*	10
			B	12
70	20	1210	*A*	10
			B	12
			C	15
			D	20
100	25	1810	*A*	10
			B	12
			C	15
			D	20

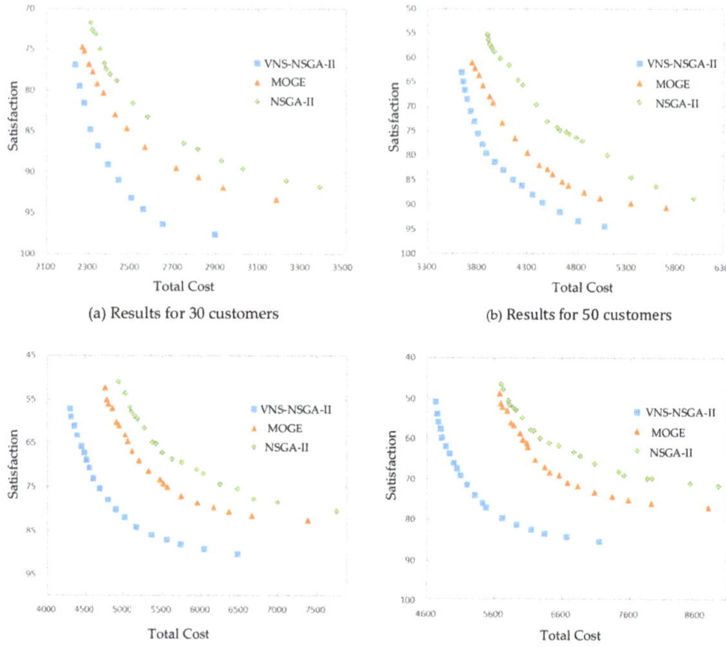

(a) Results for 30 customers

(b) Results for 50 customers

(c) Results for 70 customers

(d) Results for 100 customers

Figure 6. Pareto-optimal front comparison of different customer sizes.

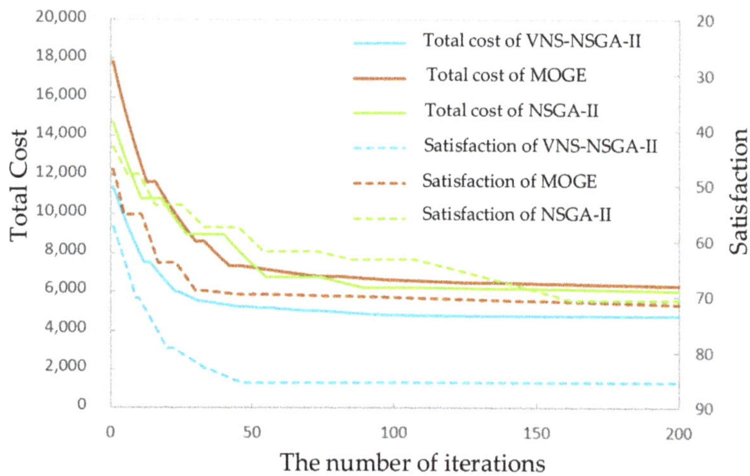

Figure 7. Comparison of convergence of algorithms.

Table 3. Experimental results of algorithm comparison.

Number of Customers	Total Demand	VNS-NSGA-II		NSGA-II		MOGE		Compared with NSGA-II		Compared with MOGE	
		TTC	Satisfaction	TTC	Satisfaction	TTC	Satisfaction	TTC Reduction %	Satisfaction Increases	TTC Reduction %	Satisfaction Increases
30	520	2212.50	96.64	2301.80	90.36	2260.71	91.86	3.88%	6.28	2.13%	4.79
50	860	3608.27	93.44	3844.34	87.57	3681.60	88.77	6.14%	5.87	1.99%	4.66
70	1210	4246.75	88.44	4865.36	79.52	4687.99	80.96	12.71%	8.92	9.41%	7.49
100	1810	4689.05	84.51	5674.60	70.48	5642.77	75.51	17.37%	14.03	16.90%	8.99
Mean	1100	3689.14	90.76	4171.52	81.98	4068.27	84.27	10.03%	8.77	7.61%	6.48

(1) Comparison of Pareto-optimal fronts

Figure 6 shows the Pareto-optimal front results obtained by three algorithms under four different data sizes. Each chart's horizontal coordinate shows the total cost, while the vertical coordinate measures customer satisfaction. Each point in the graph corresponds to a solution in the Pareto-optimum front determined by the corresponding algorithm. As illustrated in Figure 6, the solutions provided by the VNS-NSGA-II algorithm described in this paper dominate all other algorithms' solutions at various customer scales. The VNS-NSGA-II algorithm offers a more uniform solution distribution and fewer concentrated solutions, which means that the Pareto-optimal solution set obtained by VNS-NSGA-II is of better quality.

Table 3 shows the results of algorithm comparison experiments under four different data scales. The optimal value of a single optimization objective is found from the Pareto-optimal front obtained by each algorithm for comparison. Note that the total cost and satisfaction in the same algorithm with the same data size may belong to two different solutions. As shown in Table 3, the VNS-NSGA-II algorithm developed in this paper achieved the optimal results for both optimization objectives across all scales of datasets. The total cost is lowered by 10.03% when compared to the NSGA-II algorithm, and satisfaction is increased by 8.77 points. The total cost is decreased by 7.61% when compared to the MOGE algorithm, and satisfaction is increased by 6.48 points. In terms of the total cost, the VNS-NSGA-II algorithm outperforms the other two algorithms significantly. The greater the scale of the data, the more saved cost would be. When the customer size is 100, the VNS-NSGA-II algorithm can achieve an improvement of double digits in result values of total cost and satisfaction compared with the other two algorithms. The level of satisfaction optimization is also proportional to the customer scale. The above-mentioned results reflect that VNS-NSGA-II is superior and more appropriate for solving large-scale problems. The VNS-NSGA-II algorithm can produce better solutions compared with the NSGA-II algorithm and the MOGE algorithm The excellent performance of the VNS-NSGA-II algorithm can be mainly attributed to the two following reasons:

1. The VNS-NSGA-II algorithm is adaptive to the probability of crossover and mutation, which means that it can adjust the probability of crossover and mutation dynamically based on fitness, evolutionary algebra and the number of unchanged individuals during the evolution process, thereby minimizing the destruction of good solutions and ensuring population diversity. The adaptive function enhances the algorithm's search capability and prevents premature convergence.
2. The variable neighborhood search operators in VNS-NSGA-II reduce the possibility of the algorithm falling into the local optimum. Three mature neighborhood structures in variable neighborhood search operators increase the diversity of neighborhood space. The neighborhood space diversity is proportional to the offspring diversity. Greater neighborhood space diversity also represents the easier identification of the global optimal solution [47].

(2) Comparison of convergence

The comparison of algorithm convergence results is depicted in Figure 7. This experiment was conducted using a dataset of 100 customers. The horizontal coordinate of Figure 7 represents the number of iterations of the algorithm, the left vertical coordinate represents total cost, and the right vertical coordinate represents customer satisfaction. The solid line represents the optimal values of the total cost obtained by different algorithms, while the dotted line represents the optimal values of customer satisfaction obtained by different algorithms. Figure 7 shows the differences in convergence among the three algorithms. The total cost and satisfaction of the VNS-NSGA-II algorithm converge rapidly, the total cost converges after 142 iterations, and the satisfaction converges after 49 iterations. The MOGE and NSGA-II algorithms did not converge within 200 iterations. The VNS-NSGA-II algorithm has a faster convergence speed and can generate better values for objective functions.

3.2. Analysis of Optimal Solution Selection

The primary purpose of the method suggested in this paper is to develop a clear and effective distribution scheme for fresh agricultural products. Thus, after generating the Pareto-optimal solution set using the VNS-NSGA-II algorithm, it is important to select a solution and decode it as a distribution plan using the TOPSIS method. In this experiment, the dataset of 100 customers is used as an example, and the TOPSIS method is used to select the most suitable solution from the Pareto-optimal front. Figure 8 illustrates the Pareto-optimal front of 100 customers. The horizontal coordinate of Figure 8 represents the total cost, and the vertical coordinate represents customer satisfaction. Each point in the graph represents a solution in the Pareto-optimal front. Table 4 shows the process of evaluating each solution in Figure 8 using the TOPSIS method. As indicated in Table 4, the TOPSIS method is utilized to determine the S_i^+, S_i^- and C_i values for each solution. The filtered solution is compared to the solution with the ideal index values, as shown in Table 5. Note, however, that CC in Table 5 is the cost of carbon emissions and RC is the cost of refrigeration. The data in Tables 4 and 5 will be analyzed in detail next.

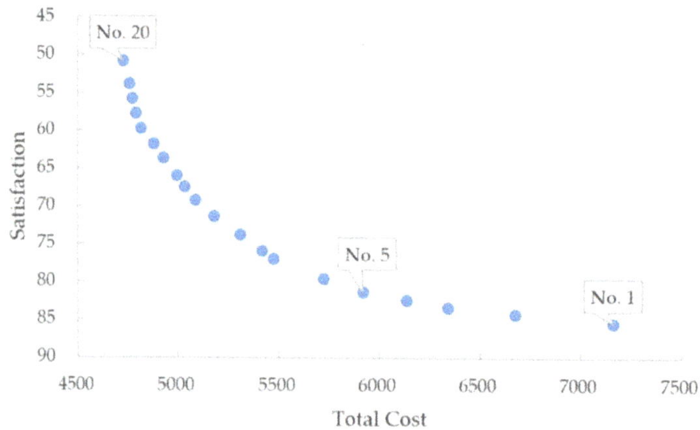

Figure 8. Distribution of 100 customers' Pareto-optimal front.

Table 4. Evaluation of results using the TOPSIS approach.

No.	TTC	Satisfaction	S_i^+	S_i^-	Closeness C_i
1	7159.35	85.33	0.15324	0.07745	0.33572
2	6669.93	84.23	0.14917	0.06207	0.29384
3	6339.67	83.36	0.14683	0.05212	0.26196
4	6132.93	82.36	0.14382	0.04670	0.24512
5	5916.87	81.31	0.14107	0.04191	0.22905
6	5720.60	79.56	0.13560	0.04074	0.23101
7	5470.51	76.94	0.12784	0.04415	0.25670
8	5413.11	75.86	0.12433	0.04738	0.27591
9	5304.79	73.74	0.11767	0.05464	0.31709
10	5176.35	71.30	0.11072	0.06388	0.36585
11	5083.69	69.12	0.10476	0.07280	0.41003
12	5029.63	67.43	0.10022	0.07997	0.44381
13	4990.14	65.90	0.09622	0.08657	0.47360
14	4924.46	63.60	0.09103	0.09657	0.51477
15	4877.52	61.75	0.08737	0.10469	0.54507
16	4813.10	59.75	0.08457	0.11347	0.57296
17	4787.05	57.70	0.08147	0.12251	0.60062
18	4770.79	55.69	0.07903	0.13145	0.62454
19	4755.34	53.77	0.07760	0.13995	0.64331
20	4724.00	50.77	0.07745	0.15324	0.66428

Table 5. Comparison of selected solution and optimal value.

No.	TTC	CC	RC	Number of Vehicles Used	Average Loading Rate	Satisfaction
1	7159.35	11.65%	34.13%	22	43.67%	85.33
5	5916.87	7.79%	26.46%	15	59.72%	81.31
20	4724.00	5.83%	16.51%	12	74.92%	50.77

We need to choose an appropriate solution based on the data in Table 4. Given that the objective function in this paper is to reduce the total cost and TOPSIS minimizes satisfaction, the smallest value should be chosen from the final closeness degree C_i. As a result, solution No. 5 should be chosen as the final delivery strategy. Figure 8 illustrates the location of solution No. 5 in the distribution of 100 customers' Pareto-optimal front. Solution No. 1 has the highest satisfaction value, while solution No. 20 has the lowest total cost value. While solution No. 5 provides 4.02 points less satisfaction than solution No. 1, the total cost is lowered by 17.4%. Although the total cost of solution No. 5 is 20.2% greater than that of solution No. 20, the satisfaction value of solution No. 5 is 30.54 points greater than that of solution No. 20. Taken together, solution No. 5 is an equilibrium solution in the middle of the weight setting's extreme values.

In Table 5, the total cost of solution No. 5 is compared with the total cost of solution No. 1 and solution No. 20, which include the proportion of carbon emission cost to the total cost, the proportion of refrigeration cost to the total cost, the number of vehicles used, and the average load rate and satisfaction. Solution No. 1 has the highest customer satisfaction but also the highest total cost, and solution No. 20 has the lowest total cost but the worst customer satisfaction. The total cost and customer satisfaction of solution No. 5 are between solutions No. 1 and No. 20. It can be determined that the number of vehicles used must grow, while the average loading rate must drop in order to boost customer satisfaction. Additionally, more vehicles will lead to increased carbon emissions and cooling costs, and refrigeration costs will increase as well; on the other hand, to reduce total costs, the number of vehicles will be reduced, the average loading rate will be increased, and consequently, fewer delivery vehicles will delay the delivery of produce from the distribution center, lowering carbon emissions and cooling costs.

From the foregoing study, it is clear that the TOPSIS method can well screen out the suitable solutions from the Pareto-optimal solution set.

3.3. Optimisation of the Fresh Agricultural Products Distribution Routes for the e-Commerce Business in the Sample

In order to verify the effectiveness of the TDSDGVRPMTW model in reality, this study uses a case in Shanghai to compare and analyze the differences between the three delivery strategies. The current common distribution strategy for fresh agricultural products is to distribute all types of products in one vehicle at one temperature without split delivery or to use multi-compartment vehicles for multi-temperature distribution. According to the two delivery strategies and the mathematical model proposed in this paper, the two delivery strategies can be modeled as a time-dependent green vehicle routing problem with a time windows model (TDGVRPTW) and a time-dependent multi-compartment green vehicle routing problem with a time windows model (TDMCGVRPTW). Among them, the TDMCGVRPTW model adopts the method from Reed et al. [3]. Therefore, the three strategies involved in the comparison are TDGVRPTW, TDMCGVRPTW and TDSDGVRPMTW.

We used the distribution data from an e-commerce business of fresh agricultural products in Shanghai. The data of one distribution center and 24 customers are shown in Table 6, including the locations of the distribution center, individual customers, demand, types of demand and time windows. Note, however, that the latitude and longitude of the customer's location are converted to X/Y coordinates for convenience. Types represent the types of fresh agricultural products that customers need. All types of fresh agricultural products in a refrigerated vehicle are loaded, and a certain temperature for delivery is set. The storage temperature of the vehicle is determined by the lowest temperature required among all types of fresh products. This is the company's current distribution strategy. The corresponding cost is calculated based on the company's actual distribution schemes and the constraints of the mathematical model established in this paper. According to the real vehicle data from the company, the total number of available vehicles is adjusted to $Q_L = 15$ and the maximum capacity is adjusted to $Q_K = 1500$. Other than that, all other parameters remain the same as in Section 2.6. Table 7 shows the relevant data for each model. Note that TC in Table 7 is the travel cost, FC is the fixed cost, and SC is the service cost. The relevant data of the TDGVRPTW model in Table 7 is the calculation result. The data corresponding to the TDMCGVRPTW model and the TDSDGVRPMTW model in Table 7 are obtained by using the VNS-NSGA-II algorithm and TOPSIS method proposed in this paper. Note, however, that Gap_mc in Table 7 refers to the difference between the related data of TDMCGVRPTW and TDGVRPTW, and Gap_sd refers to the difference between the related data of TDSDGVRPMTW and TDGVRPTW. Figure 9 is a comparison of all data of the three models. Each column in the figure represents the total cost of a model, where different colors represent different itemized costs. Broken lines represent customer satisfaction for different models. The scale on the left of the vertical coordinate corresponds to the value of total cost, while the scale on the right of the vertical coordinate corresponds to the value of satisfaction. The data presented in Table 8 can be used to analyze the different choices made by companies on the basis of different strategies when they have to give up part of the customer orders or the number of vehicles is limited. Table 8 shows the quantity and on-time rate of delivery of each type of agricultural product under three different strategies. On-time delivery means that the vehicle completes the delivery service for the customer within the customer's time window, under the circumstance of which the customer satisfaction is greater than 0. Note, that PD refers to the number of items delivered punctually. Proportion means the percentage of the quantity of products delivered punctually in the total demand for that type of product. Refrigeration cost per minute represents the refrigeration cost per minute for preserving the corresponding type of product. Total demand refers to the total demand of all customers ordering the corresponding type of product.

Table 6. Relevant data of distribution center and 24 customers.

No.	*X*	*Y*	Demand (kg)	Ready Time	Due Time	Types
0	31	47	0	5:00	17:30	
1	37	61	500	6:30	8:30	*A*, *B*
2	31	29	350	5:30	8:30	*A*
3	51	57	600	6:00	9:00	*A*
4	51	32	900	12:00	15:00	*B*, *C*
5	11	42	1200	8:00	12:00	*A*, *D*
6	21	42	100	15:00	16:00	*B*
7	16	62	250	6:00	11:00	*A*, *C*
8	6	55	400	7:30	10:00	*A*, *B*
9	51	72	800	9:00	10:30	*C*
10	26	72	750	11:00	11:30	*B*
11	16	77	600	7:30	9:30	*D*
12	46	47	800	6:00	8:00	*A*, *C*
13	26	37	1150	10:00	11:30	*B*, *D*
14	11	22	1000	5:00	7:00	*A*, *C*
15	26	17	400	14:30	16:30	*A*, *D*
16	6	32	900	5:30	7:00	*C*
17	1	42	150	6:00	8:00	*A*
18	16	52	500	7:00	8:30	*D*
19	11	72	800	6:30	8:00	*B*, *C*
20	41	77	450	8:30	9:30	*A*
21	41	32	500	7:30	10:30	*B*, *C*
22	41	22	850	13:00	14:30	*A*
23	51	17	1450	7:00	12:00	*A*, *D*
24	61	47	150	9:00	11:30	*B*

Table 7. Comparison of three delivery strategies.

	TTC	TC	FC	SC	RC	CC	Satisfaction
TDGVRPTW	2680.95	1194.32	260.00	72.00	861.88	292.75	76.36
TDMCGVRPTW	2587.27	1317.44	260.00	117.00	594.55	298.28	68.15
TDSDGVRPMTW	2525.17	1246.67	220.00	174.00	603.89	280.60	87.04
Gap_mc	−93.68	123.12	0.00	45.00	−267.33	5.53	−8.21
Gap_sd	−155.78	52.35	−40.00	102.00	−257.98	−12.15	10.68

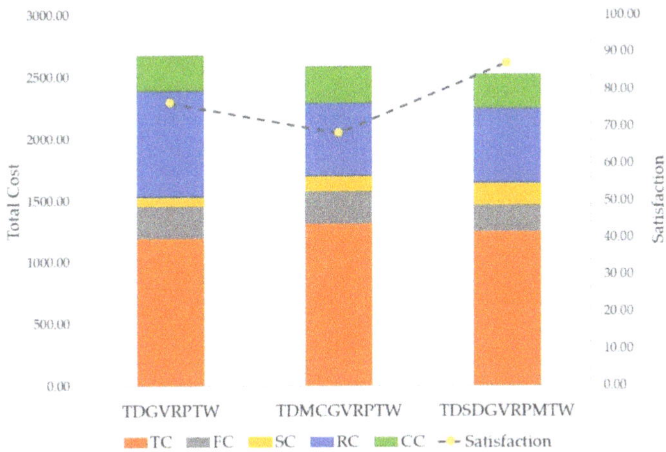

Figure 9. Total cost and satisfaction comparison of different models.

Table 8. The quantity of each type of product delivered punctually under three strategies.

	A		B		C		D	
	PD	Proportion	PD	Proportion	PD	Proportion	PD	Proportion
TDGVRPTW	4450	78.21%	3060	71.66%	2795	75.95%	1245	65.18%
TDMCGVRPTW	3525	61.95%	2730	63.93%	2645	71.88%	1435	75.13%
TDSDGVRPMTW	5290	92.97%	3770	88.29%	3020	82.07%	1375	71.99%
Refrigeration cost per minute	0.65		0.68		1.02		1.13	
Total Demand	5690		4270		3680		1910	

As shown in Table 7 and Figure 9, the proportion of each cost in the total cost is roughly the same despite the different strategies taken, which shows that the delivery schemes produced by the method with two different models in this study are feasible. TDGVRPTW and TDMCGVRPTW have the same fixed cost, which means that the two strategies use the same number of vehicles. Compared with the *RC* of TDGVRPTW, the *RC* of TDMCGVRPTW and TDSDGVRPMTW has been greatly reduced, which shows that both the multi-compartment distribution strategy and the split delivery with a single compartment distribution strategy can greatly reduce the refrigeration cost.

However, it can be found that there are many differences between the three distribution strategies. As shown in Table 7 and Figure 9, the total cost and satisfaction of TDSDGVRPMTW are optimal among the three delivery strategies. The total cost of TDSDGVRPMTW is 5.81% lower than that in TDGVRPTW, and the satisfaction of TDSDGVRPMTW is 10.68 points higher than that in TDGVRPTW. In other words, the distribution strategy of split delivery with a single compartment proposed in this paper can reduce the total cost and improve satisfaction. For the total cost, *FC*, *RC* and *CC* of TDSDGVRPMTW are all lower than those in TDGVRPTW. Although TDSDGVRPMTW is higher than TDGVRPTW in terms of *TC* and *SC*, the small difference will not trigger a large increase in *TTC*. Comparing TDMCGVRPTW with TDGVRPTW, it can be found that although the *TTC* of TDMCGVRPTW is 3.49% lower than that of TDGVRPTW, the customer satisfaction of TDMCGVRPTW is 8.21 points lower than that of TDGVRPTW. This means that the strategy reduced total costs at the expense of satisfaction. The *RC* of TDMCGVRPTW is lower than those of TDGVRPTW, but the *TC*, *SC* and *CC* are higher than those of TDGVRPTW. A high *CC* means that more carbon emissions are generated, which is not conducive to environmental protection.

As shown in Table 8, customers have the largest demand for *A* and the smallest demand for *D*. *A* has the lowest refrigeration cost, while *D* has the highest refrigeration cost. Three different delivery strategies show three different ways of allocating shipping capacity. The on-time rates of delivery of TDGVRPTW to *A*, *B*, and *C* are similar, which shows that the strategy will choose not to deliver the products with the lowest demand and the highest refrigeration cost when the number of vehicles is limited and some customer orders must be abandoned. TDMCGVRPTW has lower on-time rates of delivery for A and B than it does for *C* and *D*. This suggests that the strategy will choose to not deliver products that are in greater demand and have lower refrigeration costs when the number of vehicles is limited. This result is consistent with the findings of Hsu and Chen [2]. However, this strategy produces very low satisfaction. TDSDGVRPMTW has lower on-time rates of delivery for *C* and *D* than for *A* and *B*, which suggests that the strategy will not deliver products with lower demand and higher refrigeration costs when the number of vehicles is limited. The on-time rate of delivery of this strategy for each product is much higher than that of TDGVRPTW, which is why the satisfaction of this strategy is higher than that of TDGVRPTW.

From the above analyses of results shown in this experiment, it can be concluded that the distribution strategy represented by TDSDGVRPMTW is most suitable for the real-world case presented in this study. There are many differences between the experimental results of TDSDGVRPMTW and TDMCGVRPTW. Comparing the results of this experiment with research on multi-compartment vehicle routing problems [2–4], it can be found that the reasons for these differences are as follows:

1. Split delivery would allow customers ordering multiple agricultural products to be served by multiple vehicles, meaning that each vehicle would serve more customers and travel longer routes, which leads to higher travel costs and service costs. However, split delivery keeps each product at the optimum temperature for transport, which greatly reduces refrigeration costs and lowers the total cost.
2. Split delivery allows each vehicle to deliver a smaller number of products to customers, which leads to a shorter service time that makes the vehicle more likely to finish each delivery and leave the customer within the optimal service time window, resulting in higher customer satisfaction. Another reason for the high level of satisfaction is that TDSDGVRPMTW chooses to refuse orders with products of small quantity and prioritizes serving customers ordering products of large quantity.
3. Divide a vehicle's compartment into multiple sections. Each of them transports one type of agricultural product with different optimum temperatures as needed. Although this can reduce refrigeration costs, if a customer's demand for a certain type of agricultural product exceeds the capacity of the divided compartment, multiple deliveries are required to meet the customer's demand for this type of agricultural product and more vehicles are needed, leading to high travel costs and carbon emissions.
4. Customers have multiple time windows to choose and therefore vehicles have more opportunities to arrive at locations and complete services during a certain time window. The advantages of multiple time windows over a single time window will be leveraged, especially when a customer needs to be served by vehicles multiple times. One customer in TDMCGVRPTW needs the service of multiple vehicles, however, each customer has only one time window, and consequently many vehicles arrive at the location of the customer out of the time window, leading to very low customer satisfaction. Another reason for the low satisfaction is that TDMCGVRPTW

chooses to refuse many orders for items that are in high demand, leaving many customers unserved.

5. Therefore, the number of vehicle compartments, the capacity of each compartment, and the number of product types demanded by customers are the factors that determine why TDMCGVRPTW chooses not to deliver products that are in high demand and TDSDGVRPMTW chooses the opposite. The compartment of the vehicle in TDMCGVRPTW is divided into four parts. When a customer's demand for a certain product exceeds the compartment capacity, multiple vehicles are required to deliver the same product to that customer. When the number of vehicles needed to serve the customer exceeds the number of types of products the customer orders, the service cost is too high and the order will be refused by TDMCGVRPTW. Therefore, under this circumstance, the advantages of TDSDGVRPMTW over TDMCGVRPTW can be observed. In terms of product types, TDSDGVRPMTW needs fewer vehicles to complete the distribution to the customer with lower service costs, thus, this customer order will not be refused. When the number of such customers is large, TDSDGVRPMTW would naturally become the best strategy.

4. Conclusions

This paper presents a study on the split delivery vehicle routing problem that arises in the distribution of fresh agricultural products. A mathematical model considering travel cost, fixed cost, service cost, refrigeration cost, carbon emission cost and customer satisfaction was developed to find an optimal solution for the problem. In this paper, time-varying road network constraints are added to the model and multiple time windows are set for each customer. To solve this problem, a variable neighborhood search combined with the non-dominated sorting genetic algorithm II (VNS-NSGA-II) and techniques for order preference by similarity to an ideal solution (TOPSIS) are proposed and applied. In the stage of genetic evolution, adaptive functions are used to dynamically adjust the probability of crossover and mutation. Moreover, the variable neighborhood search operators are added to enhance the search abilities of the algorithm. After the Pareto-optimal front is obtained, the TOPSIS method is used to screen out the solutions that meet the needs.

In this paper, the VNS-NSGA-II algorithm, MOGE algorithm and NSGA-II algorithm are compared in terms of the Pareto-optimal front and convergence of the algorithm using Solomon's benchmark of different sizes. For benchmark instances with customer sizes of 30, 50, 70, and 100, the VNS-NSGA-II algorithm can obtain a better Pareto-optimal front than NSGA-II and MOGE. In terms of the average total cost, VNS-NSGA-II is 10.03% lower than NSGA-II, and 7.61% lower than MOGE. In terms of satisfaction, VNS-NSGA-II is 8.77 points higher than NSGA-II and 6.48 points higher than MOGE. In terms of convergence, the total cost of the VNS-NSGA-II algorithm converges after 142 iterations, while the satisfaction converges after 49 iterations, and the number of iterations is smaller than the other two algorithms. It is proved that the VNS-NSGA-II algorithm has better search efficiency and the Pareto-optimal front is appropriate to be applied in the case of this study. The optimal solution selection experiment also proves that the TOPSIS method can select the appropriate solution from the Pareto-optimal front. Finally, the results of a real-world case show that the TDSDGVRPMTW solution proposed in this paper is better than the existing solutions of TDGVRPTW and TDMCGVRPTW. The total cost of the TDSDGVRPMTW solution was 5.81% lower than the existing solution and the satisfaction was 10.68 points higher. Although the TDMCGVRPTW solution is 3.49% lower than the existing solution in terms of the total cost, satisfaction is 8.21 points lower than the original solution.

The contributions of this paper include the first multi-objective optimization model of TDSDGVRPMTW for fresh agricultural product distribution and propose the VNS-NSGA-II algorithm to find the Pareto-optimal front and select the appropriate solution with the TOPSIS method. The experiment verified the advantages of this method and found that the model of TDSDGVRPMTW can effectively reduce the cost of fresh agricultural product distribution and improve customer satisfaction. Enterprises can use the method proposed in this paper to model and select from different distribution strategies, so as to find the optimal one that suits their needs.

Comparing the experimental results in this study with those from Chen et al. [4], Reed et al. [3] and Hsu and Chen [2] can provide some insights into distribution management. The customer's demand, the number of types of products that the customer orders, and the capacity of each compartment in the multi-compartment delivery system, are the major factors that determine whether to use delivery by use of a multi-compartment vehicle or split delivery by a single compartment vehicle. As shown in the experiment of Reed et al. [3], it is reasonable to use multi-compartment vehicles for delivery when the customer requires several types of products and the customer's demand for a certain product does not exceed the capacity of each compartment. The experiment of a real case in this study shows that it is reasonable to split delivery with a single compartment vehicle when a customer requires a few types of products and the customer demand for a certain product exceeds the capacity of each compartment. The experiments in this paper also found that loading all types of fresh agricultural products in a vehicle and setting one single temperature for distribution is worse than the above two strategies.

On the other hand, this study has some limitations. In this paper, the calculation of carbon emissions does not consider the road slope, and it is not accurate enough since it only considers the vehicle speed, load weight and travel distance. Although the VNS-NSGA-II algorithm proposed has a good performance, there is still room for further optimization. For example, some heuristics are added in the initial population generation stage to improve the quality of the initial population. In this paper, the TOPSIS method is selected to screen suitable solutions from the solution set and to represent the effectiveness of this method. However, there are many multi-attribute decision-making methods, such as the elimination and choice expressing reality (ELECTRE) method, the preference ranking organization method for enrichment evaluations (PROMETHEE) method, and the analytical hierarchal process (AHP) method, and so on. In future, these methods can be applied to compare with the TOPSIS methods and to find more suitable ones for future research on TDSDGVRPMTW. In addition, it was found that the number of types of products needed by customers, the quantity of each ordered product, and the capacity of each compartment in a multi-compartment vehicle were factors that have an impact on what delivery strategy should be selected. Furthermore, how each factor affects the selection of delivery strategies is worth further study.

Author Contributions: D.W.: conceptualization, funding acquisition, supervision, methodology and writing—original draft; C.W.: formal analysis, validation, resources, software and editing. All authors have read and agreed to the published version of the manuscript.

Funding: This research was funded by the China Education Ministry of Humanities and Social Science Research Youth Fund project (No. 18YJCZH192), the Applied Undergraduate Pilot Project for Logistics Management of Shanghai Ocean University (No. B1-5002-18-0000), the Open Project Program of Artificial Intelligence Key Laboratory of Sichuan province (No. 2015RYJ01), and the Social Science Project in Hunan province (No. 16YBA316).

Data Availability Statement: All experimental data in this paper come from: https://people.idsia.ch//~luca/macs-vrptw/problems/welcome.htm (accessed on 7 April 2022).

Conflicts of Interest: The authors declare no conflict of interest.

References

1. iResearch Institute. Research Report on Fresh Food E-Commerce Industry in China. 2021. Available online: https://www.iresearch.com.cn/Detail/report?id=3776&isfree=0 (accessed on 7 April 2022).
2. Hsu, C.; Chen, W. Optimizing fleet size and delivery scheduling for multi-temperature food distribution. *Appl. Math. Model.* **2014**, *38*, 1077–1091. [CrossRef]
3. Reed, M.; Yiannakou, A.; Evering, R. An ant colony algorithm for the multi-compartment vehicle routing problem. *Appl. Soft Comput.* **2014**, *15*, 169–176. [CrossRef]

4. Arora, R.; Kaushik, C.; Arora, R. Multi-objective and multi-parameter optimization of two-stage thermoelectric generator in electrically series and parallel configurations through NSGA-II. *Energy* **2015**, *91*, 242–254. [CrossRef]
5. Osvald, A.; Stirn, L. A vehicle routing algorithm for the distribution of fresh vegetables and similar perishable food. *J. Food Eng.* **2008**, *85*, 285–295. [CrossRef]
6. Wen, M.; Krapper, E.; Larsen, J.; Stidsen, T. A multilevel variable neighborhood search heuristic for a practical vehicle routing and driver scheduling problem. *Networks* **2011**, *58*, 311–322. [CrossRef]
7. Stellingwerf, H.; Groeneveld, L.; Laporte, G.; Kanellopoulos, A.; Bloemhof, J.; Behdani, B. The quality-driven vehicle routing problem: Model and application to a case of cooperative logistics. *Int. J. Prod. Econ.* **2021**, *231*, 107849. [CrossRef]
8. Yao, B.; Chen, C.; Song, X.; Yang, X. Fresh seafood delivery routing problem using an improved ant colony optimization. *Ann. Oper. Res.* **2019**, *273*, 163–186. [CrossRef]
9. Zhao, Z.; Li, X.; Zhou, X. Optimization of transportation routing problem for fresh food in time-varying road network: Considering both food safety reliability and temperature control. *PLoS ONE* **2020**, *15*, e0235950. [CrossRef]
10. Hsiao, Y.; Chen, M.; Lu, K.; Chin, C. Last-mile distribution planning for fruit-and-vegetable cold chains. *Int. J. Logist. Manag.* **2018**, *29*, 862–886. [CrossRef]
11. Dror, M.; Trudeau, P. Split delivery routing. *Nav. Res. Logist. NRL* **1990**, *37*, 383–402. [CrossRef]
12. Dror, M.; Trudeau, P. Savings by split delivery routing. *Transp. Sci.* **1989**, *23*, 141–145. [CrossRef]
13. Ambrosino, D.; Sciomachen, A. A food distribution network problem: A case study. *IMA J. Manag. Math.* **2006**, *18*, 33–53. [CrossRef]
14. Yoshizaki, H.T.Y. Scatter search for a real-life heterogeneous fleet vehicle routing problem with time windows and split deliveries in Brazil. *Eur. J. Oper. Res.* **2009**, *199*, 750–758.
15. Wu, D.; Wu, C. TDGVRPSTW of fresh agricultural products distribution: Considering both economic cost and environmental cost. *Appl. Sci.* **2021**, *11*, 10579. [CrossRef]
16. Frizzell, W.; Giffin, W. The split delivery vehicle scheduling problem with time windows and grid network distances. *Comput. Oper. Res.* **1995**, *22*, 655–667. [CrossRef]
17. Silva, M.; Subramanian, A.; Ochi, S. An iterated local search heuristic for the split delivery vehicle routing problem. *Comput. Oper. Res.* **2015**, *53*, 234–249. [CrossRef]
18. Yang, W.; Wang, D.; Pang, W.; Tan, A.; Zhou, Y. Goods consumed during transit in split delivery vehicle routing problems: Modeling and solution. *IEEE Access* **2020**, *8*, 110336–110350. [CrossRef]
19. Sin, C.; Haugland, D. A tabu search heuristic for the vehicle routing problem with time windows and split deliveries. *Comput. Oper. Res.* **2004**, *31*, 1947–1964.
20. Desaulniers, G. Branch-and-price-and-cut for the split-delivery vehicle routing problem with time windows. *Oper. Res.* **2010**, *58*, 179–192. [CrossRef]
21. Archetti, C.; Bouchard, M.; Desaulniers, G. Enhanced branch and price and cut for vehicle routing with split deliveries and time windows. *Transp. Sci.* **2011**, *45*, 285–298. [CrossRef]
22. Salani, M.; Vacca, I. Branch and price for the vehicle routing problem with discrete split deliveries and time windows. *Eur. J. Oper. Res.* **2011**, *213*, 470–477. [CrossRef]
23. Bianchessi, N.; Irnich, S. Branch-and-cut for the split delivery vehicle routing problem with time windows. *Transp. Sci.* **2019**, *53*, 442–462. [CrossRef]
24. Luo, Z.; Qin, H.; Zhu, W.; Lim, A. Branch and price and cut for the split-delivery vehicle routing problem with time windows and linear weight-related cost. *Transp. Sci.* **2017**, *51*, 668–687. [CrossRef]
25. McNabb, E.; Weir, D.; Hill, R.; Hall, S. Testing local search move operators on the vehicle routing problem with split deliveries and time windows. *Comput. Oper. Res.* **2015**, *56*, 93–109. [CrossRef]
26. Li, J.; Qin, H.; Baldacci, R.; Zhu, W. Branch-and-price-and-cut for the synchronized vehicle routing problem with split delivery, proportional service time and multiple time windows. *Transp. Res. Part E Logist. Transp. Rev.* **2020**, *140*, 101955. [CrossRef]
27. Wang, H.; Du, L.; Ma, S. Multi-objective open location-routing model with split delivery for optimized relief distribution in post-earthquake. *Transp. Res. Part E Logist. Transp. Rev.* **2014**, *69*, 160–179. [CrossRef]
28. Vahdani, B.; Veysmoradi, D.; Noori, F.; Mansour, F. Two-stage multi-objective location-routing-inventory model for humanitarian logistics network design under uncertainty. *Int. J. Disaster Risk Reduct.* **2018**, *27*, 290–306. [CrossRef]
29. Goodarzi, H.; Tavakkoli-Moghaddam, R.; Amini, A. A new bi-objective vehicle routing-scheduling problem with cross-docking: Mathematical model and algorithms. *Comput. Ind. Eng.* **2020**, *149*, 106832. [CrossRef]
30. Hickman, A.; Hassel, D.; Joumard, R. *Methodology for Calculating Transport Emissions and Energy Consumption*; The National Academies of Sciences, Engineering, and Medicine: Washington, DC, USA, 1999; pp. 1–362.
31. Malandraki, C.; Daskin, M. Time dependent vehicle routing problems: Formulations, properties and heuristic algorithms. *Transp. Sci.* **1992**, *26*, 185–200. [CrossRef]
32. Malandraki, C.; Dial, B. A restricted dynamic programming heuristic algorithm for the time dependent traveling salesman problem. *Eur. J. Oper. Res.* **1996**, *90*, 45–55. [CrossRef]
33. Ichoua, S.; Gendreau, M.; Potvin, J. Vehicle dispatching with time-dependent travel times. *Eur. J. Oper. Res.* **2003**, *144*, 379–396. [CrossRef]
34. Fleischmann, B.; Gietz, M.; Gnutzmann, S. Time-varying travel times in vehicle routing. *Transp. Sci.* **2004**, *38*, 160–173. [CrossRef]

35. Wang, Y.; Zhang, J.; Guan, X.; Xu, M.; Wang, Z.; Wang, H. Collaborative multiple centers fresh logistics distribution network optimization with resource sharing and temperature control constraints. *Expert Syst. Appl.* **2021**, *165*, 113838. [CrossRef]
36. Wang, X.; Wang, M.; Ruan, J.; Li, Y. Multi-objective optimization for delivering perishable products with mixed time windows. *Adv. Prod. Eng. Manag.* **2018**, *13*, 321–332. [CrossRef]
37. Wang, H.; Li, W.; Zhao, Z.; Wang, Z.; Li, M.; Li, D. Intelligent distribution of fresh agricultural products in smart city. *IEEE Trans. Ind. Inform.* **2021**, *18*, 1220–1230. [CrossRef]
38. Xu, H.; Fan, W.; Wei, T.; Yu, L. An Or-opt NSGA-II algorithm for multi-objective vehicle routing problem with time windows. In Proceedings of the 2008 IEEE International Conference on Automation Science and Engineering, Arlington, VA, USA, 23 August 2008; pp. 309–314.
39. Wang, Y.; Zhang, S.; Guan, X.; Peng, S.; Wang, H.; Liu, Y.; Xu, M. Collaborative multi-depot logistics network design with time window assignment. *Expert Syst. Appl.* **2020**, *140*, 112910. [CrossRef]
40. Wang, Y.; Zhang, J.; Assogba, K.; Liu, Y.; Xu, M.; Wang, Y. Collaboration and transportation resource sharing in multiple centers vehicle routing optimization with delivery and pickup. *Knowl. Based Syst.* **2018**, *160*, 296–310. [CrossRef]
41. Pacciarelli, D.; Løkketangen, A.; Mandal, K.; Hasle, G. A memetic NSGA-II for the bi-objective mixed capacitated general routing problem. *J. Heuristics* **2015**, *21*, 359–390.
42. Fan, H.; Zhang, Y.G.; Tian, P.J.; Cao, Y.; Ren, X.X. Dynamic vehicle routing problem of heterogeneous fleets with time-dependent networks. *Syst. Eng. Theory Pract.* **2022**, *42*, 455–470, (In Chinese with English Abstract). [CrossRef]
43. Sánchez-Oro, J.; López-Sánchez, A.D.; Colmenar, J.M. A general variable neighborhood search for solving the multi-objective open vehicle routing problem. *J. Heuristics* **2020**, *26*, 423–452. [CrossRef]
44. Solomon, M.M. Algorithms for the vehicle routing and scheduling problems with time window constraints. *Oper. Res.* **1987**, *35*, 254–265. [CrossRef]
45. Liu, C.; Kou, G.; Zhou, X.; Peng, Y.; Sheng, H.; Alsaadi, F.E. Time-dependent vehicle routing problem with time windows of city logistics with a congestion avoidance approach. *Knowl. Based Syst.* **2019**, *188*, 104813. [CrossRef]
46. Zulvia, E.; Kuo, J.; Nugroho, Y. A many-objective gradient evolution algorithm for solving a green vehicle routing problem with time windows and time dependency for perishable products. *J. Clean Prod.* **2020**, *242*, 118428. [CrossRef]
47. Jiang, Z.; Chen, Y.; Li, X.; Li, B. A heuristic optimization approach for multi-vehicle and one-cargo green transportation scheduling in shipbuilding. *Adv. Eng. Inform.* **2021**, *49*, 101306. [CrossRef]

agriculture

MDPI

Article

A Conceptual Model for Development of Small Farm Management Information System: A Case of Indonesian Smallholder Chili Farmers

Henriyadi Henriyadi *, Vatcharaporn Esichaikul and Chutiporn Anutariya

ICT Department, School of Engineering and Technology, Asian Institute of Technology, Khlong Luang District, Pathum Thani 12120, Thailand; vatchara@ait.ac.th (V.E.); chutiporn@ait.ac.th (C.A.)
* Correspondence: st118501@ait.ac.th; Tel.: +66-63-376-1015

Abstract: Farm Management Information Systems (FMIS) assists farmers in managing their farms more effectively and efficiently. However, the use of FMIS to support crop cultivation is, at the present time, relatively expensive for smallholder farmers. Due to some handicaps, providing an FMIS that is suitable for small-holder farmers is a challenge. To analyze this gap, this study followed 3 steps, namely: (1) identified commodity and research area, (2) performed Farmers' Information Needs Assessment (FINA), and (3) developed the conceptual model using the Soft System Methodology. Indonesian smallholder chili farmers are used as a case study. The most required information of smallholder' farmers was identified through a qualitative questionnaire. Despite this, not all identified information needs could be accurately mapped. Thus, this indicates the need for a new FMIS conceptual model that is suitable for smallholder farmers. This study proposes an FMIS conceptual model for farm efficiency that incorporates five layers, namely farmers' information needs, data quality assessment, data extraction, SMM (split, match and merge), and presentation layer. SMM layer also provides a method to comprehensively tackle three main problems in data interoperability problems, namely schema heterogeneity, schema granularity, and mismatch entity naming.

Keywords: farm management information system; farmers' information needs assessment; soft system methodology; smallholder farmers; conceptual model; Indonesian chili farmers

Citation: Henriyadi, H.; Esichaikul, V.; Anutariya, C. A Conceptual Model for Development of Small Farm Management Information System: A Case of Indonesian Smallholder Chili Farmers. *Agriculture* **2022**, *12*, 866. https://doi.org/10.3390/agriculture12060866

Academic Editor: Dimitre Dimitrov

Received: 28 April 2022
Accepted: 13 June 2022
Published: 15 June 2022

1. Introduction

1.1. Background

The Farm Management Information System (FMIS) is a tool to assist farmers in managing their farms more effectively and efficiently. FMIS is a system that deals with the accuracy of data, optimization of the use of available resources, and processes by using advanced technologies for cultivating the farm [1]. Some researchers propose approaches to improving functionalities, such as improving management systems' functionality, interoperability, database inter-networking, and improving software architecture [2]. In the primary studies, 14 FMIS features appeared more than 7 times and 11 FMIS barriers appeared 3 times or more [3]. By accurately using FMIS, farmers can manage their farms more effectively and efficiently [4]. The main characteristics of existing FMISs are tailor-made applications that offer advanced functionality, focus on large farms, and concentrate solely on the specific needs of the users [3]. Moreover, over 75% of existing FMIS applications require a dedicated desktop computer to operate [5], rendering the current FMIS application expensive, especially for smallholder farmers. Therefore, providing an FMIS at an affordable price for smallholder farmers is challenging [6].

We should pay attention to smallholder farmers when developing an FMIS application due to some reasons. At present, there are about 570 million farms globally, of which, more than 475 million are smallholder farmers [7,8]. Smallholder farmers have common characteristics, including: (a) occupying a farm smaller than 2 ha [7], (b) the use of traditional product market chains, (c) the use of family labor on the farm, (d) have on-farm activities as their only source of income, (e) employ the traditional farming system, (f) have limited financial support, and (g) operate without a form management system [9]. Thus, the application of FMIS by smallholder farmers is impossible considering their lack of knowledge as well as the costs and constraints associated with it. Providing an FMIS application that is suitable for smallholder farmers at an affordable price is a herculean task [10]. This study aimed to answer the main question: "how to develop an FMIS conceptual model that is applicable for smallholder farmers?" There are 3 main goals in this research, namely (1) identification of the smallholder farmers' information needs, (2) mapping information need into the existing FMIS conceptual model to find the gap between the existing conceptual model with the information needed, and (3) develop an FMIS conceptual model for smallholder farmers.

1.2. Previous Work

The Farmers' Information Needs Assessment (FINA) is a common method for identifying farmers' information needs. Various studies using FINA have already been conducted [11–19]. In this method, the collection of information required is grouped based on some criteria, making it more understandable [11,12,17]. However, the literature study conducted did not find any relevant research that uses the agribusiness subsystem in grouping information. In addition, no research was found to have explicitly used a qualitative approach in data collection, even though it generates various benefits over the quantitative approach [9]. For farmers' information need assessment, the qualitative approach promises some benefits such as obtaining an in-depth understanding of what information is most needed by farmers and finding indigenous information that was often unthinkable before.

Furthermore, many researchers have presented various perspectives concerning FMIS. Some authors highlighted its technical aspects [1,20] while others underscored the non-technical aspects [5,21,22]. These discussions, however, have proven to be restricted as they failed to consider smallholder farmers. To address this gap, this study proposes a conceptual model for Small Farm Management Information System (sFMIS) that is relatively distinctive compared to the existing FMIS conceptual model.

The development of the sFMIS conceptual model adheres to three principles. Firstly, it only provides the functionalities required by smallholder farmers to reduce the development cost. Secondly, it optimizes the use of open external data sources to reduce operational costs. Finally, it is available as a mobile-based application to reduce equipment expenditure.

However, deploying sFMIS can be challenging and presents many problems. One main drawback is understanding the information needed by farmers. Handling data interoperability problems relating to the use of external data sources is another challenge. Some problems associated with data interoperability are schema heterogeneity, schema granularity, entity naming mismatch, and data type mismatch [10]. Much research to date has tried to address the data interoperability problems associated with the usage of external data sources. There are three main problems in data interoperability at the schema level, namely schema heterogeneity, granularity data, and inconsistency field naming. Some researchers tried to tackle the problems by using the ontology matching approach. AgreementMaker [23], COMA++ [24], Cupid [25], Falcon-AO [26], and S-Match [27] are the most commonly used and discussed ontology matching approaches in the literature. Other researchers, on the other hand, tried to address the data interoperability problems using the database approach [28–30]. Despite these initiatives, no study so far has integrated all three obstacles that may arise in using external sources comprehensively.

Another thing that should be considered in application development is the adoption of new technology or application to target users. A new application is useless if it is not adopted by the target users. Indeed, introducing a new application to small farmers is not an easy challenge. Some aspects play an important role in adopting the new application, on the adopters' side, namely: technologies' technical features, users' perceptions (farmers and farm employees) of innovation attributes, and users' characteristics such as age, education level, and existing computer skills [31].

Indonesia is an agricultural country. There are 27,682,117 agricultural households in Indonesia and approximately 10,104,682 of them work in the horticulture sub-sector [32]. However, from the literature study conducted, no research has been found related to the use of FMIS in supporting farming in Indonesia. Several earlier studies focused on certain aspects of the FMIS ecosystem. For example, some studies focus on monitoring plant growth [33], climate and planting calendars [34], water management [35], the use of drones [36], and related product marketing [37]. In addition, there is no mobile application (Android) that offers the use of FMIS as an agribusiness ecosystem. Existing applications are related to cultivation (SIPINDO, MyAgri, Lumbungin, Digitani), pest control (drtania, Plantix), marketing (tanihub, sayurbox), and financing (iGrow).

2. Methodology

The methodology used in this study is a combination of several methods, namely the purposive sampling method in selecting the commodity and research area, simple random sampling in selecting respondents, Farmers' Information Needs Assessment (FINA) in identifying information mostly needed by farmers, and Soft System Methodology (SSM) in developing the conceptual model. The block diagram of the methodology is presented in Figure 1, as explained as follows:

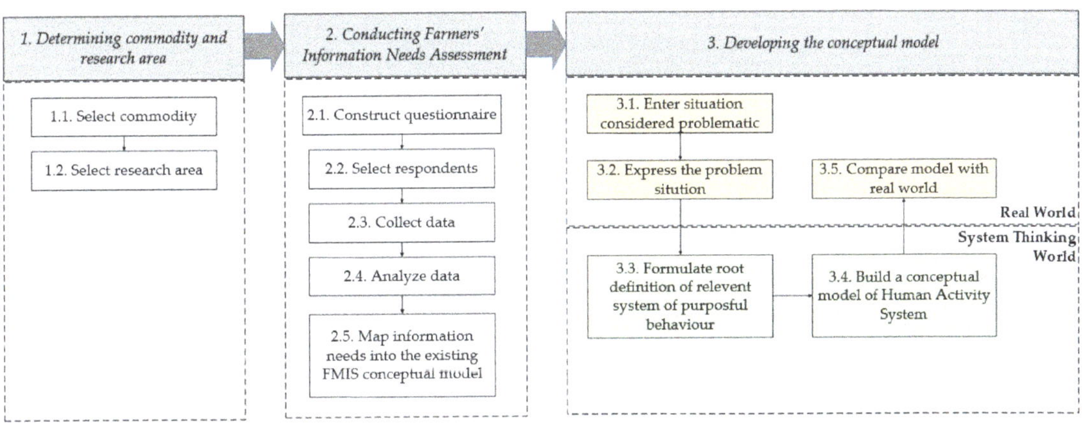

Figure 1. Research Methodology.

2.1. Determining Commodity and Research Area

The first step was determining the commodity and research area. Since the scope of the agricultural sector is extensive, various commodities or different agroecosystems require distinct handling methods. However, no method is suitable for all. Therefore, this study employed the case study approach. There are two activities in this step, namely the select commodity and the select research area.

2.1.1. Select Commodity

The purposive sampling method was applied to select the commodity. The commodity selection is based on two important considerations: it is a seasonal crop with a life cycle of fewer than six months, and it is traditionally cultivated by the majority of smallholder farmers in Indonesia.

Chili (*Capsicum annum* L.) was used as a case study because it is one of the strategic commodities in Indonesia. The need for chili in Indonesia continues to grow with the increase in income and population [38]. Additionally, chili farming ensures high profits in a relatively short time. Chili plants are ready to be harvested from the age of 3–4 months and the harvesting can be performed once a week until the plants are 6–7 months old [39]. Moreover, chili cultivation has a benefit to cost (B/C) ratio of 2.4 [39]. Despite the promising high profits, chili farming in Indonesia confronts a diverse range of challenges. Most Indonesian chili farmers are smallholder farmers, with each farmer cultivating only less than 0.5 hectares. They face a lack of information on new cultivation technology and in handling pests and diseases that often damage chili crops problems. They also struggle with high price fluctuations resulting from changes in supply and demand, and limitations in production capital and access to financing sources.

2.1.2. Select Research Area

The selection of research area started with selecting a province and was followed by selecting a district and sub-district. The purposive sampling method was employed. To select a province as a case study, statistical data analysis of the plantation area and average annual contribution to national chili production were considered. Thus, the West Java province was chosen based on the data obtained from the Central Bureau of Statistics [32,33]. The districts in this study, on the other hand, were selected based on three criteria: harvest area, distance to the research location, and willingness of the key informant. Thus, Sukabumi and Bandung Barat Districts were chosen as the research areas.

2.2. *Conducting Farmers' Information Needs Assessment*

The following step is assessing the farmers' information needed. This step includes four activities, namely: constructing a qualitative questionnaire, selecting respondents and conducting data collection, analyzing data, and mapping information needs into the existing FMIS conceptual model.

2.2.1. Construct a Qualitative Questionnaire

The first step in FINA was to construct a semi-structured qualitative questionnaire to elicit in-depth responses from the selected respondents in the study. The questionnaire was divided into two sections: the farmers' characteristics, and the farmers' information needs. The respondents' characteristics section consists of three groups of information: respondent's profile, ownership of the mobile phone, and a willingness to try a new application. Moreover, the farmers' information needs were categorized into five aspects following the agribusiness subsystems, namely: pre-planting, planting, harvesting, marketing, and supporting.

2.2.2. Select Respondents and Conduct Data Collection

The respondents were selected using simple random sampling. A total of 50 farmers were selected as the respondents, of which 27 farmers were from the Suntenjaya, a sub-district of Bandung Barat district, and 23 farmers from the Caringin and Kadudampit, a subdistrict of Sukabumi district.

The data collection was started with an explanation of the aim of the study and guidelines on how to fill out the questionnaire. The process continued with filling out the questionnaire by the respondents. In the end, an in-depth discussion with the key informant and some farmer representatives was carried out to collect more data.

2.2.3. Analyze the Data

The farmers' profiles were analyzed using the distribution frequency method. The analysis process started with grouping the data based on the "interval" or "bin" that had already been defined. The Data Analysis tools provided by Microsoft Excel were used to calculate the distribution frequency for each internal data.

The farmers' information needs were examined using word frequency analysis. Firstly, all collected data were divided into a list of words or terms using the text preprocessing method. In detail, text preprocessing consists of five consecutive processes: splitting into words, tokenizing, finding the root of the word, dropping unrelated words or terms, and grouping similar terms. After that, the frequency for each word/term is conducted by computing the occurrence frequency for each word or term and sorting the occurrence frequency from highest to lowest score.

2.2.4. Map Information Needs into the Existing FMIS functionality

The activities in this step include determining FMIS conceptual model as a reference, mapping into the existing reference model, and finding the gap between information needs with the reference model. All activities were executed manually. This study uses the FMIS conceptual model by Sorensen [40] as a reference conceptual model. The output produced in this activity formed the basis for deciding whether to apply the existing conceptual model or to propose a new one.

2.3. *Developing the Conceptual Model*

The next step is developing the conceptual model using Soft System Methodology (SSM). SSM is a cyclical learning system that utilizes different human activities to investigate the actors in the real-world problem situation, how they perceive that situation, and their readiness [41]. The main aim of this step is to decide on the appropriate activity, taking into account the perceptions, judgments, and values of various actors [42]. Although SSM originally consists of seven stages [41], it is not necessary to follow all of the phases [43]. For this study, only the first five stages of SSM were adopted to develop the model.

2.3.1. Identify the Existing Problem Situation

This first stage in SSM is to find out the problem situation and understand what the system is. Any possible problems that may be encountered were identified. This process was conducted through desk study and group discussion. The output of this step is all problems that may arise with the system that need to develop.

2.3.2. Convert the Problem Situation into a Structured Problem

The following stage is converting the problem situation into becoming structured problem. This was executed by organizing the unstructured problems that were already identified into structured problems. The mnemonic CATWOE (Customer, Actors, Transformation Process, Worldview, Owners, Environmental constraint) analysis method was employed to organize. The result of the CATWOE analysis was translated into a "Rich Picture", presenting a whole picture of the developed system.

2.3.3. Formulate Root Definition of Relevant System

The structured problems were analyzed to find the system under investigation using a root definition approach. A root definition is a sentence that describes the ideal system: What does the system do? How does it function? What is its purpose?

2.3.4. Build a Conceptual Model of the Human Activity System

The main purpose of this stage is to produce a model of what the system should execute. Data gathered from earlier works and previous models are used as references in the development of the conceptual model. The outcome is an FMIS conceptual model for smallholder farmers with Indonesian chili farmers as a case study.

2.3.5. Compare the Conceptual Model with the Identified Problem Situation

The final stage ensures that the identified problems have been addressed in the conceptual model, using a diligent mapping process for each part of the conceptual model into each identified problem.

3. Results

This study uses chili commodities as a case study. The 50 smallholder farmers from Bandung Barat and Sukabumi District, West Java province, were selected as respondents. Quantitative analysis of the respondents' characteristics indicates that the target respondents from the study are following the objectives of the study. Moreover, through the word frequency analysis of the questionnaire, the most information needed by smallholder farmers was identified. Additionally, through qualitative methods and in-depth discussion, some information that was not thought of before was found. However, not all identified information needs could be precisely mapped into the existing FMIS conceptual model. Therefore, a new FMIS conceptual model for smallholder farmers was proposed. Different from the existing conceptual model, the proposed model focuses on utilizing as much as possible external data sources, can handle data interoperability problems that may occur, and provides Android as an application interface platform.

3.1. *Farmers' Information Needs*

3.1.1. Analyzing the Respondent's Characteristics

Three groups of information in respondents' characteristics were included in the analysis, namely: respondents' profile, mobile phone ownership, and a willingness to try a new application. The descriptive analysis was used to analyze the respondents' characteristics. The analysis showed that the majority of respondents are smallholder farmers, they have a mobile phone and are willing to install a new application under some conditions.

Respondents' Profiles

The basic demographic features of respondents are shown in Table 1. The respondents had a mean age of about 38 years old and were mostly composed of low-level educated individuals, smallholder farmers, and renters of farmland for their cultivation. Approximately 78 percent of the respondents cultivated rented farmland, and 20% cultivated on their owned farmland. Additionally, the respondents cultivate on land with an average area of roughly 0.6 hectares, in a range between 0.1 and 2 hectares. However, the respondents had an average of 14 years of on-farm experience, in the range of 1 to 35 years.

Table 1. Demographic characteristics of respondents.

Variable	Total Respondents $n = 50$	p-Value
Age (mean ± SD, range (in years old))	37.6 ± 9.7 (16–57)	<0.01
Education Level (n, %)		
Primary school	25 (50%)	
Secondary School	16 (32%)	<0.01
High school	8 (16%)	
Bachelor	1 (2%)	
Experience (mean ± SD, range, in years)	13.5 ± 9.2 (1–35)	<0.01
Land ownership (n, %)		
Owned	10 (20%)	
Rental	39 (78%)	<0.01
Owned and rental	1 (2%)	
Cultivation area (mean ± SD, range, in Ha)	0.57 ± 0.48 (0.1–2)	<0.01

Mobile Phones Ownership

The mobile phone ownership of the respondents is shown in Table 2. It was found that 72% of the respondents owned a mobile phone, mostly an Android phone. In addition, approximately 48 percent of respondents (or 77 percent of those using an Android phone) subscribed to a monthly internet subscription, with slightly more than half of them (58%) spending over Rp. 50,000 (US $4).

Table 2. Mobile phone ownership of smallholder chili farmers in Sukabumi and Bandung Barat districts.

Variable	Criteria/Range	Frequency	Percentage
Mobile phone ownership	No	14	28%
	Yes	36	72%
	Total	50	100%
Mobile phone operating system	Android	31	62%
	Feature phone	5	10%
	No phone	14	28%
	IOS	0	0%
	Total	50	100%
Subscription to an Internet package	No	7	14%
	Yes	24	48%
	No phone	14	28%
	Feature phone	5	10%
	Total	50	100%
Average expenditure on monthly data package (in IDR)	0–25,000	3	6%
	25,001–50,000	7	14%
	50,001–75,000	6	12%
	75,001–100,000	8	16%
	No phone	14	28%
	Not support	5	10%
	Total	50	100%

Willingness to Try a New Application.

Another important aspect of the respondent's profile is their willingness to try a new application. It was found that 97.84 percent of the respondents considered trying new applications if they met certain criteria, such as the application supporting their agricultural farming activities, the application providing direct discussion to experts/extension workers, and the application providing facilities for marketing their product. Analysis of data also revealed three main factors influencing farmers to try new applications: ease of installation and use; benefits they obtain; new experiences in using technology.

3.1.2. Analyzing the Farmers' Information Needs

The specific information needs of the farmers were assessed based on their responses using a word frequency method. There were 1298 pieces of information derived from splitting, tokenization, and finding the root of the word. The details of all identified information were grouped based on the similarity terms into 32 types of information needed are presented in Appendix A Table A1.

Furthermore, all of the required information was sorted based on the occurrence frequency to find the top ten information mostly needed by smallholder farmers. The in-depth discussion led to additional required information. All identified farmers' information needed is presented in Table 3.

Table 3. The smallholder farmers' information needs in Sukabumi and Bandung Barat Districts.

Source of Data	Information Needed	Occurrence Frequency
Word frequency analysis	cultivation technology	233
	market price	128
	agricultural financing	106
	land preparation	97
	consultation	96
	market demand	68
	handling pest and disease	65
	another region with the same crop	60
	seed description	59
	weather forecast	54
In-depth discussion	farmland location	n.a
	Farmland owner	n.a
	the existing crop that is being cultivated	n.a
	financial record keeping	n.a
	recording their cultivation activities	n.a

The "n.a" in occurrence frequency indicates that the information needs are obtained from the in-depth discussion with representative farmers process and do not obtained from the qualitative questionnaire that the respondents filled in. The in-depth discussion was conducted with the "key informant" or "pioneer' farmer" and several senior farmer representatives, namely farmers who have more than 10 years of farming experience. With in-depth discussion find information or idea that was not thought of before.

3.1.3. Mapping Farmers' Information Needs into Existing FMIS Functionalities

Matching and mapping each identified information need with the reference conceptual model was performed manually. The result of the mapping process presented in Figure 2 indicated that some information needs could not be mapped exactly on the referenced FMIS conceptual model [40]. This demonstrates that not all functionalities provided in the FMIS conceptual model required smallholder farmers; thus, establishing the importance of a new conceptual model for the farmers to fully benefit from FMIS.

3.2. *Develop a Conceptual Model*

The result of the farmers' information needs assessment was used to develop the conceptual model. The proposed conceptual model consists of five layers, namely: farmers' information needs layer, assess the data sources quality layer, data extraction layer, split-match-merge layer, and presentation/user interface layer. The detailed processes of developing the conceptual model are as follows.

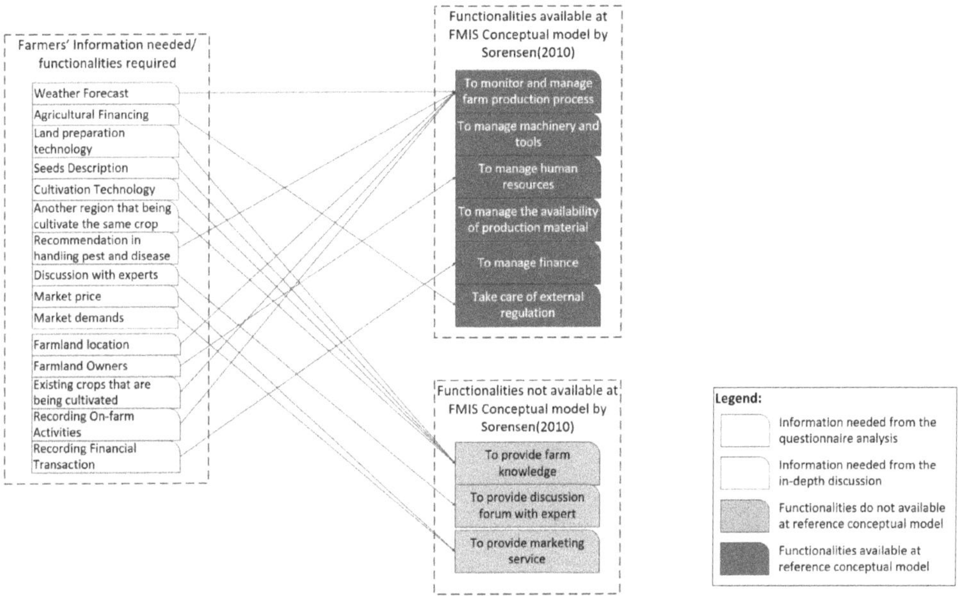

Figure 2. Mapping smallholder farmers' information needs into the FMIS Conceptual Model.

3.2.1. Identifying the Existing Problem Situation

As explained in the background section, providing an FMIS to smallholder farmers has various challenges, including requiring a comprehensive understanding of the information preferred by farmers and handling data interoperability problems associated with the use of external data sources. Based on these challenges, several critical questions are considered in providing sFMIS:

- Where do the data sources for each piece of information come from? Is this data available online?;
- What is the quality of each candidate's data source? Are all external data sources eligible for extraction, transformation, and loading?;
- Data sources are available in many formats, what method is used to extract each data source into a temporary database?;
- What is the process of transforming and loading data from the temporary database into the application database? How the algorithm could tackle the data interoperability problems that may arise in transforming and loading data process?;
- How can information be presented to users in an easy, inexpensive, and user-friendly way?

3.2.2. Converting the Problem Situation into a Structured Problem

The unstructured problems identified earlier are organized. When using external data sources, the following functionalities should be provided:

- External data source quality assessment;
- Data extraction for each eligible candidate's external data source into the temporary database;
- Data loading and transformation can handle data interoperability problems that may arise.
- A friendly User Interface (UI).

Furthermore, the CATWOE method was employed, and the following items were obtained:

- Customer: the primary actor of this model is the farmer, and the secondary actors are traders, experts/extension workers, and local government officers;
- Actors: the primary actors of this system are external websites that supplied data to support the android application. Whereas secondary actors are a group of users that interact with the android application, such as farmland owners, farmers, traders, and other data providers;
- Transformation process: in collecting and inputting data, manually inputting data transformed into an automatic process through extracting and loading data from many external data sources;
- Worldview: external data sources that can potentially be reused by the system to help farmers decide on aspects related to their farm;
- Owners: the primary owner of this system is the researcher who develops the system, while the secondary owners are the organizations who implement and manage the system;
- Environmental constraints: the primary constraints in developing the systems are the quality of data provided by an external website and access rights to external data sources. Whereas the secondary constraints are quality of infrastructure. Minor constraints are the quality of network or Internet infrastructure when collecting data from external data sources.

Moreover, to have a whole view of the system, the result of CATWOE analysis was drawn into a "Rich Picture" as shown in Figure 3. The core element of this rich picture is an sFMIS android application with four main customers/targeted users, namely farmers, farmland owners, traders, and experts/extension workers. The android application has support data from the application database through an API (Application Programming Interface) service platform. There are two data sources for the application database, namely manual data entry through the application e-form and data as a result of the extract, transform, and load (ETL) process from the temporary database. The ETL process also handles the data interoperability problem that may occur; whereas the temporary database itself is a container of the data extraction process from many external data sources.

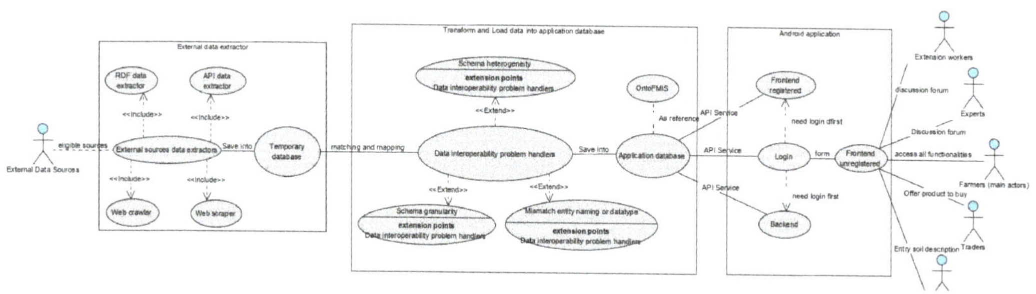

Figure 3. A rich picture in UML diagram format of the small Farm Management Information System (sFMIS).

3.2.3. Formulating the Root Definitions of Relevant Systems

The next stage is formulating a root definition for the relevant system performed by answering the three main questions as follows:

- What does the system do? The system can collect data from many external websites and conduct the assessment of the quality of data sources, extract data, transform them and load them into the application storage using the Android application;
- How does it function? The system will perform some functions, starting with identifying candidate sources of data, collecting data from many external websites, assessing the quality of the data sources, extracting, transforming, and loading data into the

application storage, and providing an android application as the interface between system and users;

- What is its purpose? The system should be easy to use by smallholder farmers. Moreover, this system should provide information deemed necessary by users. Additionally, the system should have the capability to interact with smallholder farmers, especially in managing their crop production process;

3.2.4. Building a Conceptual Model of the Human Activity System

After understanding the root definitions for the system, the next stage is creating a conceptual model. All of the data collected in the earlier steps were compiled and analyzed to develop the conceptual model. The result of developing the sFMIS conceptual model for Indonesian Chili Farmers as presented in Figure 4 consists of five layers, namely: (a) farmers' information needs layer, (b) assess the data quality layer, (c) data extraction layer, (d) split, match and merge layer and € presentation/user interface layer. The detailed explanation for each layer is as follows:

a Farmers' Information Needs Layer

This layer consists of the list of farmers' information needs is the result from the FINA as shown in Table 3. All of the information needs are a combination of the result of two analysis methods, word frequency analysis and an in-depth discussion summary with the key informant and two senior farmers.

b The data quality assessment layer

The second layer of the model involves assessing the data quality of all candidate data sources. The first activity in this layer is to identify the candidate data sources for each functionality. All identified candidate data sources for each sFMIS functionality are shown in Appendix B Table A2.

The candidate data sources for each functionality were retrieved by "googling" related keywords and other sources of information. Among 15 functionalities required, 9 functionalities found candidate external data sources and 6 functionalities required manual data entry. Furthermore, the process continued with assessing the quality of data sources. This is a fundamental process because data from various external sources have a variety of formats, platforms, levels of detail, and ownership models. Several researchers have proposed assessment dimensions to evaluate data quality [44–49]. For this study, eight of the most significant assessment dimensions from a combination of several references are presented in Table 4. The weighting score for each dimension was calculated using AHP (Analytical Hierarchy Process) method with a consistency index (CI) = 0.060395782 and consistency ratio = 0.042833888. The detailed scoring criteria for each dimension are also shown.

Table 4. The eight dimensions of data quality assessment in Small Farm Management Information System.

No.	Dimensions	Weighting Score	Scoring Criteria				
			5	6	7	8	9
1	Accessibility	0.28	protected	login and sent via email	login and download file	free with key	free access
2	License	0.22	copyright	limited free for registered user	free for registered user	free limited service	free
3	Source reliability	0.17	personal blog/others	others' company	Other organization	well-known company	Government/ international org
4	Connectedness	0.13	others	pdf	html	XLS/csv	API/RDF
5	Accuracy	0.09	very low	low	medium	high	very high

Table 4. *Cont.*

No.	Dimensions	Weighting Score	Scoring Criteria 5	6	7	8	9
6	Completeness	0.06	20%	40%	60%	80%	100%
7	Format Consistency	0.04	not use standard, inconsistent	not use standard, inconsistent	not use standard, consistent	use standard, inconsistent	use standard, consistent
8	Timeliness	0.02	never	seldom	sometime	often	always

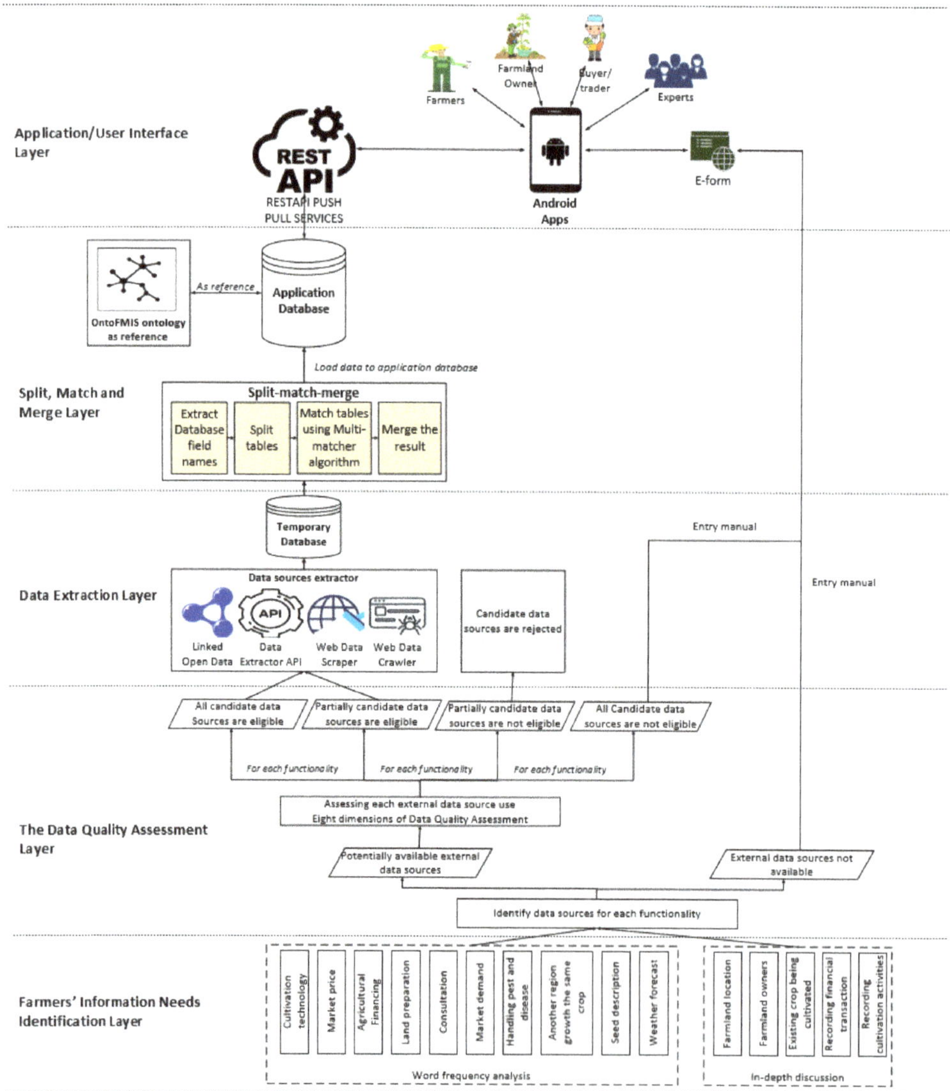

Figure 4. The small Farm Management Information System Conceptual Model for Indonesian Chili Farmers.

Moreover, assessing the data quality for each candidate's external data source uses a multi-criteria decision-making (MCDM). Appendix C Table A3 presented the details assessment score for each candidate's external data source. Among 35 candidate external data sources, 26 were accepted and eligible as external data sources, and 9 were rejected. However, some accepted data sources require manual intervention before they could be used, such as seed description and handling pest disease. Additionally, three types of candidate external data sources were rejected: those with un-supported data format, granularity data, and access-rights problems. Most of the data coming from the android application were rejected due to technical constrains.

c Data extraction layer

The data extraction layer consists of four data extraction methods depending on the format of the data provided. The first one is Linked Open Data (LOD), a tool used to data extract the external data sources that are available in RDF format. The second is the data extractor API, a tool used to extract data from the external data sources that provide the API service. The third method is a web crawler, which draws out the external data source that is available on static pages. The last one, the web scraper, extracts the external data source that is available on dynamic pages. The output of data extraction processes will then be saved in temporary storage for the next layer to transform and load the data. The flowchart to select the data extraction method based on the data provided by external data sources is presented in Figure 5.

d Split, match, and merge layer

The following layer in the conceptual model involves splitting, matching, and loading the data. This layer is the most important in this conceptual model since the quality of the information provided to the users depends on it. Data are loaded from the temporary database into the application database, as illustrated in Figure 6.

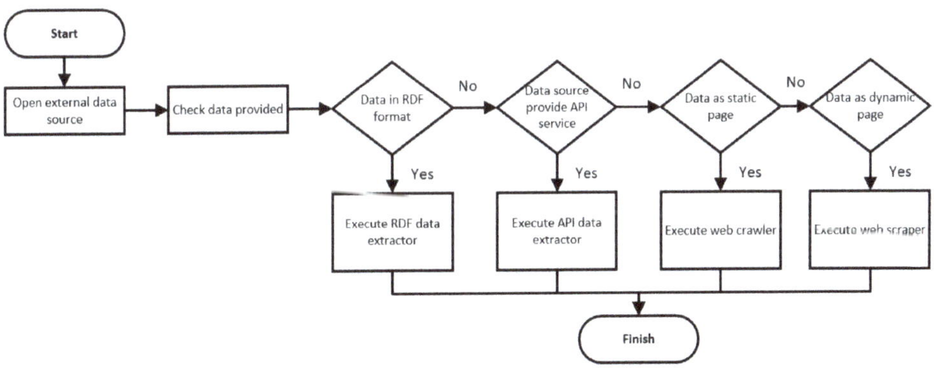

Figure 5. The flowchart in selecting data extractor tools.

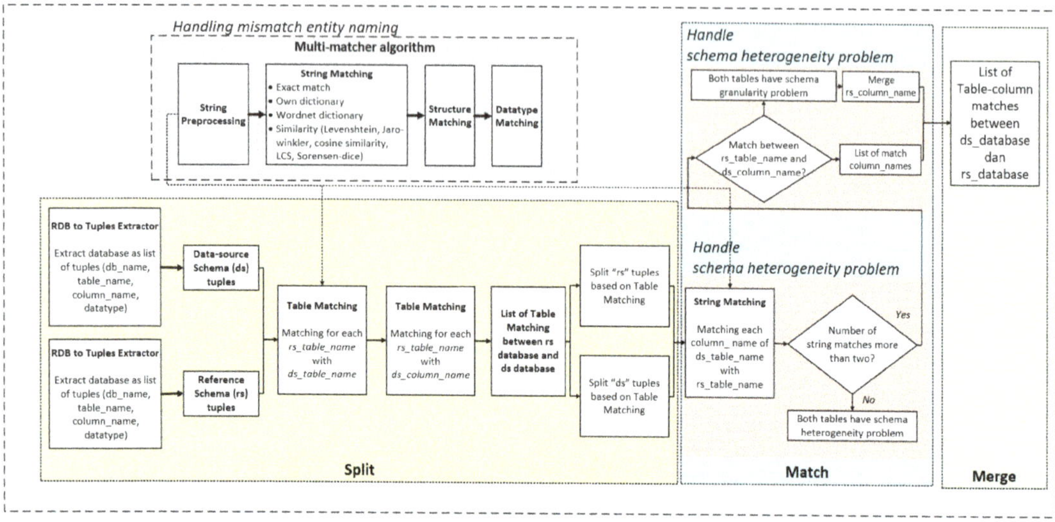

Figure 6. The details of the split, match and merge method with a multi-matcher algorithm.

The "split-match-merge method with multi-matcher algorithm" [10] was employed for this step. This layer consists of a series of activities. The first activity is converting both application and data source databases into a set of tuples. The application database developed referred to the OntoFMIS (http://103.169.28.91/ontofmis/, accessed on 22 March 2022), an ontology for small farm management information systems. Furthermore, the following activity is class matching between data source table_name and reference table_name. The next activity is is extracting each schema based on the result of class matching. For each class matching, the process continued with column name matching and datatype matching. Moreover, this layer also handles data interoperability problems that may arise, such as schema heterogeneity, the granularity of data, mismatch data type, and mismatch field naming. The output of this layer is an application database that is ready for Android applications to supply data to users.

The split-match-merge method is a modification of the method proposed by Wickham [50]. This method is aimed at making the matching and mapping process run effectively and efficiently. Through this method, the matching and mapping process only executes one-to-one matching between temporary database table structure application database table structure. While the multi-matcher algorithm is the heart of this model. The quality of data presented to the users depends on the proses inside the multi-matcher algorithm. A combination of hybrid and composite from many matcher algorithms was used to construct this algorithm. All data interoperability problems that may arise are tackled using this algorithm. Matcher algorithms that executed cascade in this paper are as follows:

- Exact-matcher: a matcher to find whether two words exactly match;
- Word synonym: a matcher to check whether two strings are synonyms. A total of 3 sources were used as of dictionary, namely (1) Wordnet in English, (2) Wordnet in the Indonesian Language, and (3) creating own dictionary;
- Similarity matcher: a matcher to find the similarity of two terms. This study uses a hybrid model as a combination of five string-similarity algorithms, namely: (1) Levenshtein, (2) Jaro-Winkler, (3) Cosine similarity, (4) Longest Common Substring (LCS), and (5) Sorensen-dice. The result for each algorithm was normalized and following with calculation the average similarity with the acceptance score is 0.9.

Moreover, the evaluation performance of the 'split-match-merge method with multi-matcher algorithm" to tackle the three data interoperability uses the precision, recall, and accuracy indicators. The result of evaluating the algorithm to handle each data interoperability problem is as follows:

- Mismatch entity naming. The mismatch entity naming is conducted by matching each entity naming of the temporary database with all entity naming of the application database. There are 236 terms extracted from the temporary database and 251 terms were extracted from the application database. Table 5 present the calculation confusion matrix for handling the mismatch entity naming problem.

Table 5. Calculation TP, TN, FP, and FN in handling mismatch entity naming.

		Manual Checking	
		True	**False**
A split-match-merge method with a multi-matcher algorithm	Positive	TP = 138	FP = 18
	Negative	TN = 59,068	FN = 4

The accuracy, precision, and recall calculation consecutively as follows

$$\text{Accuracy} = (\text{TP} + \text{TN})/(\text{TP} + \text{FP} + \text{FN} + \text{TN})$$
$$= 0.88$$
$$\text{Precision} = \text{TP}/(\text{TP} + \text{FP})$$
$$= 0.97$$
$$\text{Recall} = \text{TP}/(\text{TP} + \text{FN})$$
$$= 0.94$$

- Schema heterogeneity problem. The handling schema heterogeneity problem is executed through two steps, matching the table name, and matching the entity naming for each matched table. Table 6 presents the calculation of the confusion matrix in handling the schema heterogeneity problem.

Table 6. Calculation TP, TN, FP, and FN in handling schema heterogeneity problem.

		Manual Checking	
		True	**False**
A split-match-merge method with a multi-matcher algorithm	Positive	TP = 186	FP = 6
	Negative	TN = 6874	FN = 8

The accuracy, precision, and recall calculation in a row are as follows

$$\text{Accuracy} = (\text{TP} + \text{TN})/(\text{TP} + \text{FP} + \text{FN} + \text{TN})$$
$$= 0.98$$
$$\text{Precision} = \text{TP}/(\text{TP} + \text{FP})$$
$$= 0.97$$
$$\text{Recall} = \text{TP}/(\text{TP} + \text{FN})$$
$$= 0.96$$

- Schema granularity problem. The handling schema granularity problem is executed by matching the entity naming of the table of the application database into the table name of the temporary database and matching the entity naming of the table of the temporary database into the table name of the application database. The confusion matrix in handling schema heterogeneity problem is presented in Table 7.

Table 7. Calculation TP, TN, FP, FN in handling schema granularity problem.

		Manual Checking	
		True	**False**
A split-match-merge method with a multi-matcher algorithm	Positive	TP = 88	FP = 5
	Negative	TN = 30,139	FN = 12

The accuracy, precision, and recall calculation successively as follows

$$\text{Accuracy} = (\text{TP} + \text{TN})/(\text{TP} + \text{FP} + \text{FN} + \text{TN})$$

$$= 0.99$$

$$\text{Precision} = \text{TP}/(\text{TP} + \text{FP})$$

$$= 0.946$$

$$\text{Recall} = \text{TP}/(\text{TP} + \text{FN})$$

$$= 0.88$$

- Handling data interoperability problems comprehensively. Besides the performance in handling data interoperability problems individually, the algorithm performance evaluation is also conducted in handling data interoperability problems comprehensively. Table 6 present the result of the confusion matrix of TP, TN, FP, and FN.

  The accuracy, precision and recall calculation consecutively as follows

$$\text{Accuracy} = (\text{TP} + \text{TN})/(\text{TP} + \text{FP} + \text{FN} + \text{TN})$$

$$= 0.998$$

$$\text{Precision} = \text{TP}/(\text{TP} + \text{FP})$$

$$= 0.969$$

$$\text{Recall} = \text{TP}/(\text{TP} + \text{FN})$$

$$= 0.959$$

e Presentation/user interface layer

The top layer of this model is the presentation layer which consists of two main components. The first component is the API RESTful service server. This service is used the Android applications utilizing information provided by the application database. This component provides a push and pull service for storing and providing data. The second component is an android application which acts as an interface between the system and the users. Besides providing information, it also provides an e-form to enter data into the system. There are five target users for this Android application, namely: smallholder farmers as the main users, and four groups of the user as a supporting system, including experts, extension workers, traders, and soil specialists. Expert and extension workers have a role to answer all questions or requests for information raised by farmers. The trader's role is to submit information into the application if they need to buy chili stock owned by farmers or offer to buy chili from farmers, whereas a soil specialist is a person who appointed by authorized institutions to provide information related to the characteristics of farmland managed by farmers.

3.2.5. Comparing the Conceptual Model with the Identified Problem Situation

This stage aims to ensure that the conceptual model fully considers all of the problem situations that were identified. The first layer of the conceptual model relates to a list of farmers' information needs. Moreover, the second layer deals with the functionality of assessing the quality of the external data source of each candidate whereas the third layer handles the extraction of data from each eligible candidate's external data source into the temporary database. The fourth layer transforms and loads data into the application database. The last layer of the conceptual model deals with the functionality of a front-end application. Based on these facts, we can conclude that the model considers all of the identified problems.

4. Discussion

FMIS helps farmers to manage their farms effectively and efficiently. However, the existing FMIS application is relatively expensive for smallholder farmers [6,51]. This study proposed a new conceptual model of FMIS that is to satisfy the smallholder farmers' information needs [18]. The proposed conceptual model consists of five consecutive layers. To make an FMIS that is suitable for smallholder farmers, the development of the sFMIS conceptual model adheres to three principles. Firstly, it only provides the functionalities required by smallholder farmers to reduce the development cost. Secondly, it optimizes the use of open external data sources to reduce operational costs. Finally, it is available as a mobile-based application to reduce equipment expenditure.

What distinguishes it from the existing conceptual model is identifying farmers' information needs using qualitative approaches as the first layer of the model. This layer is important because the application aimed to provide the information needed by smallholder farmers. The use of a semi-structured qualitative questionnaire in this study revealed information needs that were not considered during the research process such as regions that grew the same crop and long-term weather forecast. The mapping of the farmers' information needs showed that not all information could be represented accurately with the functionalities of the FMIS conceptual model. Several modifications to the existing FMIS conceptual model are required to meet the needs of smallholder farmers. This also proof that a new FMIS conceptual model that applies to smallholder farmers is essential.

The second layer promotes the different methods on how to assess the quality of external data sources [45,47]. The eight dimensions that are equipped with assessment criteria for each dimension will make it easier for users to apply the developed conceptual model. Moreover, the third layer focus on the data extraction process from all eligible external data sources. Data extraction from many external data sources conducted in many ways depending on the data provided by external data sources. The REST API is used as the standard method to extract data. The web-crawling and web-scraping method is used to extract data if the external data sources do not provide API service.

The split, match and merge method are the most important layers of the conceptual model. The quality of data provided to end-users depends on the quality of data transforming in this step. This study uses a "split-match-merge method with multi-matcher algorithm" [10] to comprehensively address the three main data interoperability problems. Using a different similarity dictionary that is constantly evolving is the key factor that distinguishes the algorithm from others. Moreover, the algorithm employs a multi-matchers algorithm for term matching to improve the matching result. The algorithm that can tackle three main data interoperability problems is a differentiator from the current FMIS conceptual model. The evaluation of the algorithm performance using accuracy, precision, and recall indicators have shown that the algorithm could tackle the data interoperability problems very well. The calculation of the confusion matrix indicates that True Negative (TN) has a score which is too high compared with other indicators. This is happening because each term extracted from the temporary database only has potential matching with only one term extracted from the application database. On the other hand, the total iteration is equal to all terms from the temporary database multiplied by all terms extracted from the application database.

The top layer is an Android application as an interface between the application and users. Choosing an Android application rather than a desktop-based application because the mobile performed better on user adoption, engagement, and retention [52]. Moreover, the usage of mobile phones will reduce the need for buying a desktop computer for running the application [5].

The proposed small farm management information system conceptual model was developed, explicitly based on the case study of chili farmers in West Java taking into account distinct requirements of external data sources for different commodities or different regions. Therefore, when applying the model to other commodities, some layers should be modified. Firstly, the information required for farmers' information needs assessment layer must be reestablished. Additionally, assessing the candidate external data source layer need to be reclassified. Another modification is required for the keywords in the dictionary used for improved results in transforming and loading data into the application database layer. Lastly, the functionality for SIMUSTI must be changed and adapted to the application layer.

5. Conclusions

The results from conducting the farmers' information need assessment produced a list of information most required by smallholder farmers. However, not all of the required information could specifically be mapped into the reference conceptual model of FMIS. This means that the smallholder farmers studied do not require some of the functionalities provided in the FMIS conceptual model. On the other hand, some functionalities required by smallholder farmers are not facilitated by the existence of the FMIS conceptual model proving that the requirement to develop an FMIS differs for smallholder farmers. This study, therefore, proposes a new FMIS conceptual model for smallholder farmers termed small Farm Management Information System (sFMIS). The provision of functionalities that meet the needs of smallholder farmers and the use of external data based on the data interoperability model are the most distinguishing features of sFMIS compared to the existing Farm Management Information System (FMIS). Therefore, the sFMIS ensures better farm management for smallholder farmers at affordable prices in application development and less operational costs. The sFMIS model uses Indonesian chili farmers as a case study and consists of five layers that can be applied to other commodities, with some modifications.

Author Contributions: Conceptualization, H.H., V.E. and C.A.; data collection, H.H.; methodology, H.H.; supervision, V.E. and C.A.; writing—original draft, H.H.; editing and correction, V.E. and C.A. All authors have read and agreed to the published version of the manuscript.

Funding: This research was fully supported by the Sustainable Management of Agricultural Research and Dissemination (SMARTD) Project (World Bank P117243)—Indonesian Agency for Agricultural Research and Development—Ministry of Agriculture.

Institutional Review Board Statement: Not applicable.

Informed Consent Statement: Not applicable.

Data Availability Statement: OntoFMIS, an ontology for small farm management information systems is temporarily available at: http://103.169.28.91/ontofmis/, accessed on 22 March 2022. Other data presented in this study are available on-demand from the first author at (st118501@ait.ac.th).

Acknowledgments: We thank our colleagues from the Indonesian Agency for Agricultural Research and Development (IAARD) and the Asian Institute of Technology (AIT), who provided insight and expertise that greatly assisted the research and prototype development.

Conflicts of Interest: The authors declare no conflict of interest. Funders have no role during the research process of this manuscript. The research process starts with the research design, collection, analysis, interpretation of data, writing a script, and publishing of the results.

Appendix A

Table A1. The Frequency of Occurrence of Any Information Needed by Respondents in Sukabumi and Bandung Barat Districts.

No.	Agribusiness Sub-System	Information Needed	Occurrence Frequency
1	Pre-planting	agricultural financing	106
		land preparation	97
		another region with the same crop	60
		seed description	59
		machinery description	46
		agricultural machinery	23
		agricultural insurance	15
		seed production technology	9
		soil characteristic	6
		seed availability	5
		fertilizer and seed subsidies	3
2	Planting	cultivation technology	233
		handling pest and disease	65
		weather forecast	54
		seed recommendation	4
		labor availability	2
3	Harvesting	packaging	22
		storage technology	19
		grading	5
		yield processing	3
		Warehouse	2
4	Marketing	market price	128
		market demand	68
		marketing	42
		transportation	41
5	Support	consultation	96
		training	40
		assistance	20
		regulation	11
		management	10
		technical support	4
	Total		1298

Appendix B

Table A2. Candidate Data Sources for Each Small Farm Management Information System (sFMIS) Functionality.

No.	Functionalities	Candidate Data Sources (If Available)
1	Farmland location	Google Map, Open Street Map
2	Weather forecast	Open Weather Map, BMKG, World Bank
3	On-farm activities	Manual data entry
4	Handling pest and disease	Opete, IAARD, ICHORT, MyAgri, SIPINDO
5	Existing crops that are being cultivated	Manual data entry
6	Farmland owners' profile	Manual data entry
7	Financial transaction recording	Manual data entry
8	Agricultural financing	KUR (Kredit Usaha Rakyat)/people's business credit from Ministry Coordinator of Finance, Google News
9	Land preparation technology	Cyber extension, IAARD, Youtube
10	Seeds description	PPVT, IAARD, MyAgri, DBVaritas
11	Cultivation technology	Cyber extension, IAARD, Youtube, Repositori Publikasi, SIPINDO, ITani, Digitani
12	Another region cultivating the same crop	Generate by application
13	Consultation	Manual data entry
14	Market demand	Manual data entry
15	Market price	Toko tani BKP, PIHPS, Shopee, Bukalapak, Sayurbox, Tanihub, Info pangan Jakarta

Appendix C

Table A3. Data Quality Assessment Based on Eight Dimensions Criteria.

No.	Functionality	Candidate External Data Source	Access w = 0.28	Lice w = 0.22	Sour w = 0.17	Conn w = 0.13	Accu w = 0.09	Comp w = 0.06	Cons w = 0.04	Time w = 0.2	Total Score	Decision ≥8
1	Farmland location	Open Street Map	2.52	1.76	1.53	1.04	0.81	0.48	0.36	0.18	8.68	accepted
		Google Map	2.52	1.98	1.53	1.17	0.81	0.54	0.36	0.18	9.09	accepted
2	Weather forecast	Forecast—OWN	2.24	1.76	1.53	1.17	0.81	0.54	0.32	0.18	8.55	accepted
		BMKG	2.52	1.54	1.53	1.04	0.81	0.42	0.32	0.16	8.34	accepted
		World Bank	2.24	1.76	1.53	1.17	0.81	0.42	0.36	0.16	8.45	accepted
3	Agricultural Financing	Ministry Coordinator of Finance	2.52	1.98	1.53	0.91	0.81	0.54	0.28	0.18	8.75	accepted
		News—Google search	2.52	1.98	1.53	0.91	0.81	0.54	0.28	0.18	8.75	accepted
4	Land Preparation	IAARD	2.52	1.98	1.53	0.91	0.72	0.54	0.32	0.18	8.70	accepted
		Youtube	2.52	1.98	1.53	0.78	0.81	0.48	0.32	0.16	8.58	accepted
		Cyber extension	2.52	1.98	1.53	0.91	0.81	0.54	0.32	0.16	8.77	accepted

Table A3. *Cont.*

No.	Functionality	Candidate External Data Source	Access w = 0.28	Lice w = 0.22	Sour w = 0.17	Conn w = 0.13	Accu w = 0.09	Comp w = 0.06	Cons w = 0.04	Time w = 0.2	Total Score	Decision ≥8
5	Cultivation technology	IAARD	2.52	1.98	1.53	0.91	0.81	0.48	0.32	0.16	8.71	accepted
		Youtube	2.24	1.98	1.36	1.17	0.72	0.54	0.32	0.18	8.51	accepted
		Cyber extension	2.52	1.98	1.53	0.91	0.81	0.54	0.32	0.16	8.77	accepted
		Repositori publikasi	2.52	1.98	1.53	0.91	0.81	0.54	0.32	0.16	8.77	accepted
		Sipindo	1.40	1.10	1.36	0.65	0.81	0.42	0.20	0.16	6.10	rejected
		Itani	1.40	1.10	1.53	0.65	0.81	0.42	0.20	0.16	6.27	rejected
		Digitani	1.40	1.10	1.53	0.65	0.81	0.42	0.20	0.16	6.27	rejected
6	Seed description	PPVT	1.40	1.10	1.53	0.65	0.81	0.48	0.32	0.10	6.39	rejected
		DBVaritas—DG of Horticulture	2.52	1.98	1.19	0.91	0.72	0.48	0.28	0.16	8.24	accepted
		MyAgri	1.40	1.10	1.53	0.65	0.81	0.42	0.20	0.16	6.27	rejected
7	Handling Pest disease	IAARD	2.52	1.98	1.53	0.91	0.81	0.36	0.32	0.16	8.59	accepted
		OPETE	2.52	1.98	1.19	0.91	0.81	0.36	0.32	0.16	8.25	accepted
		MyAgri	1.40	1.10	1.53	0.65	0.81	0.42	0.20	0.16	6.27	rejected
		Sipindo	1.40	1.10	1.36	0.65	0.81	0.42	0.20	0.16	6.10	rejected
		Expert system—ICHORD	2.52	1.98	1.53	0.91	0.81	0.36	0.32	0.14	8.57	accepted
8	Market price	Toko tani—BKP	2.52	1.98	1.53	0.91	0.72	0.42	0.28	0.16	8.52	accepted
		PIHPS	2.52	1.98	1.53	0.91	0.72	0.42	0.28	0.16	8.52	accepted
		Shopee	2.52	1.98	1.19	0.91	0.72	0.54	0.28	0.16	8.30	accepted
		Bukalapak	2.52	1.98	1.19	0.91	0.72	0.54	0.28	0.16	8.30	accepted
		Sayurbox	2.52	1.98	1.19	0.91	0.72	0.48	0.28	0.16	8.24	accepted
		Tanihub	2.52	1.98	1.19	0.91	0.72	0.42	0.28	0.16	8.18	accepted
		info pangan jakarta	2.52	1.98	1.53	0.78	0.63	0.42	0.24	0.16	8.26	accepted

Note: Access: Accessibility; Lice: License; Sour: Source reliability; Conn: Connectedness; Accu: Accuracy Comp: Completeness Cons: Format Consistency; Time: Timeless.

References

1. Burlacu, G.; Costa, R.; Sarraipa, J.; Jardim-Goncalves, R.; Popescu, D. A conceptual model of Farm Management Information System for Decision Support. In *Technological Innovation for Collective Awareness Systems*; DoCEIS 2014. IFIP Advances in Information and Communication Technology; Springer: Berlin/Heidelberg, Germany, 2014; Volume 423. [CrossRef]
2. Fountas, S.; Carli, G.; Sørensen, C.G.G.; Tsiropoulos, Z.; Cavalaris, C.; Vatsanidou, A.; Liakos, B.; Canavari, M.; Wiebensohn, J.; Tisserye, B. Farm management information systems: Current situation and future perspectives. *Comput. Electron. Agric.* **2015**, *115*, 40–50. [CrossRef]
3. Tummers, J.; Kassahun, A.; Tekinerdogan, B. Obstacles and features of Farm Management Information Systems: A systematic literature review. *Comput. Electron. Agric.* **2019**, *157*, 189–204. [CrossRef]
4. Husemann, C. The Model of Farm Management Information System: A Case-Study of Diversified German Farm Model. *DETUROPE Cent. Eur. J. Reg. Dev. Tour.* **2012**, *4*, 76–90. [CrossRef]
5. Fountas, S.; Sorensen, C.G.; Tsiropoulos, Z.; Cavalaris, C.; Liakos, V.; Gemtos, T. Farm machinery management information system. *Comput. Electron. Agric.* **2015**, *110*, 131–138. [CrossRef]
6. Kaloxylos, A.; Groumas, A.; Sarris, V.; Katsikas, L.; Magdalinos, P.; Antoniou, E.; Politopoulou, Z.; Wolfert, S.; Brewster, C.; Eigenmann, R.; et al. A cloud-based farm management system: Architecture and implementation. *Comput. Electron. Agric.* **2014**, *100*, 168–179. [CrossRef]
7. Graeub, B.E.; Chappell, M.J.; Wittman, H.; Ledermann, S.; Kerr, R.B.; Gemmill-Herren, B. The State of Family Farms in the World. *World Dev.* **2016**, *87*, 1–15. [CrossRef]
8. Nguo, J.; Mwangi, S.; Melly, S. *Family Farmers: Feeding the World, Caring for the Earth*; FAO-The United Nations: Rome, Italy, 2014; pp. 1–3.
9. Calcaterra, E. *Defining Smallholders Suggestions for a RSB Smallholder Definitions*; Aidenvironment: Amsterdam, The Netherland, 2013.

10. Henriyadi, B.; Esichaikul, V.; Anutariya, C. Split-Match-Merge Method with Multi-matcher Algorithm to Handle Data Interoperability Problems in Small Farm Management Information System. In Proceedings of the 17th International Conference on Computing and Information Technology (IC2IT 2021), Bangkok, Thailand, 13–14 May 2021; Volume 1, pp. 1–12.
11. Bachhav, N.B. Information Needs of the Rural Farmers: A Study from Maharashtra, India: A Survey. 2012. Paper 866. Available online: https://digitalcommons.unl.edu/libphilprac/866/ (accessed on 14 April 2022).
12. Meitei, L.S.; Purnima, T. Farmers information needs in rural Manipur: An assessment. *Ann. Libr. Inf. Stud.* **2009**, *56*, 35–40.
13. Yusuf, S.F.G.; Masika, P.; Ighodaro, D.I. Agricultural Information Needs of Rural Women Farmers in Nkonkobe Municipality: The Extension Challenge. *J. Agric. Sci.* **2013**, *5*, 107. [CrossRef]
14. Daramola, C.F.; Adebo, T.I.; Adebo, G.M. Challenges and Information Needs Assessment of Dry Season Vegetable Farmers in Akure Metropolis, Ondo State. *IOSR J. Agric. Vet. Sci. Ver. I* **2016**, *9*, 52–58. [CrossRef]
15. Subash, S.; Gupta, J.; Babu, G.P. Information Needs Assessment and Prioritization of Dairy Farmers. *J. Krishi Vigyan* **2015**, *4*, 51. [CrossRef]
16. Babu, S.C.; Glendenning, C.J.; Asenso-Okyere, K.; Govindarajan, S.K. *Farmers' Information Needs and Search Behaviors: Case Study in Tamil Nadu*; IFPR: New Delhi, India, 2012.
17. Elly, T. Agricultural information needs and sources of the rural farmers in Tanzania A case of Iringa rural district. *Libr. Rev.* **2013**, *62*, 266–292. [CrossRef]
18. Naveed, M.A.; Anwar, M.A. Agricultural information needs of Pakistani farmers. *Malays. J. Libr. Inf. Sci.* **2013**, *18*, 13–23.
19. John, O.; Wakilu, O.; Olateju, A. Agricultural information needs of farmers in Lagos. *Int. J. Agric. Sci. Res.* **2013**, *2*, 116–123.
20. Husemann, C.; Novkovic, N.; Novković, N. Farm management information systems: A case study on a German multifunctional farm. *Ekon. Poljopr.* **2014**, *61*, 441–453. [CrossRef]
21. Wolfert, S.; Ge, L.; Verdouw, C.; Bogaardt, M.J. Big Data in Smart Farming—A review. *Agric. Syst.* **2017**, *153*, 69–80. [CrossRef]
22. Kruize, J.W.; Wolfert, J.; Scholten, H.; Verdouw, C.N.; Kassahun, A.; Beulens, A.J.M. Original papers A reference architecture for Farm Software Ecosystems. *Comput. Electron. Agric.* **2016**, *125*, 12–28. [CrossRef]
23. Cruz, I.F. AgreementMaker: Efficient Matching for Large Real-World Schemas and Ontologies. In Proceedings of the VLDB '09, Lyon, France, 24–28 August 2009; pp. 1586–1589.
24. Aumueller, D.; Do, H.-H.; Massmann, S.; Rahm, E. COMA++—Schema and ontology matching with COMA. In Proceedings of the 2005 ACM SIGMOD International Conference on Management of Data, Baltimore, MA, USA, 14–16 June 2005; p. 906. [CrossRef]
25. Madhavan, J.; Bernstein, P.A.; Rahm, E.; Bernstein, P.A. Generic Schema Matching with Cupid. *VLDB* **2001**, *10*, 49–58. [CrossRef]
26. Hu, W.; Qu, Y. Falcon-AO: A practical ontology matching system. *Web Semant.* **2008**, *6*, 237–239. [CrossRef]
27. Shvaiko, P.; Giunchiglia, F.; Yatskevich, M. Semantic Matching with S-Match. In *Semantic Web Information Management: A Model-Based Perspective*; de Virgilio, R., Giunchiglia, F., Tanca, L., Eds.; Springer: Berlin/Heidelberg, Germany, 2010; pp. 1–20. ISBN 978-3-642-04328-4.
28. Jain, S.; Tanwani, S. Schema matching technique for heterogeneous web database. In Proceedings of the 2015 4th International Conference on Reliability, Infocom Technologies and Optimization (ICRITO) (Trends and Future Directions), Noida, India, 2–4 September 2015; Volume 2. [CrossRef]
29. Rachman, M.A.F.; Saptawati, G.A.P. Database integration based on combination schema matching approach (case study: Multi-database of district health information system). In Proceedings of the2017 2nd International Conferences on Information Technology, Information Systems and Electrical Engineering (ICITISEE), Yogyakarta, Indonesia, 1–2 November 2017; pp. 430–435. [CrossRef]
30. Do, H.H. *Schema Matching and Mapping-Based Data Integration*; Interdisciplinary Center for Bioinformatics and Department of Computer Science University of Leipzig Germany: Leipzig, Germany, 2006.
31. Giua, C.; Materia, V.C.; Camanzi, L. Management information system adoption at the farm level: Evidence from the literature. *Br. Food J.* **2021**, *123*, 884–909. [CrossRef]
32. Purnawan, E.; Brunori, G.; Prosperi, P. Small Family Farms; A Perspective from Indonesia, Challenges and Investment. *No. Dec.* **2020**. [CrossRef]
33. Masriwilaga, A.A.; Munadi, R.; Rahmat, B. Wireless Sensor Network for Monitoring Rice Crop Growth. *MESA Tek. Mesin Tek. Elektro* **2018**, *5*, 47–52.
34. Fitriana, G.F.; Prasetyo, N.A.; Engineering, S.; Java, C.; Sprint, M. Rice Planting Calendar Application Development using Scrum. *IJCCS Indones. J. Comput. Cybern. Syst.* **2022**, *16*, 169–180. [CrossRef]
35. Srihartanto, E.; Widodo, S. The Potency of the Rice Crop Index Development through Adjustment of Agroclimate and Water Management Situated in Rainfed Field Gunungkidul. *Agromet* **2020**, *34*, 75–88. [CrossRef]
36. Hakim, V.A.A.; Wibowo, A.; Wibowo, H.; Bhandralia, A.; Arya, R.; Panda, S.N.; Ahuja, S.; Fitriana, G.F.; Prasetyo, N.A.; Engineering, S.; et al. Analisa Pengembangan Drone Penyemprotan Hama Tanaman Dengan Jenis Nosel Dan Ketinggian Untuk Mengetahui Luas Semprotan. *Int. J. Adv. Appl. Sci.* **2019**, *5*, 64–69. [CrossRef]
37. Bhandralia, A.; Arya, R.; Panda, S.N.; Ahuja, S. Polyhouse Agricultural Marketing System Using Big Data Hadoop. *Int. J. Adv. Appl. Sci.* **2016**, *5*, 78–84. [CrossRef]
38. Yanuarti, A.R.; Afsari, M.D. *Profil Komoditas Barang Kebutuhan Pokok dan Barang Penting Komoditas Cabai*; Direktorat Jenderal Perdagangan Dalam Negeri Kementerian Perdagangan: Jakarta, Indonesia, 2016; p. 68.

39. *Direktorat Kredit BPR dan UKM Bank Indoensia PPUK-Budidaya Cabai Merah*; Bank Indonesia: Jakarta, Indonesia, 2014.
40. Sørensen, C.G.; Fountas, S.; Nash, E.; Pesonen, L.; Bochtis, D.; Pedersen, S.M.; Basso, B.; Blackmore, S.B. Conceptual model of a future farm management information system. *Comput. Electron. Agric.* **2010**, *72*, 37–47. [CrossRef]
41. Checkland, P. Information systems and systems thinking: Time to unite? *Int. J. Inf. Manag.* **1988**, *8*, 239–248. [CrossRef]
42. Checkland, P.; Poulter, J. Soft System Methodology. In *Systems Approaches to Managing Change: A Practical Guide*; Reynolds, M., Holwell, S., Eds.; Springer: Berlin/Heidelberg, Germany, 2010; pp. 191–242. ISBN 978-1-84882-809-4.
43. Mehregan, M.R.; Hosseinzadeh, M.; Kazemi, A. An application of Soft System Methodology. *Procedia-Soc. Behav. Sci.* **2012**, *41*, 426–433. [CrossRef]
44. Zaveri, A.; Rula, A.; Maurino, A.; Pietrobon, R.; Lehmann, J. Quality Assessment for Linked Data: A Survey. *Semant. Web* **2012**, *1*, 63–93. [CrossRef]
45. Vetrò, A.; Canova, L.; Torchiano, M.; Minotas, C.O.; Iemma, R.; Morando, F. Open data quality measurement framework: Definition and application to Open Government Data. *Gov. Inf. Q.* **2016**, *33*, 325–337. [CrossRef]
46. Assaf, A.; Senart, A. Data Quality Principle in the Semantic Web. In Proceedings of the 2012 IEEE Sixth International Conference on Semantic Computing, Palermo, Italy, 19–21 September 2012.
47. Veiga, A.K.; Saraiva, A.M.; Chapman, A.D.; Morris, P.J.; Gendreau, C.; Schigel, D.; Robertson, T.J. A conceptual framework for quality assessment and management of biodiversity data. *PLoS ONE* **2017**, *12*, e0178731. [CrossRef] [PubMed]
48. Batini, C.; Cappiello, C.; Francalanci, C.; Maurino, A. Methodologies for data quality assessment and improvement. *ACM Comput. Surv.* **2009**, *41*, 1–52. [CrossRef]
49. Cai, L.; Zhu, Y.; Weiskopf, N.G.; Weng, C.; Webzell, S. Methods and dimensions of electronic health record data quality assessment: Enabling reuse for clinical research. *J. Am. Med. Inform. Assoc.* **2013**, *20*, 144–151. [CrossRef]
50. Wickham, H. The Split-Apply-Combine Strategy for Data Analysis. *J. Stat. Softw.* **2011**, *40*, 1–29. [CrossRef]
51. Murakami, E.; Saraiva, A.M.; Ribeiro, L.C.M.; Cugnasca, C.E.; Hirakawa, A.R.; Correa, P.L.P. An infrastructure for the development of distributed service-oriented information systems for precision agriculture. *Comput. Electron. Agric.* **2007**, *58*, 37–48. [CrossRef]
52. Wang, W.; Reani, M. The rise of mobile computing for Group Decision Support Systems: A comparative evaluation of mobile and desktop. *Int. J. Hum. Comput. Stud.* **2017**, *104*, 16–35. [CrossRef]

agriculture

Article

GrapeNet: A Lightweight Convolutional Neural Network Model for Identification of Grape Leaf Diseases

Jianwu Lin [1], Xiaoyulong Chen [2], Renyong Pan [1], Tengbao Cao [1], Jitong Cai [1], Yang Chen [1], Xishun Peng [1], Tomislav Cernava [3] and Xin Zhang [1,*]

1 College of Big Data and Information Engineering, Guizhou University, Guiyang 550025, China; ljw971121@163.com (J.L.); panry198@163.com (R.P.); 15283673634@163.com (T.C.); gs.jtcai21@gzu.edu.cn (J.C.); cy52cv2022@163.com (Y.C.); pxs19970921@163.com (X.P.)
2 College of Tobacco Science, Guizhou University, Guiyang 550025, China; chenxiaoyulong@sina.cn
3 Institute of Environmental Biotechnology, Graz University of Technology, 8010 Graz, Austria; tomislav.cernava@tugraz.at
* Correspondence: xzhang1@gzu.edu.cn

Abstract: Most convolutional neural network (CNN) models have various difficulties in identifying crop diseases owing to morphological and physiological changes in crop tissues, and cells. Furthermore, a single crop disease can show different symptoms. Usually, the differences in symptoms between early crop disease and late crop disease stages include the area of disease and color of disease. This also poses additional difficulties for CNN models. Here, we propose a lightweight CNN model called GrapeNet for the identification of different symptom stages for specific grape diseases. The main components of GrapeNet are residual blocks, residual feature fusion blocks (RFFBs), and convolution block attention modules. The residual blocks are used to deepen the network depth and extract rich features. To alleviate the CNN performance degradation associated with a large number of hidden layers, we designed an RFFB module based on the residual block. It fuses the average pooled feature map before the residual block input and the high-dimensional feature maps after the residual block output by a concatenation operation, thereby achieving feature fusion at different depths. In addition, the convolutional block attention module (CBAM) is introduced after each RFFB module to extract valid disease information. The obtained results show that the identification accuracy was determined as 82.99%, 84.01%, 82.74%, 84.77%, 80.96%, 82.74%, 80.96%, 83.76%, and 86.29% for GoogLeNet, Vgg16, ResNet34, DenseNet121, MobileNetV2, MobileNetV3_large, ShuffleNetV2_×1.0, EfficientNetV2_s, and GrapeNet. The GrapeNet model achieved the best classification performance when compared with other classical models. The total number of parameters of the GrapeNet model only included 2.15 million. Compared with DenseNet121, which has the highest accuracy among classical network models, the number of parameters of GrapeNet was reduced by 4.81 million, thereby reducing the training time of GrapeNet by about two times compared with that of DenseNet121. Moreover, the visualization results of Grad-cam indicate that the introduction of CBAM can emphasize disease information and suppress irrelevant information. The overall results suggest that the GrapeNet model is useful for the automatic identification of grape leaf diseases.

Keywords: convolutional neural network; residual block; attention mechanism; grape leaf disease

Citation: Lin, J.; Chen, X.; Pan, R.; Cao, T.; Cai, J.; Chen, Y.; Peng, X.; Cernava, T.; Zhang, X. GrapeNet: A Lightweight Convolutional Neural Network Model for Identification of Grape Leaf Diseases. *Agriculture* **2022**, *12*, 887. https://doi.org/10.3390/agriculture12060887

Academic Editor: Dimitre Dimitrov

Received: 10 May 2022
Accepted: 18 June 2022
Published: 20 June 2022

1. Introduction

Grapes are one of the most popular fruits in the world and also the main raw material for the production of wine, thus the yield and quality of grapes are of substantial economic value [1]. However, grape leaves are susceptible to various diseases that are influenced by the weather as well as the environment, and mainly caused by fungi, viruses, and bacteria. If the diseased leaves of grapes are not effectively controlled, the disease spreads to the whole plant, thereby affecting the quality and yield of grapes. In the early days, grape leaf disease identification was mainly conducted by means of classic phytopathology [2];

however, manual identification is time-consuming and labor-intensive. The growing area of land used for grape production makes manual identification methods unreliable. Therefore, automatic identification of grape leaf disease is of great significance for the future development of grape production [3].

With the rapid development of computer technology, a new visual recognition method based on machine learning [4], has been employed for disease recognition. Using machine learning methods to identify crop diseases generally involves three steps: spot segmentation, feature extraction, and classifier recognition [5]. Majumdar et al. extracted wheat disease characteristics and used artificial neural networks (ANNs) to classify diseases, achieving an accuracy of 85% [6]. Guru et al. presented a novel algorithm for extracting lesion areas and applying the probabilistic neural network (PNN) to classify seedling diseases such as anthracnose and frog-eye spots on tobacco leaves, achieving an accuracy of 88.59% [7]. Rumpf et al. accomplished early disease identification of sugar beets using support vector machine (SVM) and hyperspectral techniques [8]. Their experiments showed an accuracy of 97% for healthy and diseased leaves of sugar beets. Moreover, Padol et al. used the SVM classification technique to detect and classify grape leaf diseases [9]. First, the diseased region is identified using segmentation by K-means clustering, and then useful features are extracted. Finally, SVM classification is used to classify the categories of grape leaf diseases, achieving an accuracy of 88.89%. The abovementioned results indicate that it is feasible to use machine learning to identify crop diseases. However, its cumbersome steps lead to low recognition efficiency, and the artificially extracted features are subject to a certain degree of subjectivity, resulting in low recognition accuracy.

Convolutional neural network (CNN) models have been widely used in various application fields, such as facial recognition [10] and license plate detection [11]. The CNN models use sliding window extraction to automatically extract image features and then use fully connected layers for classification to implement an end-to-end disease detection model. Recently, CNN models were used to detect and identify crop diseases instead of traditional machine learning methods [12]. Liu et al. proposed a novel recognition approach based on an improved CNN model for the diagnosis of grape leaf diseases [13]. In this approach, a dense connectivity strategy was introduced to encourage feature reuse and strengthen feature propagation. Finally, a new CNN model named DICNN was built and trained from scratch and achieved an accuracy of 97.22%. Tang et al. proposed a novel method based on a lightweight CNN applying the channel-wise attention mechanism. ShuffleNetV1 and ShuffleNetV2 were chosen as the backbones [14]. The results showed that the proposed model achieved a best trained accuracy of 99.14%, and the model size was only 4.2 MB. Mohanty et al. used GoogLeNet to identify plant disease images from PlantVillage. After GoogLeNet was trained with two methods of training from scratch and isomorphic transfer learning, the accuracy rates were 98.36% and 99.35%, respectively [15]. Pandian et al. proposed a CNN model for image-based plant leaf disease identification using data augmentation and hyperparameter optimization techniques [16]. The results show that the model achieved an accuracy of 98.41% and illustrate the importance of data augmentation techniques and hyperparameter optimization techniques. Chan et al. proposed an early diagnosis method for apple tree leaf diseases based on a deep CNN [17]. The CNN combines DenseNet and Xception, using global average pooling to replace fully connected layers. It achieved an overall accuracy of 98.82% in identifying apple tree leaf diseases. Gao et al. proposed a dual-branch, efficient, channel attention (DECA)-based crop disease recognition model, and the recognition accuracy of the model was 86.65%, 99.74%, and 98.54% on the datasets of PlantVillage, AI Challenger 2018, and Cucumber disease, respectively [18]. Chen et al. introduced the Location-wise Soft Attention mechanism to the pre-trained MobileNetV2 [19]. Furthermore, a two-phase progressive strategy was executed for model training. The experimental results showed that the average accuracy of the model was 99.71% on the open-source dataset. Zeng et al. proposed a lightweight dense-scale network (LDSNet) for corn leaf disease identification under field conditions [20]. The accuracy of the optimized model on the test data was 95.4%. Kamal et al. proposed a novel deep-separable convolution block, through which MobileNet was able to construct just a few parameters with an accuracy of 98.34% on the PlantVillage dataset [21].

The findings of the abovementioned studies confirm that CNN models have advantages in crop disease identification. However, the objectives of these studies were based on the classification of different disease categories. The classification of different stages of specific diseases was so far neglected. However, accurate identification of different symptom stages of a distinct disease has potential value in modern agriculture. The objective of this study was to examine three defined grape leaf diseases. To that end, the symptoms of the same grape disease were divided into two stages (general symptoms and severe symptoms). We present a well-designed CNN model to provide a novel method for the identification of grape leaf diseases. The main objections of this study were as follows:

(1) Proposing a lightweight CNN model, named GrapeNet, based on residual feature fusion block (RFFB) modules and convolutional block attention modules (CBAMs) [22], for the identification of different symptom stages for specific grape diseases.
(2) Implementing ablation experiments and visualization of results of the model to verify the effectiveness of the RFFB modules and the CBAM modules, respectively.
(3) Comparing GrapeNet with other classical network models to verify the performance advantages of GrapeNet.

2. Materials and Methods

2.1. Image Acquisition

In this study, we obtained seven types of grape leaves in the AI challenger 2018 dataset, for a total of 2850 grape leaf images, including 2456 in the training set and 394 in the test set. Representative images are shown in Figure 1. As the same disease is divided into general and serious symptoms, the inter-class variance in the dataset is small. Therefore, it is challenging for a CNN model to identify the disease accurately.

Figure 1. Examples of grape leaves. (**a**) Grape healthy leaf (GH). (**b**) Grape black rot fungus with general symptoms (BRF_G). (**c**) Grape black rot fungus with serious symptoms (BRF_S). (**d**) Grape black measles fungus with general symptoms (BMF_G). (**e**) Grape black measles fungus with serious symptoms (BMF_S). (**f**) Grape leaf blight fungus with general symptoms (LBF_G). (**g**) Grape leaf blight fungus with serious symptoms (LBF_S).

2.2. Image Preprocessing

The number of grape disease samples was limited, and the number of samples of different categories was not evenly distributed. To reduce overfitting during model training and enhance the generalization ability of the model, the dataset had to be expanded. The following operations were carried out. First, we redivided the training set into a training set and validation set in the ratio of 9:1 and performed data augmentation on the new training set through some operations such as rotation, color enhancement, contrast enhancement, and Gaussian noise. Some of the expanded images are shown in Figure 2. Thereafter, the validation set and test set did not need to be expanded. The validation set was used to verify whether the model training fits, and the test set was used to test the performance of the model. The sample distribution before and after augmentation is shown in Table 1.

Figure 2. Some of the expanded images. (**a**) Original image. (**b**) Image expanded by Gaussian noise. (**c**) Image expanded by contrast enhancement. (**d**) Image expanded by horizontal flip. (**e**) Image expanded by a rotation of 90 degrees counter-clockwise. (**f**) Image expanded by a rotation of 60 degrees counter-clockwise.

Table 1. Sample distribution before and after augmentation. As the number of gaps between samples was too large in the training set, we iteratively augmented the dataset with fewer samples to ensure a comparable number of samples per class.

	Sample Distribution before Augmentation			**Sample Distribution after Augmentation**		
Class	Training set	Validation set	Test set	Training set	Validation set	Test set
Grape healthy leaf (GH)	265	29	42	2650	29	42
Grape black rot fungus with general symptoms (BRF_G)	343	38	54	2058	38	54
Grape black rot fungus with serious symptoms (BRF_S)	416	46	66	2496	46	66
Grape black measles fungus with general symptoms (BMF_G)	453	50	74	2718	50	74

Table 1. *Cont.*

	Sample Distribution before Augmentation			Sample Distribution after Augmentation		
Grape black measles fungus with serious symptoms (BMF_S)	378	41	59	2268	41	59
Grape leaf blight fungus with general symptoms (LBF_G)	55	6	9	1980	6	9
Grape leaf blight fungus with serious symptoms (LBF_S)	567	63	90	3402	63	90
Total	2477	273	394	17,572	273	394

2.3. *GrapeNet Model Framework*

In this study, we propose a lightweight CNN model named GrapeNet for grape leaf disease recognition. GrapeNet extracts rich grape leaf disease features by using the RFFB modules while introducing attention mechanisms to focus on useful disease features and enhance the ability to identify grape leaf diseases. The network structure of the GrapeNet model is shown in Figure 3. It consists of convolutional layers, residual blocks, RFFB modules, CBAM modules, an adaptive average pooling layer, and a classifier. First, the image size is resized to 224 × 224 when the image is input to the network model before. Then, a convolutional layer with a stride of 2 and a convolutional kernel size of 3 are used to extract shallow feature information such as the contour and color of the grape leaves. Third, alternate structures of residual blocks, RFFB modules, and CBAM modules are used to deepen the network structure while improving the model's ability to extract disease features, thereby improving the recognition accuracy. Next, the remaining convolutional layers are used to integrate the high-dimensional feature information. Finally, the adaptive average pooling layer integrates the shape of the feature map to 1 × 1 × 1280. The classifier (the fully connected layer) adopts SoftMax for the classification of the extracted features.

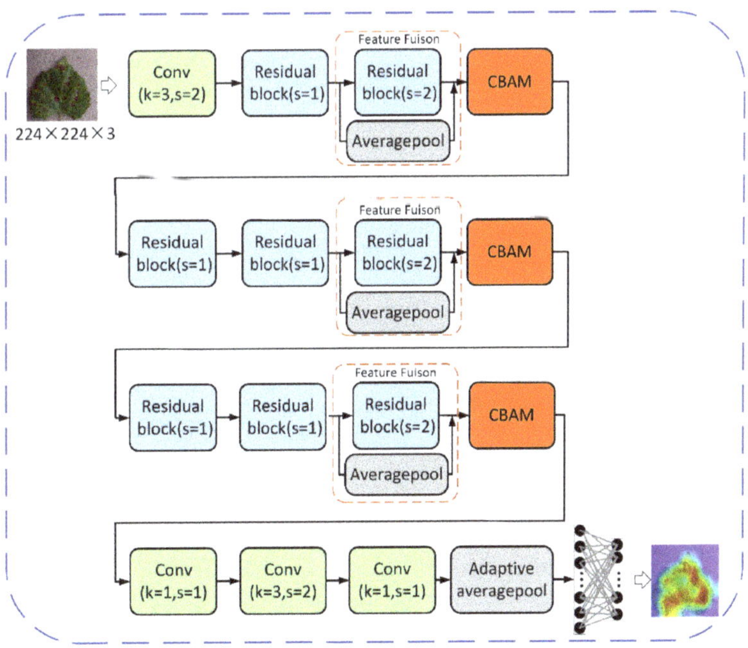

Figure 3. *Cont.*

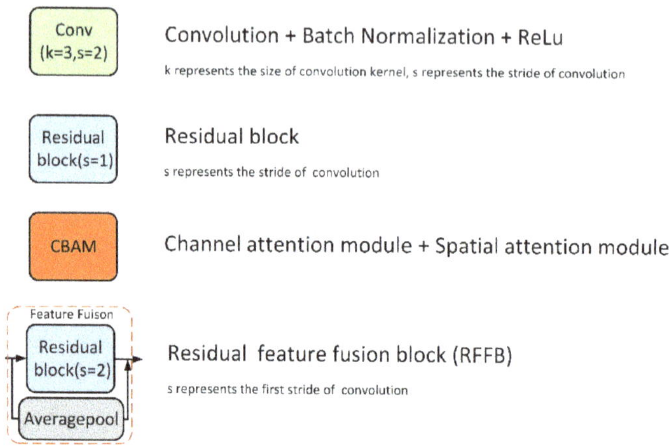

Figure 3. The network structure of the GrapeNet model. It consists of convolutional layers, residual blocks, residual feature fusion block (RFFB) modules, convolutional block attention modules (CBAMs), an adaptive average pooling layer, and a classifier.

2.4. RFFB Module

The residual block is a network structure proposed in the ResNet model. It mainly solves the problem of network degradation caused by the deep structure of the network model through residual learning [23]. He et al. proposed two types of residual blocks in ResNet34. As shown in Figure 4, Figure 4a represents the residual block when the stride is 1. The feature maps of the input and output are added by a skip connection. Figure 4b represents the residual block when the stride is 2. The input feature map is first subjected to a convolution operation with a stride of 2 and a convolution kernel size of 1, and is then added to the output feature map by a skip connection. When designing the GrapeNet model, we found that the residual block when the stride was 2 lost some detailed features, which made the model unable to capture more useful feature information. Therefore, we designed an RFFB module based on this residual block. To preserve the feature information to the greatest extent, we discarded the method of adding by skip connections in the residual module and adopted the method of concatenating by skip connection while using average pooling to replace the convolution operation with a kernel size of 1 on the shortcut branch. In this way, the parameters of the module can be reduced, and more disease characteristics can be preserved. The structure of an RFFB module is shown in Figure 5.

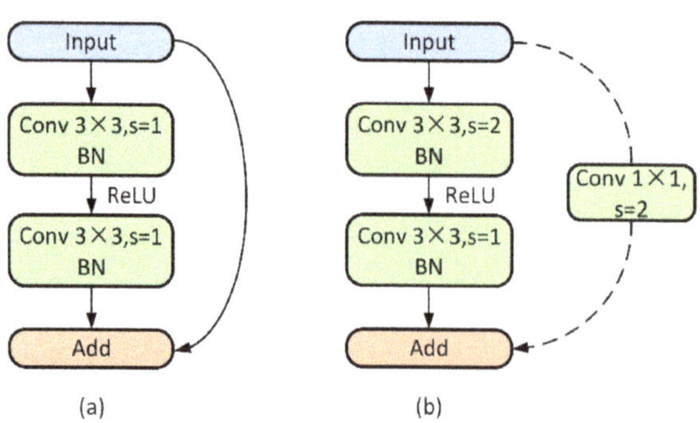

Figure 4. Structures of the residual blocks. (**a**) The residual block when the stride is 1. (**b**) The residual block when the stride is 2.

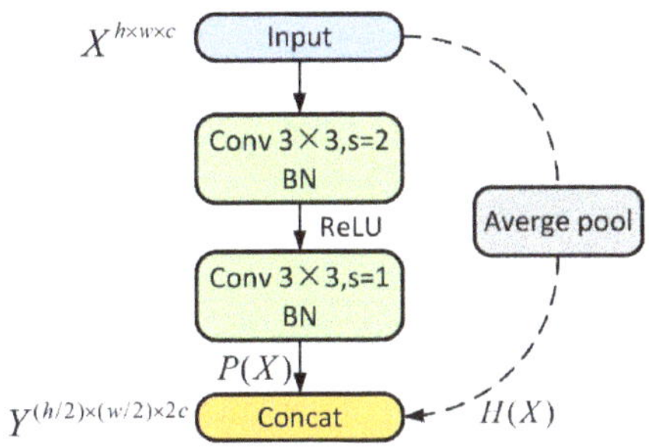

Figure 5. The structure of the RFFB module.

Given an input $X^{h \times w \times c}$, where h, w, and c denotes the height, width, and number of channels of the feature map, respectively. $P(X)$ is the feature map obtained by the convolution operation, which can be expressed as:

$$P(X) = Conv(X) \tag{1}$$

$H(X)$ is the feature map obtained by the average pooling operation, which can be written as:

$$H(X) = Avergepool(X) \tag{2}$$

The final output is obtained by concatenating the high-dimensional feature map $P(X)$ and the high-resolution feature map $H(X)$, thereby yielding:

$$Y^{(h/2) \times (w/2) \times 2c} = Concat\{P(X); H(X)\} \tag{3}$$

The RFFB module prevents the loss of feature information during down sampling by concatenating different forms of feature maps, which retains rich disease feature information, and increases the feature dimension so that the model can more accurately identify grape disease leaves.

2.5. CBAM Module

CNNs can obtain a large amount of useless information when extracting features, including background information and noise. This useless information greatly affects the effect of disease identification. The attention mechanism can ensure that the network focuses on useful feature information, suppresses the background and noise, and improves the recognition accuracy. In this study, the CBAM module was introduced into the network model so that the network can highlight the disease information of grape leaves. The structure of the CBAM module in the GrapeNet model is shown in Figure 6. After the RFFB module extracts a large amount of feature information, the CBAM module assigns different weights to different feature information; for example, it assigns more weight to disease information and assigns less weight to the background and noise. Finally, the residual block integrates the obtained information.

Figure 6. The structure of the CBAM module in the GrapeNet model.

Figure 7 shows the network structure of the CBAM module. This module first extracts the channel information of the feature map through spatial channel attention and then extracts the spatial information through spatial attention. Therefore, the CBAM module can be divided into two sub-modules: the channel attention module and spatial attention module.

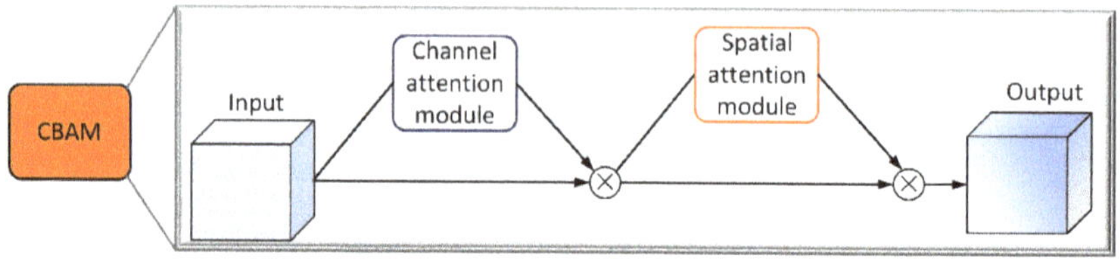

Figure 7. The network structure of the CBAM module.

The network structure of the channel attention module is shown in Figure 8. It mainly weights the channel information of the input feature map and highlights the channels with disease information.

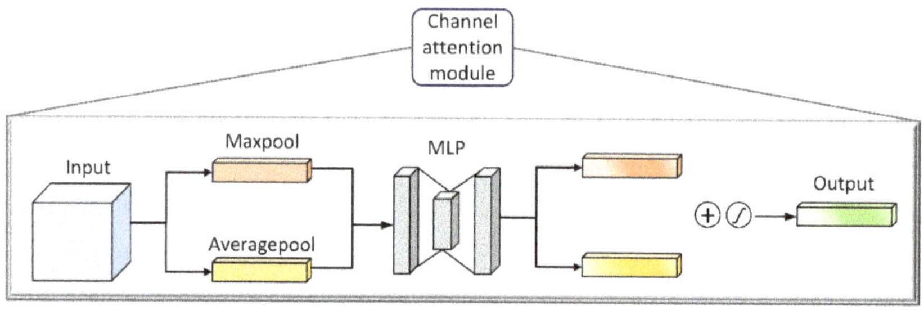

Figure 8. The network structure of the channel attention module.

The input feature map is **F**, and the output feature map can be expressed as:

$$\mathbf{M_c}(\mathbf{F}) = \sigma(\mathbf{W}_1(\mathbf{W}_0(\mathbf{F}^{\mathbf{c}}_{\mathbf{avg}}) + \mathbf{W}_1(\mathbf{W}_0(\mathbf{F}^{\mathbf{c}}_{\mathbf{max}})) \quad (4)$$

where σ denotes the sigmoid function. The MPL weights ($\mathbf{W}_1$ and $\mathbf{W}_0$) are shared for both inputs and the ReLU activation function.

The network structure of the spatial attention module is shown in Figure 9. It highlights the diseased area of interest in the feature map by weighting the spatial information of the input feature map. The input feature map is **F**, and the formula can be written as:

$$\mathbf{M_c}(\mathbf{F}) = \sigma(f^{7\times7}(Concat\left[\mathbf{F}^{\mathbf{s}}_{\mathbf{avg}};\mathbf{F}^{\mathbf{s}}_{\mathbf{max}}\right])) \quad (5)$$

where $f^{7\times7}$ represents a convolution operation with the kernel size of 7×7, and σ denotes the sigmoid function.

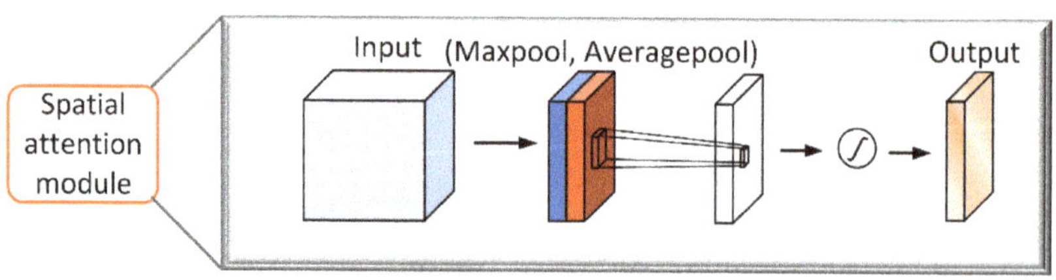

Figure 9. The network structure of the spatial attention module.

2.6. Evaluation Indexes

In this study, we used accuracy, precision, recall, and F1-score as evaluation indicators. The formulas are as follows:

$$accuracy = \frac{TP + TN}{TP + FP + TN + FN} \quad (6)$$

$$precision = \frac{TP}{TP + FP} \quad (7)$$

$$recall = \frac{TP}{TP + FN} \quad (8)$$

$$F1 - score = \frac{2TP}{2TP + FP + FN} \quad (9)$$

where *TP* is the number of true-positive samples, *FP* is the number of false-positive samples, *FN* is the number of false-negative samples, and *TN* is the number of true-negative samples.

2.7. Experimental Environment and Hyperparameter Setting

The experimental environment is shown in Table 2. The hyperparameters were set as follows. The cross-entropy loss function (CE) was used as the loss function, and the Adam optimizer [24] was used to optimize the model. The initial learning rate and batch size during training were set to 0.0001 and 64, respectively. The number of iterations was 120.

Table 2. Experimental environment.

Name	Parameter
CPU	Intel(R) Xeon(R) W-2235
GPU	NVIDIA GeForce RTX 2080Ti
System	Windows 10
Programming language	Python 3.8.8
Deep learning framework	Pytorch 1.6.0

3. Results

3.1. The Impact of Data Augmentation on the Model

Figure 10 shows each epoch of the GrapeNet model with data augmentation and without it. We found that the training loss of the GrapeNet model with data augmentation dropped faster than that of the GrapeNet model without data augmentation, and the average accuracy of the model with data augmentation on the validation set was higher than that of the GrapeNet model without data augmentation on the validation set. This indicates that data augmentation can increase the diversity of data, reduce model overfitting, and enable the model to have better recognition ability. Moreover, the accuracy of the GrapeNet model with data augmentation was 86.25% in the test set, which is 4% higher than that of the GrapeNet model without data augmentation. This also indicates that the data augmentation method enables the model to have a higher generalization ability.

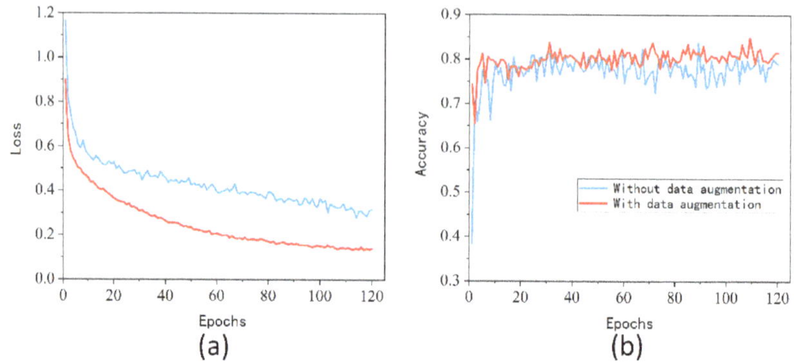

Figure 10. Each epoch of the GrapeNet model. (**a**) Training loss. (**b**) Validation accuracy.

3.2. *Ablation Experiment*

To verify the effectiveness of the RFFB module and CBAM module in the GrapeNet model, we performed ablation experiments on the test set. The obtained results are shown in Table 3. We found that the network model with the RFFB module achieved an accuracy of 82.49%. The accuracy was improved by 3.05% compared to the accuracy of the network model with no module introduced. We also found that the accuracy of the network model with the CBAM module was improved by 2.79% compared to that of the network model that does not introduce any module; the accuracy was 82.23%. The introduction of the RFFB module and the CBAM module did not add too many parameters to the network model. Finally, the network model GrapeNet, in which both modules were introduced simultaneously, achieved the best performance in terms of accuracy, precision, recall, and F1-score values (86.29%, 77.76%, 88.43%, and 79.05%, respectively). This indicates that the RFFB module and CBAM module can effectively enhance the identification of grape leaf disease.

Table 3. Results of ablation experiments.

RFFB	CBAM	Accuracy	Recall	Precision	F1-Score	Param (M)
-	-	0.7944	0.7569	0.7372	0.7413	2.05
√	-	0.8249	0.7738	0.7878	0.7756	2.14
-	√	0.8223	0.7658	0.7884	0.7689	2.05
√	√	0.8629	0.7776	0.8843	0.7905	2.15

3.3. *Visual Comparison of Output Feature Maps*

To demonstrate the effect of the RFFB module on the network model, we visualized the output feature maps of GrapeNet without the RFFB module, and GrapeNet with the RFFB module. As shown in Figure 11, the network model extracted the texture, color, and edge of grape leaf diseases in the first several convolutional layers. With the deepening of the network structure, the extracted feature information gradually became abstract feature information. We found that the abstract information of GrapeNet with RFFB module was richer than that of GrapeNet without the RFFB module; this is because the output of the average pooling operation and residual down sampling is concatenated in the RFFB module, and avoids the loss of a large number of detailed features, thus improving the model's capability to identify grape leaf diseases.

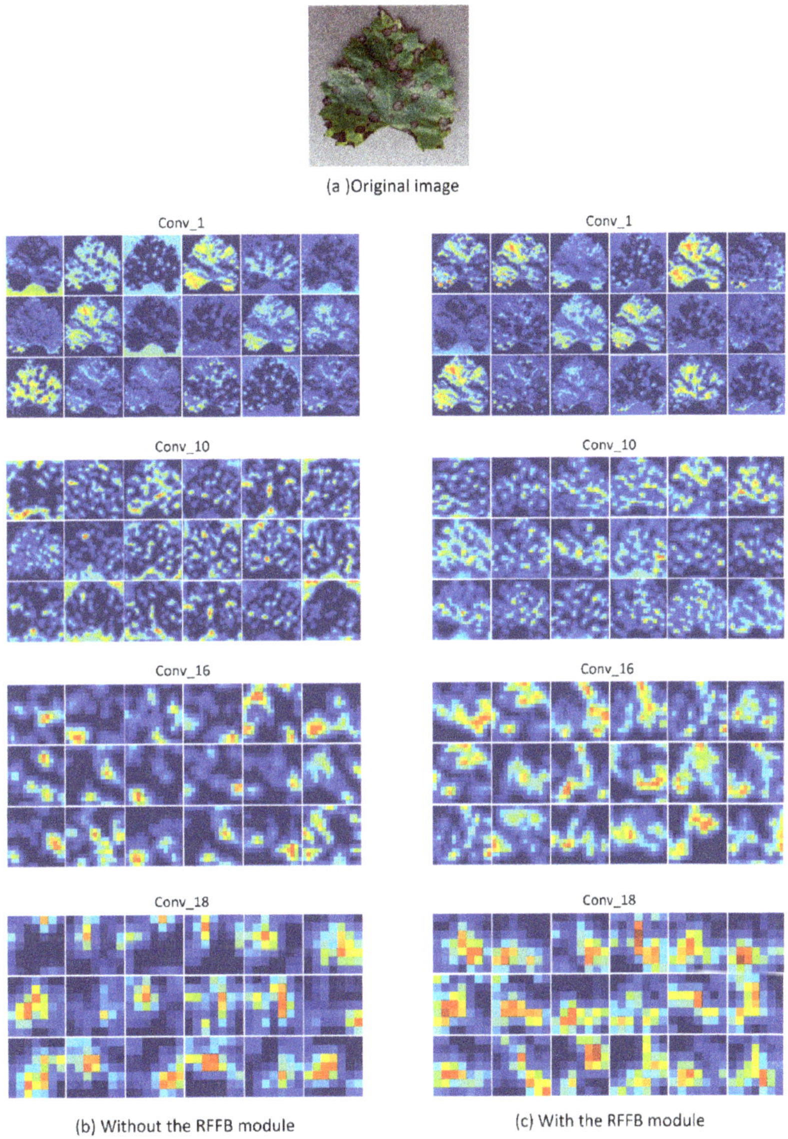

Figure 11. Visualization results of the output feature maps of the convolution layer. (**a**) The original image. (**b**) The visualization results of the output feature maps of GrapeNet without the RFFB module. (**c**) The visualization results of the output feature maps of GrapeNet with the RFFB module.

3.4. Comparison of Results of Different Attention Mechanisms

Table 4 shows the comparison of results of different attention mechanisms (the Squeeze-and-Excitation (SE) module, the Coordinate Attention (CA) module, and the CBAM module) on the network. None of the three attention models were found to produce an excessive number of parameters. However, the GrapeNet model with the introduction of the CBAM module had the highest accuracy of 86.29%, making it 0.76% more accurate than the GrapeNet model with the introduction of the SE module and 0.76% more accurate than the GrapeNet model with the introduction of the CA module. This result indicates that introducing the CBAM module can make the GrapeNet model better focus on the disease region and reduce the influence of useless features, thereby improving the accuracy of grape leaf disease identification.

Table 4. Comparison of results of different attention mechanisms.

Attention Mechanism	Accuracy	Recall	Precision	F1-Score	Param (M)
SE [25]	0.8553	0.8012	0.8111	0.8053	2.15
CA [26]	0.8553	0.8206	0.8267	0.8217	2.15
CBAM	0.8629	0.7776	0.8843	0.7905	2.15

To show the regions of interest of the network model, Grad-cam [27] was used to visualize the class activation maps of the model using different attention mechanisms. As shown in Figure 12, the first row included the image of grape black rot fungus with serious symptoms (BRF_S). We found that the region of disease captured using the SE module and using the CA module was incomplete; however, the region of disease was captured intact using the CBAM module. For the image of grape black measles fungus with general symptoms (BMF_G) in the second row, the disease area was accurately located using the CBAM module, the disease area was not accurately captured using the SE module, and only a part of the disease area was captured using the CA module. This is because the CBAM module highlights the region of interest along the spatial and channel directions, thereby capturing more complete information about the grape leaf disease. In summary, the obtained results show that the introduction of the CBAM module can focus on the disease information and filter the background information.

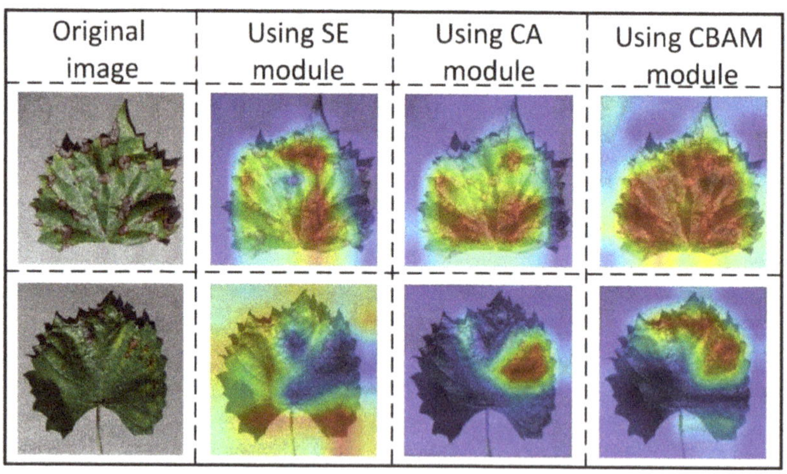

Figure 12. Visualization of results using different attention mechanisms.

3.5. Comparison of Identification Results with Classical CNN Models

Next, we aimed to verify that our proposed GrapeNet model has some advantages compared with classical CNN models such as heavyweight network models, VGG16, EfficientNet, ResNet, DenseNet, GoogLeNet and lightweight networks, MobileNetV2, MobileNetV3, and ShuffleNet. The results of the grape leaf disease test set are shown in Table 5. Furthermore, Table 5 also shows the training time of the nine models. The GrapeNet model achieved good performance on the test set for all evaluation metrics. It achieved a maximum accuracy of 86.29%, the number of parameters was 2.15 million, and the training time was 101 min. Compared to the DenseNet121 model, which was the most accurate of the classical networks, the GrapeNet model yielded a 1.52% improvement in accuracy, a decrease of 4.81 million parameters, and a two times shorter training time. Compared to the ShuffleNetV2_x1.0 model, which had the fewest parameters and minimum training time among the classical networks, the GrapeNet model had a 5.33% improvement in accuracy, although the number of parameters increased by 0.89 million and the training time increased. Therefore, the GrapeNet model is a lightweight CNN model. It can achieve a balance between accuracy, the number of parameters, and training time, demonstrating the potential for grape leaf disease recognition.

Table 5. Identification results of the nine CNN models.

Model	Accuracy	Recall	Precision	F1-Score	Param (M)	Training Time (mins)
GoogLeNet [28]	0.8299	0.7521	0.8069	0.7601	5.98	107
Vgg16 [29]	0.8401	0.7761	0.7817	0.7777	134.29	254
ResNet34 [23]	0.8274	0.7617	0.77	0.762	21.29	108
DenseNet121 [30]	0.8477	0.7845	0.8357	0.7972	6.96	206
MobileNetV2 [31]	0.8096	0.7327	0.7572	0.74	2.23	98
MobileNetV3_large [32]	0.8274	0.7479	0.7818	0.7569	4.21	84
ShuffleNetV2_×1.0 [33]	0.8096	0.7455	0.7472	0.7424	1.26	64
EfficientNetV2_s [34]	0.8376	0.7738	0.8241	0.7865	20.19	290
GrapeNet	0.8629	0.7776	0.8843	0.7905	2.15	101

The confusion matrix for the nine models is shown in Figure 13. It was evident that the different periods of manifestation of the same grape leaf disease were the biggest factors affecting the recognition effectiveness of the network models. For grape black rot fungus (BRF), the DensNet121 model accurately classified the highest number of samples, 92, and the GrapeNet model accurately classified the next highest number of samples, 90. For both grape black measles fungus (BMF) and grape leaf blight fungus (LBF), the GrapeNet model achieved the highest number of true positive samples, 116 and 92, respectively. This indicates that our proposed GrapeNet model has better identification performance in different periods of the same grape leaf disease.

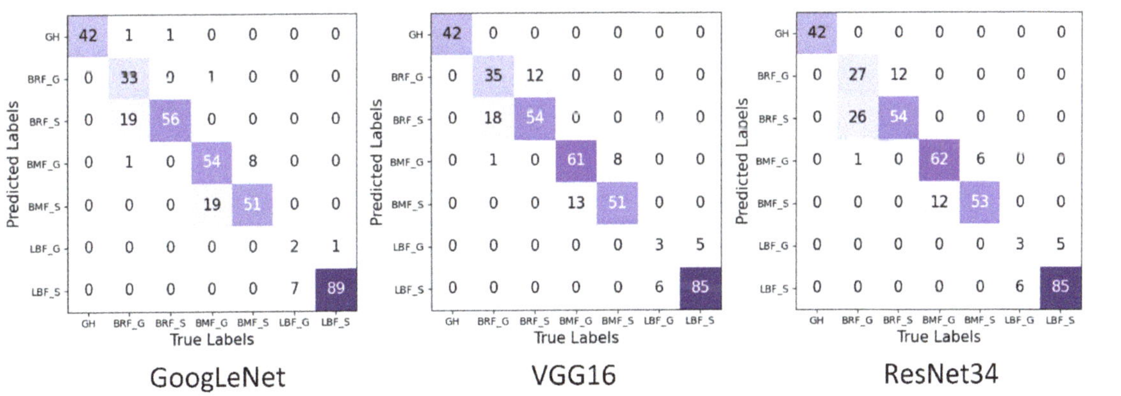

Figure 13. *Cont.*

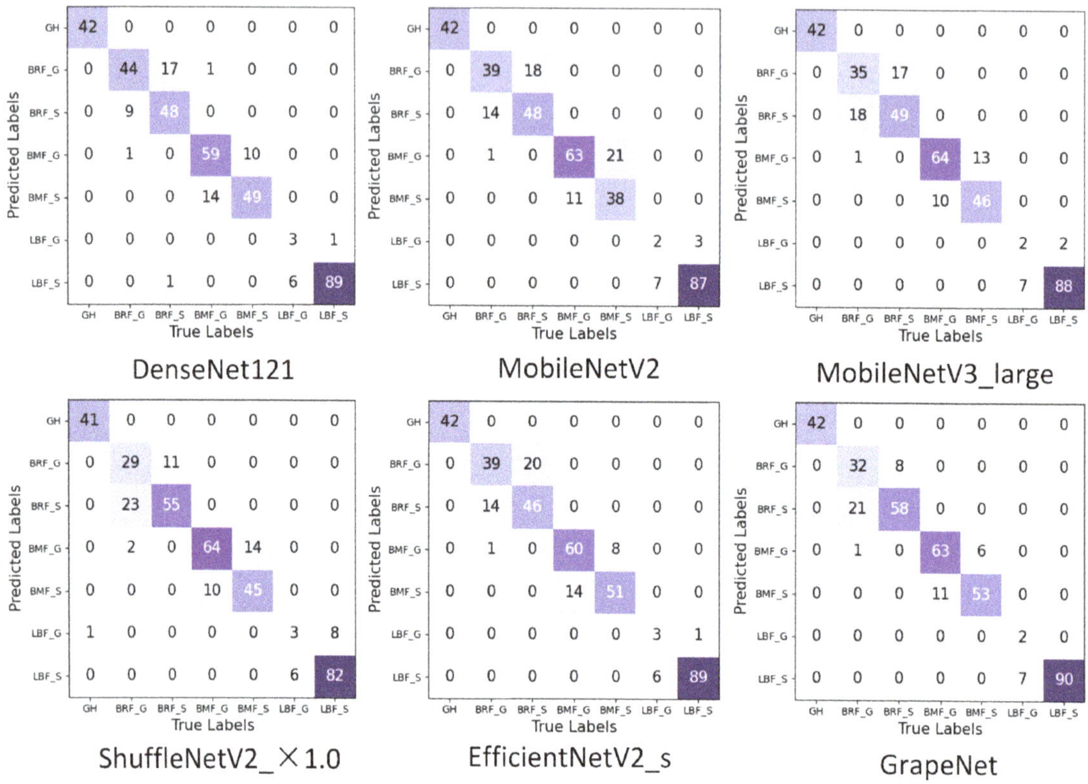

Figure 13. The confusion matrix for the nine models.

4. Discussion

Crop disease is a major threat affecting the safety of global agricultural production. Therefore, it is necessary to utilize new technologies for crop disease identification. Among deep learning methods, the CNN model has the advantages of high speed and high accuracy. It has been widely applied to identify crop diseases [14,16].

In this study, we compared our experimental results with some of the existing literature on crop disease identification. Zhao et al. proposed a deep CNN that combines the SE modules for the identification of tomato diseases [35]. The results suggested that the introduction of the SE module improved the accuracy of ResNet50 by 4.26%. Bao et al. proposed a CNN model called SimpleNet for the identification of wheat diseases [36]. The experiment results showed that the introduction of the CBAM modules improved the accuracy of the benchmark model by 3%. These results demonstrated that the attention mechanism is an effective module for improving the accuracy of crop disease identification. In our approaches, the accuracy of the network model with the CBAM module was improved by 2.79% compared to the accuracy of the network model with no module introduced. Perhaps, it is due to our dataset having a small intra-class variance, which requires fine-grained features for classification for the model. Subsequently, with the increasing difficulty of the classification task, the improved accuracy of the attention mechanism was reduced. Furthermore, the existing models ignored the loss of feature information during feature extraction [18,19]. Thus, we designed a module to reduce the loss of feature information, namely the RFFB module. With the introduction of the RFFB module on the benchmark, the accuracy of the model was improved by 6.85%. The model can retain more feature information while emphasizing them by the combination between the RFFB module and the attention mechanism. The obtained result suggested that a single attention module is not enough to obtain the best performance of the network model. Hence, we should pay more attention to the loss of feature information to design new models.

Here, GrapeNet was designed for the specific task of grape leaf disease identification based on the above findings. It has a simplified architecture that deepens the network and extracts features with only a few residual blocks. It can reduce the number of parameters and provide an architecture of suitable depth and width for grape leaf disease identification. Afterwards, GrapeNet was compared with the classical models. As shown in Table 5, the disease identification accuracy of GrapeNet achieved 86.29%. The accuracy was 3.3%, 2.28%, 3.55%, 1.52%, 5.33%, 3.55%, 5.33%, and 2.53% higher than GoogLeNet, Vgg16, ResNet34, DenseNet121, MobileNetV2, MobileNetV3_large, ShuffleNetV2_×1.0, and Efficient-NetV2_s. Furthermore, the parameters of GrapeNet amounted to only 2.15 million. Although ShuffleNetV2_×1.0 has a lower number of parameters, its accuracy is 5.33% lower compared to GrapeNet. The results have clearly demonstrated the advantages of the proposed GrapeNet for grape leaf disease identification. Our approach can be deployed in smart mechanical devices or mobile devices to facilitate the rapid identification of grape leaf diseases, thus saving time and labor costs. Moreover, GrapeNet can be used not only for grape leaf diseases, but may have potential research value for other crop diseases with different symptom stages.

5. Conclusions

In this study, a lightweight CNN named GrapeNet was developed for the identification of grape leaf disease. To reduce the loss of disease features, an RFFB module was designed by concatenating the average pooled feature map before the input of the residual block and the output feature map of the residual block. In addition, the CBAM module was included between the RFFB module and the residual block to emphasize the disease features of interest and reduce the influence of redundant information. On the seven-class grape leaf disease test set, the GrapeNet model achieved 86.29%, 77.76%, 88.63%, and 79.05% in the accuracy, recall, precision, and F1-score, respectively. Furthermore, the GrapeNet model only included 2.15 million parameters. The GrapeNet model achieved an excellent balance between high accuracy and low parameter size. The results demonstrated that GrapeNet has high potential for identification of grape leaf diseases on mobile and embedded devices. However, the study has a specific shortcoming. The overly simple background in the grape disease images may lead to a decrease in the recognition accuracy of the model in relevant environments. In the future, we will continue to optimize our model for grape leaf disease recognition under field conditions.

Author Contributions: Conceptualization, X.Z. and J.L.; methodology, X.Z. and X.C.; software, R.P.; validation, X.Z., J.L. and R.P.; formal analysis, J.L. and T.C. (Tengbao Cao); investigation, J.C.; data curation, Y.C. and T.C. (Tengbao Cao); writing—original draft preparation, J.L., R.P. and T.C. (Tomislav Cernava); writing—review and editing, X.Z., T.C. (Tomislav Cernava), and X.C.; visualization, X.P.; supervision, X.Z and X.C.; project administration, X.Z. and X.C.; funding acquisition, X.C. All authors have read and agreed to the published version of the manuscript.

Funding: This research was funded by National Key Research and Development Plan Key Special Projects, grant number 2021YFE0107700, National Nature Science Foundation of China, grant numbers 61865002 and 31960555, Guizhou Science and Technology Program, grant number 2019-1410, and Outstanding Young Scientist Program of Guizhou Province, grant number KY2021-026. In addition, the study received support by the Program for Introducing Talents to Chinese Universities, 111 Program, grant number D20023.

Institutional Review Board Statement: Not applicable.

Informed Consent Statement: Not applicable.

Data Availability Statement: Not applicable.

Conflicts of Interest: The authors declare no conflict of interest.

References

1. Peng, Y.; Zhao, S.Y.; Liu, J.Z. Fused Deep Features-Based Grape Varieties Identification Using Support Vector Machine. *Agriculture* **2021**, *11*, 869. [CrossRef]
2. Ji, M.; Zhang, L.; Wu, Q. Automatic grape leaf diseases identification via UnitedModel based on multiple convolutional neural networks. *Inf. Process. Agric.* **2020**, *7*, 418–426. [CrossRef]
3. Singh, V.; Misra, A. Detection of plant leaf diseases using image segmentation and soft computing techniques. *Inf. Process. Agric.* **2017**, *4*, 41–49. [CrossRef]
4. Zhang, Z.; Khanal, S.; Raudenbush, A.; Tilmon, K.; Stewart, C. Assessing the efficacy of machine learning techniques to characterize soybean defoliation from unmanned aerial vehicles. *Comput. Electron. Agric.* **2022**, *193*, 106682. [CrossRef]
5. Jaisakthi, S.; Mirunalini, P.; Thenmozhi, D. Grape leaf disease identification using machine learning techniques. In Proceedings of the 2019 International Conference on Computational Intelligence in Data Science (ICCIDS), Vatsala, Australia, 21–23 February 2019; pp. 1–6.
6. Majumdar, D.; Kole, D.K.; Chakraborty, A.; Majumder, D.D. An integrated digital image analysis system for detection, recognition and diagnosis of disease in wheat leaves. In Proceedings of the Third International Symposium on Women in Computing and Informatics, Kerala, India, 10–13 August 2015; pp. 400–405.
7. Guru, D.; Mallikarjuna, P.; Manjunath, S. Segmentation and classification of tobacco seedling diseases. In Proceedings of the Fourth Annual ACM Bangalore Conference, Bangalore, India, 25–26 March 2011; pp. 1–5.
8. Rumpf, T.; Mahlein, A.K.; Steiner, U.; Oerke, E.C.; Dehne, H.W.; Plümer, L. Early detection and classification of plant diseases with support vector machines based on hyperspectral reflectance. *Comput. Electron. Agric.* **2010**, *74*, 91–99. [CrossRef]
9. Padol, P.B.; Yadav, A.A. SVM classifier based grape leaf disease detection. In Proceedings of the 2016 Conference on Advances in Signal Processing (CASP), Pune, India, 9–11 June 2016; pp. 175–179.
10. Martins, P.; Silva, J.S.; Bernardino, A. Multispectral Facial Recognition in the Wild. *Sensors* **2022**, *22*, 4219. [CrossRef]
11. Khan, I.R.; Ali, S.T.A.; Siddiq, A.; Khan, M.M.; Ilyas, M.U.; Alshomrani, S.; Rahardja, S. Automatic License Plate Recognition in Real-World Traffic Videos Captured in Unconstrained Environment by a Mobile Camera. *Electronics* **2022**, *11*, 1408. [CrossRef]
12. Orchi, H.; Sadik, M.; Khaldoun, M. On Using Artificial Intelligence and the Internet of Things for Crop Disease Detection: A Contemporary Survey. *Agriculture* **2022**, *12*, 9. [CrossRef]
13. Liu, B.; Ding, Z.; Tian, L.; He, D.; Li, S.; Wang, H. Grape leaf disease identification using improved deep convolutional neural networks. *Front. Plant Sci.* **2020**, *11*, 1082. [CrossRef]
14. Tang, Z.; Yang, J.; Li, Z.; Qi, F. Grape disease image classification based on lightweight convolution neural networks and channelwise attention. *Comput. Electron. Agric.* **2020**, *178*, 105735. [CrossRef]
15. Mohanty, S.P.; Hughes, D.P.; Salathé, M. Using Deep Learning for Image-Based Plant Disease Detection. *Front. Plant Sci.* **2016**, *7*, 1419. [CrossRef] [PubMed]
16. Pandian, J.A.; Kanchanadevi, K.; Kumar, V.D.; Jasińska, E.; Goňo, R.; Leonowicz, Z.; Jasiński, M. A Five Convolutional Layer Deep Convolutional Neural Network for Plant Leaf Disease Detection. *Electronics* **2022**, *11*, 1266. [CrossRef]
17. Chao, X.; Sun, G.; Zhao, H.; Li, M.; He, D. Identification of Apple Tree Leaf Diseases Based on Deep Learning Models. *Symmetry* **2020**, *12*, 1065. [CrossRef]
18. Gao, R.; Wang, R.; Feng, L.; Li, Q.; Wu, H. Dual-branch, efficient, channel attention-based crop disease identification. *Comput. Electron. Agri.* **2021**, *190*, 106410. [CrossRef]
19. Chen, J.; Zhang, D.; Suzauddola, M.; Zeb, A. Identifying crop diseases using attention embedded MobileNet-V2 model. *Appl. Soft Comput.* **2021**, *113*, 107901. [CrossRef]
20. Zeng, W.; Li, H.; Hu, G.; Liang, D. Lightweight dense-scale network (LDSNet) for corn leaf disease identification. *Comput. Electron. Agri.* **2022**, *197*, 106943. [CrossRef]
21. Kamal, K.; Yin, Z.; Wu, M.; Wu, Z. Depthwise separable convolution architectures for plant disease classification. *Comput. Electron. Agric.* **2019**, *165*, 104948.
22. Woo, S.; Park, J.; Lee, J.Y.; Kweon, I.S. Cbam: Convolutional block attention module. In Proceedings of the European Conference on Computer Vision (ECCV), Munich, Germany, 8–14 September 2018; pp. 3–19.
23. He, K.; Zhang, X.; Ren, S.; Sun, J. Deep residual learning for image recognition. In Proceedings of the IEEE Conference on Computer Vision and Pattern Recognition (CVPR), Las Vegas, NV, USA, 26 June–1 July 2016; pp. 770–778.
24. Kingma, D.P.; Ba, J. Adam: A method for stochastic optimization. *arXiv* **2014**, arXiv:1412.6980.
25. Hu, J.; Shen, L.; Sun, G. Squeeze-and-excitation networks. In Proceedings of the IEEE Conference on Computer Vision and Pattern Recognition (CVPR), Salt Lake City, UT, USA, 18–22 June 2018; pp. 7132–7141.
26. Hou, Q.; Zhou, D.; Feng, J. Coordinate attention for efficient mobile network design. In Proceedings of the 2021 IEEE/CVF Conference on Computer Vision and Pattern Recognition (CVPR), Nashville, TN, USA, 20–25 June 2021; pp. 13708–13717.
27. Selvaraju, R.R.; Cogswell, M.; Das, A.; Vedantam, R.; Parikh, D.; Batra, D. Grad-cam: Visual explanations from deep networks via gradient-based localization. In Proceedings of the IEEE International Conference on Computer Vision (ICCV), Venice, Italy, 22–29 October 2017; pp. 618–626.
28. Szegedy, C.; Liu, W.; Jia, Y.; Sermanet, P.; Reed, S.; Anguelov, D.; Erhan, D.; Vanhoucke, V.; Rabinovich, A. Going deeper with convolutions. In Proceedings of the IEEE Conference on Computer Vision and Pattern Recognition (CVPR), Boston, MA, USA, 7–12 June 2015; pp. 1–9.

29. Simonyan, K.; Zisserman, A. Very deep convolutional networks for large-scale image recognition. *arXiv* **2014**, arXiv:1409.1556.
30. Huang, G.; Liu, Z.; Van Der Maaten, L.; Weinberger, K.Q. Densely connected convolutional networks. In Proceedings of the IEEE Conference on Computer Vision and Pattern Recognition, Honolulu, HI, USA, 21–26 July 2017; pp. 4700–4708.
31. Sandler, M.; Howard, A.; Zhu, M.; Zhmoginov, A.; Chen, L.-C. Mobilenetv2: Inverted residuals and linear bottlenecks. In Proceedings of the IEEE Conference on Computer Vision and Pattern Recognition, Salt Lake City, UT, USA, 18–23 June 2018; pp. 4510–4520.
32. Howard, A.; Sandler, M.; Chu, G.; Chen, L.-C.; Chen, B.; Tan, M.; Wang, W.; Zhu, Y.; Pang, R.; Vasudevan, V.; et al. Searching for MobileNetV3. In Proceedings of the 2019 IEEE/CVF International Conference on Computer Vision (ICCV 2019), Seoul, Korea, 27 October–2 November 2019; pp. 1314–1324.
33. Ma, N.; Zhang, X.; Zheng, H.-T.; Sun, J. Shufflenet v2: Practical guidelines for efficient cnn architecture design. In Proceedings of the European Conference on Computer Vision (ECCV), Munich, Germany, 8–14 September 2018; pp. 116–131.
34. Tan, M.; Le, Q.V. Efficientnetv2: Smaller models and faster training. *arXiv* **2021**, arXiv:2104.00298.
35. Zhao, S.; Peng, Y.; Liu, J.; Wu, S. Tomato Leaf Disease Diagnosis Based on Improved Convolution Neural Network by Attention Module. *Agriculture* **2021**, *11*, 651. [CrossRef]
36. Bao, W.; Yang, X.; Liang, D.; Hu, G.; Yang, X. Lightweight convolutional neural network model for field wheat ear disease identification. *Comput. Electron. Agric.* **2021**, *189*, 106367. [CrossRef]

agriculture

MDPI

Article

An Aquatic Product Price Forecast Model Using VMD-IBES-LSTM Hybrid Approach

Junhao Wu, Yuan Hu, Daqing Wu * and Zhengyong Yang

College of Economics and Management, Shanghai Ocean University, Shanghai 201306, China
* Correspondence: dqwu@shou.edu.cn; Tel.: +86-021-6190-0856

Abstract: Changes in the consumption price of aquatic products will affect demand and fishermen's income. The accurate prediction of consumer price index provides important information regarding the aquatic product market. Based on the non-linear and non-smooth characteristics of fishery product price series, this paper innovatively proposes a fishery product price forecasting model that is based on Variational Modal Decomposition and Improved bald eagle search algorithm optimized Long Short Term Memory Network (VMD-IBES-LSTM). Empirical analysis was conducted using fish price data from the Department of Marketing and Informatization of the Ministry of Agriculture and Rural Affairs of China. The proposed model in this study was subsequently compared with common forecasting models such as VMD-LSTM and SSA-LSTM. The research results show that the VMD-IBES-LSTM model that was constructed in this paper has good fitting results and high prediction accuracy, which can better explain the seasonality and trends of the change of China's aquatic product consumer price index, provide a scientific and effective method for relevant management departments and units to predict the aquatic product consumer price, and have a certain reference value for reasonably coping with the fluctuation of China's aquatic product market price.

Keywords: aquatic products price forecast; VMD; IBES; LSTM; hybrid model

Citation: Wu, J.; Hu, Y.; Wu, D.; Yang, Z. An Aquatic Product Price Forecast Model Using VMD-IBES-LSTM Hybrid Approach. *Agriculture* **2022**, *12*, 1185. https://doi.org/10.3390/agriculture12081185

Academic Editor: Giuseppe Timpanaro

Received: 26 June 2022
Accepted: 27 July 2022
Published: 9 August 2022

1. Introduction

Aquatic products play an important role in China's fishery economic development and international market competition. As a specific reflection of the fishery production cost and the relationship between the supply and demand of aquatic products, aquatic product price is not only related to the production and sales of enterprises and economic interests, but also related to China's macroeconomic policies. In recent years, with the rapid development of China's economy and the continuous advancement of urbanization, residents' demand for high-quality and safe aquatic products such as abalone and shark fin has gradually increased [1–3]. Meanwhile, China's aquatic product market structure is increasingly in line with the needs of the world [4]. The accurate forecast of aquatic product price can make aquiculturists understand the changing trend of market in time, and then rationally plan the aquaculture structure, and realize the maximization of aquaculture benefits. At the same time, the price forecast provides a scientific basis for the government to make relevant industry policies, and strive to make full use of resources and promote the healthy and sustainable development of aquaculture. In addition, the price of aquatic products can also play a certain reference role in consumers' choice.

This study takes the prices of five common fishery products, including crucian carp, grass carp, and carp, as the object of study. The reasons are shown below. Changes in aquatic product prices affect a large number of Chinese people. According to the China Fisheries Economic Statistics Bulletin for 2021, in 2021, the value of fishery output was 224347.724 million dollars, accounting for 51.1% of China's total annual fishery economic output. Secondly, the total population that was engaged in fishery fishing was 16,342,400. However, in other countries, due to the relatively small proportion of people that are engaged in fishing, the impact of price changes on fishermen is relatively small. Therefore, it is very necessary to establish the prediction model of aquatic product price.

The innovations in this paper are shown below (1) Introducing variational modal decomposition (VMD), which has great advantages in dealing with non-stationary and non-linear time series [5]. The VMD model can decompose the original time series into several sub-series, which enlarges the details in the time series data and makes the fluctuation of sub-series smoother than the original series, which can improve the prediction accuracy of each decomposed sub-model. Therefore, this study introduced variational modal decomposition to decompose the fish price series in order to improve the accuracy of the model. (2) Based on the characteristics that the bald eagle search optimization algorithm has strong optimization-seeking ability and requires fewer parameters to be set, but easily falls into local optimality, levy flight and Tent mapping are introduced to improve the bald eagle search algorithm, and the improved bald eagle search optimization algorithm is utilized to optimize the parameters of the long short-term memory network(LSTM) to improve the accuracy of the model. (3) A comparison with common prediction models was made and the outcome indicated that compared with other popular prediction models, the *RMSE*s of the VMD-IBES-LSTM model that was proposed in this paper on the five fish price test sets are 0.480, 0.214, 0.288, 0.58, and 0.68, respectively, which is much lower than other models.

2. Literature Review

Scholars have proposed a large number of methods for price prediction. Huang et al. [6] used coal prices, which are highly non-linear and non-stationary, as the subject of their study. First, VMD was used to decompose the coal price dataset. Subsequently, GARCH and LSTM models were used to forecast each IMF component, respectively. The results show that the model has the smallest error compared to other econometric and deep-learning models. Lin et al. [7] constructed a VMD-AR-IBILSTM-ELMAN model using the prices of gas and coal from 1 December 2009 to 30 November 2020 as the study object. Subsequently, it was compared with 16 models such as GRU, VMD-GRU, and others, and the outcome indicated that the *MSE* of the VMD-AR-IBILSTM-ELMAN model on the gasoline and coal price datasets was 0.0106 and 0.649, respectively, which was much lower than the other models. Sun et al. [8] developed an EMD-VMD-LSTM model using the carbon prices of eight carbon markets in China, including Beijing and Fujian, as the object of study. First, EMD was used to decompose the carbon price dataset into a number of IMF components. Subsequently, based on the high volatility of IMF1, VMD was used to perform a secondary decomposition of IMF1. Finally, LSTM was introduced to forecast the individual IMF components. The results show that the robustness and accuracy of the model is optimal compared to EMD-LSTM, LSTM and other models. Liang et al. [9] constructed an ICEEMDAN-LSTM-CNN-CBAM model using the price of gold as the object of study. First, ICEEMDAN was used to decompose the gold price dataset into individual IMF components, and subsequently, the LSTM-CNN-CBAM model was used to forecast the individual IMF components. Ultimately, the ICEEMDAN-LSTM-CNN-CBAM model is compared with 11 common models such as LSTM and CNN-LSTM. The outcome indicated that the accuracy of the models is greatly improved after the signal processing approach is adopted. In addition, the accuracy of the ICEEMDAN-LSTM-CNN-CBAM model is significantly better than the other models. Huang et al. [10] constructed a VMD-LSTM-MW model using crude oil prices from January 1994 to July 2018 as the study object and subsequently compared it with ELM, ARIMA, and other models. The results showed that the *MAPE* of the VMD-LSTM-MW model was 0.46, which was much lower than the other models. Liu et al. [11] constructed a VMD-LSTM model using the non-ferrous metal prices for each trading day from 2 June 2006 to 21 March 2019 as the study object. First, VMD was used to decompose it into a number of IMF components. Subsequently, each IMF component was used as an input to the LSTM and the test set was predicted. The outcome showed that the *RMSE*s of the VMD-LSTM model on the zinc, copper, and aluminum price datasets are 4.79, 7.48, and 2.83, respectively, which are much lower than those of the ARIMA and LSTM models. Rezaei et al. [12] constructed CEEMD-CNN-LSTM and EMD-CNN-LSTM models using four groups of stock prices from January 201 to September 2019 as the subjects of their study. The results indicated that the CEEMD-CNN-LSTM outperformed the other models in terms of accuracy.

Aquatic products are a crucial component of agricultural products and the fluctuation of aquatic product price is characterized by nonlinear, non-stationary, and periodicity [13]. Its price forecasting research is less explored compared to other agricultural products. Nam and Sim [14] proposed an ARMA model with different parameters based on the price of abalone with different shell sizes and then used Diebold–Mariano to test the model. The test results indicated an increase in the accuracy of the improvement. Mustapa et al. [15] used an autoregressive integrated moving average model (ARIMA) to predict the prices of ten kinds of fish and vegetables in Malaysia based on data from the network. Hasan et al. [16] used the ARIMAX model to forecast catfish prices and the results indicated that the model has high predictive accuracy for both in-sample and out-of-sample. Gordon [17] used the ARDL/Bounds model to forecast lobster prices. Guillen et al. [18] used the Hedonic model to analyze eight aquatic products (e.g., cod, red mullet, flounder, etc.) in Spain from 2000 to 2013 and the results showed that the prices of aquatic products depend on the economic cycle. Khanh Nguyen et al. [19] used the ARIMAX model to forecast the price of catfish based on the uncertainty of the profitability of the catfish industry. They concluded that Vietnam people have laid little attention to the need for a sustainable and comprehensive action plan for animal-based aquatic product exports. This could have negative impacts on many aspects such as the environment and the economy.

In recent years, with the rapid development of machine learning technology, scholars have gradually applied various machine learning techniques to fish price prediction. Li Hongwei et al. [20] used wavelet function to replace the excitation function in the BP neural network to predict the price of perch and validated the model with three fish prices in ULUNGU lake. Duan, et al. [21] used the genetic algorithm to optimize the SVR model and predicted the prices of fish such as Mandarin Fish. The model was then compared with the BP neural network model and the SVR model. The results proved that the prediction accuracy of the model was significantly improved and superior compared to other models. Bloznelis [22] selected ARIMA, ANN, and KNN models to predict the price of salmon. The results showed that KNN had the highest prediction accuracy for salmon prices.

In summary, although there has been much research for price prediction [23,24], laying the foundation for this study, there is still room for further improvement: (1) A single prediction method is easily affected by the fluctuation of fish price series, resulting in lower prediction accuracy. (2) The selection of model parameters, such as the number of nodes in the implicit layer, the number of training times, and the initial learning rate in the long short-term memory network (LSTM), will have a great impact on the fitting ability of the model, and an unreasonable setting of parameters will not lead to satisfactory prediction results.

The rest of this paper is organized as follows. In Section 3, We introduce variational modal decomposition (VMD), bald eagle search algorithm (BES), improved bald eagle search optimization algorithm (IBES), long short-term memory network (LSTM), VMD-IBES-LSTM model framework, and the evaluation index criterion. In Section 4, the results of the numerical experiments are analyzed and compared with the relevant literature and also the reasons for the similarities and differences are discussed. Finally, conclusions and future research directions are given in Section 5.

3. Materials and Methods

3.1. Variational Modal Decomposition

Variational modal decomposition (VMD) is a new adaptive signal decomposition method that was proposed by Dragomiretskiy et al. [25]. It works by decomposing a multi-component signal into multiple single-component AMF signals, and then decomposing the original signal into several IMF components by solving a constrained variational problem, which has powerful non-linear and non-smooth signal processing capability. It can minimize the impact of fish price data on the prediction results due to high volatility and strong nonlinearity, etc. Compared with other decomposition methods such as EEMD, it can also solve the residual noise problem.

3.2. Bald Eagle Search Algorithm (BES)

Malaysian scholars proposed the BES algorithm in 2020 [26], which is a novel meta-heuristic algorithm with strong optimal solution search capability, and thus has received extensive research and attention from scholars in various countries. The algorithm simulates the predation behavior of bald eagles on salmon. In the process of predation on salmon, the bald eagle will firstly select a search space based on the distance of individuals and populations to salmon, flying towards a specific area; secondly, they search the water within the selected search space until a suitable prey is found; and finally the bald eagle will gradually change the altitude of its flight and dive downwards rapidly to successfully capture a prey item such as salmon from the water.

The bald eagle search algorithm is modeled on the behavior of a bald eagle hunting for prey in three stages:

- Select the search space.

The bald eagle randomly selects an area to search and subsequently makes a judgement about the number of prey items to find the best location. The equation for updating a bald eagle's position is:

$$p_{i,new} = p_{best} + \alpha \times r(p_{mean} - p_i) \tag{1}$$

In the Equation (1), p_{best} is the best position that is determined by the current bald eagle search down. p_{mean} is the average position of the bald eagle at the end of the previous search. r is a random number between 0 and 1, and α is the parameter, which takes values in the range 1.5 to 2.

- Search for prey in a selected space.

To speed up the search process to find the prey, the bald eagle flies in a spiral form, so the polar equation of the spiral is adopted here for the position update. The relevant equation is shown below:

$$\theta(i) = \alpha \times \pi \times rand \tag{2}$$

$$r(i) = \theta(i) + R \times rand \tag{3}$$

$$xr(i) = r(i) \times \sin(\theta(i)) \tag{4}$$

$$yr(i) = r(i) \times \cos(\theta(i)) \tag{5}$$

$$x(i) = xr(i)/\max(xr) \tag{6}$$

$$y(i) = yr(i)/\max(yr) \tag{7}$$

where $\theta(i)$ is the polar angle of the spiral equation, $r(i)$ is the polar diameter of the spiral equation, α, R are both parameters controlling the trajectory of the spiral, the ranges of variation are (0.5),$(0.5, 2)$, $x(i)$, and $y(i)$ are the position of the bald eagle in polar coordinates, the range of values are $(-1, 1)$, The formula for updating the location of the bald eagle is shown below:

$$p_{i,new} = p_i + x(i) \times (p_i - p_{mean}) + y(i) \times (p_i - p_{i+1}) \tag{8}$$

- Swoop to capture prey.

After the first and second steps, the bald eagle swoops from its optimal position towards its target. This is still represented here using the polar equation:

$$\theta(i) = \alpha \times \pi \times rand \tag{9}$$

$$r(i) = \theta(i) \tag{10}$$

$$xr(i) = r(i) \times \sinh(\theta(i)) \tag{11}$$

$$yr(i) = r(i) \times \cosh(\theta(i)) \tag{12}$$

$$x(i) = xr(i)/\max(xr) \tag{13}$$

$$y(i) = yr(i)/\max(yr) \tag{14}$$

During the dive, the equation for updating the position of the bald eagle is:

$$\delta_x = x_1(i) \times (p_i - c_1 \times p_{mean}) \tag{15}$$

$$\delta_y = y_1(i) \times (p_i - c_2 \times p_{best}) \tag{16}$$

$$p_{i,new} = rand \times p_{best} + \delta_x + \delta_y \tag{17}$$

where: c_1 denotes the intensity of the bald eagle's movement towards the optimal position and c_2 denotes the intensity of the bald eagle's movement towards the central position. The range of c_1 and c_2 is between 1 and 2.

3.3. Improved Bald Eagle Search Algorithm (IBES)

To address the problem that the bald eagle search algorithm is prone to slow convergence and falls easily into local optima when dealing with problems [20], the overall performance of the Bald eagle search algorithm is improved by two main strategies.

(1) Population initialization strategy based on Tent mapping.

The more uniformly the initial population is distributed in the search space, the more beneficial it is to improve the algorithm's optimization-seeking efficiency and the accuracy of the solution [27–29]. The initialized population of the traditional bald eagle search algorithm is randomly generated, which cannot guarantee that the initial population individuals are uniformly distributed in the solution space, and the over-concentration of the population may result in local optimality. Chaotic sequences have better randomness, regularity as well as ergodicity. Compared with other mappings, the Tent mapping generates a more balanced distribution of sequences. the expression of the Tent mapping is:

$$x_{t+1} = \left\{ \begin{array}{c} \frac{x_t}{u} 0 \le x_t \le u \\ \frac{1-x_t}{1-u} u \le x_t \le 1 \end{array} \right\} \tag{18}$$

The most uniform distribution series can be produced when u = 0.5. The distribution density at this point is insensitive to changes in the parameters, which is the most typical Tent mapping. The formula is:

$$x_{t+1} = \left\{ \begin{array}{c} 2x_t 0 \le x_t \le 0.5 \\ 2(1 - x_t) 0.5 \le x_t \le 1 \end{array} \right\} \tag{19}$$

The equation for population X is shown below:

$$X = X_{\min} + x_t \times (X_{max} - X_{\min}) \tag{20}$$

where: X_{max} and X_{min} are the upper and lower bound of the search, respectively.

(2) Local search strategy based on levy flight.

Levy flight was proposed by P. Levy in 1937, initially to describe the activity of a population of organisms. Through the study of the wandering foraging behavior of a population of organisms, a Levy flight with a combination of long and short distance jumps was gradually formed. Levy flight has been widely used in many fields. The position update formula of Levy flight is:

$$x_i^{t+1} = x_i^t + \alpha \oplus Levy(\lambda) \tag{21}$$

where: α is the random step size, $\oplus$ is the dot product, Levy is the random search path that fits the Levy distribution and meets the following constraints:

$$Levy(\lambda) = \frac{\varphi u}{|v|^{1/2}} \tag{22}$$

where: u and v follow standard normal distribution and λ = 1.5:

$$\varphi = [\frac{\Gamma(1+\lambda)\sin(\pi\lambda/2)}{\Gamma[(1+\lambda)/2]\lambda 2^{(\lambda-1)/2}}]^{1/\lambda} \tag{23}$$

The steps of the improved bald eagle search algorithm are shown below:

1. Set the parameters of the algorithm. The parameters that are initially set are mainly the number of populations, the maximum number of iterations T, the dimension D, the upper limit UB, and the lower limit LB of the solution space.
2. Initialize the population using the Tent mapping strategy, the maximum number of iterations and other parameters.
3. Calculate the fitness values and rank them.
4. Select the search space.
5. Search for prey in the selected space.
6. Swoop to capture prey.
7. Calculate fitness values and update bald eagle position.
8. Calculate the inertia weighting factors and use roulette wheel selection to levy flight variation on the selected individual sparrows.
9. Determine whether the stopping condition is satisfied. Exit and output the results if the stopping condition is met, otherwise, repeat Steps 2–9.

3.4. LSTM

Recurrent neural networks were introduced in the 1980s, as a popular algorithm in deep-learning. Compared to deep-learning networks (DNNs), their recurrent network structure allows them to make full use of the sequence information in the sequence data itself, and, therefore, have many advantages in dealing with time series, and the ability to correct errors that is achieved through back propagation and gradient descent algorithms. However, there are also many problems such as gradient disappearance or gradient explosion due to the increase in the number of network layers with time. Therefore, Hochreiter et al. [30] proposed a long-short term memory network in 1997. Figure 1 shows the topology of a long-short term memory network.

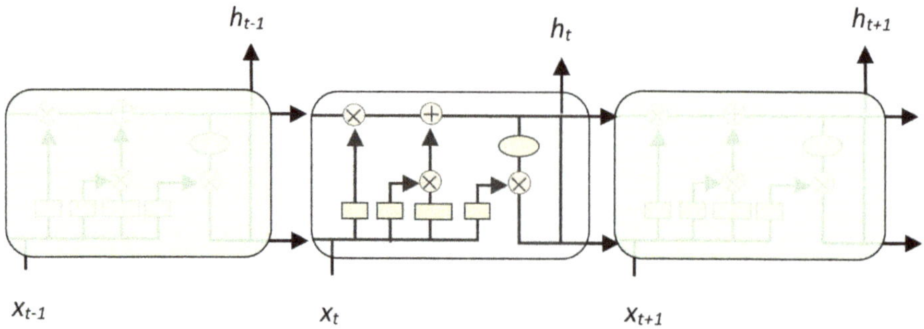

Figure 1. Structure of LSTM.

Unlike traditional recurrent neural networks that rewrite memories at each time step, LSTMs save the important features that they learn as long-term memories and selectively retain, update, or forget the saved long-term memories according to the learning process, while features that are always given little weight in multiple iterations are considered short-term memories and are eventually forgotten by the network. This mechanism allows important feature information to be passed on over iterations, allowing the network to perform better in classification tasks with long-dependent samples. In recent years, LSTM has been applied extensively in time series prediction [31–33], predicting key parameters of nuclear power plants [34], wind speed prediction [35,36], financial price trends [37], language processing [38], etc. The LSTM model has made a series of improvements on the basis of RNN neurons. It adds a transmission unit state in the RNN hidden layer and is controlled by three gating units: forgetting gate, input gate, and output gate.

3.5. *Hybrid Model of Aquatic Products Prices Forecasting*

The model that is proposed in this study is divided into four main parts as follows

- The data pre-processing stage. First, the original aquatic product time series with strong non-linearity is decomposed into a series of IMF components using variational modal decomposition.
- Optimization based on IBES. The time window step, the number of hidden layer units, the learning rate and the number of training times in the LSTM model are used as the optimization objects of the improved bald eagle search algorithm, and the parameters of the IBES algorithm (maximum number of iterations, number of populations, upper limit, lower limit, etc.) are initialized.
- Use the hybrid model to do prediction and evaluation. The LSTM network model is constructed using the optimal time window step, number of hidden layer units, learning rate and training times, and the model is trained and subsequently predicted for the test set.
- The prediction results that are obtained from the sub-series test set are added by simple linear summation to obtain the final prediction results. The flow chat is shown below (Figure 2).

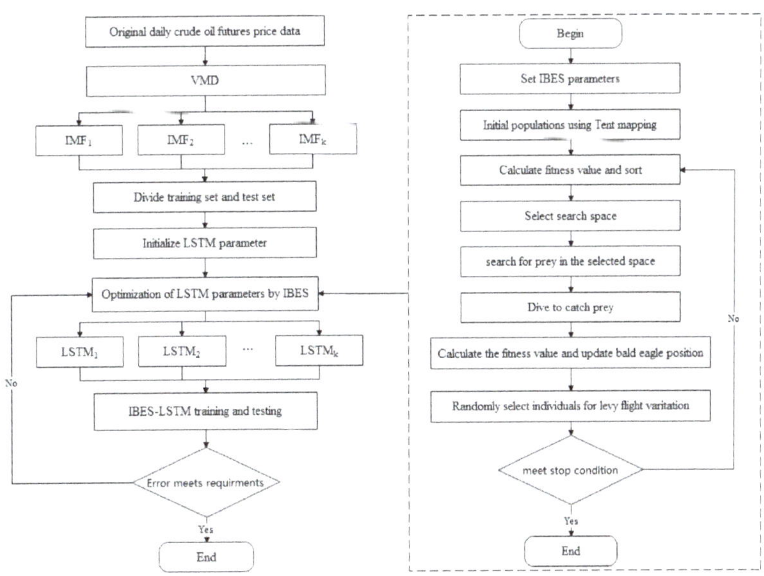

Figure 2. Flow chart of VMD-IBES-LSTM model.

3.6. Error Evaluation Criteria

In this study, four metrics, namely mean square error (*MSE*), root mean square error (*RMSE*), mean absolute error (*MAE*), and mean percentage error (*MAPE*), are chosen as the basis for judging the predictive performance of the model. *MAE* is used to measure the mean absolute error between the predicted and actual values. *RMSE* is used to measure the deviation between the predicted and actual values, which is sensitive to outliers, while *MAPE* is used to measure the average relative error between the predicted and actual values. The formulae for calculating each indicator are shown below:

$$MSE = \frac{1}{N}\sum_{t=1}^{N}(y_t - \overline{y_t})^2 \tag{24}$$

$$RMSE = \sqrt{\frac{1}{N}\sum_{t=1}^{N}(y_t - \overline{y_t})^2} \tag{25}$$

$$MAE = \frac{1}{N}\sum_{t=1}^{N}|(y_t - \overline{y_t})| \tag{26}$$

$$MAPE = \frac{1}{N}\sum_{t=1}^{N}\left|\frac{y_t - \overline{y_t}}{y_t}\right| \tag{27}$$

4. Computational Experiments and Analyses

4.1. Data Source and Descriptive Analysis

In this study, five common aquatic products, grass carp, crucian carp, carp, white chub, and big scallop were selected as the research objects. The time was from the 52nd week of 2012 to the 44th week of 2021. The data update cycle was once a week, with a total of 452 groups of data. The data are mainly from the market and Information Department of the Ministry of Agriculture and Rural Areas of China. The images of each dataset are shown below (Figure 3).

This paper first used SPSS26 software (Armonk, NY, USA) to conduct descriptive statistics on the data and the results are shown in Table 1.

Table 1. Descriptive statistics of the data set.

	Minimum Value	Maximum Value	Mean	Standard Deviation
Grass carp	11.73	22.29	14.2	2.08
Crucian carp	12.99	25.22	16.1	2.34
Carp	10.64	18.98	12.25	1.57
White Chub	6.43	12.53	7.63	1.23
Big Scallop	18.04	45.01	33.7	4.9

The Pearson correlation coefficient was then measured using SPSS26 software to screen the input data for the prediction model and determine the correlation results between the different data (Table 2).

Table 2. Correlation coefficients for datasets.

	Grass Carp	Crucian CARP	Carp	White Chub	Big Scallop
Grass Carp	1	0.83	0.78	0.81	0.43
Crucian carp		1	0.73	0.78	0.56
Carp			1	0.79	0.47
White Chub				1	0.45
Big Scallop					1

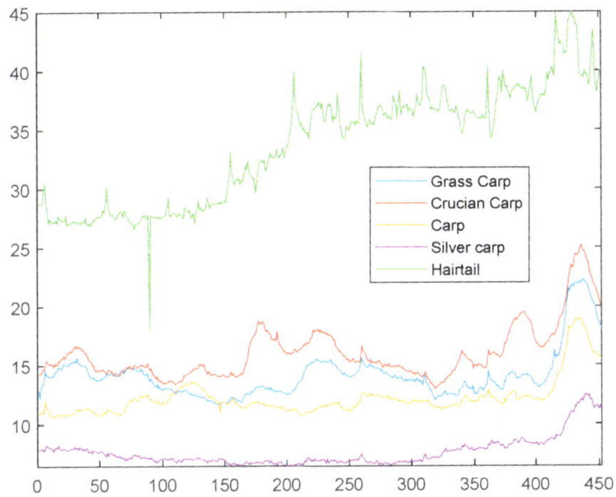

Figure 3. Dataset images.

4.2. VMD Results

In the VMD parameters, if the K value is too small, some important information of the original data may be ignored, resulting in insufficient prediction accuracy. If the K value is too large, the central frequencies of neighboring modal components may be close to each other, which may lead to problems such as mode repetition or the generation of additional noise, adversely affecting the achievement of high prediction accuracy. Therefore, it is crucial to determine the correct value of the modal number k. In this paper, to determine the k-value, the original data is first automatically decomposed into multiple modes by EMD, and then the K-value is determined according to the number of modes that are decomposed by the EMD algorithm adaptively. This method can effectively improve the efficiency of parameter selection [39–41].

VMD parameters are set as follows: The VMD decomposition layers of grass carp and crucian carp price data set are seven layers. The VMD decomposition layer of carp, white chub and big scallop price data set is eight layers. The convergence tolerance 10^{-6}. The empirical modal decomposition (EMD) results for each dataset are shown below (Figures 4 and 5).

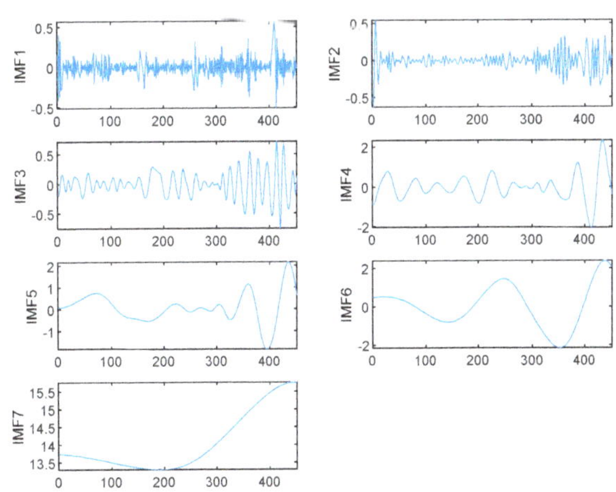

Grass carp price

Figure 4. *Cont.*

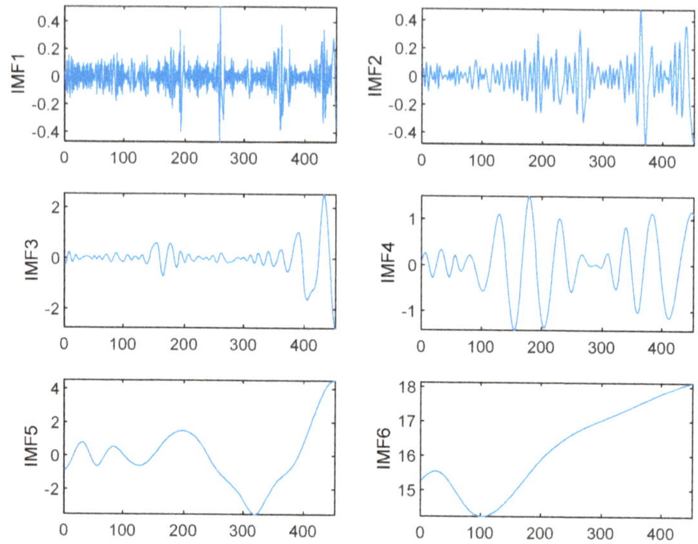

Crucian Carp price

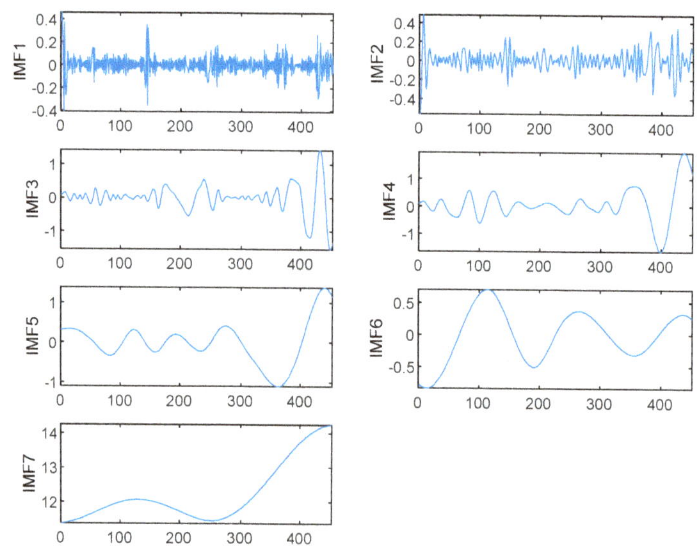

Carp price

Figure 4. *Cont.*

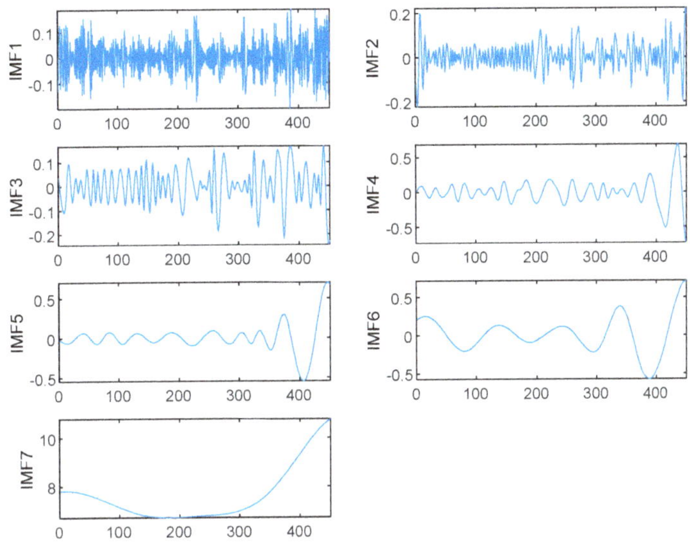

White chub price

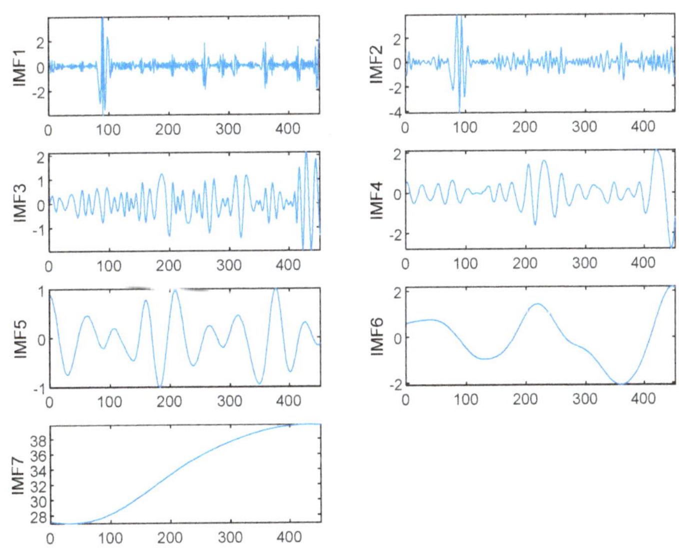

Big Scallop price

Figure 4. The outcome of EMD for each aquatic product price dataset. The variational modal decomposition (VMD) results for each dataset are shown below.

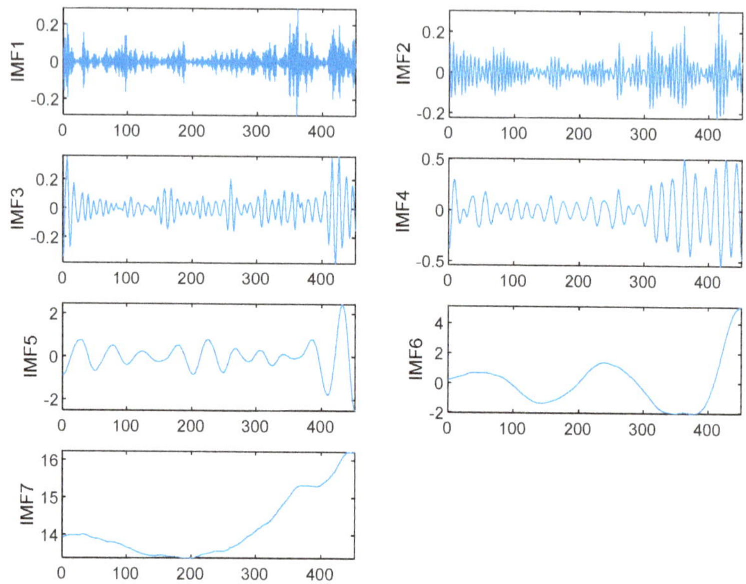

Grass carp price

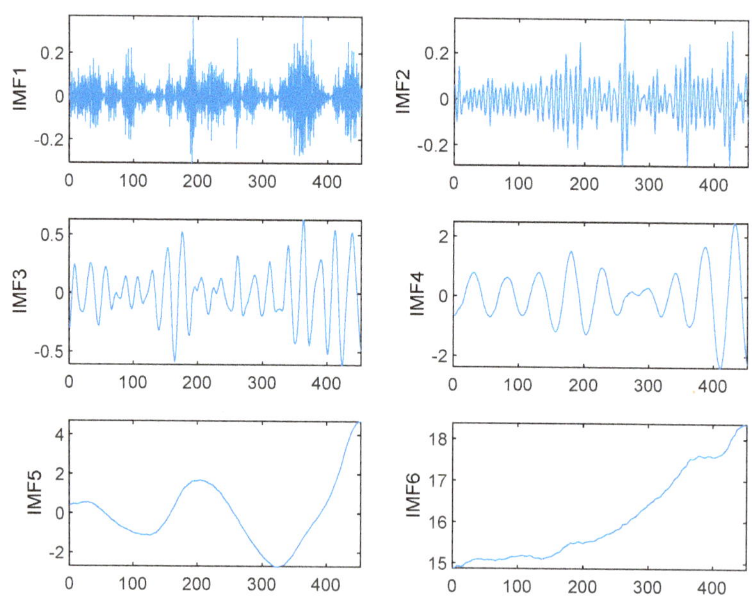

Crucian carp price

Figure 5. *Cont.*

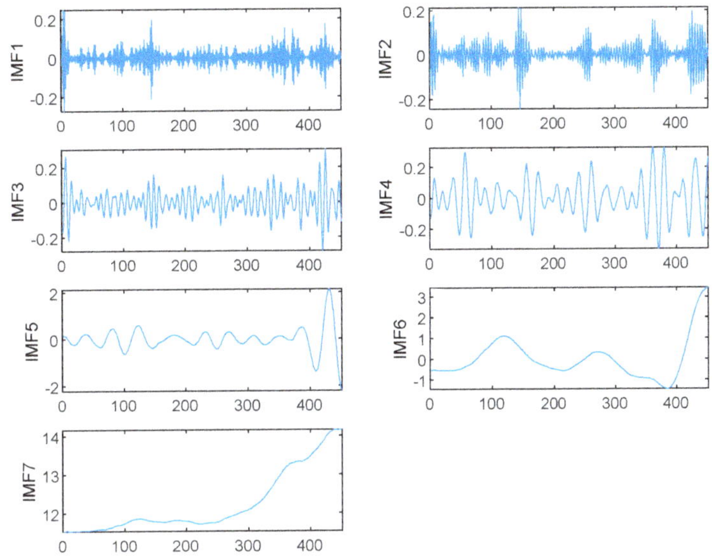

Carp price

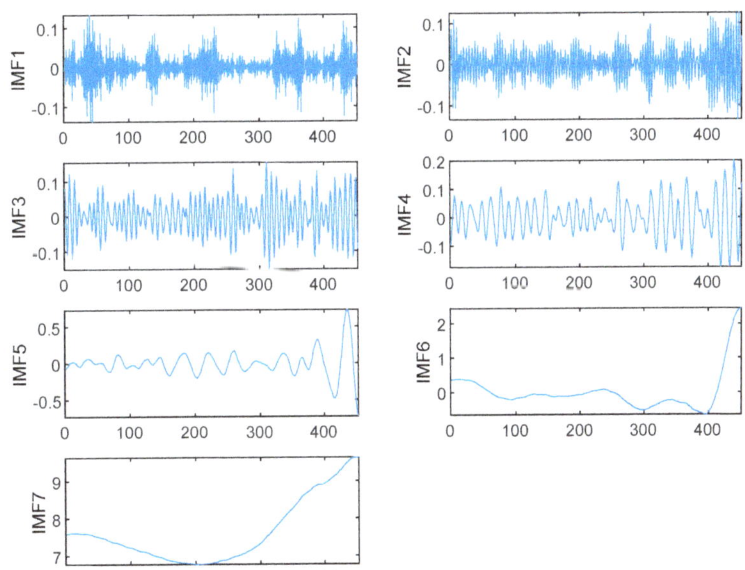

White chub price

Figure 5. *Cont.*

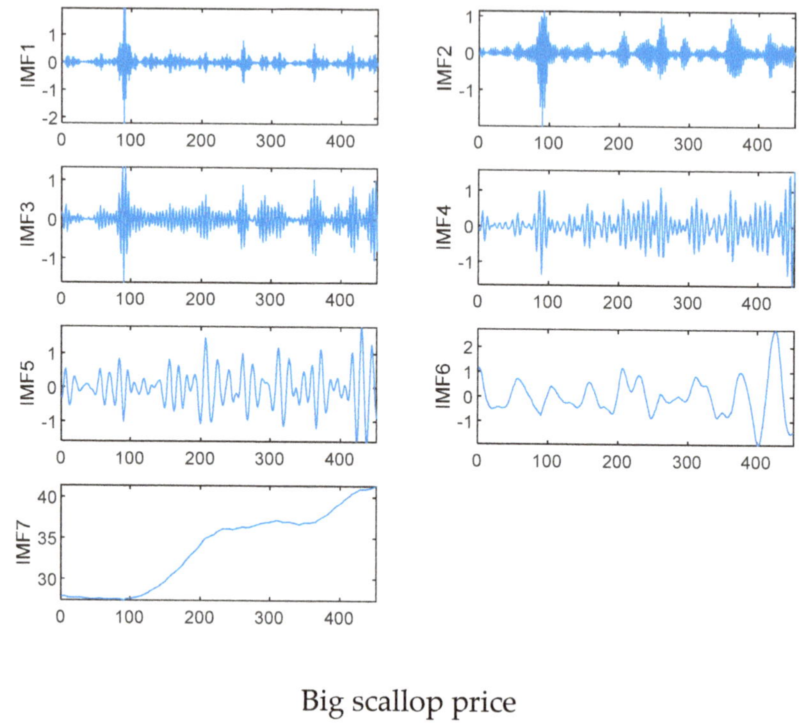

Figure 5. The outcome of VMD for each aquatic product price dataset.

From the above figure we can see that the IMF components in the upper layers are strongly non-linear and unstable during the decomposition of EMD and VMD. Therefore, it is extremely crucial to predict the IMF components in the upper layers precisely. In addition, the EMD algorithm has the problem of modal mixing in the decomposition process, which greatly affects the subsequent prediction accuracy, while the application of the VMD algorithm can effectively tackle this problem.

4.3. Aquatic Product Price Forecasting Results Based on VMD-IBES-LSTM Model

The parameter settings of the improved bald eagle search algorithm are as follows: the number of bald eagles is 5, the dim is 4, the range of learning rate is [0.001, 1], the number of neurons is [10, 500], and the maximum number of iterations is 100. Set the first 367 groups as the training set and the last 41 groups as the test set. The fitting results of each IMF price component are shown in the Figures 6–10 below.

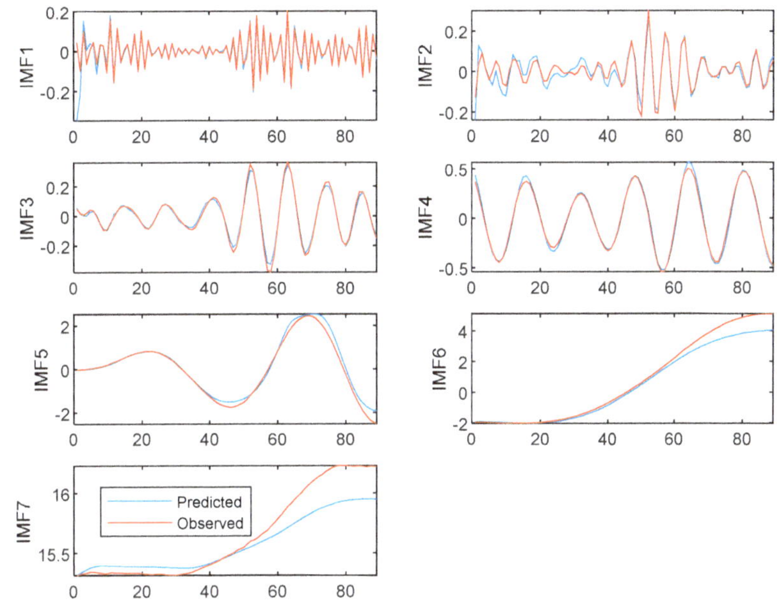

Figure 6. Fitting results of IMF components in grass carp price dataset.

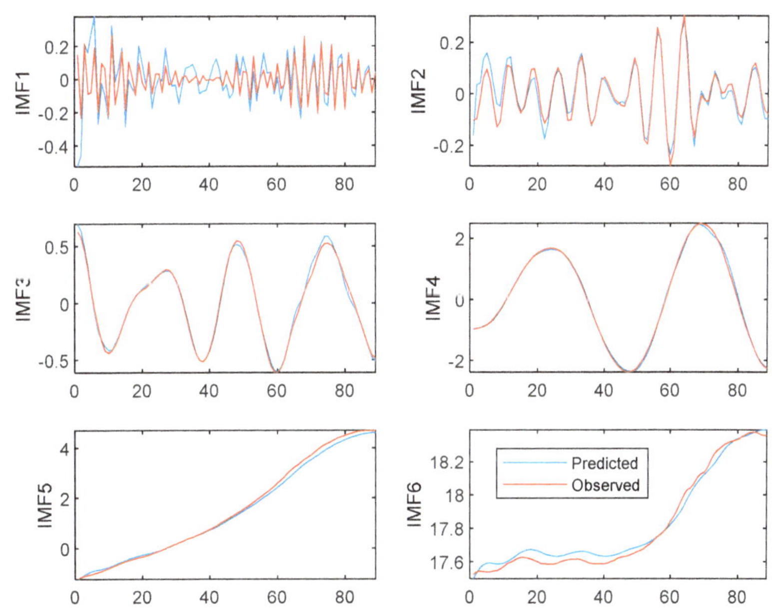

Figure 7. Fitting results of IMF components of crucian carp price dataset.

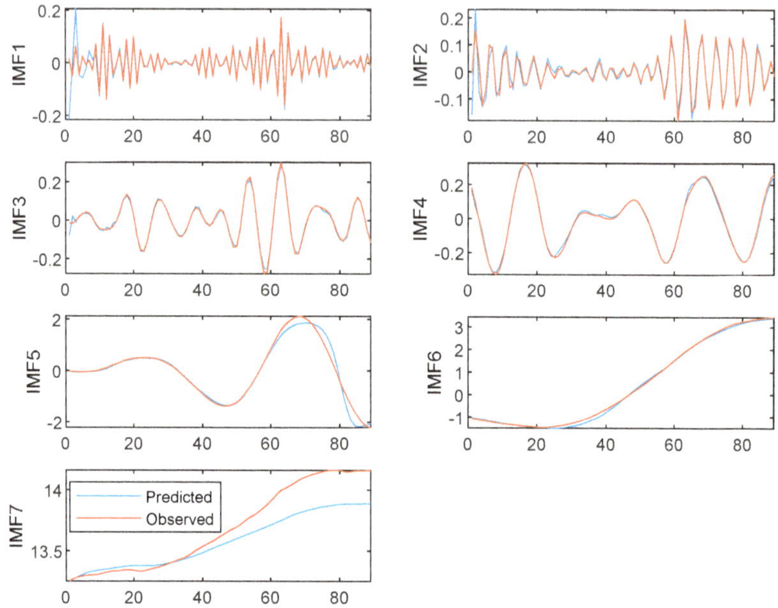

Figure 8. Results of fitting each IMF component to the carp price dataset.

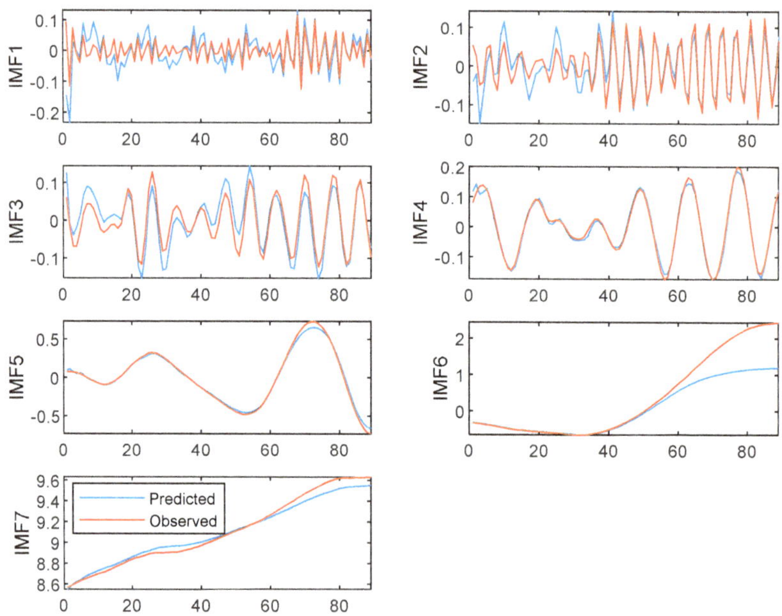

Figure 9. Fitting results for each IMF component of the white chub price dataset.

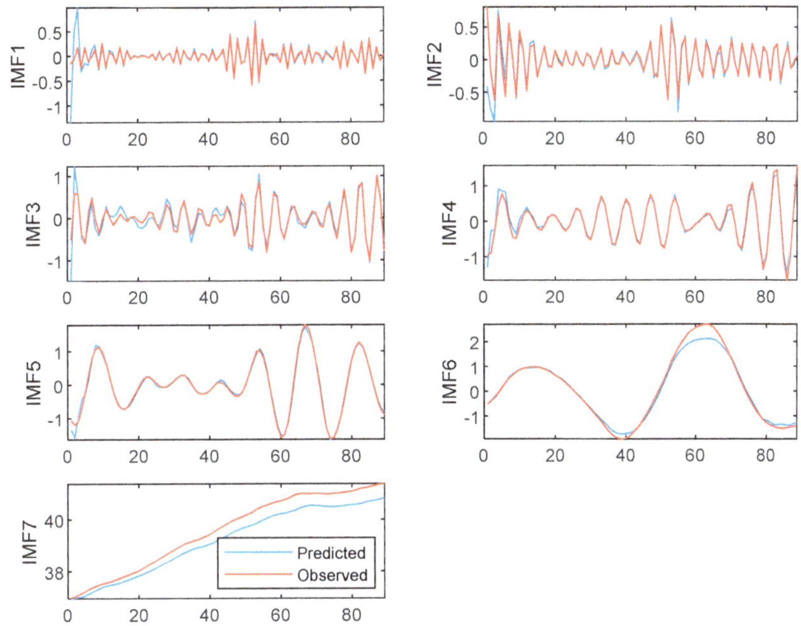

Figure 10. Fitting results for each IMF component of the big scallop price dataset.

After summarizing each IMF, the regression images of each dataset are as follows (Figure 11):

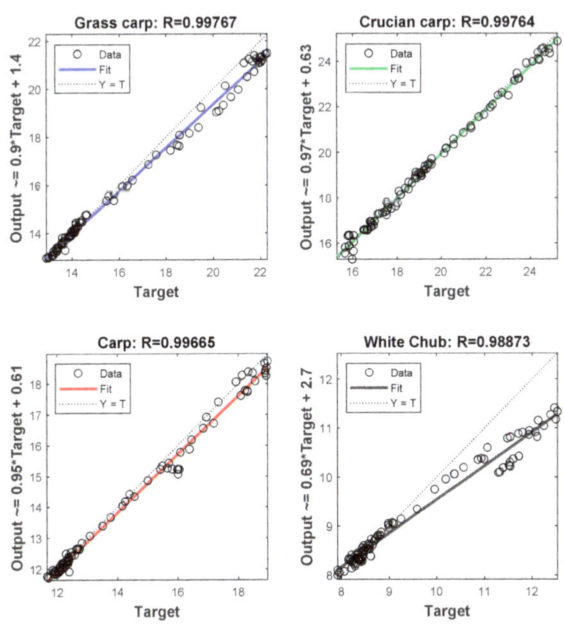

Figure 11. *Cont.*

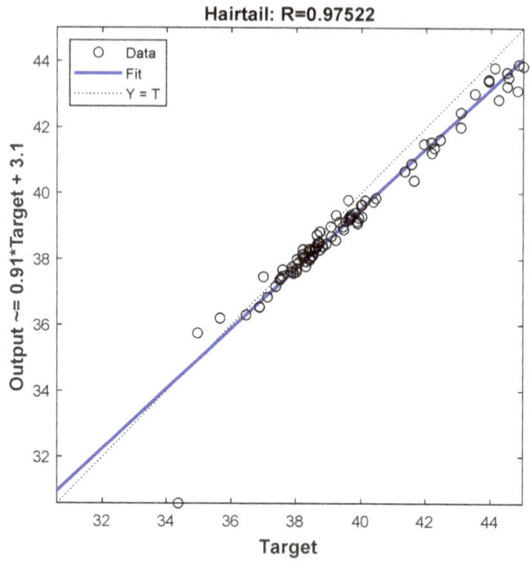

Figure 11. Data regression image of each aquatic product price test set.

It can be seen from the above figure that the goodness of fit of the VMD-IBES-LSTM model on Grass carp, crucian carp, carp, white chub, and big scallop price is 0.99767, 0.99764, 0.99665, 0.98873, and 0.97522, respectively, which indicated that the VMD-IBES-LSTM model fits best on the grass carp price dataset, while the fit is relatively average on the big scallop price dataset.

The error analysis of each IMF component is shown in the Table 3 below.

Table 3. Error analysis table.

		MSE	*RMSE*	*MAE*	*MAPE*
Grass Carp	IMF1	0.0020	0.0447	0.0150	0.3840
	IMF2	0.0010	0.0316	0.0240	1.2040
	IMF3	0.0003	0.0173	0.0140	0.2390
	IMF4	0.0001	0.0100	0.0273	0.3378
	IMF5	0.0440	0.2098	0.1401	0.2438
	IMF6	0.2473	0.4973	0.3250	0.1741
	IMF7	0.0235	0.1532	0.1136	0.0071
Crucian carp	IMF1	0.0130	0.1140	0.0731	6.2650
	IMF2	0.0011	0.03317	0.0250	1.112
	IMF3	0.0009	0.0300	0.0213	0.4643
	IMF4	0.0049	0.0700	0.0499	0.1003
	IMF5	0.0138	0.1175	0.0948	0.0842
	IMF6	0.0014	0.0374	0.0330	0.0019
Carp	IMF1	0.0010	0.0316	0.0121	0.5859
	IMF2	0.0007	0.0265	0.0146	1.1712
	IMF3	0.0001	0.0100	0.0074	0.5499
	IMF4	0.0002	0.0140	0.0100	0.3814
	IMF5	0.0333	0.1824	0.1044	0.2633
	IMF6	0.0058	0.07622	0.0589	0.0724
	IMF7	0.0269	0.1640	0.1227	0.0088

Table 3. *Cont.*

		MSE	*RMSE*	*MAE*	*MAPE*
White chub	IMF1	0.0017	0.0412	0.0297	3.2284
	IMF2	0.0009	0.0300	0.0239	0.9942
	IMF3	0.0010	0.0316	0.0285	1.0449
	IMF4	0.0001	0.0100	0.0085	0.1973
	IMF5	0.0009	0.0300	0.0212	0.3164
	IMF6	0.2717	0.5212	0.2926	1.9635
	IMF7	0.0026	0.0510	0.0427	0.0046
Hairtail	IMF1	0.0293	0.1712	0.0590	0.9651
	IMF2	0.0254	0.1594	0.0629	0.4858
	IMF3	0.0267	0.1634	0.1055	1.8694
	IMF4	0.0150	0.1225	0.0741	0.6865
	IMF5	0.0060	0.0775	0.0513	0.2352
	IMF6	0.0378	0.1944	0.1277	0.3146
	IMF7	0.1669	0.4085	0.3791	0.0095

4.4. Comparative Results and Discussion

To test and validate the effectiveness and superiority of the model that is proposed in this study. In this research, we selected the EMD-VMD-LSTM model [8], VMD-LSTM model [11], CEEMD-CNN-LSTM [12] model, MOGWO-LSSVM model [42], and the model that was proposed in this paper for comparison

- LSTM: The learning rate is 0.3%, the number of iterations is 100, the number of cells in the hidden layer is 200, the solver is set to adam, the gradient threshold is set to 1, and the mini-batch size is 32.
- CNN-LSTM: the maximum number of iterations is 100 and the learning rate is 0.003.
- BiLSTM: the number of iterations is 100. The number of cells in the hidden layer is 200. The learning rate is 0.3%.
- MOGWO: We set the smoothness and accuracy of the model as the objective function. Set the size of the repository to 30, the maximum number of iterations to 100, and the number of grey wolves to 100. Set the grid expansion parameter to 0.1. Set the leader Selection Pressure Parameter to 4.
- The error analysis for each comparison model is shown in the Table 4 as below.

Table 4. Comparison of various models in prediction performance.

		VMD-IBES-LSTM	LSTM	BPNN	CNN-LSTM	BiLSTM	CNN-BiLSTM	Bayes-BiLSTM	VMD-LSTM	EMD-VMD-LSTM	MOGWO-LSSVM	CEEMD-CNN-LSTM
Grass carp	*MSE*	0.230	4.260	5.700	7.540	20.970	8.539	3.507	1.430	1.460	1.358	6.281
	RMSE	0.480	2.064	2.390	2.750	4.570	2.922	1.873	1.196	1.208	1.165	2.506
	MAE	0.016	0.065	5.040	0.080	0.180	0.088	0.058	0.041	0.040	0.037	0.110
	MAPE	0.320	1.310	1.690	1.690	3.340	1.804	1.168	0.801	0.804	0.739	1.993
Crucian carp	*MSE*	0.046	0.880	2.750	5.860	10.670	5.717	0.142	0.160	0.710	1.716	2.921
	RMSE	0.214	0.940	1.660	2.420	3.270	2.391	0.377	0.400	0.843	1.310	1.709
	MAE	0.008	0.036	1.460	0.070	0.160	0.071	0.014	0.015	0.028	0.049	0.053
	MAPE	0.157	0.720	1.330	1.570	3.070	1.582	0.276	0.304	0.600	1.013	1.157
Carp	*MSE*	0.083	1.320	5.040	4.390	9.540	3.714	3.577	0.086	0.709	1.343	4.889
	RMSE	0.288	1.150	2.240	2.100	3.100	1.927	1.891	0.293	0.842	1.159	2.211
	MAE	0.012	0.044	3.900	0.080	0.120	0.068	0.118	0.013	0.028	0.041	0.112
	MAPE	0.190	0.730	1.500	1.290	2.000	1.168	1.296	0.200	0.596	0.700	1.751
White chub	*MSE*	0.240	1.630	6.020	3.600	6.360	3.577	0.282	0.253	0.242	1.589	2.403
	RMSE	0.489	1.280	2.450	1.900	2.520	1.891	0.531	0.503	0.492	1.261	1.550
	MAE	0.032	0.091	1.200	0.120	0.230	0.118	0.034	0.033	0.039	0.082	0.118
	MAPE	0.350	0.960	2.160	1.300	2.310	1.296	0.365	0.362	0.318	0.887	1.227
Big Scallop	*MSE*	0.470	2.820	4.540	7.420	142.700	7.107	6.520	1.326	0.814	3.665	5.108
	RMSE	0.680	1.680	2.130	2.720	11.950	2.666	2.55	1.152	0.902	1.914	2.260
	MAE	0.012	0.030	1.690	0.050	0.260	0.049	0.042	0.020	0.016	0.035	0.049
	MAPE	0.490	2.220	1.630	2.100	10.670	2.016	1.740	0.794	0.65	1.426	1.972

5. Discussion

The fitting images of VMD-IBES-LSTM model and other models that were proposed in this study are as follows (Figure 12).

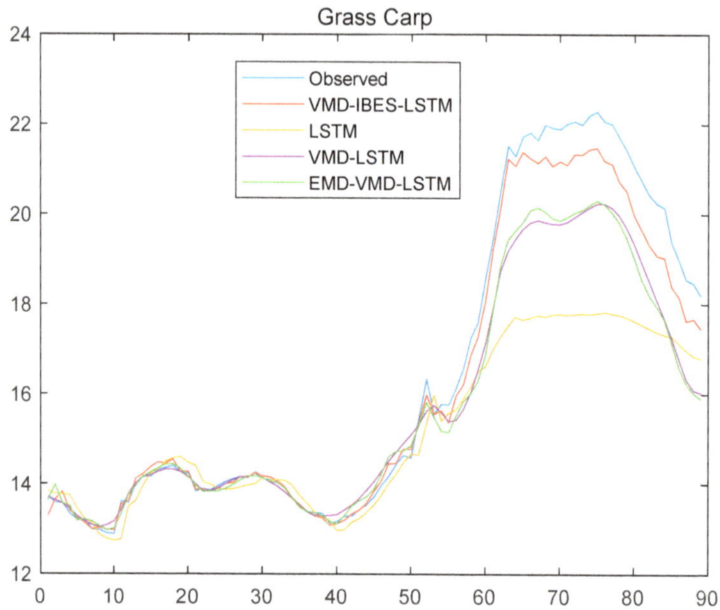

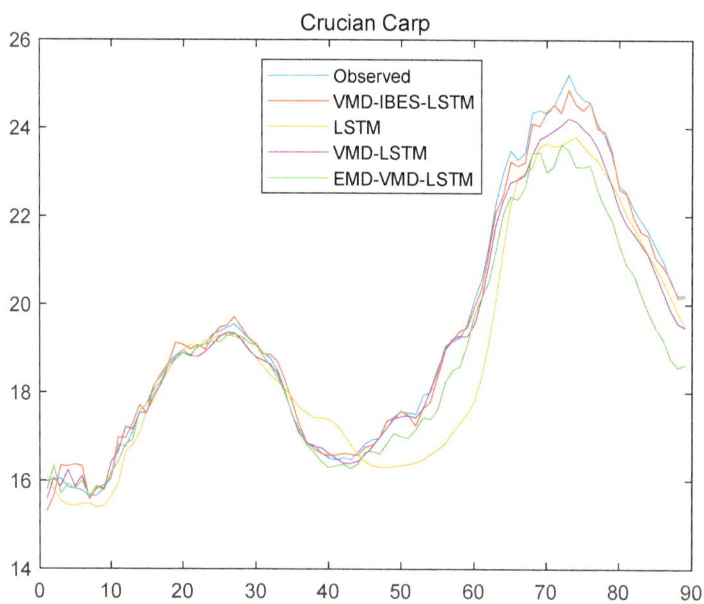

Figure 12. *Cont.*

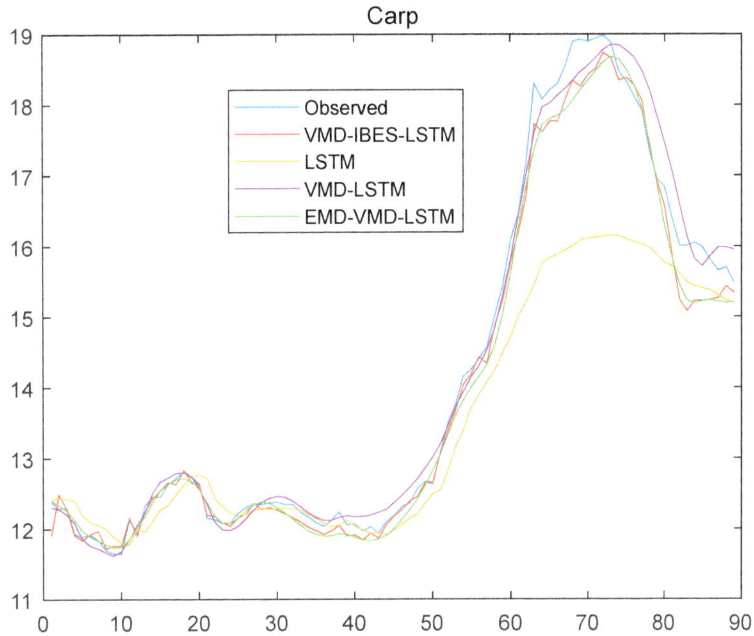

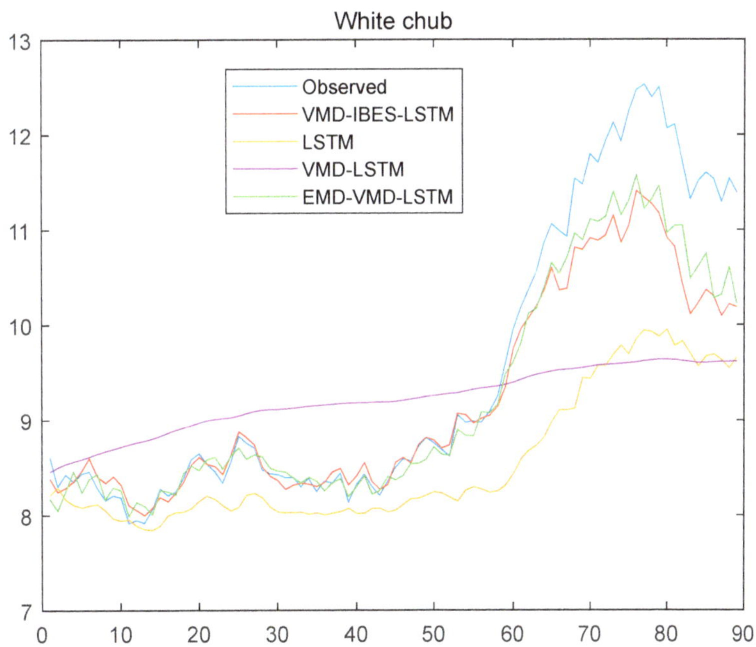

Figure 12. *Cont.*

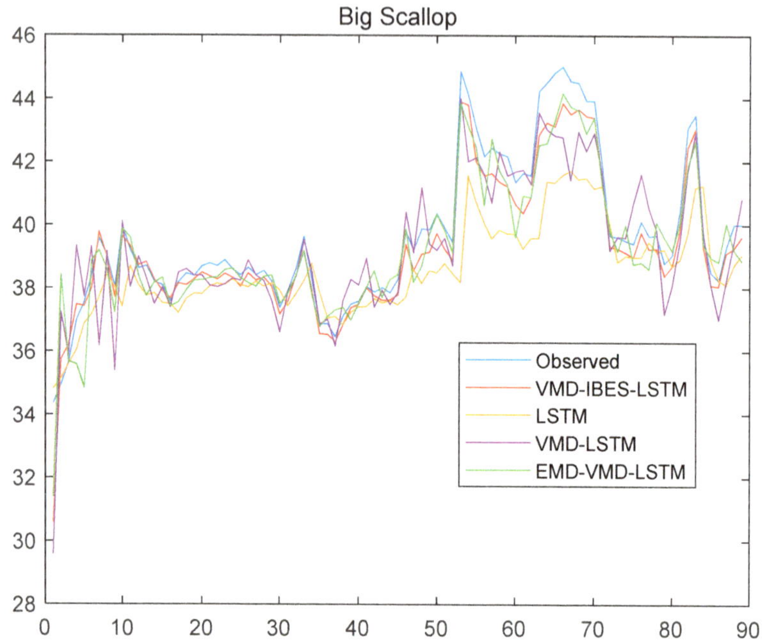

Figure 12. Fitting images of each model.

In the beginning, the models fit well because the fluctuations in grass carp, crucian carp, carp, and chub prices were small. However, with the passage of time, the fit of the single model (LSTM) becomes progressively worse, especially at the extremes. Taking the grass carp and carp price dataset as an example, it is clear that the prediction accuracy of the single model (LSTM) is poor at the extremes, which indicates the limitations of the single model in dealing with highly non-linear and non-stationary time series. Models that are based on the "decomposition-prediction-integration" idea such as the VMD-IBES-LSTM model, VMD-LSTM model, and EMD-VMD-LSTM model are less affected and have a better overall fit to the original dataset. To a higher degree, this indicated that compared to traditional single models (LSTM, BPNN, BiLSTM) and models that are based on the "feature extraction-prediction" idea (CNN-LSTM model, CNN-BiLSTM model), models that are based on the "decomposition-prediction-integration" models showed significant improvements in robustness and model accuracy.

In addition, from the above data results, it can be seen that the hybrid models that are based on the idea of "decomposition-prediction-integration" (e.g., VMD-LSTM) have better performance in terms of *MSE*, *RMSE*, and *RMSE* for each dataset. The *MSE*, *RMSE*, *MAPE*, and *MAE* of these hybrid models (e.g., NAR, LSTM, BPNN) are all smaller than those of the individual models (e.g., LSTM, BPNN), demonstrating that these hybrid models are significantly more accurate in prediction than the individual models. The reason for this analysis may lie in the fact that signal decomposition techniques can effectively solve problems such as prediction difficulties that are caused by the non-smoothness of time series data, significantly reducing the complexity of the data, and providing the possibility of improving the prediction accuracy of the models.

From the above figure, it can be seen that the VMD-IBES-LSTM model proposed in this study outperforms VMD-LSTM, CNN-BiLSTM, Bayes-BiLSTM and other models on each dataset. The improvement in the accuracy of the model is mainly in the following aspects.

(1) The variational modal decomposition is used to decompose the fish price dataset into several IMF components, which can effectively reduce the non-linearity and non-smoothness of the dataset and improve the accuracy of the model.
(2) An improved optimization algorithm is used to optimise each hyperparameter of the LSTM. In LSTM, the choice of parameters has an extremely important impact on the accuracy of the model. In previous studies, although scholars have proposed models that are based on the idea of "decomposition-prediction-integration" that can effectively improve the accuracy of prediction, less attention has been paid to the selection of parameters. The bald eagle search algorithm has received a lot of attention because of its advantages such as strong merit-seeking ability and difficulty in falling into local optimality. Thus, in this study, an improved bald eagle search algorithm was selected to optimize the hyperparameters of the LSTM.
(3) A long short-term memory network was used to predict the individual IMF components. Compared with the traditional recurrent neural network, LSTM can effectively overcome the problem of gradient explosion and gradient disappearance with an increasing number of iterations of the traditional recurrent neural network. Thus, the VMD-IBES-LSTM model that was proposed in this study is effective and competitive.

Based on the above model and results, we would like to make the following policy recommendations.

1. Data collection standards should be improved. The cross-fertilization of agricultural data and information technology should be strengthened and a unified system of data collection standards, including data collection and storage, should be established. A unified standard system should be formed through standardized data types, classifications, storage, interfaces, etc.
2. Information analysis should be strengthened. First, we should actively study the analysis models, both to strengthen the study of the adaptability of existing models, such as time series, autoregressive models, moving average models, mixed autoregressive-moving average models, vector autoregressive models, etc., and to study new models that are based on existing data. Secondly, we should strengthen the cross analysis of different disciplines. Agricultural monitoring and early warning is a multidisciplinary field that requires both economic analysis and the integrated use of information technology, computer technology, database technology, and agricultural technology. Third, special analyses should be carried out in conjunction with different target groups. It is necessary to take full account of the economic operation of agriculture, taking into account reasonable fluctuations in the prices of agricultural products, as well as the returns of producers and the benefits to consumers. At the same time, technology is actively used to liberate manpower and increase labor productivity.
3. The construction of the agricultural Internet of Things should be strengthened to improve the efficiency of agricultural products production. The agricultural Internet of Things is the application of Internet of Things technology in agricultural production, It is a specific application of agricultural production, management, and services, using various types of sensing devices to collect information about the agricultural production process, logistics of agricultural products, and animals and plants. It uses various sensing devices to collect information about the agricultural production process, the logistics of agricultural products, and the plants and animals themselves, and to transmit them through wireless sensor networks, mobile communication wireless networks, and the Internet. The information that was obtained is fused and processed through wireless sensor networks, mobile communication wireless networks, and internet transmission, and finally, through intelligent operation terminals, the process monitoring, scientific decision-making, and real-time services are realized for the pre-production, production, and post-production of agricultural products.

However, the method that was proposed in this study can be improved in the following aspects:

(1) In this study, LSTM is used to predict fish prices. While there are many improved versions of LSTM, including Bi-LSTM, Adaptive Neuro-Fuzzy Inference System (ANFIS), etc., the above methods can be compared with the model that was proposed in this study.
(2) The improved bald eagle search algorithm that was proposed in this study can also be combined with other optimisation algorithms (e.g., gravitational search algorithm, etc.) to subsequently optimize the parameters of the machine learning prediction model.
(3) There are still some errors in the accuracy of the model that was proposed in this paper. The reasons for this are mainly the following. Price fluctuations of aquatic products are closely related to a variety of factors, such as the supply and demand of aquatic products, policy changes, consumer preferences, etc. In subsequent studies, consideration can be given to adding the above-mentioned influencing factors to further improve the accuracy of the model.

6. Conclusions

As an essential resource, the price trend of aquatic products has a crucial impact on economic and social development. To address the non-linear and non-stationary characteristics of aquatic product prices, this paper proposes a new hybrid VMD-IBES-LSTM model for fish price forecasting and compares with VMD-LSTM and other models. The results indicated that the VMD-IBES-LSTM model outperforms the other listed models in *MSE*, *RMSE*, *MAE*, and *MAPE* indicators. Ultimately, based on the above model, we put forward three policy recommendations.

However, the model that was proposed in this study can still be improved in the following aspects. (1) Consider other improved versions of the LSTM, such as Bi-LSTM and GRU, for comparison testing. (2) It can be combined with other optimization algorithms to verify whether the accuracy of the model has been improved. (3) The inclusion of factors that are closely related to aquatic product prices can be considered to further improve the prediction accuracy of the model. On 31 January 2020, the World Health Organization listed the epidemic situation of novel coronavirus as a public health event of international concern. Cities in China adopted the strategy of "closing cities" and isolation. The aquatic product trade fell into a stagnant state. Changes in the external factors in the short-term led to a decline in the prediction accuracy of this model. Some studies also show that the prediction accuracy of the SARIMA model decreases with time, which is more accurate when predicting the values of the next three to six periods, but when the prediction range exceeds six periods, the simulation effect becomes worse, and the prediction error gradually increases. The same conclusion has been reached in the actual prediction process in this paper. It can be seen from the comparison between the actual value and the predicted value in Table 4 that the relative error of prediction gradually increases with the passage of time. The short-term changes of the internal and external factors also need to re-evaluate their parameters regularly according to the constantly updated data, so as to improve the accuracy of model prediction.

Author Contributions: J.W.: Writing—original draft, investigation, visualization, writing—review & editing. Z.Y.: Methodology, supervision, validation. Y.H.: Writing—original draft, diagram and flowchart preparation, writing-review & editing. D.W.: Conceptualization, formal analysis, resources. Z.Y.: Investigation, visualization. All authors have read and agreed to the published version of the manuscript.

Funding: This research was funded by the China Education Ministry of Humanities and Social Science Research Youth Fund project (No. 18YJCZH192). Ministry of Finance and Ministry of agriculture and rural areas: national special fund for the construction of modern agricultural industrial technology system "Industrial Economic Research on national marine fish industrial technology system" (No. CARS-47-G29). Major project of National Social Science Fund "Research on the development strategy of China's deep blue fishery under the background of accelerating the construction of a marine power" (No. 21 & ZD100).

Institutional Review Board Statement: Not applicable.

Informed Consent Statement: Not applicable.

Data Availability Statement: All experimental data in this paper come from Price information system of national agricultural products wholesale market in China: http://pfsc.agri.cn/#/indexPage (accessed on 4 April 2022).

Acknowledgments: Thanks to the computing science center of Shanghai Ocean University for its support for scientific research.

Conflicts of Interest: The authors declare no conflict of interest.

References

1. Fabinyi, M.; Liu, N. The social context of the chinese food system: An ethnographic study of the beijing seafood market. *Sustainability* **2016**, *8*, 244. [CrossRef]
2. Sun, Y.; Lian, F.; Guo, W.Y.; Yang, Z.Z. The spatial evolution and optimization of supply channels for marine products consumed in China. *Marit. Policy Manag.* **2022**, *8*, 1–23. [CrossRef]
3. Fabinyi, M. Historical, cultural and social perspectives on luxury seafood consumption in China. *Environ. Conserv.* **2012**, *39*, 83–92. [CrossRef]
4. Miao, M.; Liu, H.; Chen, J. Factors affecting fluctuations in China's aquatic product exports to Japan, the USA, South Korea, Southeast Asia, and the EU. *Aquacult. Int.* **2021**, *29*, 2507–2533. [CrossRef]
5. Wang, R.; Li, C.; Fu, W.; Tang, G. Deep learning method based on gated recurrent unit and variational mode decomposition for short-term wind power interval prediction. *IEEE Trans. Neural Netw. Learn. Syst.* **2019**, *31*, 3814–3827. [CrossRef]
6. Huang, Y.; Dai, X.; Wang, Q.; Zhou, D. A hybrid model for carbon price forecasting using GARCH and long short-term memory network. *Appl. Energy* **2021**, *285*, 116485. [CrossRef]
7. Lin, Y.; Lu, Q.; Tan, B.; Yu, Y. Forecasting energy prices using a novel hybrid model with variational mode decomposition. *Energy* **2022**, *246*, 123366. [CrossRef]
8. Sun, W.; Huang, C. A novel carbon price prediction model combines the secondary decomposition algorithm and the long short-term memory network. *Energy* **2020**, *207*, 118294. [CrossRef]
9. Liang, Y.; Lin, Y.; Lu, Q. Forecasting gold price using a novel hybrid model with ICEEMDAN and LSTM-CNN-CBAM. *Expert Syst. Appl.* **2022**, *206*, 117847. [CrossRef]
10. Huang, Y.; Deng, Y. A new crude oil price forecasting model based on variational mode decomposition. *Knowl.-Based Syst.* **2021**, *213*, 106669. [CrossRef]
11. Liu, Y.; Yang, C.; Huang, K.; Gui, W. Non-ferrous metals price forecasting based on variational mode decomposition and LSTM network. *Knowl.-Based Syst.* **2020**, *188*, 105006. [CrossRef]
12. Rezaei, H.; Faaljou, H.; Mansourfar, G. Stock price prediction using deep learning and frequency decomposition. *Expert Syst. Appl.* **2021**, *169*, 114332. [CrossRef]
13. Duran, N.M.; Maciel, E.D.S.; Galvao, J.A.; Savay-da-Silva, L.K.; Sonati, J.G.; Oetterer, M. Availability and consumption of fish as convenience food—Correlation between market value and nutritional parameters. *Food Sci. Technol.-Brazil* **2018**, *37*, 65–69. [CrossRef]
14. Nam, J.; Sim, S. Forecast accuracy of abalone producer prices by shell size in the republic of korea: Modified diebold–mariano tests of selected autoregressive models. *Aquacult. Econ. Manag.* **2017**, *22*, 474–489. [CrossRef]
15. Mazliana, M.; Raja, R.P.; Ho, M.K. Forecasting Prices of Fish and Vegetable using Web Scraped Price Micro Data. *Int. J. Rec. Eng.* **2019**, *7*, 251–256.
16. Hasan, M.R.; Dey, M.M.; Engle, C.R. Forecasting monthly catfish (ictalurus punctatus.) pond bank and feed prices. *Aquacult. Econ. Manag.* **2019**, *23*, 86–110. [CrossRef]
17. Gordon, D.V. A short-run ARDL-bounds model for forecasting and simulating the price of lobster. *Mar. Resour. Econ.* **2020**, *35*, 43–63. [CrossRef]
18. Guillen, J.; Maynou, F. Characterisation of fish species based on ex-vessel prices and its management implications: An application to the spanish mediterranean. *Fish. Res.* **2015**, *167*, 22–29. [CrossRef]
19. Nguyen, H.T.K.; Thu, T.T.N.; Lebailly, P.; Azadi, H. Economic challenges of the export-oriented aquaculture sector in Vietnam. *J. Appl. Aquac.* **2019**, *31*, 367–383. [CrossRef]
20. Li, H.; Gao, X.; Cheng, K. The application of wavelet neural network in prediction of the fish price. *Appl. Mech. Mater.* **2014**, *687*, 1945–1949. [CrossRef]
21. Duan, Q.; Zhang, L.; Wei, F.; Xiao, X.; Wang, L. Forecasting model and validation for aquatic product price based on time series GA-SVR. *Trans. Chin. Soc. Agric. Eng.* **2017**, *33*, 308–314.
22. Bloznelis, D. Short term salmon price forecasting. *J. Forecast.* **2018**, *37*, 151–169. [CrossRef]
23. Yuan, H.; Chen, Y.; Ju, J. A CBR Based Prediction Method for Web Aquatic Products Prices. *Int. J. Comput. Int. Sys.* **2007**, 195–200.
24. Shi, J.; Leau, Y.B.; Li, K.; Park, Y.J.; Yan, Z. Optimization and Decomposition Methods in Network Traffic Prediction Model: A Review and Discussion. *IEEE Access* **2020**, *8*, 202858–202871. [CrossRef]
25. Dragomiretskiy, K.; Zosso, D. Variational mode decomposition. *IEEE Trans. Signal Process.* **2013**, *62*, 531–544. [CrossRef]

26. Alsattar, H.A.; Zaidan, A.A.; Zaidan, B.B. Novel meta-heuristic bald eagle search optimisation algorithm. *Artif. Intell. Rev.* **2020**, *53*, 2237–2264. [CrossRef]
27. Angayarkanni, S.A.; Sivakumar, R.; Ramana Rao, Y.V. Hybrid Grey Wolf: Bald Eagle search optimized support vector regression for traffic flow forecasting. *J. Amb. Intel. Hum. Comp.* **2021**, *12*, 1293–1304. [CrossRef]
28. Li, L.L.; Liu, Z.F.; Tseng, M.L.; Zheng, S.J.; Lim, M.K. Improved tunicate swarm algorithm: Solving the dynamic economic emission dispatch problems. *Appl. Soft Comput.* **2021**, *108*, 107504. [CrossRef]
29. Chen, S.; Wang, S. An Optimization Method for an Integrated Energy System Scheduling Process Based on NSGA-II Improved by Tent Mapping Chaotic Algorithms. *Processes* **2020**, *8*, 426. [CrossRef]
30. Hochreiter, S.; Schmidhuber, J. Long short-term memory. *Neural Comput.* **1997**, *9*, 1735–1780. [CrossRef]
31. Zhang, J.; Zhu, Y.; Zhang, X.; Ye, M.; Yang, J. Developing a long short-term memory (LSTM) based model for predicting water table depth in agricultural areas. *J. Hydrol.* **2018**, *561*, 918–929. [CrossRef]
32. Wang, X.; Wang, Y.; Yuan, P.; Wang, L.; Cheng, D. An adaptive daily runoff forecast model using VMD-LSTM-PSO hybrid approach. *Hydrol. Sci. J.* **2021**, *66*, 1488–1502. [CrossRef]
33. Fang, Z.; Wang, Y.; Peng, L.; Hong, H. Predicting flood susceptibility using LSTM neural networks. *J. Hydrol.* **2021**, *594*, 125734. [CrossRef]
34. Chen, Y.; Lin, M.; Yu, R.; Wang, T. Research on simulation and state prediction of nuclear power system based on LSTM neural network. *Sci. Technol. Nucl. Install.* **2021**, *2021*, 8839867. [CrossRef]
35. Fukuoka, R.; Suzuki, H.; Kitajima, T.; Kuwahara, A.; Yasuno, T. Wind Speed Prediction Model Using LSTM and 1D-CNN. *J. Signal Process.* **2018**, *22*, 207–210. [CrossRef]
36. Ehsan, M.A.; Shahirinia, A.; Zhang, N.; Oladunni, T. Wind Speed Prediction and Visualization Using Long Short-Term Memory Networks (LSTM). In Proceedings of the 10th International Conference on Information Science and Technology (ICIST), Bath, London, and Plymouth, UK, 9–15 September 2020; pp. 234–240.
37. Troiano, L.; Villa, E.M.; Loia, V. Replicating a Trading Strategy by Means of LSTM for Financial Industry Applications. *IEEE Trans. Ind. Inform.* **2018**, *14*, 3226–3234. [CrossRef]
38. Sundermeyer, M.; Schlüter, R.; Ney, H. LSTM Neural Networks for Language Modeling. In Proceedings of the Thirteenth Annual Conference of the International Speech Communication Association, Portland, OR, USA, 9–13 September 2012.
39. Wu, S.; Feng, F.; Zhu, J.; Wu, C.; Zhang, G. A method for determining intrinsic mode function number in variational mode decomposition and its application to bearing vibration signal processing. *Shock. Vib.* **2020**, *2020*, 8304903. [CrossRef]
40. Li, Y.; Li, Y.; Chen, X.; Yu, J. Research on ship-radiated noise denoising using secondary variational mode decomposition and correlation coefficient. *Sensors* **2017**, *18*, 48. [CrossRef]
41. Zhang, K.; Cao, H.; Thé, J.; Yu, H. A hybrid model for multi-step coal price forecasting using decomposition technique and deep learning algorithms. *Appl. Energy* **2022**, *306*, 118011. [CrossRef]
42. Bai, L.; Liu, Z.; Wang, J. Novel hybrid extreme learning machine and multi-objective optimization algorithm for air pollution prediction. *Appl. Math. Model.* **2022**, *106*, 177–198. [CrossRef]

agriculture

Article

IoAT Enabled Smart Farming: Urdu Language-Based Solution for Low-Literate Farmers

Sehrish Munawar Cheema [1,*], Muhammad Ali [2], Ivan Miguel Pires [3,*], Norberto Jorge Gonçalves [4], Mustahsan Hammad Naqvi [1] and Maleeha Hassan [5]

1 Department of Computer Science, University of Management and Technology, Sialkot 54770, Pakistan
2 Department of Software Engineering, The Superior University, Lahore 54600, Pakistan
3 Instituto de Telecomunicações, Universidade da Beira Interior, 6200-001 Covilhã, Portugal
4 Escola de Ciências e Tecnologia, University of Trás-os-Montes e Alto Douro, Quinta de Prados, 5001-801 Vila Real, Portugal
5 Department of Nutritional Sciences, University of Sialkot, Sialkot 51310, Pakistan
* Correspondence: sehrish.munawar@skt.umt.edu.pk (S.M.C.); impires@it.ubi.pt (I.M.P.)

Abstract: The agriculture sector is the backbone of Pakistan's economy, reflecting 26% of its GPD and 43% of the entire labor force. Smart and precise agriculture is the key to producing the best crop yield. Moreover, emerging technologies are reducing energy consumption and cost-effectiveness for saving agricultural resources in control and monitoring systems, especially for those areas lacking these resources. Agricultural productivity is thwarted in many areas of Pakistan due to farmers' illiteracy, lack of a smart system for remote access to farmland, and an absence of proactive decision-making in all phases of the crop cycle available in their native language. This study proposes an internet of agricultural things (IoAT) based smart system armed with a set of economical, accessible devices and sensors to capture real-time parameters of farms such as soil moisture level, temperature, soil pH level, light intensity, and humidity on frequent intervals of time. The system analyzes the environmental parameters of specific farms and enables the farmers to understand soil and environmental factors, facilitating farmers in terms of soil fertility analysis, suitable crop cultivation, automated irrigation and guidelines, harvest schedule, pest and weed control, crop disease awareness, and fertilizer guidance. The system is integrated with an android application 'Kistan Pakistan' (prototype) designed in bilingual, i.e., 'Urdu' and 'English'. The mobile application is equipped with visual components, audio, voice, and iconic and textual menus to be used by diverse literary levels of farmers.

Keywords: smart farming; precision agriculture; IoT; sensor network; semi-literate farmers; interactive interface; User Interface (UI); Android apps

Citation: Cheema, S.M.; Ali, M.; Pires, I.M.; Gonçalves, N.J.; Naqvi, M.H.; Hassan, M. IoAT Enabled Smart Farming: Urdu Language-Based Solution for Low-Literate Farmers. *Agriculture* 2022, *12*, 1277. https://doi.org/10.3390/agriculture12081277

Academic Editor: Dimitre Dimitrov

Received: 19 July 2022
Accepted: 20 August 2022
Published: 22 August 2022

1. Introduction

Agriculture is considered the base for human living because it is the primary food source and plays a crucial role in the global economy. Pakistan is 79.6 million km^2 and is home to a population of 192 million. The contribution of the agricultural sector to gross domestic product (GDP) in Pakistan gradually decreased to 19.3% in the year 2020–2021 from 22.04% previously recorded in 2019 and generating employment opportunities for 38.5% of Pakistan's labor force and valuable foreign exchange for the country [1–3]. It supports the manufacturing and services sectors of the economy by providing backward-forward linkages in inputs-outputs markets and the most significant consumer of household durables. Therefore, our agriculture sector can be considered an economic activity in the country [4]. Farmers are facing issues in the agriculture sector, so it's significant to research, develop of latest mechanisms, and adopt new practices to enhance production. Pressure on the agricultural system will increase with the continuing expansion of the human population.

Many areas of Pakistan are trailing in agricultural productivity due to a lack of farmers' awareness, timely access to crucial information, and proactive decision-making [5,6]. It is vital for human development in these underdeveloped areas to utilize information and communication technologies (ICTs), artificial intelligence techniques, machine learning (ML), and deep learning (DL) to make such information more readily accessible to farmers, significantly increasing crop production [7–13]. Climate change and shortage of agricultural resources are also significant concerns for the downfall of agricultural performance resulting in food insecurity [14–17]. This lets farmers hamper soil with intensified pesticides, which affect agricultural practices in a harmful manner. Finally, fields remain barren [18–22]. These are reasons for crop failure, lower production due to diseases, unpredictable climate change, and loss of soil fertility [23,24].

In this scenario, the traditional agriculture trends are insufficient to increase agricultural growth. Agriculture is also out of the reach of less conventional technologies. In this context, digital agriculture, automation, and precision farming, now termed smart farming, have arisen as new scientific fields that use intense techniques to drive agricultural productivity while minimizing its environmental impact [25–27]. Data generated by smart farming operations is provided by various sensors that enable a better understanding of the operational environment (interaction of dynamic conditions of the crop, soil, weather, and environmental factors) and the operation itself, leading to more accurate and timely decision-making [28,29]. Variability in climate and labor shortage is increasing continuously, providing better insights for agricultural machinery automation. Remote monitoring technologies facilitate farmers to access every inch of the farmlands by creating virtual fences to monitor, detect and protect crops in real-time [30–32]. IoT-based technologies allow farmers, among other things, to gather data on plants' environmental conditions like climate change, soil fertility level, humidity, temperature, and light intensity to monitor fields and farms remotely. These technologies assist farmers in having know-how and status of crops anywhere and anytime [33–36].

In Pakistan, most farmers have android phones [37–41] but are regrettably underutilized. Our preliminary literature study compelled us to work to facilitate farmers for agriculture automation, recommendations, and guidelines in their local language, i.e., Urdu, by using the internet of agricultural things (IoAT), also known as agricultural internet of things (Ag-IoT) and artificial intelligence technologies with transliteration and voice-speech support in the local language. IoAT is the network of complex and diverse agricultural objects that compute, process, and recommend solutions intelligently based on data generated from every connectable thing [42].

Previous research shows that using graphical cues, audio, speech, and video in mobile interfaces helps low-literates better adapt [43,44]. Field study experiences reflect that low-literates feel more challenged in understanding and interpreting textual information than their literate peers [45]. In our research, we try to accommodate such users by introducing audio, speech, Urdu language support, and an interactive graphical interface. Researchers also talked about the improvements in information dissemination systems for less literate farmers via different means [46–50]. However, none considered interface design and Urdu language-based real-time updates about agricultural guidelines using the android application and user preferences. Our choice of Urdu in this work was made by observing that 87% of farmers preferred Urdu as a medium of information dissemination [51]. Our research is an extended form of [52–54] and primarily focuses on developing an IoT-based and user-friendly system with these utilities.

1.1. Rationale

This section reflects the findings and an evaluation report on information found in the literature relevant to our research domain. It represents the overview of different approaches used by other researchers. Integrating the Internet of Things into the agricultural system has led to the internet of Agricultural things (IoAT) and advanced computing techniques. The researchers applied this to obtain maximum benefits and also to improve the production of agriculture, artificial intelligence, and IoT [55]. The agriculture domain is experiencing new evolution and revolution motivated by cloud technology, IoT, Edge and fog computing, sensors, IoT, and big data [56]. A proposal was presented for agriculture applications by investigating integrated platforms, including cloud computing, IoT, and data mining techniques [57]. An IoT-based smart agricultural system was developed using deep learning combined with a cloud environment comprised of four layers: data collection, edge computing, data transmission, and cloud computing layer [58]. A scalable network-based architecture was proposed to monitor and regulate agricultural farms in rural areas using IoT-based wifi, long-distance network, and fog computing [59]. Agricultural data analytics employed with IoT has transformed from specific crops to any kind of crop. The developed system could support various applications, from controlling and monitoring the crops to promoting them to market [60].

Literacy is the ability to read and write simple statements [61]. Illiteracy, low education, and computer illiteracy are significant concerns in developing countries like Pakistan. Studies indicate that user interface (UI) would be designed differently for literate, low-literate, and illiterate users. A user interface should also consider the cultural context, such as language and images. The non-textual interface is more user-friendly than the textual one for illiterate users [62].

The inability to read and write and the illiteracy of small farmers make them vulnerable to various workers and cause human health risks [63]. Previous research inferred that complex hierarchy and multi-screens become difficult for low-literates to understand helpful information, so the visuals, audio, video, speech, icons, and images are a better approach to passing complex data and information to mobile users [64]. The research findings shed light on some user interface (UI) design guidelines for illiterate and semi-literate users that can help take advantage of information and communication technologies ICT [51]. The most powerful design factors that should be incorporated into a user interface (UI) for low-literate users are localization and graphics [65]. An android application with audio, textual, and visual components was designed for farmers with diverse literacy levels. It could facilitate them regarding vital weather information [45]. Pakistani farmers typically rely on traditional sources of information, which could be a reason for their information deficiency. Data analysis indicated that farmers had diverse demographic conditions, but primary among them is the ordinary level of education (52.4% illiterate). A high level of information deficiency was observed among farmers regarding fertilizers application, seed rate, disease diagnosis, pests, and insects' identification, and a medium level of lack in information regarding the selection of varieties, harvesting, and pets' management was observed [66]. Providing information access to low-literate, linguistic minority, tech-shy, handicapped and marginalized users using speech-based services is a viable solution. These services were made the national weather hotline of Pakistan [50]. A survey data revealed that farmers in the Vehari district of Pakistan have a low literary rate and less technical knowledge. They are unable to read agricultural instructions, unaware of pesticides persistence and toxicity (73%), unable to diagnose cotton pests and diseases (86%), and unable to decide which crop to grow on cotton adjacent farms (100%) [67]. The research was conducted to study knowledge, attitude, and practices regarding pesticide usage by vegetable growers in three districts; Dadu, Larkana, and Shikarpur of Sindh, Pakistan. Results show that most vegetable growers (40.90%) have low primary education literacy, and 27.27% possess a middle pass. That's why most growers are unfamiliar with pest and insect damage indications and the safe handling of pesticides [68]. Pakistani farmers' awareness of the damaging effects of different pesticides can lead to integrated and smart pest control and management [69]. Research findings reflect farmers' behavior and a low tendency towards reading the labels of the pesticides due to low education, advanced age, usage of too technical language, illegible fonts, and unclear texts [70]. In [71], the authors developed a basic interactive voice response (IVR) system for agro-information dissemination, such as fertilizer, pesticide information, and weather forecast. In terms of usability and information extraction, their study reflected that simple menu-based navigation interfaces are relatively easy to use and understand.

A remote agricultural monitoring platform was proposed in [72] after a detailed literature study. Cyber security-based precision farming conceptual architecture was presented in [73] for the frost prediction in peach production by analyzing data captured by sensors implanted around an orchard. IoT-based precision farming comprises multiple control and monitoring applications like monitoring water needs according to climate conditions, analyzing soil patterns, monitoring crops disease and pest attacks, and assessing optimum time for planting, harvesting, and tracking [74,75].

AquaAgro offers IoT and Artificial Intelligence (AI) enabled solutions for precision farming. Using a software or app embedded hardware, the predictions will be made for the Irrigation scheduling, Fertilizer requirement, Pest attack prediction, and Plant disease detection. The essential four services that AquaAgro provides are irrigation scheduling, Fertilizer requirement, Pest attack prediction, and Plant disease detection. They have received an overwhelming response from the people [76]. An android mobile application named 'Mentha Mitra' was developed with an interactive interface with bilingual (Hindi and English) for menthol mint growers [77]. Android application provides scientific e-advisories on crop-related diseases, high-yield varieties, pests, insects, and improved distillation units.

An IoT-based wireless sensor network (WSN) framework was proposed to monitor crops smartly by analyzing environmental factors [78]. In [79], the authors utilized the benefits of IoT for the implementation of precision agriculture by sensing required parameters from the field and making suitable decisions such as activation and deactivation of irrigation valves. Parameters include soil moisture, temperature and light intensity, etc. Sensors could also send the gathered data to the cloud, and an Android application was developed to access these parameters. An expert IoT-based system relies on the stored knowledge base and real-time data for farmer recommendations [80]. This system will help in proactive and reactive tasks to a minimum the loss of water. Farooq et al. [81] performed a comprehensive literature study on state-of-the-art techniques in smart farming. They discussed agriculture networks, platforms, architecture, and topologies to help farmers to enhance the corps' productivity. This survey paper shows that Government and many other stakeholders are interested in deploying IoT in Pakistan's agriculture field. To increase agricultural productivity, the authors suggested that collaboration between allied and agriculture activities can be built by integrating big data into climate-smart agriculture with resource utilization [82].

In [83], the authors are more concerned about the water supply to the plants. They proposed a system in which a farmer can water the plants with a push of a button on his phone when he is out of the station. Machine learning algorithms and radio frequency identification (RFID) tags detect and measure moisture and humidity. Internationally, many studies [12,84–86] have been conducted to improve agricultural processes based on soil fertility level, crops, weather patterns, and fertilizers. These studies used IoT, Global Position Systems (GPS), Global information systems (GIS), Wireless Sensor Networks (WSN), and many machine learning techniques. The implementation of studies results in increased profitability and self-sufficiency. IoT enabled decision support systems based on real-time farm sensors data, improving the water consumption by crops [87–89]. Authors in [90] provided the real-time farm data, weather, and crops data to a Penman-Monteith and crop-coefficient model to produce recommendations about irrigation schedules. An intelligent approach for diagnosing crop disease was proposed in [91], capable of working with android devices equipped with fuzzy decision-making at the backend. The system interacts with farmers in their native language of Urdu for crop disease diagnosing. 'Padi2U' is an android application developed for farmers to manage paddy fields. It provides guidelines related to paddy varieties, planting schedule, pest, disease, weed, weather forecast, and yield information in their native language 'Malay' [92]. In [93], the authors developed an application named 'BLYNK' to control the IoT-based hardware remotely. The purpose of 'BLYNK' was to automate irrigation and fertilizer supply to farms. Their results reflect approximately 50% water saving and a 35% increase in yield. Irrigation monitoring and automatic control systems were developed using fuzzy decision support to generate a moisture content distribution map of soil and enhance affectivity [53,94–98].

Soil having an essential quantity of macro and micro-nutrients would be capable of cultivating different crops. The soil's lack of significant nutrients (Nitrogen, Phosphorus, and Potassium) declines crops' cultivation, growth, and yield. To increase crop production, the suitability of a specific crop to be planted can be recommended by exploiting the soil's macronutrients [99–103]. Soil pH level is the major parameter for measuring soil macronutrients (N, P, and K) and some of the micronutrients [104–108]. Smart agro farms [109] use solar power and a low-cost smart system, a perfect combination of IoT, data mining, and Android application. The system monitors and extracts a farm's environmental factors such as soil moisture, humidity, and weather and temperature parameters via data mining modules, and provides optimized guidance regarding crop cultivation, irrigation, and weather forecast in the English language.

1.2. Objectives and Hypotheses

The proposed system obtains agricultural data through implanted IoT sensors, such as pH, soil moisture, humidity, and temperature. The Internet plays a mediatory role in communication and data exchange. We integrated agricultural data acquired from implanted IoT devices with the cloud platform. Data is processed in a decision-making system based on learning prediction rules in conjunction with a rule-based engine. Generally, a farmer requires guidelines, even from the crop selection phase to the harvesting stage. As presented in Figure 1, to facilitate low literate farmers at each of these steps in their native language, we performed the following research objectives:

- Investigated traditional techniques and systems with different agricultural interfaces to find a research gap.
- Design and develop an interface in an easy-to-use format and Urdu for low-literate farmers to facilitate their awareness and guidelines in their native language.
- Design and develop a mechanism for measuring soil fertility of specific land to recommend suitable crops according to soil fertility using fuzzy logic.
- Provide crops cultivation schedule, crop harvest schedule, automated irrigation process, and watering guidelines to farmers.
- Facilitate farmers concerning guidelines for weeds and their eradication, pest attacks, and awareness of best pesticides, crop diseases, and suitable fertilizers.

Figure 1. Agriculture Cycle.

2. Materials and Methods

Previous studies indicate that most farmers are unfamiliar with the latest practices of agriculture as they are not facilitated with new technologies to access agricultural information and thus rely on traditional methods to grow their crops. Related studies indicate that there is no such smart system providing an interactive interface to a low literate or illiterate farmer and guidelines from the crop selection phase to the harvesting stage. Significant barriers to accessing modern information systems are the low literacy of farmers, the non-availability of local-language information systems, and systems with fewer features. Our research identified that it is essential to equip rural and semi-literate or illiterate farmers with updated information through ICT, IoT, Edge Computing, Cloud Computing, and Machine learning techniques, and provide them guidance in almost every phase of the crop cycle. It is necessary to develop Urdu-language-based information smart systems to enhance farmers' comprehension, crop production, and sustainable agriculture.

Proposed System Design and Architecture

The overall design and architecture of the solution proposed to cover smart agriculture are depicted in Figure 2. It comprises three layers: crop (edge) layer, fog computing layer, data analytics, and smart management at the cloud layer. The edge and cloud layers are designed to be deployed respectively at local crop premises and remote data servers. The intermediate fog computing layer comprises a set of virtualized control modules in the form of Network Function Virtualization (NFV) nodes that can be initiated along the network path from the farm facilities to the cloud layer. NFV is a way to virtualize network services, for example, firewalls, routers, and load balancers that have traditionally been run on proprietary hardware. The intermediate fog layer increases the versatility of deployed solutions and connectivity performances with the edge layer. At the crop premises, suitable sensors like humidity, temperature, soil moisture sensor, light intensity, pH sensor, and actuators like water pumps, valves, and activation of devices for smart farming automation are deployed and connected with wireless nodes as shown in Figure 2. Sensors' data is captured at the edge layer through wireless nodes and transmitted to the fog layer. This layered architecture lets atomic operations requiring high reliability and low latency between sensors and actuators to be processed at the fog layer, such as executing irrigation mandates for a specific time interval. The fog layer subsystem comprises the farm's operative control like irrigation, farm monitoring, energy management, etc. The fog layer is responsible for data fusion and aggregation to offload analytics functions that are usually performed. The fog layer control modules are virtualized through NFV techniques that communicate with edge nodes via IoT protocols like constrained application protocol (CoAP) and MQ telemetry transport (MQTT). As depicted in Figure 2 cloud layer serves as an interface between users and the core platform. At this layer, crops current status and configuration parameters are maintained. Any change in configuration parameters triggers the control actions to be managed at fog subsystems.

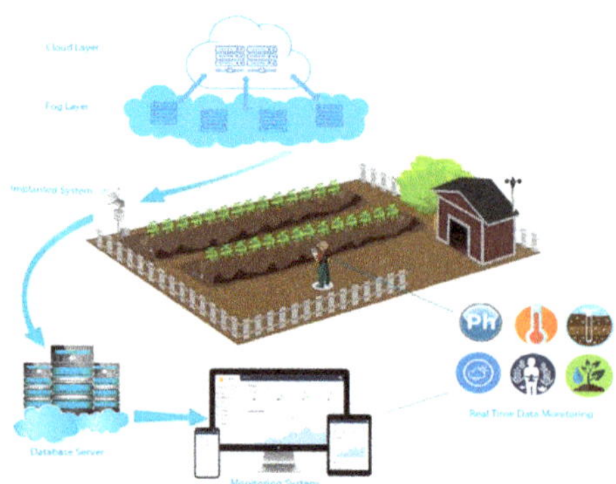

Figure 2. System Design and architecture.

3. Results and Discussion

3.1. System Components and Expected Outcomes

The proposed system gets real-time values from sensors implanted in farmland. The controller grabs data from sensors and transmits it to a cloud server, where data analysis is performed to match predefined conditions and the current state of crops. After mapping the requirements and data, the analysis system performs suitable actions via actuators. Our system provides access to an Android application for farmer facilitation with the following main features.

3.1.1. Soil Nutrient Analysis

With increased emphasis on precision agriculture, economics, and the environment, soil analysis is a tool to determine areas where adequate and excessive fertilization has occurred. Soil analysis is also used to monitor past fertility practices to changes in a field's nutrient status. Nutrient availability can be impacted by soil chemical and physical properties.

In determining soil nutrient contents, soil pH analysis is one parameter. Soil pH refers to the acidity and alkalinity of soil measured on a logarithmic scale; thus decrease in 1 unit of pH value causes an increase in acidity by a factor of 10. Small changes in pH values have significant consequences. Table 1 represents the range values defined for soil pH.

Table 1. Soil pH range values.

pH Level	Range Values
<3.5	Ultra-Acidic
3.6–3.9	Extremely Acidic
4–5.5	Strong acidic
5.6–6	Medium acidic
6.1–6.5	Slightly acidic
6.6–7	Very Slightly acidic
7.1–7.5	Very Slightly alkaline
7.6–8	Slightly alkaline
8.1–8.5	Medium alkaline
8.6–10	Strongly alkaline

Measuring the acidity and alkalinity of soil is essential for analyzing the number of macro-nutrients present in the soil, particularly nitrogen (N), potassium (K), and phosphorus (P). Crops need these macro-nutrients in their growth, thrive, and combat diseases. Removal of bases from the soil due to harvested crops, leaching, and acidic residual left in soil due to fertilizers causes an increase in acidity of the soil. Soil acidity affects crops and plants in many ways, such as whether the surface pH is very high or too low, when the efficacy of herbicides and chemical reactions may be affected. Soil analysis is the best way to check pH levels, and maintaining at least a pH of 6.0 is a realistic goal. When soil pH is very low (acidity is high) following conditions occur:

- Soluble metals, especially Manganese and Aluminum, may be toxic.
- The population of organisms and their activities accountable for transforming N, P, and S to plant-available forms may be reduced.
- Deficiency of Calcium. The soil's cation exchange capacity (CEC) is low.
- Symbiotic N fixation in legume crops is significantly impaired. The symbiotic association entails a narrower range of soil reactions than does the growth of plants not relying on 'N' fixation.
- Acidic soil with less organic matter is poorly aggregated and has poor tilt.

- The availability of mineral elements in soil may be affected. Association between soil pH and nutrient availability to plants can be depicted in Figure 3. The wider the blue bar, the greater the nutrient availability. For example, for a pH range of 5.5–7.5, the availability of P is highest and drops below 5.5. If the soil pH is 6, an amount of P applied to it will be more available than if the same amount is used in soil with a pH less than 5.5. Soil with high pH (>7.4) reduces several nutrients such as Fe, Mn, Zn, and P, which is not economical for growing agronomic crops.

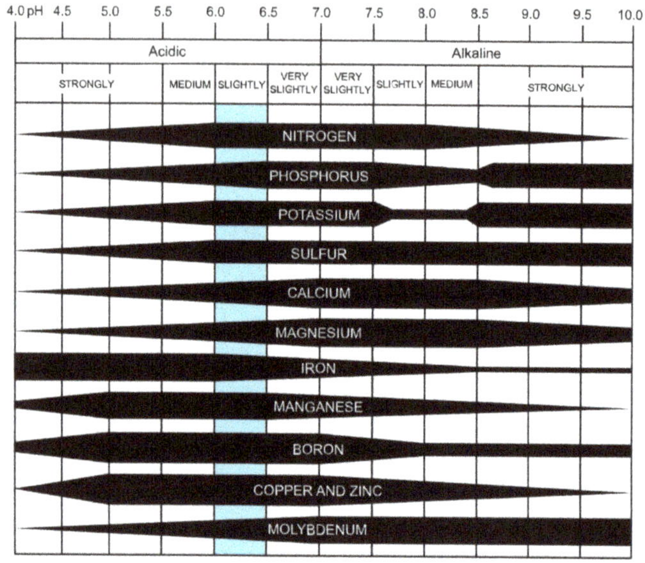

Figure 3. Chart representing availability of soil nutrients in terms of soil pH [110].

In relatively large amounts, soil provides nitrogen, potassium, phosphorus, calcium, magnesium, and sulfur. These are known as macronutrients. Soil supplies iron, boron, manganese, copper, molybdenum, and zinc in relatively small amounts, often called micronutrients. Plant nutrition is difficult to understand entirely because of the variation between different species of plants or individuals of a given clone.

Macronutrients are essential for plant growth and an excellent overall plant state. The primary macronutrients are nitrogen (N), phosphorus (P), and potassium (K). Nitrogen is a principal constituent of several essential plant substances necessary for plant development, energy metabolism, and protein synthesis. Phosphorus is involved in vital plant processes. Unlike other macronutrients, potassium is not included in the composition of essential metabolism components. Still, it substantially occurs in all plant parts for enzyme activities. Soil pH sensor and soil moisture sensor measure the soil characteristics frequently so that a farmer can monitor the status of crops in a healthy range in real-time and remotely. We can predict a specific value for nitrogen (N), phosphorus (P), and potassium (K), as Table 2. represents some ideas about these relations.

Table 2. Soil pH and corresponding estimation of N, P, and K.

pH Range	Nitrogen (N)	Phosphorus (P)	Potassium (K)
0–3.9	0%	0%	0%
4–4.5	2%	5%	2%
4.5–5	50%	20%	35%
5–5.5	100%	35%	50%
5.5–6	100%	45%	70%
6–6.5	100%	55%	100%
6.5–7.0	100%	100%	100%
7	100%	100%	100%
7–7.5	100%	100%	100%
7.5–8	100%	70%	2%
8–8.5	75%	20%	2%
8.5–9	65%	100%	100%
9–9.5	50%	100%	100%
9.5–10	2%	100%	100%

3.1.2. Crops Recommendation

The recommendation system proceeds based on a decree made by a fuzzy logic-based decision support system. Fuzzy logic is the key concept for decision-making systems and characterizes each object of a set by a degree of member functions from the interval [0,1]. The membership function defines the degree of similarity of an object to the fuzzy subset. Fuzzification is the method of allocating a system's numerical input to fuzzy sets with some degree of membership. The fuzzy system decides by considering predefined conditions and real-time data captured by sensors implanted on a specific farm. A fuzzy decision system is integrated with the controller to recommend suitable crops that can be cultivated on farmland based on available soil nutrients in the soil. Finally, real-time data is processed on the server, and a list of suitable crops is directed to the farmer's mobile app, as shown in Figure 4, where the farmer can select any crop to cultivate.

A fuzzy set S with parameters (U, i) where U is the universe of discourse and 'i' denotes the interval of U, i.e., i:U —> [0,1]. 'e' elements can signify a fuzzy set S ordered pairs. This universe of discourse is characterized by a membership function mS(e) that depicts the probability of belonging of 'e' to 'S' as shown in Equation (1):

$$S = \{(e, mS\ (e), e \in U)\} \tag{1}$$

The proposed fuzzy logic system design has four main components: fuzzifier, rule base, inference engine, and unfuzzified, represented in Figure 5. The fuzzifier converts crisp inputs to fuzzy sets. Rules are depicted as a group of if-then statements provided by an expert or acquired from data. The inference engine combines the rules and membership function to produce a fuzzy output.

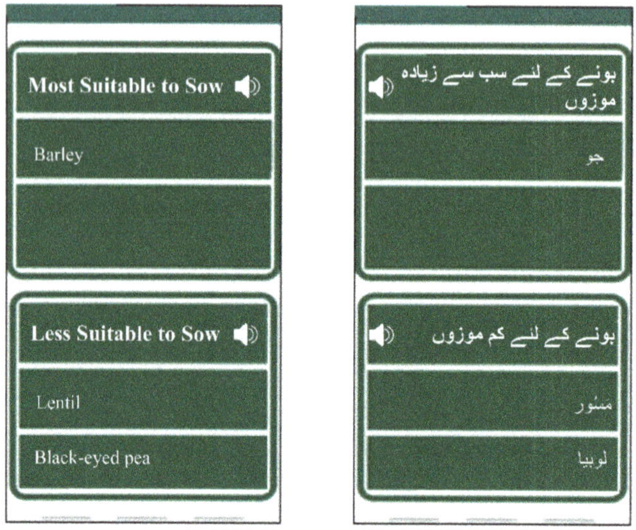

Figure 4. List of Recommended Crops.

The fuzzy logic system starts by fuzzing input variables. Later, the inference engine takes the decision based on if-then rules, membership functions, and fuzzy logic operators, i.e., "and", "or". The fuzzy inference maps input variables that are the pH level of soil, temperature, humidity, and season to fuzzy output by considering a fuzzy inference system that infers results based on fuzzy logic. Defuzzification evaluates the outcome from an input rule set provided as if-then statements. These rules are then stored in a knowledge base of the proposed system. Following is a brief description of the proposed algorithm.

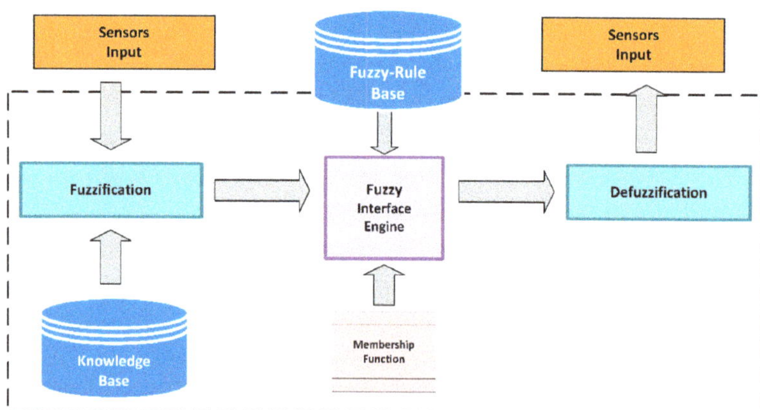

Figure 5. Fuzzy logic System for crop recommendation.

Algorithm 1: A Fuzzy Logic System

```
Input: RealTimePh, phMin, phMax, currentDate
Output: cropDetails [ ]
fetchSensorPh()
return RealTimePh
For row in TimeframOfCrop
If (CurrentDate > CultivationStartTime) && (CurrentDate < CultivationEndTime)
cropDetails [ ] = fetchCropDetail(CultivationStartTime, CultivationEndTime)
end
For row in ph Table
if (RealTimePh > phMin) && (RealTimePh < phMax)
cropDetails [ ] = showCropDetail (phMin, phMax)
end
Else
Print error
"No crop can be cultivated in these environmental conditions"
end
```

3.1.3. Land Preparation and Cultivation

A well-prepared land plays a vital role in providing the important nutrients to crops in weeds control and is suitable for sowing the seeds. A structured soil is required for ventilation and root penetration. The proposed system gets real-time data from the sensors implanted in the farms and recommends a list of crops most suitable for cultivating specific fields. From the suggested list, the farmers can choose any crop to sow. After the crop selection phase, systems provide guidelines for land preparation along with a list of appropriate fertilizers to prepare the soil for a specific crop. It also provides a cultivation schedule (suitable season) and cultivation method for each particular crop. All guidance is provided in text and voice to make the interface rural farmer-friendly, as shown in Figure 6.

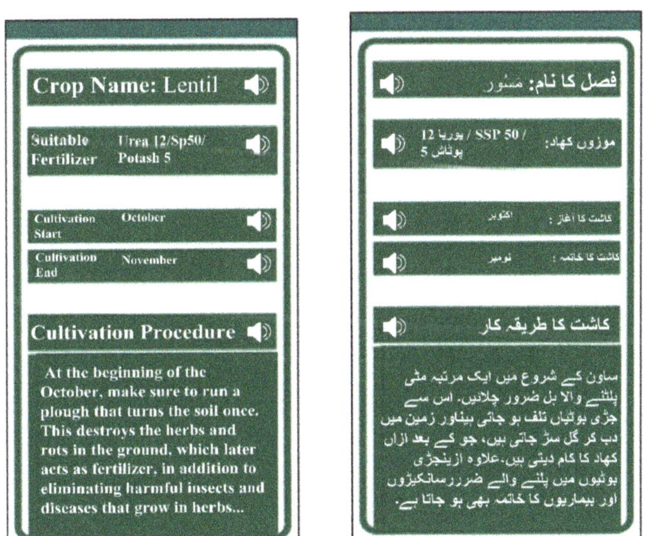

Figure 6. Land Preparation and Cultivation.

3.1.4. Irrigation

The system transmits the input from deployed IoT devices in a specific farm to an underlying irrigation calculation algorithm (ICA) illustrated in Figure 7, which recommends the irrigation scheduling for a particular farm. An android application interface is presented to the farmer to monitor the farm parameters and to get feedback on the irrigation requirement. The whole process is controlled by an irrigation control module in the fog computing layer.

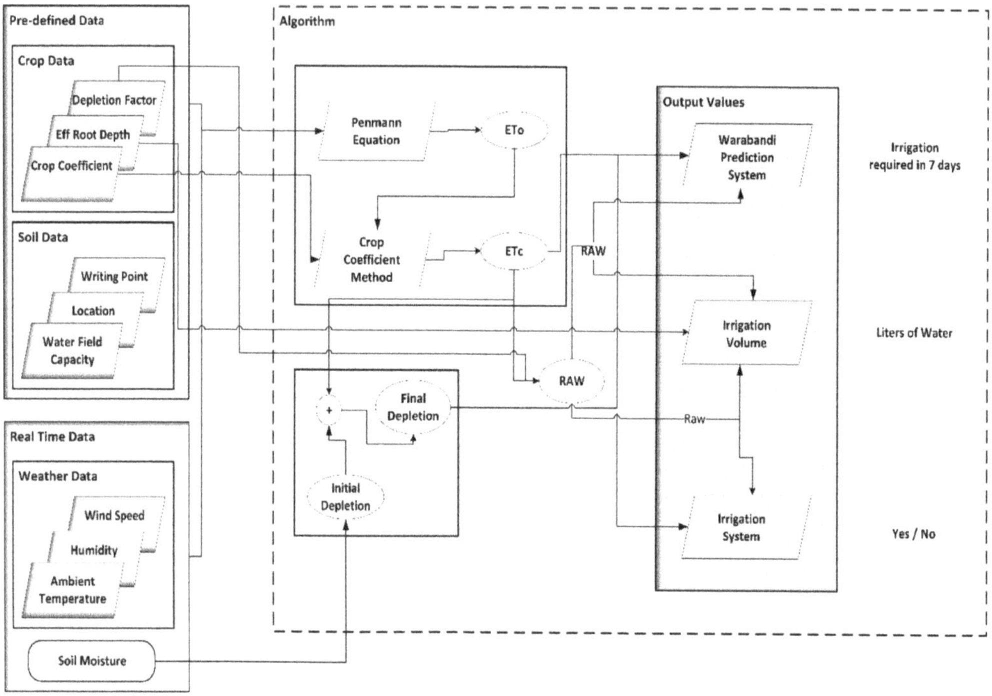

Figure 7. Irrigation Calculation Algorithm.

The irrigation calculation algorithm (ICA) determines whether irrigation is required or not and calculates the volume of irrigation needed. ICA operates on two types of data: real-time data collected by sensors and predetermined static data such as crop and soil data. Ambient temperature, humidity, and soil moisture measured by sensors are dynamic as they change hourly. Real-time data also incorporate average wind data (m/s) to calculate ET_0 from an online source [111]. Crop data comprises crop coefficient, depletion factor, and adequate root depth. Soil data contains soil category, water capacity, wilting point, and location. Location data further comprises the latitude and longitude of specific farms.

Food and Agriculture Organization (FAO) [112] recommends an essential condition that ICA evaluates daily to calculate irrigation decisions for a particular farm and crop. If $D_{r,i} \geq RAW$, there is a need for irrigation, here $D_{r,i}$ is the root zone depletion or final depletion at the end of an ith day, and RAW is readily available water or amount of water in the root zone measured in 'mm'. To ensure proper crop growth and avoid water stress, RAW must be maintained above final depletion ($D_{r,i}$). If the above condition is good, the RAW value and total farm area are used to compute the necessary irrigation volume. RAW is calculated using ET_c (depletion value) and predefined crop data. Every day depletion value increases due to crop evapotranspiration that cause an irrigation need if it increases than RAW. Depletion before evapotranspiration, called initial shortage, and lack after evapotranspiration, represented as the final deficit, are calculated using average soil moisture, water capacity, and adequate root depth daily. For optimistic irrigation, measuring the water amount a crop loses and requires for a specific duration is essential. Every crop type and soil has different water requirements; however, water loss occurs due to evaporation from the soil surface and plant transpiration. Evapotranspiration is a combination of evaporation and transpiration. Evapotranspiration 'ET_0' be determined by real-time and predefined variables such as humidity, wind speed, latitude, and altitude. ET_0 and ET_c can be computed using the Penman-Monteith model and crop coefficient, respectively.

Penman-Monteith Method:

The Penman-Monteith Equation (2) is an effective way to compute reference evapotranspiration (ET_0)

$$ET_0 = \frac{0.408\Delta(Rn - G) + Y\frac{900}{T+273}u_2(es - ea)}{\Delta + Y(1 + 0.34u_2)} \tag{2}$$

where Rn is net radiation at the surface and computed from publicly available libraries that apply an estimation formula named metabolic [113] and FAO [114], the values of maximum temperature, minimum temperature, longitude, and latitude are used to calculate Rn. 'G' is the soil heat flux, the amount of thermal energy that transfers through the soil surface per unit of time. As the ICA measures ET_0 every 24h, the value of soil heat flux is so tiny that it can be neglected; thus, $G \approx 0$. u_2 is the wind speed (m/s) measured by an anemometer placed at the height of 2 m above ground level. u_2 can be computed by Equation (3).

$$u_2 = u_z \frac{4.87}{\ln(67.8z - 5.42)} \tag{3}$$

where 'z' is the elevation (m) above sea level. Saturation vapor pressure (es) required in equation (1) is computed from Equation (4).

$$es = \frac{e0(Tmax) + e0(Tmin)}{2} \tag{4}$$

where 'T' is the temperature (°C) and $e0$ (T) is the saturation vapor pressure at air temperature T (kPa), represented in Equation (5).

$$e0\ (T) = 0.6108 \exp\left[\frac{17.27T}{T + 273.3}\right] \tag{5}$$

ea is the actual vapor pressure in Equation (1) is computed by Equation (6)

$$ea = \frac{e0(Tmax)\frac{RHmax}{100} + e0\left(Tmin\frac{RHmin}{100}\right)}{2} \tag{6}$$

where 'T' is the temperature (°C). 'Δ' in Equation (1) is the vapor pressure curve computed by Equation (7).

$$\Delta = \frac{4098\left[0.618 \exp\left(\frac{17.27T}{T+237.3}\right)\right]}{(T + 237.3)2} \tag{7}$$

where 'T' is the temperature (°C). 'Υ' in Equation (1) is the psychometric constant represented in Equation (8)

$$\Upsilon = 0.665 \times 10^{-3}P \tag{8}$$

where 'P' is the atmospheric pressure (mb) computed by Equation (9).

$$P = 101.3\left(\frac{293 - 0.0065z}{293}\right)^{5.26} \tag{9}$$

where 'z' is the sea level (m) altitude.

Crop Coefficient:

The evapotranspiration (ET_0) calculated by the Penman-Monteith Equation (1) is used to compute reference evapotranspiration (ET_C). As every crop has different evapotranspiration, thus Penman-Monteith equation assigns 'ET_0' to every crop type. The 'ET_c' crop coefficient approach can be used as equation (10).

$$ET_c = K_c\,ET_0 \tag{10}$$

where 'K_c' is the crop coefficient which varies from crop to crop and their growth stages.

ICA Outputs:

The irrigation calculation algorithm (ICA) provides flexibility for the farmer with multiple options regarding irrigation parameters and user application interface in their native language. Some farmers need irrigation output in terms of volume, such as gallons or liters in acre per inch, whereas some need output in terms of time. ICA facilitates farmers with various output parameters as per their requirements. For example, if a crop in some specific farm needs 1000L of water, then the system transforms 1000L, whether the output in time, volume, and acre per inch. The system adjusts the output, calculates how much time or acre per inch equals 10L of water, and presents the correct output amount to the farmer. Therefore, our proposed solution can work on any farm in Pakistan with varying output parameter requirements.

3.1.5. Crops Disease Prevention and Cure

For ease of the user, the proposed system provides guidelines about diseases and prevention and cure methods for cultivated crops. This feature enables the farmer to take precautionary steps to avoid any illness before any disease occurs. Moreover, in case of any disease symptom found, the farmer can cure that disease with the help of disease cure methods provided by the proposed system, as shown on the app screen in Figure 8.

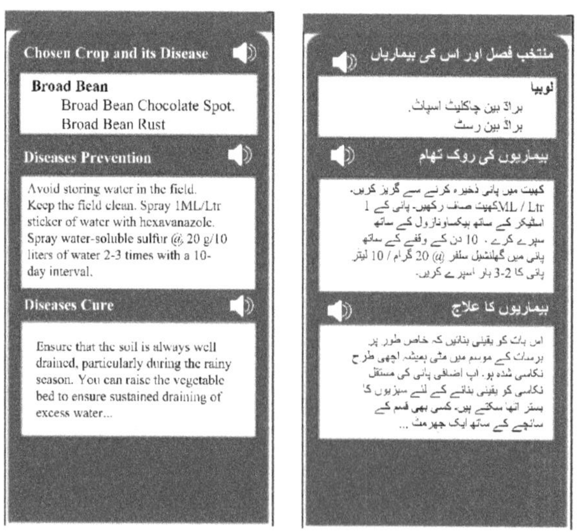

Figure 8. Crops' Disease Prevention and Cure.

3.1.6. Pest and Weed Control

Pests are harmful organisms that threaten crops' existence, spread diseases in crops, and cause destruction. On the other hand, weeds are plants that grow where and when they are not needed and compete with crops for nutrients, space, light, and water. Weeds and pests increase production costs, decrease the overall yield, and affect crop quality, so getting rid of them is important to maintain quality and yield. They become a big challenge if not controlled correctly at the right time because they cause severe damage to the crop. Our system aims to protect crops from economic damage by insects, plant pathogens, weeds, pests, and other harmful organisms while reducing reliance on hazardous pesticides. The system provides farmers with authoritative and up-to-date information about each crop's weeds and pests. It provides guidelines for controlling pest attacks and weed eradication methods, as shown in Figures 9 and 10, respectively.

Figure 9. Pests attack control guidelines.

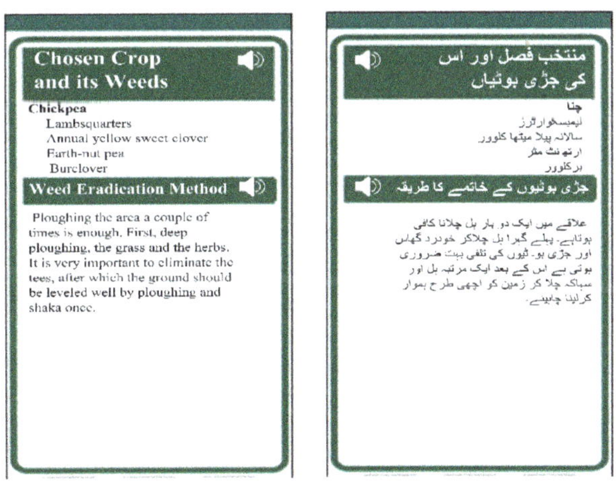

Figure 10. Weeds eradication methods.

3.1.7. Fertilizing

Fertilizers have become a vital part of farming nowadays. Whether there is a need for weed eradication or to increase production, both farmers must use fertilizer. So, it is essential to choose a suitable fertilizer to fulfill the requirements. The concentration of macro and micronutrients varies season by season, so we cannot show the same crop every season. In the same way, we cannot use the same fertilizer every time. The selection of fertilizer depends upon the crops' requirements that the farmer may fulfill or the purpose they have to achieve. If the goal is to eradicate the weeds, the farmer should use some specific fertilizers for a particular weed. Suppose the requirement is to enhance crop growth and production. In that case, the fertilizer selection depends upon the nature of the crop as the native farmers are low-literate and less aware of choosing the right fertilizer. Thus, the proposed system "Kisan Pakistan" provides accurate guidance in terms of relevant fertilizers along with weed eradication support. The system suggests suitable fertilizers for different types of weeds and the crops' growth, along with usage guidance in the native and English languages, as shown in Figure 6. This makes it much easy for native and low-literate farmers to solve their issues without acquiring help from any external entity.

3.1.8. Harvesting and Storing

Harvesting and storing are critical phases in the agriculture cycle because if these are done correctly, they provide high-quality products resulting in high income. So right way of harvesting maximizes the yield and reduces crop fatalities. The proposed system makes it convenient for the farmer by providing the best harvesting schedule for each recommended crop and harvesting methods, as shown in Figure 11.

Figure 11. Harvesting and storing guidelines.

3.2. Discussion

This research was conducted on a small-scale farm of 2 acres in Sialkot, Pakistan. Of the two, one acre was controlled by the farmer (farm A), where they applied traditional farming techniques. The remaining one acre, farm B, was controlled by our proposed smart system integrated with sensors and IoT techniques. The system recommends different suitable crops to be cultivated according to the soil analysis, i.e., 6 pH level for farm B. Farms A and B were cultivated with the same crop. Regarding the irrigation module, we compared the water usage on both farm A and farm B. Farm A was irrigated by farmers who used conventional estimations for irrigation time and volume. Farm B was irrigated using decisions made by the Irrigation Calculation Method (ICA) as a function of real-time data supplied by IoT devices deployed on the farm. Table 3. highlights the total irrigation volume consumed in farms A and B. It can be depicted that farm A, using conventional farming methods, consumed 48,569 L of irrigation water, and farm B, using the proposed solution, utilized 22,779 L of irrigation water, which resulted in 25,790 L of water saved, approximately 53%. The data for the detailed irrigation schedule for both farms are also plotted in Figure 12 to illustrate the water usage efficiency in the proposed solution.

Table 3. Irrigation Statistics.

Water Consumption (Farm A)	Water Consumption (Farm B)	Water Saving (L)	Water Saving (%)
48,569	22,779	25,790	53

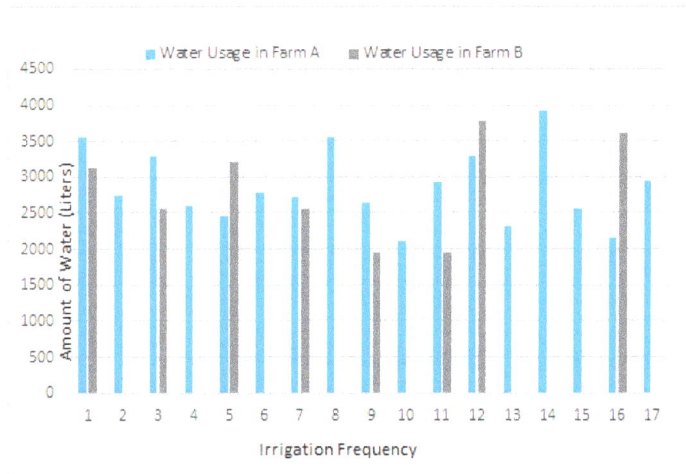

Figure 12. Irrigation frequency and water usage efficiency.

We implanted the proposed smart system on a small-scale farm. Results show that if we add more sensors and IoT devices, the proposed model has the flexibility to be implemented on medium to large farms. The system incorporates Edge, fog, and cloud computing with IoT devices which offers low latency, high bandwidth, less energy consumption, and real time analytics that make it more efficient. Currently, the system incorporates data of major crops in Pakistan, but by involving more crop data from other global regions, the system could be implemented on farms with more crops.

Our research covered a wide range of previously proposed models, papers, and studies. All these researches and studies were thoroughly read and understood, their domain of interest, their architecture, the pros and cons, and the features added in their proposed studies. After critically evaluating many studies on smart agriculture, some crucial information about related studies is provided in Table 4. Readers can obtain an overview and comparison of the previous work done by researchers, practitioners, authors, and technologists related to our research contributions.

A few limitations are incorporated in this study. We could not conduct the yield analysis of crops cultivated on farm B. Concerning soil analysis, we could also involve more soil sensors, such as NPK sensors, for better fertility measurements. System recommendations address only major crops to be grown in Pakistan. In the future, we will incorporate more crop data for different global regions. This study was carried out when most regions were on lockdown, with restrictions on movements within Pakistan. We are intended to conduct the qualitative usability test of the android application 'Kisan Pakistan' among farmers in the future. We will perform experiments proposed system on large-scale farm lands to measure and improve its performance in the future.

Table 4. Comparison summary of related studies vs. proposed solution.

Study	Smart Solution	Interface for Semi-literate	Medium (Language)	Proposed Features and Guidelines								
				Soil Analysis	Crop Cultivation	Land Preparation	Irrigation	Crop Disease	Pest and Weed Control	Fertilizer	Harvest	Weather Forecast
[45]		Yes	Urdu	No	No	No	No	No	No	No	No	Yes
[53]	Yes	No	English	No	No	No	Yes	No	No	No	No	No
[54]	Yes	No	English	No	Yes	No	Yes	No	No	No	No	No
[111]	Yes	No	Urdu	No	No	No	No	Yes	No	No	No	No
[50]	No	Yes	N/A	No	No	No	No	No	No	No	No	Yes
[71]	No	Yes	N/A	No	No	No	No	No	Yes	Yes	No	Yes
[77]	Yes	No	Hindi, English	No	No	No	No	Yes	Yes	Yes	No	No
[109]	Yes	No	English	Yes	Yes	No	Yes	No	No	No	No	Yes
[92]	Yes	No	Malay	No	Yes	No	No	Yes	Yes	No	No	Yes
[93]	Yes	No	English	Yes	No	No	Yes	Yes	No	Yes	No	No
Proposed Solution	Yes	Yes	Urdu	Yes	Yes	Yes	Yes	Yes	Yes	Yes	Yes	Yes

4. Conclusion

Agriculture is the backbone of Pakistan. It is necessary to ensure its sustainable growth over the years. We studied traditional trends followed by farmers and investigated why productivity lags. The key barriers are information inadequacies, lack of information systems for illiterates or less-literates, and lack of a system that provides guidance at every stage of the crop cycle. This study was carried out to provide a smart advisory system for illiterate and semi-literate farmers of Pakistan that could provide them guidance from crop selection to the harvest stage phase. In this research work, we built a cost-effective smart system equipped with multiple sensors and devices related to the internet of things (IoT) technologies. We also developed an android application named 'Kistan Pakistan' that allows illiterate and low-literate farmers to manage their farms remotely. The interface of the android application is interactive due to its visual, audio, voice, and iconic components. The proposed solution is applicable globally as all information and guidelines are disseminated in both the 'Urdu' and 'English' languages. Edge-cloud computing delivers more accurate guidelines in less time and in almost every phase of the agricultural cycle, increasing productivity and making the agricultural ecosystem more robust. We experimented on a small-scale farm, but the results reflect that it will be efficient for medium to large-scale fields.

Author Contributions: Conceptualization, S.M.C.; methodology, S.M.C.; software, S.M.C.; validation, M.H.N. and M.H.; formal analysis, S.M.C.; investigation, S.M.C.; writing—original draft preparation, S.M.C., M.A., I.M.P., N.J.G., M.H.N. and M.H.; writing—review and editing, S.M.C., M.A., I.M.P., N.J.G., M.H.N. and M.H.; funding acquisition, I.M.P. All authors have read and agreed to the published version of the manuscript.

Funding: This work is funded by FCT/MEC through national funds and, when applicable, co-funded by the FEDER-PT2020 partnership agreement under the project UIDB/50008/2020.

Institutional Review Board Statement: Not applicable.

Informed Consent Statement: Not applicable.

Data Availability Statement: Not applicable.

Acknowledgments: This article is based upon work from COST Action IC1303-AAPELE—Architectures, Algorithms, and Protocols for Enhanced Living Environments and COST Action CA16226–SHELD-ON—Indoor living space improvement: Smart Habitat for the Elderly, supported by COST (European Cooperation in Science and Technology). COST is a funding agency for research and innovation networks. Our actions help connect research initiatives across Europe and enable scientists to develop their ideas by sharing them with their peers. It boosts their research, career, and innovation. More information is available at www.cost.eu (accessed on 20 August 2022).

Conflicts of Interest: The authors declare no conflict of interest.

References

1. Zaman, S.B.; Ishaq, M.; Niazi, M.A. Contribution of Agriculture Sector in Economic Growth of Pakistan: An Empirical Analysis. *J. Appl. Econ. Bus. Stud.* **2021**, *5*, 103–120. [CrossRef]
2. Jatoi, F.Z. Agriculture in Pakistan and it's impact on Economic growth. *SSRN Electron. J.* **2020**. [CrossRef]
3. 02-Agriculture.pdf. Available online: https://www.finance.gov.pk/survey/chapters_21/02-Agriculture.pdf (accessed on 18 July 2022).
4. Khan, Z.A.; Koondhar, M.A.; Khan, I.; Ali, U.; Tianjun, L. Dynamic linkage between industrialization, energy consumption, carbon emission, and agricultural products export of Pakistan: An ARDL approach. *Environ. Sci. Pollut. Res. Int.* **2021**, *28*, 43698–43710. [CrossRef] [PubMed]
5. Saqib, S.E.; Arifullah, A.; Yaseen, M. Managing farm-centric risks in agricultural production at the flood-prone locations of Khyber Pakhtunkhwa, Pakistan. *Nat. Hazards* **2021**, *107*, 853–871. [CrossRef]
6. Yaseen, M.; Shahzad, M.S.; Khan, F.U.; Luqman, M.; Saleem, U.; Nasir, S. Public Sector Advisory Services for Rice Productivity: A Case Study of Farmers' Awareness in Tehsil Shakargarh of Pakistan. *Sarhad J. Agric.* **2021**, *38*, 229–237. [CrossRef]
7. Sansa-Otim, J.; Nsabagwa, M.; Mwesigwa, A.; Faith, B.; Owoseni, M.; Osuolale, O.; Mboma, D.; Khemis, B.; Albino, P.; Ansah, S.O.; et al. An Assessment of the Effectiveness of Weather Information Dissemination among Farmers and Policy Makers. *Sustainability* **2022**, *14*, 3870. [CrossRef]

8. Salik, M.H.; Tanwir, F.; Saboor, A.; Akram, M.B.; Anjum, F.; Mehdi, M.; Ashraf, I.; Naazer, M.A.; Suleman, M.; Latif, M.; et al. Role of Radio Communication and Adoption of Modern Agricultural Technology: A Study of Farmers in District Jhang, Punjab-Pakistan. *Pak. J. Agric. Sci.* **2021**, *58*, 731–738. [CrossRef]
9. Alant, B.P.; Bakare, O.O. A case study of the relationship between smallholder farmers' ICT literacy levels and demographic data w.r.t. their use and adoption of ICT for weather forecasting. *Heliyon* **2021**, *7*, e06403. [CrossRef]
10. Javaid, N. Integration of context awareness in Internet of Agricultural Things. *ICT Express* **2021**. [CrossRef]
11. Ayim, C.; Kassahun, A.; Addison, C.; Tekinerdogan, B. Adoption of ICT innovations in the agriculture sector in Africa: A review of the literature. *Agric. Food Secur.* **2022**, *11*, 22. [CrossRef]
12. Jha, K.; Doshi, A.; Patel, P.; Shah, M. A comprehensive review on automation in agriculture using artificial intelligence. *Artif. Intell. Agric.* **2019**, *2*, 1–12. [CrossRef]
13. Bannerjee, G.; Sarkar, U.; Das, S.; Ghosh, I. Artificial Intelligence in Agriculture: A Literature Survey. *Int. J. Sci. Res. Comput. Sci. Appl. Manag. Stud.* **2018**, *7*, 1–6.
14. Agesa, B.; Onyango, C.; Kathumo, V.; Onwonga, R.; Karuku, G. Climate Change Effects on Crop Production in Kenya: Farmer Perceptions and Adaptation Strategies. *Afr. J. Food Agric. Nutr. Dev.* **2019**, *19*, 14010–14042. [CrossRef]
15. Ngoune Liliane, T.; Shelton Charles, M. Factors Affecting Yield of Crops. *Agron. Clim. Change Food Secur.* **2020**. [CrossRef]
16. Ahmad, D.; Afzal, M. Impact of climate change on pastoralists' resilience and sustainable mitigation in Punjab, Pakistan. *Environ. Dev. Sustain.* **2021**, *23*, 11406–11426. [CrossRef]
17. Fahad, S.; Wang, J. Climate change, vulnerability, and its impacts in rural Pakistan: A review. *Environ. Sci. Pollut. Res.* **2019**, *27*, 1334–1338. [CrossRef]
18. Subeesh, A.; Mehta, C.R. Automation and digitization of agriculture using artificial intelligence and internet of things. *Artif. Intell. Agric.* **2021**, *5*, 278–291. [CrossRef]
19. Kumar, S.; Singh, R.; Venkatesh, A.S.; Udayabhanu, G.; Singh, T.B.N. Assessment of Potentially Toxic Elements Contamination on the Fertile Agricultural Soils Within Fluoride-Affected Areas of Jamui District, Indo-Gangetic Alluvial Plains, India. *Water Air Soil Pollut.* **2022**, *233*, 39. [CrossRef]
20. Zambito Marsala, R.; Capri, E.; Russo, E.; Bisagni, M.; Colla, R.; Lucini, L.; Gallo, A.; Suciu, N.A. First evaluation of pesticides occurrence in groundwater of Tidone Valley, an area with intensive viticulture. *Sci. Total Environ.* **2020**, *736*, 139730. [CrossRef]
21. Gupta, A.; Singh, U.B.; Sahu, P.K.; Paul, S.; Kumar, A.; Malviya, D.; Singh, S.; Kuppusamy, P.; Singh, P.; Paul, D.; et al. Linking Soil Microbial Diversity to Modern Agriculture Practices: A Review. *Int. J. Environ. Res. Public Health* **2022**, *19*, 3141. [CrossRef]
22. Kopittke, P.M.; Menzies, N.W.; Wang, P.; McKenna, B.A.; Lombi, E. Soil and the intensification of agriculture for global food security. *Environ. Int.* **2019**, *132*, 105078. [CrossRef] [PubMed]
23. Khan, N.A.; Gao, Q.; Iqbal, M.A.; Abid, M. Modeling food growers' perceptions and behavior towards environmental changes and its induced risks: Evidence from Pakistan. *Environ. Sci. Pollut. Res.* **2020**, *27*, 20292–20308. [CrossRef] [PubMed]
24. Ullah, W.; Nafees, M.; Khurshid, M.; Nihei, T. Assessing farmers' perspectives on climate change for effective farm-level adaptation measures in Khyber Pakhtunkhwa, Pakistan. *Environ. Monit. Assess.* **2019**, *191*, 547. [CrossRef] [PubMed]
25. Zhang, M.; Wang, N.; Chen, L. Sensing Technologies and Automation for Precision Agriculture. In *Women in Precision Agriculture*; Springer: Cham, Germany, 2021; pp. 35–54. [CrossRef]
26. Blasch, J.; van der Kroon, B.; van Beukering, P.; Munster, R.; Fabiani, S.; Nino, P.; Vanino, S. Farmer preferences for adopting precision farming technologies: A case study from Italy. *Eur. Rev. Agric. Econ.* **2020**, *49*, 33–81. [CrossRef]
27. Gonçalves, P.; Pedreiras, P.; Monteiro, A. Recent Advances in Smart Farming. *Animals* **2022**, *12*, 705. [CrossRef] [PubMed]
28. De Alwis, S.; Hou, Z.; Zhang, Y.; Na, M.H.; Ofoghi, B.; Sajjanhar, A. A survey on smart farming data, applications and techniques. *Comput. Ind.* **2022**, *138*, 103624. [CrossRef]
29. Bhat, S.A.; Huang, N.-F. Big Data and AI Revolution in Precision Agriculture: Survey and Challenges. *IEEE Access* **2021**, *9*, 110209–110222. [CrossRef]
30. Ullo, S.L.; Sinha, G.R. Advances in IoT and Smart Sensors for Remote Sensing and Agriculture Applications. *Remote Sens.* **2021**, *13*, 2585. [CrossRef]
31. Gao, F. Remote Sensing for Agriculture. In *Springer Remote Sensing/Photogrammetry*; Springer: Cham, Germany, 2021; pp. 7–24. [CrossRef]
32. Waleed, M.; Um, T.-W.; Kamal, T.; Usman, S.M. Classification of Agriculture Farm Machinery Using Machine Learning and Internet of Things. *Symmetry* **2021**, *13*, 403. [CrossRef]
33. Qazi, S.; Khawaja, B.A.; Farooq, Q.U. IoT-Equipped and AI-Enabled Next Generation Smart Agriculture: A Critical Review, Current Challenges and Future Trends. *IEEE Access* **2022**, *10*, 21219–21235. [CrossRef]
34. Tao, W.; Zhao, L.; Wang, G.; Liang, R. Review of the internet of things communication technologies in smart agriculture and challenges. *Comput. Electron. Agric.* **2021**, *189*, 106352. [CrossRef]
35. Mehmood, M.Z.; Ahmed, M.; Afzal, O.; Aslam, M.A.; Zoq-ul-Arfeen, R.; Qadir, G.; Komal, S.; Shahid, M.A.; Awan, A.A.; Awale, M.A.; et al. Internet of Things (IoT) and Sensors Technologies in Smart Agriculture: Applications, Opportunities, and Current Trends. *Build. Clim. Resil. Agric.* **2021**, 339–364. [CrossRef]
36. Hassan, S.I.; Alam, M.M.; Illahi, U.; Al Ghamdi, M.A.; Almotiri, S.H.; Su'ud, M.M. A Systematic Review on Monitoring and Advanced Control Strategies in Smart Agriculture. *IEEE Access* **2021**, *9*, 32517–32548. [CrossRef]

37. Khan, N.A.; Qijie, G.; Ali, S.; Shahbaz, B.; Shah, A.A. Farmers' use of mobile phone for accessing agricultural information in Pakistan: Ciência Rural 2019, 49. *Ciência Rural* **2019**, *49*. [CrossRef]
38. Khan, N.; Siddiqui, B.N.; Khan, N.; Khan, F.; Ullah, N.; Ihtisham, M.; Ullah, R.; Ismail, S.; Muhammad, S. Analyzing mobile phone usage in agricultural modernization and rural development. *Int. J. Agric. Ext.* **2020**, *8*, 139–147. [CrossRef]
39. Chhachhar, A.R.; Chen, C.; Jin, J. Mobile Phone Impact on Agriculture and Price Information among Farmers. *Indian J. Sci. Technol.* **2016**, *9*. [CrossRef]
40. Aldosari, F.; Al Shunaifi, M.S.; Ullah, M.A.; Muddassir, M.; Noor, M.A. Farmers' perceptions regarding the use of Information and Communication Technology (ICT) in Khyber Pakhtunkhwa, Northern Pakistan. *J. Saudi Soc. Agric. Sci.* **2019**, *18*, 211–217. [CrossRef]
41. Farooq, U. Revolutionising Pakistan Agriculture by Increasing the Use of Knowledge, Science and Technology and ICT. In *Building Knowledge-Based Economy in Pakistan: Learning from Best Practices*; Islamabad Policy Research Institute: Islamabad, Pakistan, 2016; pp. 112–143. ISBN 978-969-8721-49-7.
42. Lee, I.; Lee, K. The Internet of Things (IoT): Applications, investments, and challenges for enterprises. *Bus. Horiz.* **2015**, *58*, 431–440. [CrossRef]
43. Medhi, I.; Patnaik, S.; Brunskill, E.; Gautama, S.N.N.; Thies, W.; Toyama, K. Designing mobile interfaces for novice and low-literacy users. *ACM Trans. Comput. Hum. Interact.* **2011**, *18*, 1–28. [CrossRef]
44. Ujakpa, M.M.; Kristof, A.; Domingos, A.; Hashiyana, V.; Suresh, N.; Osakwe, J.O.; Iyawa, G. Farmers' Use of Mobile Devices in Developing Countries. In Proceedings of the 2021 IST-Africa Conference (IST-Africa), South Africa, 10–14 May 2021; pp. 1–7.
45. Idrees, F.; Qadir, J.; Mehmood, H.; Hassan, S.U.; Batool, A. *Urdu Language based Information Dissemination System for Low-Literate Farmers*; ACM: New York, NY, USA, 2019.
46. Smart Agriculture on Computers and Handheld Devices. *Int. J. Adv. Trends Comput. Sci. Eng.* **2021**, *10*, 1177–1182. [CrossRef]
47. Sharma, U.; Chetri, P.; Minocha, S.; Roy, A.; Holker, T.; Patt, A.; Joerin, J. Do phone-based short message services improve the uptake of agri-met advice by farmers? A case study in Haryana, India. *Clim. Risk Manag.* **2021**, *33*, 100321. [CrossRef]
48. Chengalur-Smith, I.; Potnis, D.; Mishra, G. Developing voice-based information sharing services to bridge the information divide in marginalized communities: A study of farmers using IBM's spoken web in rural India. *Int. J. Inf. Manag.* **2021**, *57*, 102283. [CrossRef]
49. Mubin, O.; Tubb, J.; Novoa, M.; Naseem, M.; Razaq, S. *Understanding the Needs of Pakistani Farmers and the Prospects of an ICT Intervention*; ACM: New York, NY, USA, 2015.
50. Qasim, M.; Zia, H.B.; Athar, A.; Habib, T.; Raza, A.A. Personalized weather information for low-literate farmers using multimodal dialog systems. *Int. J. Speech Technol.* **2021**, *24*, 455–471. [CrossRef]
51. Srivastava, A.; Kapania, S.; Tuli, A.; Singh, P. Actionable UI Design Guidelines for Smartphone Applications Inclusive of Low-Literate Users. *Proc. ACM Hum. Comput. Interact.* **2021**, *5*, 1–30. [CrossRef]
52. Sheikh, J.A.; Cheema, S.M.; Ali, M.; Amjad, Z.; Tariq, J.Z.; Naz, A. IoT and AI in Precision Agriculture: Designing Smart System to Support Illiterate Farmers. *Adv. Intell. Syst. Comput.* **2020**, *1213*, 490–496. [CrossRef]
53. Munir, M.S.; Bajwa, I.S.; Cheema, S.M. An intelligent and secure smart watering system using fuzzy logic and blockchain. *Comput. Electr. Eng.* **2019**, *77*, 109–119. [CrossRef]
54. Cheema, S.M.; Khalid, M.; Rehman, A.; Sarwar, N. Plant Irrigation and Recommender System–IoT Based Digital Solution for Home Garden. *Commun. Comput. Inf. Sci.* **2019**, *932*, 513–525. [CrossRef]
55. Chatterjee, S.; Dey, N.; Sen, S. Soil moisture quantity prediction using optimized neural supported model for sustainable agricultural applications. *Sustain. Comput. Inform. Syst.* **2020**, *28*, 100279. [CrossRef]
56. Ghorbani, M.A.; Deo, R.C.; Karimi, V.; Kashani, M.H.; Ghorbani, S. Design and implementation of a hybrid MLP-GSA model with multi-layer perceptron-gravitational search algorithm for monthly lake water level forecasting. *Stoch. Environ. Res. Risk Assess.* **2018**, *33*, 125–147. [CrossRef]
57. Morais, R.; Silva, N.; Mendes, J.; Adão, T.; Pádua, L.; López-Riquelme, J.A.; Pavón-Pulido, N.; Sousa, J.J.; Peres, E. mySense: A comprehensive data management environment to improve precision agriculture practices. *Comput. Electron. Agric.* **2019**, *162*, 882–894. [CrossRef]
58. Dewi, C.; Chen, R.-C. Decision Making Based on IoT Data Collection for Precision Agriculture. *Intell. Inf. Database Syst. Recent Dev.* **2019**, *830*, 31–42. [CrossRef]
59. Yang, Z.; Ding, Y.; Hao, K.; Cai, X. An adaptive immune algorithm for service-oriented agricultural Internet of Things. *Neurocomputing* **2019**, *344*, 3–12. [CrossRef]
60. Kale, A.P.; Sonavane, S.P. IoT based Smart Farming: Feature subset selection for optimized high-dimensional data using improved GA based approach for ELM. *Comput. Electron. Agric.* **2019**, *161*, 225–232. [CrossRef]
61. White, S. Evaluation of Articles Written about Agriculture and Comprehension of Agriculture Literacy. Master's Thesis, Tarleton State University, Stephenville, TX, USA, 2021.
62. Jan, S.; Maqsood, I.; Ahmad, I.; Ashraf, M.; Khan, F.; Imran, M. A Systematic Feasibility Analysis of User Interfaces for Illiterate Users. *Proc. Pak. Acad. Sci.* **2019**, *56*, 75–91.
63. Hussain, A.; Akhtar, W.; Jabbar, A. Risk management for small farmers in Pakistan: A review. *Pak. J. Agric. Sci.* **2022**, *59*, 247–259. [CrossRef]

64. Cuendet, S.; Medhi, I.; Bali, K.; Cutrell, E. VideoKheti: Making Video Content Accessible to Low-Literate and Novice Users. In Proceedings of the SIGCHI Conference on Human Factors in Computing Systems, Association for Computing Machinery, New York, NY, USA, 27 April 2013; pp. 2833–2842.
65. Chaudhry, S.; Muhammad, A.; Shah, A.A.; Batoo, F. Human-Computer User Interface Design for Semi-literate and Illiterate Users. *Lahore Garrison Univ. Res. J. Comput. Sci. Inf. Technol.* **2021**, *5*, 62–77. [CrossRef]
66. Hassan, G.; Ashraf, I.; Ul Hassan, N.; Ali, M.; Khalid, I.; Ashraf, E.; Raza, H.; Husnain, R.T.; Zia, S.-R.; Asghar, S. Information deficiency among farmers regarding vegetable production practices in peri-urban areas of the Punjab-Pakistan. *Int. J. Agric. Ext.* **2021**, *9*, 19–28. [CrossRef]
67. Ahmad, A.; Shahid, M.; Khalid, S.; Zaffar, H.; Naqvi, T.; Pervez, A.; Bilal, M.; Ali, M.A.; Abbas, G.; Nasim, W. Residues of endosulfan in cotton growing area of Vehari, Pakistan: An assessment of knowledge and awareness of pesticide use and health risks. *Environ. Sci. Pollut. Res.* **2018**, *26*, 20079–20091. [CrossRef]
68. Khuhro, S.N.; Junejo, I.A.; Hullio, M.H.; Hassan, M.F.; Maitlo, S.A.; Shaikh, M.A. Knowledge Attitude Practice Regarding Pesticide Application among Vegetable Growers of Dadu Canal Irrigated Areas of Northern Sindh Pakistan. *Pak. J. Agric. Res.* **2020**, *33*, 331. [CrossRef]
69. Khan, F.Z.A.; Manzoor, S.A.; Gul, H.T.; Ali, M.; Bashir, M.A.; Akmal, M.; Haseeb, M.; Imran, M.U.; Taqi, M.; Manzoor, S.A.; et al. Drivers of farmers' intention to adopt integrated pest management: A case study of vegetable farmers in Pakistan. *Ecosphere* **2021**, *12*, e03812. [CrossRef]
70. Bagheri, A.; Emami, N.; Damalas, C.A. Farmers' behavior in reading and using risk information displayed on pesticide labels: A test with the theory of planned behavior. *Pest Manag. Sci.* **2021**, *77*, 2903–2913. [CrossRef] [PubMed]
71. Riaz, W.; Durrani, H.; Shahid, S.; Raza, A.A. Ict intervention for agriculture development: Designing an ivr system for farmers in pakistan. In Proceedings of the Ninth International Conference on Information and Communication Technologies and Development, New York, NY, USA, 16 November 2017; pp. 1–5.
72. Nakutis, Z.; Deksnys, V.; Jaruevicius, I.; Marcinkevicius, E.; Ronkainen, A.; Soumi, P.; Nikander, J.; Blaszczyk, T.; Andersen, B. Remote Agriculture Automation Using Wireless Link and IoT Gateway Infrastructure. In Proceedings of the 26th International Workshop on Database and Expert Systems Applications (DEXA), IEEE, Valencia, Spain, 1–4 September 2015.
73. Brun-Laguna, K.; Diedrichs, A.L.; Chaar, J.E.; Dujovne, D.; Taffernaberry, J.C.; Mercado, G.; Watteyne, T. A Demo of the PEACH IoT-Based Frost Event Prediction System for Precision Agriculture. In Proceedings of the 13th Annual IEEE International Conference on Sensing, Communication, and Networking (SECON), IEEE, London, UK, 27–30 June 2016.
74. Swain, M.; Hashmi, M.F.; Singh, R.; Hashmi, A.W. A cost-effective LoRa-based customized device for agriculture field monitoring and precision farming on IoT platform. *Int. J. Commun. Syst.* **2020**, *34*, e4632. [CrossRef]
75. Gaikwad, S.V.; Vibhute, A.D.; Kale, K.V.; Mehrotra, S.C. An innovative IoT based system for precision farming. *Comput. Electron. Agric.* **2021**, *187*, 106291. [CrossRef]
76. Available online: https://aquaagro.smartcube.pk/ (accessed on 18 July 2022).
77. Singh, P.P.; Pandey, P.; Singh, D.; Singh, S.; Khan, M.S.; Semwal, M. 'Mentha Mitra'—An android app based advisory digital tool for menthol mint farmers. *Ind. Crops Prod.* **2020**, *144*, 112047. [CrossRef]
78. Ayaz, M.; Ammad-Uddin, M.; Sharif, Z.; Mansour, A.; Aggoune, E.-H.M. Internet-of-Things (IoT)-Based Smart Agriculture: Toward Making the Fields Talk. *IEEE Access* **2019**, *7*, 129551–129583. [CrossRef]
79. Dholu, M.; Ghodinde, K.A. Internet of Things (IoT) for Precision Agriculture Application. In Proceedings of the 2018 2nd International Conference on Trends in Electronics and Informatics (ICOEI), IEEE, Tirunelveli, India, 11–12 May 2018.
80. Shahzadi, R.; Ferzund, J.; Tausif, M.; Asif, M. Internet of Things based Expert System for Smart Agriculture. *Int. J. Adv. Comput. Sci. Appl.* **2016**, *7*, 341–350. [CrossRef]
81. Farooq, M.S.; Riaz, S.; Abid, A.; Abid, K.; Naeem, M.A. A Survey on the Role of IoT in Agriculture for the Implementation of Smart Farming. *IEEE Access* **2019**, *7*, 156237–156271. [CrossRef]
82. Reghunadhan, R. Big Data, Climate Smart Agriculture and India–Africa Relations: A Social Science Perspective. In *IoT and Analytics for Agriculture*; Pattnaik, P.K., Kumar, R., Pal, S., Panda, S.N., Eds.; Springer: Singapore, 2020; pp. 113–137. ISBN 978-981-13-9177-4.
83. Roopaei, M.; Rad, P.; Choo, K.-K.R. Cloud of Things in Smart Agriculture: Intelligent Irrigation Monitoring by Thermal Imaging. *IEEE Cloud Comput.* **2017**, *4*, 10–15. [CrossRef]
84. The National Artificial Intelligence Research and Development Strategic Plan: 2019 Update. Available online: https://www.nitrd.gov/pubs/National-AI-RD-Strategy-2019.pdf (accessed on 1 August 2022).
85. U.S. LEADERSHIP IN AI: A Plan for Federal Engagement in Developing Technical Standards and Related Tools. Available online: https://www.nist.gov/system/files/documents/2019/08/10/ai_standards_fedengagement_plan_9aug2019.pdf (accessed on 15 January 2020).
86. Artificial Intelligence in Agriculture. Available online: https://www.mindtree.com/sites/default/files/2018-04/Artificial%20Intelligence%20in%20Agriculture.pdf (accessed on 1 August 2022).
87. García, L.; Parra, L.; Jimenez, J.M.; Lloret, J.; Lorenz, P. IoT-Based Smart Irrigation Systems: An Overview on the Recent Trends on Sensors and IoT Systems for Irrigation in Precision Agriculture. *Sensors (Basel)* **2020**, *20*, 1042. [CrossRef]

88. Torres-Sanchez, R.; Navarro-Hellin, H.; Guillamon-Frutos, A.; San-Segundo, R.; Ruiz-Abellón, M.C.; Domingo-Miguel, R. A Decision Support System for Irrigation Management: Analysis and Implementation of Different Learning Techniques. *Water* **2020**, *12*, 548. [CrossRef]
89. Jiménez, A.-F.; Cárdenas, P.-F.; Jiménez, F. Intelligent IoT-multiagent precision irrigation approach for improving water use efficiency in irrigation systems at farm and district scales. *Comput. Electron. Agric.* **2022**, *192*, 106635. [CrossRef]
90. Zia, H.; Rehman, A.; Harris, N.R.; Fatima, S.; Khurram, M. An Experimental Comparison of IoT-Based and Traditional Irrigation Scheduling on a Flood-Irrigated Subtropical Lemon Farm. *Sensors (Basel)* **2021**, *21*, 4175. [CrossRef]
91. Toseef, M.; Khan, M.J. An intelligent mobile application for diagnosis of crop diseases in Pakistan using fuzzy inference system. *Comput. Electron. Agric.* **2018**, *153*, 1–11. [CrossRef]
92. Athirah, R.N.; Norasma, C.Y.N.; Ismail, M.R. Development of an android application for smart farming in crop management. In *IOP Conference Series: Earth and Environmental Science*; IOP Publishing: Bristol, UK, 2020; Volume 540, p. 012074.
93. Kumar, T.U.; Periasamy, A. IoT Based Smart Farming (E-FARM)'S. *Int. J. Recent Adv. Multidiscip. Top.* **2021**, *2*, 85–87.
94. Nabati, J.; Nezami, A.; Neamatollahi, E.; Akbari, M. An integrated approach land suitability for agroecological zoning based on fuzzy inference system and GIS. *Environ. Dev. Sustain.* **2022**. [CrossRef]
95. Orojloo, M.; Hashemy Shahdany, S.M.; Roozbahani, A. Developing an integrated risk management framework for agricultural water conveyance and distribution systems within fuzzy decision making approaches. *Sci. Total Environ.* **2018**, *627*, 1363–1376. [CrossRef]
96. Li, M.; Sui, R.; Meng, Y.; Yan, H. A real-time fuzzy decision support system for alfalfa irrigation. *Comput. Electron. Agric.* **2019**, *163*, 104870. [CrossRef]
97. Benyezza, H.; Bouhedda, M.; Rebouh, S. Zoning irrigation smart system based on fuzzy control technology and IoT for water and energy saving. *J. Clean. Prod.* **2021**, *302*, 127001. [CrossRef]
98. Bhat, S.K.; Kumar, S.S.; Krishnakumar, K.; Shaju, S.; Kumar, G.P. *Enhancing Effectivity of Automated irrigation SYSTEM Using Fuzzy Logic*; AIP Publishing: Sausalito, CA, USA, 2021.
99. Rajeswari, A.M.; Anushiya, A.S.; Fathima, K.S.A.; Priya, S.S.; Mathumithaa, N. Fuzzy Decision Support System for Recommendation of Crop Cultivation based on Soil Type. In Proceedings of the 4th International Conference on Trends in Electronics and Informatics (ICOEI) (48184), IEEE, Tirunelveli, India, 15–17 June2020.
100. Banerjee, G.; Sarkar, U.; Ghosh, I. A Fuzzy Logic-Based Crop Recommendation System. *Adv. Intell. Syst. Comput.* **2020**, *1225*, 57–69. [CrossRef]
101. Joss, B.N.; Hall, R.J.; Sidders, D.M.; Keddy, T.J. Fuzzy-logic modeling of land suitability for hybrid poplar across the Prairie Provinces of Canada. *Environ. Monit. Assess.* **2007**, *141*, 79–96. [CrossRef]
102. Zhang, L.; Cao, B.; Bai, C.; Li, G.; Mao, M. Predicting suitable cultivation regions of medicinal plants with Maxent modeling and fuzzy logics: A case study of Scutellaria baicalensis in China. *Environ. Earth Sci.* **2016**, *75*, 361. [CrossRef]
103. Wickramasinghe, C.P.; Lakshitha, P.L.N.; Hemapriya, H.P.H.S.; Jayakody, A.; Ranasinghe, P.G.N.S. Smart Crop and Fertilizer Prediction System. In Proceedings of the 2019 International Conference on Advancements in Computing (ICAC), IEEE, Malabe, Sri Lanka, 5–7 December 2019.
104. Martinez-Ojeda, C.O.; Amado, T.M.; Dela Cruz, J.C. In Field Proximal Soil Sensing For Real Time Crop Recommendation Using Fuzzy Logic Model. In Proceedings of the 2019 International Symposium on Multimedia and Communication Technology (ISMAC), IEEE, Quezon City, Philippines, 19–21 August 2019.
105. Fernando, P., Jr.; Lacatan, L. *Microcontroller-Based Soil Nutrients Analyzer for Plant Applicability using Adaptive Neuro-Fuzzy Inference System*; Mattingley Publishing Co., Inc.: Oakland, CA, USA, 2020.
106. Kapse, S.; Kale, S.; Bhongade, S.; Sangamnerkar, S.; Gotmare, Y. IOT Enable Soil Testing & NPK Nutrient Detection. *A J. Compos. Theory* **2020**, *13*, 310–318.
107. Fernández, F.G.; Hoeft, R.G. Managing Soil pH and Crop Nutrients. *Ill. Agron. Handb.* **2009**, *24*, 91–112.
108. Akande, S.; Chukwuweike, M.E.; Olaoluwa, S.S. Development of a Mechatronics System for Measuring Soil pH and approximating NPK Value. In Proceedings of the International Conference on Industrial Engineering and Operations Management, Monterrey, Mexico, 3–5 November 2021.
109. Devapal, D. Smart agro farm solar powered soil and weather monitoring system for farmers. *Mater. Today Proc.* **2020**, *24*, 1843–1854. [CrossRef]
110. File:Soil pH effect on nutrient availability.svg-Wikimedia Commons. Available online: https://commons.wikimedia.org/wiki/File:Soil_pH_effect_on_nutrient_availability.svg (accessed on 19 May 2021).
111. Gadap Weather—7, 10 & 14 Day Weather Forecast—Sindh, PK. Available online: https://www.worldweatheronline.com/gadap-weather/sindh/pk.aspx (accessed on 19 May 2021).
112. Allen, R.; Pereira, L.; Raes, D.; Smith, M. Chapter 08—ETc under soil water stress conditions. In *Crop Evapotranspiration-Guidelines for Computing Crop Water Requirements-FAO Irrigation and Drainage Paper 56*; 1998; Volume 56.
113. Meteorology and Evaporation Function Modules for Python—Meteorology and Evaporation Function Modules 1.0.1 Documentation. Available online: http://python.hydrology-amsterdam.nl/moduledoc/index.html (accessed on 19 July 2022).
114. Richards, M. PyETo—Pyeto 0.2 Documentation. Available online: https://pyeto.readthedocs.io/en/latest/# (accessed on 19 July 2022).

agriculture

MDPI

Article

A Machine Learning Model for Early Prediction of Crop Yield, Nested in a Web Application in the Cloud: A Case Study in an Olive Grove in Southern Spain

Juan J. Cubillas [1], María I. Ramos [2,*], Juan M. Jurado [3] and Francisco R. Feito [4]

1 Department Tecnologías de la Información y Comunicación Aplicadas a la Educación, Universidad Internacional de La Rioja, 26006 Logroño, Spain
2 Department Ingeniería Cartográfica, Geodésica y Fotogrametría, Universidad de Jaén, 23071 Jaén, Spain
3 Department Lenguajes y Sistemas Informáticos, Universidad de Granada, 18071 Granada, Spain
4 Department Informática, Universidad de Jaén, 23071 Jaén, Spain
* Correspondence: miramos@ujaen.es; Tel.: +34-953212372

Abstract: Predictive systems are a crucial tool in management and decision-making in any productive sector. In the case of agriculture, it is especially interesting to have advance information on the profitability of a farm. In this sense, depending on the time of the year when this information is available, important decisions can be made that affect the economic balance of the farm. The aim of this study is to develop an effective model for predicting crop yields in advance that is accessible and easy to use by the farmer or farm manager from a web-based application. In this case, an olive orchard in the Andalusia region of southern Spain was used. The model was estimated using spatio-temporal training data, such as yield data from eight consecutive years, and more than twenty meteorological parameters data, automatically charged from public web services, belonging to a weather station located near the sample farm. The workflow requires selecting the parameters that influence the crop prediction and discarding those that introduce noise into the model. The main contribution of this research is the early prediction of crop yield with absolute errors better than 20%, which is crucial for making decisions on tillage investments and crop marketing.

Keywords: machine learning; regression algorithms; web application; early prediction of crop yield

Citation: Cubillas, J.J.; Ramos, M.I.; Jurado, J.M.; Feito, F.R. A Machine Learning Model for Early Prediction of Crop Yield, Nested in a Web Application in the Cloud: A Case Study in an Olive Grove in Southern Spain. *Agriculture* **2022**, *12*, 1345. https://doi.org/10.3390/agriculture12091345

Academic Editor: Dimitre Dimitrov

Received: 19 July 2022
Accepted: 22 August 2022
Published: 31 August 2022

1. Introduction

Spain is the country with the largest olive grove area in the world, reaching 2.5 Mha, 60% of which is concentrated in the Andalusia Region, in southern Spain [1], where the climate provides ideal growing conditions for olive trees [2]. Olive growing is therefore an important economic factor for Spain, especially for Andalusia. Moreover, olive oil is known worldwide for its culinary contributions and its health benefits. In fact, olive oil is included in the Mediterranean Diet Pyramid, which underlines the importance of the foods making up the principal food groups [3].

Olive crop is undergoing constant growth on an international level due to a steady worldwide increase of the olive growing area and improvements in irrigation systems and technological advances [4,5]. In the current political–financial framework for agriculture, the new Common Agricultural Policy (CAP) reforms are geared towards achieving environmental objectives such as fighting climate change and supporting European farmers in achieving a sustainable and competitive agricultural sector by focusing on the digitalization of the olive sector. In this process, agriculture 4.0 could play a key role. This term refers to the technological revolution that characterizes the modern agricultural sector, based on the widespread sharing of digital technologies, smart farming, and knowledge-based production methods [6,7]. This technology has enormous potential, as these tools can be applied to a wide variety of farming systems and require less financial investment compared to machinery or heavy equipment. Technological improvement increases the quantity of outputs relative to the quantity of inputs and shifts the "technological frontier". This displacement translates into higher productivity.

Information technologies and Artificial Intelligence (AI) are key in multiple sectors [8,9], since they are based on the optimization of production systems and marketing and help in the decision-making process [10,11]. Machine Learning (ML), a branch of artificial intelligence, is a practical approach used in many fields, including agriculture, for several years [12]. Today, ML remains the most common and popular approach to AI in the field of agriculture and beyond [13]. The valuable knowledge shared between farmers and experts in these technologies allows valuable information to be inferred for making the right decisions in the agricultural sector and improving crop productivity and environmental sustainability, the main objectives of the new agricultural policies. In this sense, the reality for the farmer, and especially for the olive farmer, is that due to numerous reasons, his hard work does not always result in maximum crop yields. Crop yields depend on several factors, such as soil, climate, irrigation, rainfall, Pesticides, fertilizers, tillage, temperature, and the harvest of last year. The farmer or farm manager often needs to make decisions for which having advance information on future harvest would help to define the best management strategy [14]. The olive sector is not an exception, and currently the actors involved must make decisions every day related to agricultural practices and management, with the consequent economic investment that this implies in the farms (fertilizer, tillage, etc.). Additionally, the farmer or the farm manager has to make decisions about the best way to market the oil. This is an important matter that requires a comprehensive and in-depth knowledge of the current state of the farm and how it can evolve in the medium and long term. Consequently, the need to carry out an analysis of variables from different sources and nature is evident. From the point of view of optimizing the farmer's economic resources, the most useful thing for him is to know well in advance, before making the investment, what the yield of his crop will be depending on the practices he performs or does not perform during each season. This information is interesting for other professionals who are part of the olive sector in addition to being useful for farmers. As an example, advance information on harvest quantity can be key for insurance companies to know the risk of the insured property, and based on this, establish their rates. In the case of those actors with responsibility for marketing the oil, it is important to determine the best time to carry out the sale or purchase, or to establish their storage forecasts. In summary, a system capable of making an early prediction about the harvest, that is, in January, February, or even March, with a low ratio of error is key to designing a correct marketing strategy.

Crop yield prediction is one of the challenging problems in precision agriculture; however, as Xu et al. (2019) [15] indicates, this is not a trivial task. Nowadays, crop yield prediction models can estimate the actual values reasonably, but a better performance in yield prediction is still desirable [16]. Numerous authors have emphasized for years the importance of the quantitative forecasting of crop yields, considering it as a valuable tool in the support of the farmers in the olive sector [17]. There are numerous investigations about the use of long-term data series to carry out crop-forecasting technique for numerous species, also in the olive grove [18–20]. In this research, the close relationships between pollen emission and fruit production are widely studied. Nevertheless, the final fruit production is influenced by several weather and agronomic conditions during both the pre-flowering period and the time period between flowering and harvest, such as water deficit, temperature extremes, and phytopathological problems [21–23].

There are several studies in olive crop yield prediction, and most of them are based on the predictive value of pollen emission levels [16,21–28]. In these studies, basic parameters of pollen levels are taken into account, in addition to other factors of temperature, rainfall, and relative humidity. Regression analysis is performed in the last study. This model presents results with an error of 0.96% in July. It must be considered that generally, in traditional olive groves, the olives were harvested between the months of December and January, and it is in January that the tillage work begins. In other words, these models are very accurate, but reliable prediction is made after pollination, which occurs at a very advanced stage of the agricultural year: April, May, or June. Therefore, it is a very reliable model, but it provides a very late prediction (only 5 months after the olive harvest). More recently, other prediction works, such as the novel study presented by López-Bernal et al., 2021 [29], estimate a conceptual model for predicting fruit oil content. The results provide useful information for the farmer; it is about helping to establish optimal harvesting periods. However, the purpose of the prediction is not aimed at the objective set out above.

The basis of the predictive challenge is that it is used by the farmer or farm manager. Often predictive systems are implemented on applications that are unfriendly and too complex, and only their developers know how to operate them. As a result, these systems are not used. In order for these tools to be successful, it is essential that the expert system is accessible with a user-friendly interface.

Therefore, this system is additionally integrated into a web application with cloud deployment. This system is fed with public data from web services of government agencies and is freely accessible. Additionally, data provided by the farmer or farm manager himself complete the data set. The aim is to generate a predictive model that provides information at an early stage, at the beginning of the year, on the amount of olive crop that will be harvested that year. The innovation of this work lies in the early stage of the prediction and in the combination of variables downloaded from official web services, these being web services belonging to official meteorological institutions. Other essential data to generate the predictive models are the values of the amount of harvest collected in previous years. This information is crucial for all the agents involved in the olive sector, as it helps to make the right strategic decisions, both for the initial tillage phase and for the final oil marketing phase. For this reason, the predictive model has been integrated into a web application that is accessible and simple for users. The proposed method represents a disruption in relation to the most advanced methods, as previous contributions provide a harvest prediction based on pollen data and remote observation of the fruit at the stages closest to harvest. Our solution does not need these data that require field data collection campaigns and arduous processing, but rather information available on the internet and harvest quantity data collected from previous years.

Objectives and Hypotheses

In relation to the review of the state of the art, there are works related to the prediction of olive crop yield; however, there is very little research on prediction at an early stage of the agricultural season. These data are of great interest in the olive-growing sector, as they provide information in advance that allows for adequate economic and sustainable planning by all the actors involved in this sector. However, for the results of this research to be really useful, it is necessary that this prediction be transferred to the productive sector, i.e., the end user. Thus, this work proposes the design of a tool that allows this predictive model to be accessible. This research hypothesised the generation of an early prediction of olive crop yield using new models from ML algorithms. In addition, the idea is that this information should reach the farmer or person in charge of managing olive orchards, which is why the aim is to integrate this model into a tool that is easy to use for these users. Thus, this work has a main objective which is: (1) to generate an early prediction model of olive harvest yield, and as a complement to the previous objective: (2) to nest this ML model in a cloud-based web application to improve the convenience, accessibility, and use of the proposed software by the end users.

The scientific hypothesis of this work is focused on the aim of this paper, which is to predict, at an early stage, the amount of kilograms of crop that the farmer will harvest using several regression algorithms. The process takes as a starting point the hypothesis that there are several variables related to the harvest quantity, mainly the previous year's harvest and a number of climatic variables. In this sense, as described in the introduction, there are several studies that corroborate this relationship. For all these reasons, ML techniques are used in a supervised learning study, where all the predictor variables are labelled.

There are numerous regression algorithms, such as Linear Regression, Logistic Regression, Generalised Regression Model, One-Class Support Vector Machine (SVM), etc. In this study, a priori the nature of the variables is unknown and there are few training data available (we have harvest data collected over eight years); therefore, linear and non-linear regression algorithms are used, as follows:

- Linear: purely linear algorithms have a great strength due to their characteristics and simplicity, as they are calculated with a simple weighted sum of the variables:

$$y = \beta_0 + \beta_1 x_1 + \ldots + \beta_p x_p + \epsilon \quad (1)$$

The first algorithm selected in this study is Generalised Linear Models (GLM), which works mathematically as the weighted sum of the features with the mean value of the distribution assumed using the link function g, which can be chosen flexibly depending on the type of result.

$$g(EY(y \mid x)) = \beta_0 + \beta_1 x_1 + \ldots \beta_p x_p \quad (2)$$

In other words, this algorithm is an extension of the linear algorithms allowing to model linear or normal distributions and non-constant variances.

Another Linear algorithm selected is SVM, which has the great advantage of being able to be used with different Kernels. Kernels allow the data to be distributed in a hyperplane according to a function, which makes it easier for the algorithm to adapt to the nature of the data, allowing infinite transformations, Figure 1.

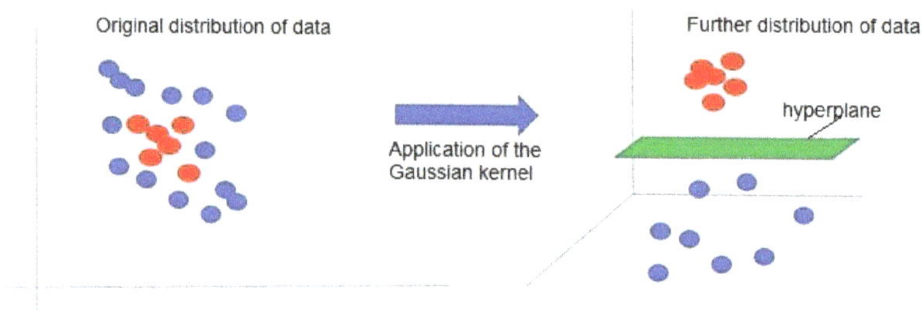

Figure 1. Graphical representation of the analysis of the distribution of variables using Gaussian kernel algorithms.

In this study, we work with SVM with Linear Kernel. When using the Linear kernel, the following transformation is performed:

$$K(x,x') = x \cdot x' \quad (3)$$

This algorithm has the advantage that it fits very well if the nature of the data is linear, and if there are many predictor variables (as in our case). It should be noted that in this algorithm, there is no upper limit on the number of predictor attributes; the only limitations are those imposed by the hardware. In this study, we have a limited set of training data, since we have a limited number of years with harvest quantity data, but we do have a large number of predictor variables.

- Non-linear: in this case, the SVM algorithm is applied with a Gaussian Kernel. This kernel applies the following transformation to the data:

$$K(x,x') = \exp(-\gamma \mid \mid x - x' \mid \mid 2) \quad (4)$$

The value of γ controls the behavior of the kernel. When it is very small, the final model is equivalent to that obtained with a linear kernel, and as its value increases the data becomes more distant, forming a Gaussian bell in the hyperplane, adapting very well when the nature of the data does not have a linear distribution.

In summary, these are the advantages of these three algorithms in this study, based on the hypothesis of a priori ignorance of the relationship between the variables with the target attribute and also taking into account that the training data we have are limited and the predictor variables are numerous. Furthermore, the complexity of these algorithms means that the relationship between the attributes used cannot be described by means of a specific equation.

2. Materials and Methods

The model implemented is based on the application of different data mining techniques to generate predictive models of the amount of olive crop that will be harvested. Data mining arises from the convergence of various disciplines, such as computer science, statistics, artificial intelligence, database technology, etc. Data mining is a logical process of finding useful information to discover useful data. Once the information and patterns are found, they can be used to make decisions for developing the business; in this case, this study seeks the relationship between the amount of olive crop harvested in a given year and multiple variables, such as the historical yield data and meteorological variables of previous years. Data mining techniques are lengthily functional to the agricultural sector [30]. It is used to examine a large dataset and establish serviceable classifications and patterns in this dataset. The overall aim of the data mining method is to extract useful information from the dataset and exchange it into an explicable structure for additional use.

The methodology followed in this work is sequenced in several phases, from the data understanding, data preparation, and generation of the models to the validation and implementation of the models. The output data of the model is the amount of olive crop, that is, the number of kg of olives that will be collected in each campaign on the farm under study. This is an unknown variable, but it can be deduced from other variables that are known and that are directly or inversely related to the target. In this sense, the study is based on a supervised analysis of data mining, where the value of an unknown variable is deduced from a few known variables.

The general flow of the methodology carried out is composed of data mining algorithms and techniques used as follows in each phase:

1. Extraction and loading of data. Meteorological data are downloaded from web servers for public use. Olive crop harvested data is provided by the owner of the farm. Both will be uploaded to the database management system.
2. First analysis of data. All data are explored and analysed using distribution techniques (histograms), with the objective of reviewing the data and cleaning those whose dispersion or variability may cause inconsistencies in the study.
3. Anomaly detection. The detection of the anomalies is implemented as a class classification algorithm, where the algorithm is able to predict, with a certain probability, whether a record of the data is typical of the distribution. The objective of this phase is to identify those cases that are not common within our information.
4. Transformations of data. In this section, both the yield data and the meteorological data are transformed, i.e., adapting formats, units, rescaled, etc, so that they could be optimally exploited by predictive models.
5. Grouping of data. The information downloaded is aggregated monthly; therefore, the rest of the data to be added to the study should be aggregated in the same way.
6. Integration of data. For our work, we have heterogeneous information that comes from different sources. The farmer indicates the annual net olive crop harvest, and meteorological information is available for the area where the farm is located. Thus, in order to carry out the data mining study, it is necessary to integrate all the information in a single source that serves as the input for the predictive models.
7. Detection of the level of influence of the input attributes on the target attribute. Before generating the model, the influence of each attribute on the target attribute (kg of

olives, the olive crop harvest) is analyzed in order to include or exclude attributes from the study based on their level of influence on the prediction.

8. Application of regression algorithms [31–33]. To perform the crop harvest prediction, different regression algorithms are tested. The goal of regression analysis is to determine the parameter values from a function that best fit a set of observational data. There are different families of regression functions and different ways of measuring error. This study uses:
   - Linear regression: this technique can be used if the relationship between variables can be approximated to a straight line. This is used to predict the value of a variable based on the value of another variable. The variable you want to predict is called the dependent variable. The variable you are using to predict the other variable's value is called the independent variable.
   - Nonlinear regression: it is a nonlinear combination of the model parameters and depends on one or more independent variables. It relates the two variables in a nonlinear, curved, relationship. The data are fitted by a method of successive approximations.

2.1. Experimental Site

This study was conducted in an olive growing farm situated in Jaén (Andalusia), which is located in the southern area of Spain, Figure 2. The perfect habitat for the cultivation of the olive tree is between latitudes 30° and 45°, so the Mediterranean climate and specifically the dry and hot summer climate of Jaén is perfect for its healthy growth and its greater use in harvest. The farm is an unirrigated olive orchard situated in 38°00′ N, 4°01′ W. The olive orchard studied is 3, 5 ha with 330 olive trees of similar properties; they all are about 25 years old and their mean fruit harvest is 60 kg/year. This is a traditional irrigated olive grove, in which the trees have two trunks and are planted at a distance of 10 m from each other. Cultivation methods include ploughing or other intensive tillage, such as harrowing to remove most of the plant residue cover. The main objective of this type of tillage is to keep the top cover of the olive grove free of weeds. After harvesting, from December to February, a 15 cm deep moldboard ploughing is carried out, which is essential to prepare the topsoil to absorb rainwater. The frequency of ploughing depends on rainfall. In summer, surface ploughing is carried out to increase the storage capacity of surface water in the soil but keep the organic substance at the surface level.

Regarding the rest of the treatment of the farm, pruning is carried out after harvesting, followed by fertilization with water-soluble fertilizers containing nitrogen, phosphorus, potassium, sulphur, magnesium, and micronutrients.

2.2. Dataset

Any predictive study requires the availability of adequate data in order to guarantee a quality prediction. Regarding the farm selected as a case study, there are eight growing seasons' yield data recorded from 2013 to 2020, Table 1. Olives were harvested in a mixed way, although the use of machinery to help the fruit fall from the tree predominates. The amount of the harvested crop was measured at the factory where the harvest was transported and stored for the farmer to collect the corresponding amount. In Spain, the weighing system of the mills must be checked regularly. The metrological systems are regulated by law and must comply with the current ISO 9001 and ISO 17025 standards.

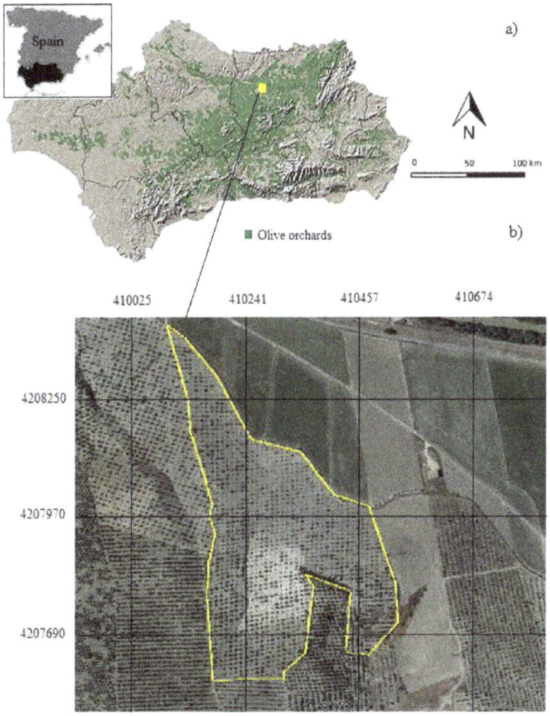

Figure 2. (**a**) Location of study area in Andalusia (Spain). (**b**) Precise location of the farm studied. The coordinates (m) are UTM, zone 30, referred to ETRS89. Source: SIGPAC web, juntadeandalucia.es, (accesed on 15 February 2022).

Table 1. Historical yield data of the olive orchard studied.

Growing Seasons	Yield (kg)
2013/2014	8697
2014/2015	7629
2015/2016	19,068
2016/2017	5755
2017/2018	10,700
2018/2019	11,475
2019/2020	11,249
2020/2021	15,071

Meteorological information has also been collected for those eight years. The meteorological data belong to the Spanish State Meteorological Agency (AEMET). The AEMET contributes to the protection of lives and property through the adequate prediction and monitoring of adverse meteorological phenomena and as support for social and economic activities in Spain through the provision of quality meteorological services. It is responsible for the planning, management, development, and coordination of meteorological activities of any kind at the national level, as well as representing it in international organizations and spheres related to Meteorology. In Spain, the State Meteorological Agency [34] offers the service AEMET OpenData. This is an API REST (Application Programming Interface. REpresentational State Transfer) through which the explicit data can be downloaded. Close to the study farm, there is a weather station that has provided meteorological data. The station is about 2 km away from the farm, which ID is 5298X. JSON (JavaScript Object Notation) files have been downloaded for each year. The metadata of these JSON files show the content of the files and the type of variables that have been used in this research. The meteorological data used are those indicated in Table 2. There are 26 variables, which are grouped by month, so there are 12 lines per each variable.

Table 2. Meteorological information included in JSON files downloaded from AEMET OpenData. ID is the name of each variable according to the JSON file.

ID	Description
fecha	Date
indicativo	ID of the weather station
p_max	Maximum daily precipitation (mm) of month/year and date
Hr	Average monthly/annual relative humidity (%)
q_max	Maximum absolute pressure monthly/yearly and date (hPa)
nw_55	No. of days of wind speed greater than or equal to 55 km/h in the month/year
q_mar	Monthly/yearly mean pressure at sea level (hPa)
q_med	Monthly/yearly mean pressure at station level (hPa)
tm_min	Mean monthly/yearly minimum temperature (degrees Celsius)
ta_max	Absolute maximum temperature of the month/year and date (degrees Celsius)
ts_min	Highest minimum temperature of the month/year (degrees Celsius)
nt_30	Number of days with maximum temperature greater than or equal to 30 degrees Celsius.
w_racha	Direction (tens of degree), speed (m/s) and date of maximum gust in month/year
np_100	No. of days of precipitation greater than or equal to 10 mm in the month/year
nw_91	No. of days of wind speed greater than or equal to 91 km/h in the month/year
np_001	No. of days of appreciable precipitation ($\geq$0.1 mm) in the month/year
w_rec	Average daily wind speed (from 07 to 07 UTC) in the month/year (km)
E	Mean monthly/yearly vapor tension (tenths hPa)
np_300	Number of days of precipitation greater than or equal to 30 mm in the month/year
p_mes	Total precipitation monthly/yearly (mm)
w_med	Monthly mean velocity elaborated from the observations of 07, 13 and 18 UTC. (km/h)
nt_00	Number of days with minimum temperature less than or equal to 0 degrees Celsius)
ti_max	Lowest maximum temperature of the month/year (degrees Celsius)
tm_mes	Average monthly/yearly average temperature (degrees Celsius)
tm_max	Average monthly/yearly maximum temperature (degrees Celsius)
q_min	Minimum monthly/yearly maximum pressure and date (hPa)

2.3. *Data Mining Techniques Used*

Data mining involves the intersection of statistics, computer science, and machine learning. The techniques used in data mining are a set of calculations that creates a model from data. In this sense, in order to create a model, first an algorithm analyzes the data provided, looking for specific types of patterns or trends. It uses the results of this analysis over many iterations to find the optimal parameters for generating the model. These parameters are then applied across the entire data set for extracting actionable patterns and detailed statistics. In this sense, the complete data analysis has been carried out with an analysis module included in the Oracle Data Mining software. Choosing the best algorithm to use for our task is a challenge. The algorithms used in each phase and the flow of this study are described below, Figure 3, justifying their selection.

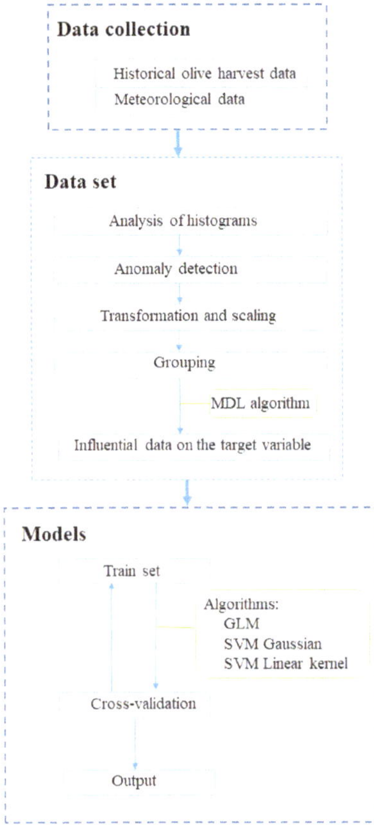

Figure 3. Flow diagram of the methodology.

As explained in Section 2.2, once the data set is available, it is essential to carry out an anomaly detection. This is to identify cases that are unusual within data that are seemingly homogeneous. Anomaly detection is important for detecting fraud, outliers, and other rare events that may have great significance but are hard to find. In this phase, a classification algorithm is used because anomaly detection can be considered as a type of classification. A one-class classifier develops a profile that generally describes a typical case in the training data. Deviation from the profile is identified as an anomaly. Specifically, in this phase, the algorithm used has been the SVM algorithm [35–37]. SVM works on the basic idea of minimizing the hypersphere of the single class of examples in training data and considers all the other samples outside the hypersphere to be outliers or out of training data distribution. This algorithm produces a prediction and a probability for each case in the scoring data. If the prediction is 1, the case is considered typical. If the prediction is 0, the case is considered anomalous. This behavior reflects the fact that the model is trained with normal data.

Another important phase in the process of generating the predictive model is the determination of the level of influence of the variables used on the target attribute. In this case, the Minimum Description Length (MDL) algorithm [38] is used. MDLprinciple is a powerful method of inductive inference, the basis of statistical modeling, pattern recognition, and machine learning. It holds that the best explanation, given a limited set of observed data, is the one that permits the greatest compression of the data. This is a supervised technique used for calculating attribute importance. MDL considers each attribute as a simple predictive model of the target class. Model selection refers to the process of comparing and ranking the single-predictor models. The implementation of this algorithm returns a value of between −1 and 1, taking a value of 1 those attributes that have the greatest relationship with the target, 0 those which have no relationship, and a negative value means that the attribute is not related to the target, and therefore can insert noise into the studio. Only those attributes with weight greater than 0 are considered in this study, discarding all those with 0 or negative values.

Finally, regression analysis algorithms have been used in order to generate the model, that is the prediction of the target, the amount of olive harvest. The main reason for using regression is that it is a data mining function that predicts numeric values along a continuum. A regression task begins with a data set in which the target values are known. A regression algorithm estimates the value of the target as a function of the predictors for each case in the data set. These relationships between predictors and target are summarized in a model, which can then be applied to a different data set in which the target values are unknown. Regression models are tested by computing various statistics that measure the difference between the predicted values and the expected values. The historical data for a regression project is typically divided into two data sets: one for building the model and the other for testing the model. It is necessary to specify that the year used for model testing is not included in the training data set. For this study, a pure linear model is used from the GLM algorithm and other models applying Support Vector Machines (SVM) with Gaussian and Linear Kernel respectively.

Linear models make a set of restrictive assumptions, where the target is normally distributed conditioned on the value of predictors with a constant variance regardless of the predicted response value. In this sense, generalized models (GLM) relax these restrictions, and for a binary response example, the response is a probability in the range [0, 1] [35,39].

SVM regression supports two kernels: The Gaussian kernel for nonlinear regression, and the linear kernel for linear regression. SVM performs well on data sets that have many attributes, even if there are very few cases on which to train the model. There is no upper limit on the number of attributes; the only constraints are those imposed by hardware. This is especially interesting for our study, since in the historical training data, particularly in the first years, some data are missing for certain meteorological variables. However, the algorithm works quite well despite this circumstance [40,41].

2.4. Web application

The development of an application that includes the predictive model requires a tool that will provide for better means of knowledge acquisition, inference mechanism, and user interface. It is a technology with a broad impact on business and industry. It offers practical use and commercial potential [42]. In this case, it is essential that the system is easily accessible to the farmer or farm manager. Therefore, our application has been deployed in the cloud, allowing the user to work with it from any device with an internet connection.

Efficient analysis, simulation, and visualization of the predictive crop model for a selected farm is needed. In such a way, the user can graphically see anticipated yield data based on real weather data and even make simulations with fictitious data entered by the user. In this way, future action plans can be drawn up according to the different circumstances that may arise.

One of the most interesting aspects of the tool is that the loading of the meteorological variable data required by the application is automatic and always up to date. The only data that the user must insert are the data on the amount of crop harvested each year. This action is performed by the user through the application by a user-friendly interface. This data will be included in the training data set that allows the predictive model to be improved, Figure 4.

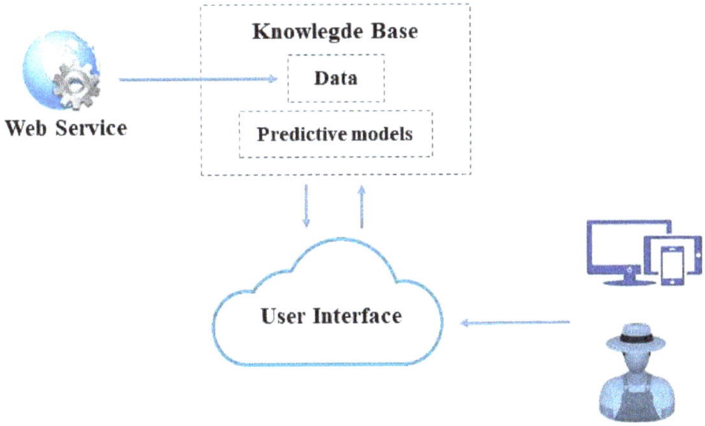

Figure 4. Schematic representation of the web application.

The development of the web application is carried out under the Oracle Database Management System 19c [43]. Oracle provides us the necessary tools for the agile development of the application: the database to manage the information, Oracle Data Mining, which is a module integrated in the database in which there are all necessary data mining algorithms, and Oracle Applications Express (APEX), which is a WEB application development module that is also implemented in the database. Apex allows us to develop WEB applications for both PC and mobile devices. The web application has been sequentially developed from the following phases: design of the model and uploading data, generating the predictive models, development of the application interface, and the design and development of the learning system.

2.4.1. Design of the Data Model and Uploading Data

The data model for our system is quite basic; we just need two tables. The first table contains historical data including the amount of crop yield harvested in the olive grove on the farm under study each growing season, meteorological data, and environmental quality. The second table stores the results and input parameters of the prediction; both tables make the knowledge base of the system, Figure 4.

There are several tools to upload the data into the database. In this case, SQL-Loader has been used. It is a tool that allows us to read text files and transfer the data into the tables. Once the information was uploaded into the auxiliary tables, it was grouped by SQL querys following the framework criteria of the study.

2.4.2. Generating the Models with Oracle Data Mining

Oracle Data Miner is an extension to Oracle SQL Developer that allows data analysts to work directly with the data within the database, explore the data graphically, build and evaluate multiple data mining models, and more. The Oracle Data Miner workflow captures and documents the user's analytical methodology and can be saved and shared with others to automate advanced analytical methodologies. It also allows the creation of predictive models that application developers can integrate into applications to automate the discovery and distribution of new business intelligence: predictions, patterns, and discoveries.

Once the data was stored, the models were generated with Oracle Data Mining. This was performed using the SQL Developer tool. First, we selected the source table and the type of model we wanted to generate, in our case, a regression. Then, we selected the algorithm to apply, and the system asked us to select the target attribute and the attributes that form part of the predictive model.

2.4.3. Development of the Application and System Learning

The second objective of this work is to integrate the prediction system into a useful application for the farmer or manager of the olive farm. As this is a user who is not an expert in Information and Communication Technologies, it is a priority that the system is implemented in a simple application with a user-friendly interface. Oracle Application Express (Oracle Apex) has been used for this purpose. It is an Oracle's primary tool for developing Web applications with SQL and PL/SQL. Using a web browser, professional Web-based applications for desktop and mobile devices can be developed. In this study, the system has been mounted on Oracle's Cloud, allowing it to be accessed from the Internet on the page https://apex.oracle.com/pls/apex/f?p=50617 (accesed on 20 February 2022). The system has the following functionalities:

9. Log-in system. In order to access the system, user registration and authorization is required, Figure 5a.
10. Main screen. There are 3 icons, Figure 5:
    a Prediction data. This accesses the screen where the user inserts the data to make a prediction. It opens a variable collection form. Figure 6a.
    b Historical Data. This allows us to visualize the result of all the predictions made by the system. It is possible to search for any prediction made by the system at any time in different formats (table, report, or graph), send it to anyone by mail, make groupings, etc. You can even make simulations from fictitious data. Figure 6b.
    c Real Yield Data. In this option, the user enters the real data on the amount of harvest harvested in that year. This data becomes part of the training data, so the system is learning through a continuous feedback process, Figure 6.

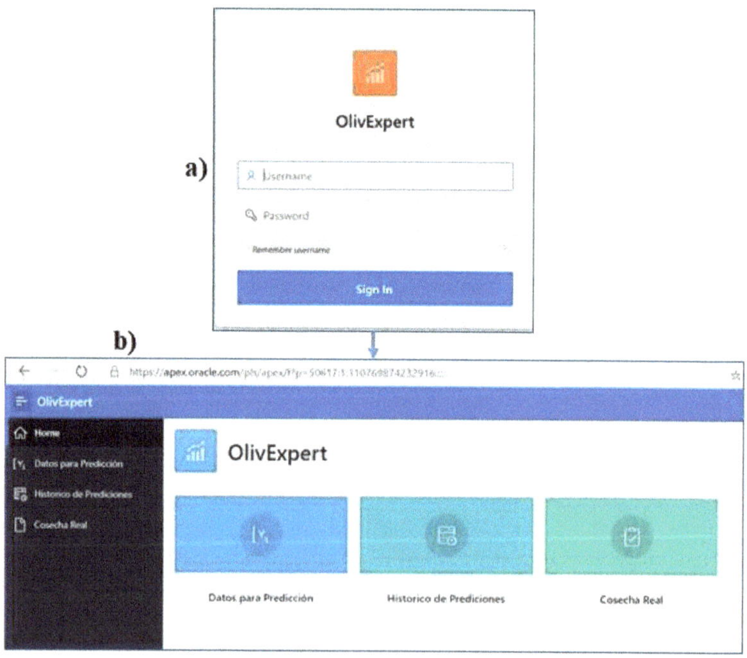

Figure 5. Interface of the web application, (**a**) login; (**b**) main menu.

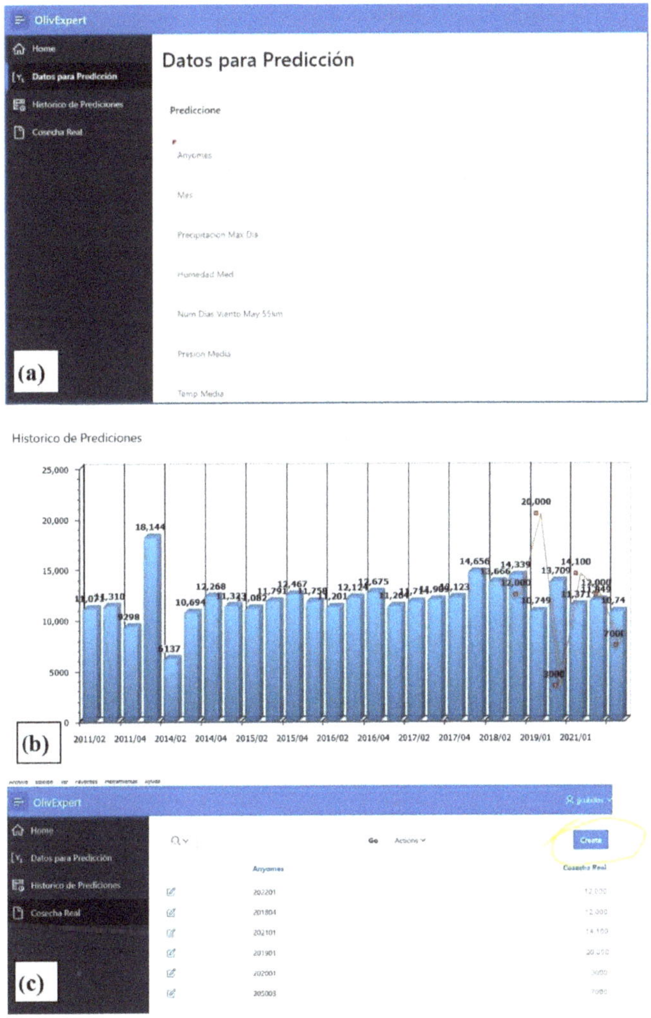

Figure 6. Interface of the web application, (**a**) prediction data entry module; (**b**) predictions visualization and graphical representation module; and (**c**) actual yield data entry module.

Finally, the screen showing the result of the web application was designed. To do this, the wizard provided by Oracle APEX was used and relied on the same results table. To improve the interface and help the manager, the application generates representative graphs of the stored data, such as the harvested data versus the prediction values for each year.

3. Results

3.1. Datasets Analysis and Data Cleaning

The idea is to use data from the first seven years to generate the model and the last one to analyze the quality of the prediction. The development of this data mining study includes the following phases: data loading, data cleaning, detection of anomalous data, generation of predictive models, and finally testing for validation. Firstly, a table was created in order to store the information extracted from the AEMET Web Services as well as the harvest quantity data supplied by the farmer. Previously loaded data in the database JSON files were converted to CSV using " | " as field delimiters. Once this conversion was carried out, they were loaded into the database. All the information was stored in a single table, denormalized.

Additionally, data were also analyzed with a one-class classification algorithm for the detection of anomalies. A one-class classifier develops a profile that generally describes a typical case in the training data. The profile deviation is identified as an anomaly. Outliers are sought in this study, such as unusual cases, because they fall outside the distribution considered normal for those data.

The algorithm shows the distribution of the data and marks with 0 abnormal data together with their probability. The results can be seen in Table 3, in which record number 9 is marked as anomalous value with a probability of 75%.

Table 3. Results of the one-class classification algorithm.

Anomaly Detection	Probability (%)
1	50.19
1	66.72
1	62.08
1	64.90
1	67.38
1	65.18
1	62.58
1	53.31
0	75.48
...	...

The anomalous value detected was removed from the study. Figure 7 shows the values of this abnormal record, and indeed, it can be confirmed that August marks a maximum temperature of 0 degrees, minimum temperature of 0 degrees, etc. It is clear that it is an outlier (that is a mistake in the data), because in Jaen, the mean temperature in August is higher than 30 degrees. This review of the table could hardly have been carried out with the naked eye. This checking and filtering of variable values ensures that the data used as training data are quality data. Consequently, the final prediction errors are attributed to the adequacy of the model/equations and not to the parameters.

In addition to the analysis of the meteorological variable data, an analysis of the yield data was carried out. When analyzing the harvest quantity data for each year, Figure 8, it was observed that the maximum and minimum values correspond to two consecutive campaigns, those of 2015 and 2016, with values of more than 19,000 kg and around 5000 kg, respectively. This is a significant difference, more than 300%. At first it could be considered that this fact could negatively affect the model testing, since it presents a significant peak jump in the distribution of the yield data. However, if the 2015 data is excluded from the training data to generate the model when testing its prediction, the model validation will give good results, but for global model purposes there will be no training data with the 2015 crop. The same argument could be applied to the 2016 crop data, so it is to be expected that although the prediction of the 2015 and 2016 crop will not be reliable in the validation check, this does not imply a deficiency of the model. The reason is that this circumstance, in the final model, is mitigated by including all years as training data, i.e., from 2013 to 2020. In this sense, as the amount of training data increases, the model will be adjusted to the different casuistry of each year.

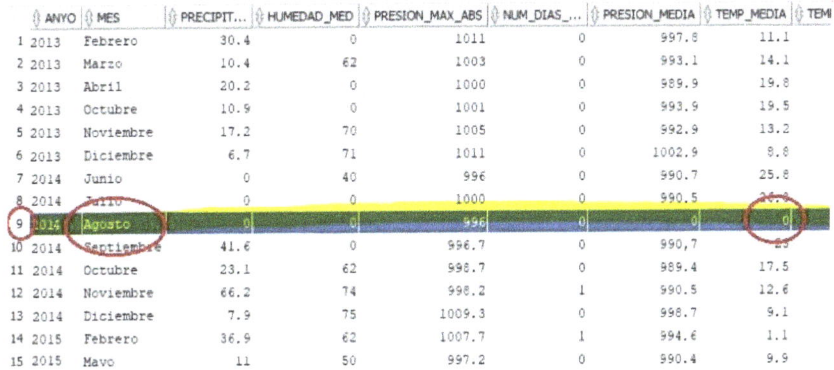

	ANYO	MES	PRECIPIT...	HUMEDAD_MED	PRESION_MAX_ABS	NUM_DIAS_...	PRESION_MEDIA	TEMP_MEDIA	TEM
1	2013	Febrero	30.4	0	1011	0	997.8	11.1	
2	2013	Marzo	10.4	62	1003	0	993.1	14.1	
3	2013	Abril	20.2	0	1000	0	989.9	19.8	
4	2013	Octubre	10.9	0	1001	0	993.9	19.5	
5	2013	Noviembre	17.2	70	1005	0	992.9	13.2	
6	2013	Diciembre	6.7	71	1011	0	1002.9	8.8	
7	2014	Junio	0	40	996	0	990.7	25.8	
8	2014	Julio	0	0	1000	0	990.5	[illegible]	
9	2014	Agosto	0	0	996	0	0	0	
10	2014	Septiembre	41.6	0	996.7	0	990,7	[illegible]	
11	2014	Octubre	23.1	62	998.7	0	989.4	17.5	
12	2014	Noviembre	66.2	74	998.2	1	990.5	12.6	
13	2014	Diciembre	7.9	75	1009.3	0	998.7	9.1	
14	2015	Febrero	36.9	62	1007.7	1	994.6	1.1	
15	2015	Mavo	11	50	997.2	0	990.4	9.9	

Figure 7. Example of anomaly detection.

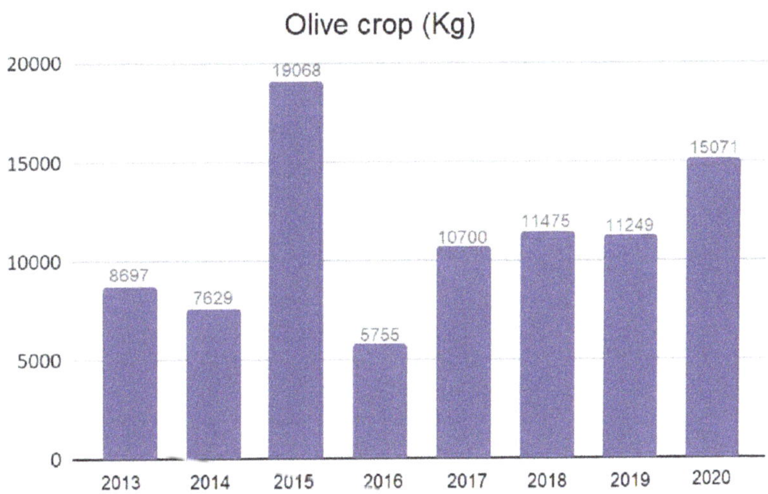

Figure 8. Yield data history of the farm under study.

3.2. Variables Influencing Olive Crop Prediction

The next phase in the workflow of this study is to determine the level of influence of each variable on the target attribute, in this case, the crop harvested in each year. The Minimum Description Length algorithm (MDL) was applied. This algorithm returns a value of between −1 and 1. Values of 0 are returned for those cases which have no relationship. For this study, only those attributes with weight greater than 0 were considered, discarding those attributes with a 0 or negative value. Once the most suitable attributes for this study were selected, they were used to generate the predictive models. Table 4 shows the results of the algorithm execution. Here it can be confirmed the variables that have been included in the model are those that have a level of influence on the target attribute above zero; those with a negative influence value have been discarded. The attribute with the greatest influence is the harvest quantity of the previous year, last year's crop. Then, the next variable in importance level is the number of days with rainfall greater than 30 L, np_300, followed by the number of days with wind of more than 55 km/h, nw_55. It is understood that this will affect pollination. It was also observed that the absolute maximum atmospheric pressure, q_max, data have a negative value and, therefore, is not related to the target attribute; this attribute was removed from the study.

Table 4. Ranking and weight assigned to each attribute.

Variable	Level of Influence on Target
Last year's crop	0.793
np_300	0.436
nw_55	0.156
nt_00	0.152
np_100	0.132
p_mes	0.121
ti_max	0.111
p_max	0.092
ta_max	0.091
e	0.073
tm_mes	0.061
nt_30	0.053
w_med	0.042
tm_max	0.041
ts_min	0.031
tm_max	0.021
nt_00	0.019
q_med	0.015
hr	0.015
np_001	0.015
np_300	0.005
month	0.004
q_max	−0.756

3.3. Models and Validation

Once variables were selected and all information was available and normalized, the model was designed. The k-fold cross validation technique was used to evaluate results in statistical analyses [44]. The technique consisted of evaluating the quality of each year's prediction by separating that year's data from the training data used for the prediction. Particularly, for this research, it was used to check the reliability of the model in terms of harvest quantity prediction of a specific year, being this excluded in the generation of the model. As previously indicated, for this study, production data from 2013 to 2020 were used. Specifically, a predictive model was generated using data from seven years, and next, data from the eighth were used to evaluate the reliability of the model; that is, the prediction of the harvest of the eighth year was estimated and then it was compared with the actual crop data of that date.

There was no dependence between the training data and the prediction; therefore, another model will be generated with the data for the years 2013, 2014, 2015, 2016, 2017, 2018, and 2020, and its reliability will be tested in the 2019 crop prediction, and so on. In this way, the model can be tested against the actual production of several years, Figure 9. Finally, a final model will be generated including all years, that is, using all training data.

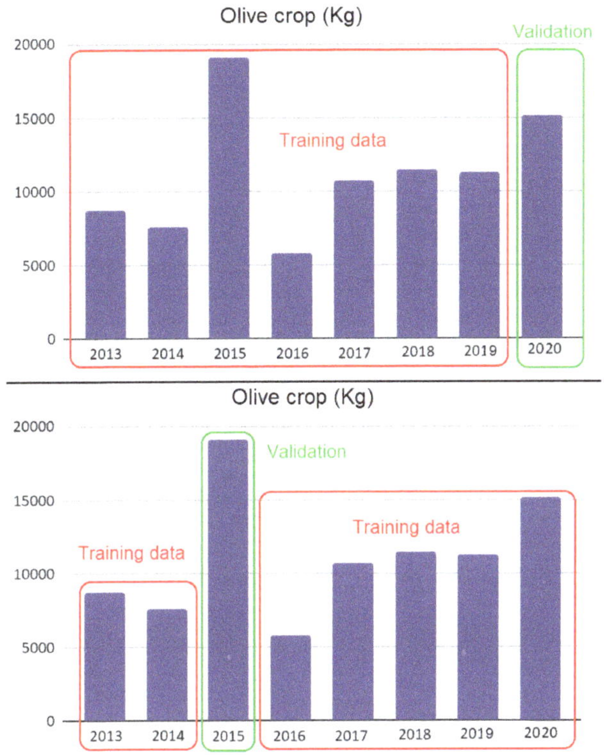

Figure 9. The k-fold cross validation technique applied to the yield data from 2013 to 2020. (**Up**) Training data used correspond from 2013 to 2019 and 2020 data are used for validation. (**Down**) Model is generated using training data from 2013 to 2020 not including 2015 data which is used for the validation process.

Another important consideration in the design of the model is the reliability of an early prediction of the crop during the harvesting year. As indicated in previous sections, the farmer must make decisions once the harvesting year has been completed, that is, from January to March, he has to make decisions regarding tilling the land, fertilizing it, pruning it, selling the oil obtained, etc. In this sense, in this study, since the meteorological data are grouped by months, a monthly prediction of harvest quantity is generated and then the quality of each prediction is assessed. Thus, for each year, 12 predictions were generated, one for each month, and the mean absolute error (MAE) of each prediction was calculated in order to determine how consistent the prediction was. The MAE is the average of all absolute errors, and it is calculated as shown in Equation (5), where n is the number of errors, Σ is the summation symbol which means "add them all up", and $|x_i - x|$ is the absolute errors.

$$\text{MAE} = \frac{1}{n}\sum_{i=1}^{n}|x_i - x| \tag{5}$$

In this way, the farmer has a series of tools in addition to his own experience to make appropriate decisions about the management of his farm.

As the predictive model generation process has been designed, a cross validation is carried out. For this purpose, several models are generated, one for each year to be checked. A pure linear model is used from the GLM algorithm and other models applying SVM with Gaussian and Linear Kernel, respectively. Once the algorithms have been tested, the one that best suits the nature of the data is taken into account. Finally, the final model is generated using this algorithm, including as training data all the harvest years from 2013 to 2020.

In this first phase of results, the models generated for cross-validation are presented. A model is generated for each year to be tested. For example, to check the year 2020, training data included are from 2013 to 2019. Those for 2020 are not included; they are used to test the effectiveness of the model. In other models, we proceed in the same way, but checking the year 2017. The model training data are: 2013, 2014, 2015, 2016, 2018, 2019 and 2020, leaving 2017 out, as it is used for testing, and so on. Three different algorithms are used: a pure linear model such as GLM and others such as SVM with Gaussian and Linear Kernel.

The Oracle Data Mining tool is used in this research to provide information about the theoretical efficiency of each generated model, Figure 10. To analyze the effectiveness of each model, we analyzed the MAE, as seen in Equation (5) and also the root mean square error (*RMSE*), The latter is the standard deviation of the residuals, prediction errors. Residuals are a measure of how far from the regression line data values are. *RMSE* is a measure of how spread out these residuals are. In other words, it indicates how concentrated the data are around the line of best fit. In Equation (6), n is the number of values predicted.

$$RMSE = \sqrt{\sum_{i=1}^{n} \frac{(Predicted_i - Actual_i)^2}{n}} \quad (6)$$

Algorithm	Mean Abs. Error	Square Root Error	Expected average value	Actual average value
GLM	35,863.81	40,734.24	25,459.60	9,747.91
SVM Gaussian	1458.41	2,072.90	10,653.42	9,747.91
SVM Linear Kernel	2,356.55	3,199.80	10,245.63	9,747.91

Figure 10. Validation of the model using different algorithms.

In the comparison of the three models, it is observed that the lowest errors, both for MAE and *RMSE*, are those obtained with the SVM model with Gaussian Kernel. Therefore, based on these data, the selection of this algorithm to generate the final model is justified. The real data for the year 2020 are used for the error comparison.

The residual plot is also analyzed to visualize graphically how the model fits the evolution of the data. Thus, it detects strange behaviors in individual data, Figure 11. Three large groups that are very far apart, whose values have an influence on the model can be detected. Although they are very distant from each other, they are not considered as outliers since the residuals are homogeneously located in the three groups. Analyzing the impact of the last year's crop variable on the model, it can be confirmed that harvests can be basically classified into three types: low harvest, medium harvest, or high harvest.

As indicated in the introduction, the optimization of agricultural resources on an olive orchard requires the farmer to make appropriate decisions in advance. In this sense, resource management during the months of January or February is crucial, since there is still no indication of what the harvest will be that year. However, the contribution of this research focuses precisely on providing reliable data to the farmer also at an early stage of the agricultural year of the olive crop. Table 5 shows the model predictions for the months of January and February of all years. Note that the prediction for 2013 is not included. This year, being the first year, its data are used as training data, but the harvest cannot be predicted because data from the previous year are not available.

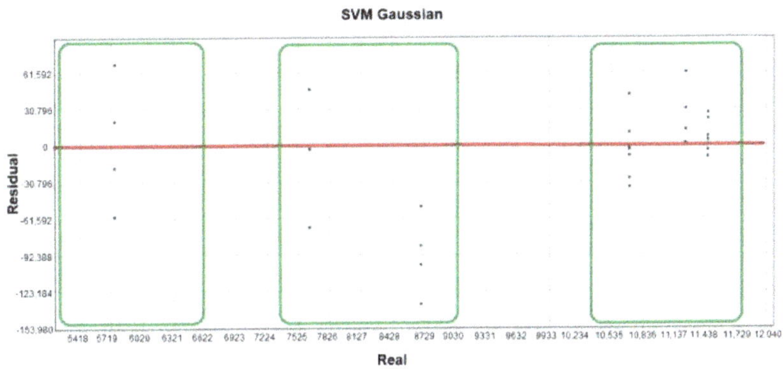

Figure 11. Distribution of the residuals using the SVM with Gaussian kernel algorithm.

Table 5. Early prediction of crop data for each year (kg/ha).

Year	Month	Yield Data	Prediction	Relative Error
2020	January	4566.18	3815.97	16.43%
	February	4566.18	4275.08	6.38%
2019	January	3408.20	3592.49	5.41%
	February	3408.20	3806.77	11.69%
2018	January	3476.67	3141.15	9.65%
	February	3476.67	3478.94	6.52%
2017	January	32.42	3881.56	19.73%
	February	32.42	3763.93	16.10%
2016	January	1743.64	2481.55	42.32%
	February	1743.64	2588.64	48.46%
2015	January	5777.19	3511.67	39.00%
	February	5777.19	5760.12	29.54%
2014	January	2308.39	2310.70	10.03%
	February	2308.39	2310.08	7.32%

However, crop prediction errors for 2015 and 2016 are less encouraging. The latter range between 48% and 39%. In this regard, it is important to remember that the yield data for these two years correspond to the maximum and minimum values, respectively, of the total training data set. Therefore, these results are to be expected in the validation study since, in order to perform the cross-validation for the years 2015 and 2016, the data for each year are not included in the training data. For this reason, the model is overestimated, and these results come out. However, this is not a problem, since these extreme data are included in the final model, and in a similar situation the regression model would fit more accurately.

Traditionally, the farm manager usually makes decisions based on the average values of previous years, which, in the case of years with extraordinary circumstances, our predictive model would be able to provide extraordinary knowledge. In this sense, considering the crop variability between maximum and minimum is more than 331%, between 2015 and 2016, having an error of 48% in January, this prediction could be useful for the farmer. In fact, cross-validation of the prediction for the months of January and February, Figure 12, confirms that the crop prediction for these two years of extreme values is closer to the actual values than the naïve model, based on averaged values.

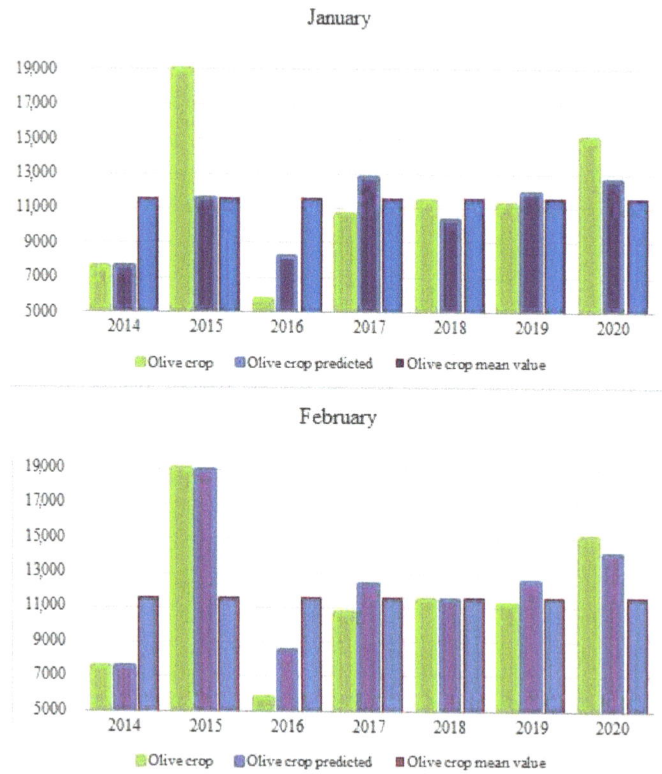

Figure 12. January and February yield data for all years studied: actual, predicted and averaged data.

Once the algorithms have been tested, taking into account that the SVM algorithm with Linear Kernel is the one that best adapts to the nature of the data that influence the target attribute, the amount of harvest was generated, which includes as training data all the harvest years, from 2013 to 2020.

4. Discussion

An early crop yield prediction in the crop season, January-February, as resulted in previous sections, is key for the farmer to make important decisions, such as choosing the type of tillage on the farm, investment in fertilizers, irrigation, or even the marketing of the oil. These decisions are highly dependent on the crop forecast. Currently there is research oriented to the generation of predictive models; however, although they are very efficient, they are not useful for the case of olive orchards. The main reason is that these models are based on the analysis of pollination [16,21–28], which occurs very late in the crop year, between May and July. Moreover, in this period, the fruit is already visible on the olive tree, so the farmer, based on his knowledge of previous years, can already have a reliable idea about the harvest; therefore, it is a good and reliable prediction, but not very practical because the economic investment in ploughing and other work on the exploitation has already been made.

The contribution of this research is that based on historical olive crop harvested data and meteorological data such as temperature, rainfall, wind, etc., crop prediction models on February have been generated, with absolute errors of less than 17% in 70% of the years used for the training data. Similar results have been obtained by other authors, but in research related to early potato harvest prediction. In this work, regression models and yield data from seven previous years have also been used, obtaining a mean absolute error of around 15%, with the same validation method [45]. Taking into account that crop variation from one year to another can vary by more than 330%, these results can be considered satisfactory. In addition, due to the design of the study, the errors obtained are maximum errors, since training data were excluded for the validation tests; however, in the final model, the training data are composed of yield data from all years, which increases the robustness and efficiency of the predictive model.

As a particular case, in the years of maximum and minimum harvest, relative errors of around 40% have been obtained, Table 5. This is a large error compared to the rest of the years; however, it is an acceptable error considering that this error is not real and is overestimated. When removing, in the cross validation, data from the years of maximum and minimum harvest, there are no training data for these harvest extremes, hence such large differences between the estimated and real values. Even so, the model fits with a relatively low error, making an acceptable prediction for the margins of error typical of such an early prediction. However, as already mentioned, these values are included in the final model, thus allowing the model to be able to predict future extreme harvests under similar circumstances.

The study confirms that the main variable that most influences the prediction of harvest quantity is the previous year's harvest. In this sense, and to demonstrate the sensitivity of the model to this variable, the harvest prediction for the growing seasons of 2015/16, 2016/17, and 2017/18 was generated by considering the actual harvest values of the previous year in the training data. The harvest data for the year before each season was then increased by 10%, and each of the three predictions was generated again. It should be noted that the rest of the variables were kept as the real values. The results can be seen in Table 6. It can be seen that decreasing the harvest of the previous year by 10% has a direct effect on the prediction; in such a case, the lower the harvest of the previous year, the higher the prediction.

Table 6. Comparison in yield data prediction (Kg) by changing the key parameter by 10%. Analysis for three selected growing seasons. (1) Prediction using real values for the key parameter and (2) prediction using the key parameter value modified by 10%.

Growing Season	Yield Data Current Season	Real Yield Data Previous Season	Prediction Current Season (1)	Changed Yield Data Previous Season	Prediction Current Season (2)
2015/16	11,475	10,700	10,367.58	9630	11,033.24
2016/17	11,249	11,475	12,564.51	10,327.5	12,664.51
2017/18	15,071	11,249	14,110.19	10,124.1	14,412.36

In this harvest prediction research, harvest data and meteorological values have been used. In fact, the Discussion of this paper focuses on these two variables as key parameters. The reason is that, as already developed in the methodology section, the MDL algorithm provided us the level of influence of each variable on the target attribute, with the amount of harvest from the previous season and rainfall being the most influential. However, in other works [46], remote sensing data have also been used. However, in this study, it was found that the estimation results change depending on different agricultural zones and temporal training settings. Nevertheless, factors influencing crop production used to be the same, the greatest weight attributed to environmental factors such as crop variety, soil type and surface cover or topography, etc.

The analysis of the influence of the variables studied on the early harvest prediction has been carried out using the MDL algorithm. The results obtained are in line with the works [47,48], which state that years of very good olive yields alternate with other in which very few kilograms of olives are harvested per tree. This alternation is not due to climatic factors, but to the fact that the high yields of the productive years interfere with the vegetative development of the tree or exhaust its reserves, and therefore, it will need time to recover and accumulate the lost resources again. This has been confirmed in our study by analyzing the variables that influence the amount of harvest, the target attribute. The MDL algorithm statistically quantifies that the most influential variable is the amount of harvest from the previous year, Table 4. The analysis of the influence of the variables studied on the early harvest prediction has been carried out using the MDL algorithm. The results obtained are in line with the works [47,49], which state that years of very good olive yields alternate with others in which very few kilograms of olives are harvested per tree. This alternation is not due to climatic factors, but to the fact that the high yields of the productive years interfere with the vegetative development of the tree or exhaust its reserves and, therefore, it will need time to recover and accumulate the lost resources again. This has been confirmed in our study by analyzing the variables that influence the amount of harvest, the target attribute. The MDL algorithm statistically quantifies that the most influential variable is the amount of harvest from the previous year, Table 4. In addition, the argument of alternation between productive crop years and crop-shortage years is also corroborated in this study from the residue analysis in Figure 11. Here, we can see how the harvest quantity data are grouped into three clearly differentiated classes: low harvest, medium harvest, or high harvest.

Secondly, another key variable in crop production is the amount of rainfall, especially that accumulated at certain times of the year. This has been confirmed by several authors [49–51], whose results showed in these works identify the highest production coincided with increased rainfall, namely in two consecutive months, during August and December. It was concluded that the impact of rainfall on olive production depends on the intensity and monthly distribution of rainfall. In this sense, our work identifies the variable "np_300" as the second most influential variable on the target attribute. This variable identifies the number of days of precipitation greater than or equal to 30 mm in the month/year, which coincides with the importance of the accumulated rainfall factor identified by the aforementioned authors.

The results obtained with respect to prediction reflect that of the three predictive algorithms used; the one that best fits the objective of this research is the SVM algorithm with Gaussian Kernel. This indicates that there is no linear relationship between the variables and the target attribute. The algorithm distributes the data in a Gaussian bell-shaped hyperplane in such a way that a plane of separation is established between the data, Figure 1. In this study, the following Gaussian Kernel configuration has been used in the SVM algorithm:

- Kernel Cache Size specifies the cache size (in bytes), which is used to store the kernels computed during the build operation. As expected, larger cache sizes generally result in faster builds. In our case, it has been configured with the value of 50 MB.
- Convergence tolerance specifies the tolerance value allowed for the generation of the model before completion. The value must be between 0 and 1. The value configured in our case was 0.001. Higher values tend to result in faster generations but less accurate models. In our study we have aimed for a value very close to 0 to ensure maximum accuracy without penalizing computational time.
- Standard deviation allows us to specify the standard deviation parameter that the Gaussian kernel uses. This parameter affects the trade-off between the complexity of the model and the ability to generalize to other data sets (overfitting and underfitting of the data). Larger standard deviation values favor under-fitting. We have left this parameter with the default setting. With this setting, the system has automatically estimated this parameter from the training data.
- Epsilon. Specifies the value of the error interval allowed in the generation of models not sensitive to epsilon. In other words, it distinguishes small errors (which are ignored) from large errors (which are not ignored). The value must be between 0 and 1. As with the previous parameter, the default setting has been used, so that it has been calculated by the system based on the training data.

- Complexity factor. Allows the determination of the complexity factor that balances model error (measured with respect to the training data) and model complexity in order to avoid over-fitting or under-fitting of the data. Higher values provide a higher penalty to errors, which means a higher risk of over-fitting the data. Smaller values provide a smaller penalty to errors and may lead to under-fitting. In our study, this value has been set as automatic and the system has determined it based on the training data.
- The normalization method specifies the method used for the continuous input and the target attribute. Z-scores, Min-Max, or None can be selected. Oracle performs the normalization automatically if none is specified. In our case, it has been left unselected.
- Active learning provides a method for handling large generated sets. With active learning, the algorithm creates an initial model based on a small sample before applying it to the entire training data set and then updates the sample and model incrementally based on the results. This cycle is repeated until the model converges on the training data or until the maximum number of allowed support vectors is reached. In our case, to speed up the calculations, this parameter has been activated.

Given the specific configuration of the parameters used in the algorithm, it could be concluded that the model generated is only valid for the case of the farm studied. For this reason, and in order to check its effectiveness at other scales, our model was tested on a much larger area. The statistical model has been applied to another crop prediction in Spain, but this is a much larger farm extension, the joint prediction at the level of all the farms belonging to the same municipal district, Lahiguera (Jaén, Spain). This includes 50% of traditional rainfed olive groves and the other 50% of irrigated olive groves. Harvest prediction was generated for the 2020/21 and 2021/22 seasons. Our model was trained with data from the 2013/14 seasons up to the 2021/22 season. As in our study, we removed the year to be predicted from the model, Table 5. The meteorological data were obtained from two stations located within the boundary of Lahiguera, and the arithmetic means of the climatic data were used. The predictions obtained had similar relative error to those of our case study, Table 7.

Table 7. Early prediction of crop data for Lahiguera district (Kg).

Growing Season	Yield Data	Prediction	Relative Error
2020/21	15,349,191.85	15,899,198.74	3.58%
2021/22	14,823,594.35	16,367,993.56	10.41%

In summary, despite the local character of the study, the results are in line with larger scale studies and with the results of other authors elsewhere, which is another way of validating our methodology. It is also an endorsement to improve our predictive model by adding more data and more farms to our training data.

5. Conclusions

All development and generation of the predictive models were performed under the cloud application without the user being aware of it. The farmer or manager using the application only selected the prediction they needed to see. They could also export the information in the desired format or simply visualize a graphical representation of the results. The application is, therefore, a powerful tool that is very accessible and very useful for decision making.

The conclusions of this study respond to the case of a type of olive grove with specific characteristics, an unirrigated olive grove where traditional tillage is practiced. Future lines of work include the application of these models in other farms with similar characteristics and in others where a different type of tillage is practiced. It also remains to be seen how these models would respond in an irrigated olive grove or in other types of olive groves, such as the growing intensive and super-intensive olive groves, where the yield data are very different from those dealt with in this study. However, the working methodology and implementation in the web application is designed. Adaptation to other crop types would be achieved by following the same workflow and adapting the predictive models according to the influence of weather data and harvested crop quantity data.

Author Contributions: J.J.C. has contributed to the design and development of the web application, the generation of the prediction models, the interpretation of results, and to the conceptualization, formal analysis, and writing—review and editing. M.I.R. has contributed to the general supervision of the paper, the writing—original draft preparation, and to the mailing and exchange of mailings with the journal. J.M.J. has contributed to the design and implementation of the web application. F.R.F. has contributed to the supervision of the products resulting from the prediction model and the analysis of results and conclusions. All authors have read and agreed to the published version of the manuscript.

Funding: This research has been partially funded through the research projects PYC20-RE-005-UJA 1381202-GEU and IEG-2021 y PREDIC_I-GOPO-JA-20-0006, which are co-financed with the European Union FEDER, Instituto de Estudios Giennenses, and the Junta de Andalucía funds. We are also grateful for the support provided by the Ministry for Ecological Transition and the Demographic Challenge, Spanish Government (AEMET, Agencia Estatal de Meteorología).

Institutional Review Board Statement: Not applicable.

Informed Consent Statement: Not applicable.

Data Availability Statement: Not applicable.

Conflicts of Interest: The authors declare no conflict of interest.

References

1. INEbase; Agriculture and Environment; Agriculture. Available online: https://www.ine.es/dyngs/INEbase/en/categoria.htm?c=Estadistica_P&cid=1254735727106 (accessed on 13 October 2020).
2. Quiroga, S.; Iglesias, A. A Comparison of the Climate Risks of Cereal, Citrus, Grapevine and Olive Production in Spain. *Agric. Syst.* **2009**, *101*, 91–100. [CrossRef]
3. Olive Oil & Health. Available online: https://www.internationaloliveoil.org/olive-world/olive-oil-health/ (accessed on 13 October 2020).
4. Moral, A.; Manuel, P.; Ruiz, F.J. *El Comportamiento Comercial Del Cooperativismo Oleícola En La Cadena de Valor de Los Aceites de Oliva En España*; Agrícola Española: Madrid, Spain, 2013; ISBN 978-84-92928-23-1.
5. Vilar, J.; Cárdenas, J.R. *Un Estudio Descriptivo de Los 56 Países Productores*; El Sector Internacional de Elaboración de Aceite de Oliva: Jaén, España, 2016.
6. Carey, M. The Common Agricultural Policy's New Delivery Model Post-2020: National Administration Perspective. *EuroChoices* **2019**, *18*, 11–17. [CrossRef]
7. The Common Agricultural Policy at a Glance. Available online: https://ec.europa.eu/info/food-farming-fisheries/key-policies/common-agricultural-policy/cap-glance_en (accessed on 13 October 2020).
8. Fleitas, N.S.; Rdoríguez, R.C.; Lorenzo, M.M.G.; Quesada, A.R. Modelo de manejo de datos, con el uso de inteligencia artificial, para un sistema de información geográfica en el sector energético. *Enfoque UTE* **2016**, *7*, 95–109. [CrossRef]
9. Juarez Ruelas, J.; Trentin, G.; Heinen, M. Determinación de Evapotranspiración de Referencia a Partir de Modelos de Inteligencia Artificial. In Proceedings of the Congreso de AgroInformática (CAI)-JAIIO 47, Buenos Aires, Argentina, 9 July 2018.
10. Ramos, M.I.; Cubillas, J.J.; Jurado, J.M.; Lopez, W.; Feito, F.R.; Quero, M.; Gonzalez, J.M. Prediction of the Increase in Health Services Demand Based on the Analysis of Reasons of Calls Received by a Customer Relationship Management. *Int. J. Health Plan. Manag.* **2019**, *34*, e1215–e1222. [CrossRef]
11. van Klompenburg, T.; Kassahun, A.; Catal, C. Crop Yield Prediction Using Machine Learning: A Systematic Literature Review. *Comput. Electron. Agric.* **2020**, *177*, 105709. [CrossRef]
12. McQueen, R.J.; Garner, S.R.; Nevill-Manning, C.G.; Witten, I.H. Applying Machine Learning to Agricultural Data. *Comput. Electron. Agric.* **1995**, *12*, 275–293. [CrossRef]
13. Ahmad, L.; Nabi, F. *AGRICULTURE 5.0 Artificial Intelligence, Iot and Machine Learning*; CRC PRESS: Boca Raton, FL, USA, 2021; ISBN 978-1-00-036441-5.
14. Beulah, R. A Survey on Different Data Mining Techniques for Crop Yield Prediction. *Int. J. Comput. Sci. Eng.* **2019**, *7*, 738–744. [CrossRef]
15. Xu, X.; Gao, P.; Zhu, X.; Guo, W.; Ding, J.; Li, C.; Zhu, M.; Wu, X. Design of an Integrated Climatic Assessment Indicator (ICAI) for Wheat Production: A Case Study in Jiangsu Province, China. *Ecol. Indic.* **2019**, *101*, 943–953. [CrossRef]

16. Filippi, P.; Jones, E.J.; Wimalathunge, N.S.; Somarathna, P.D.; Pozza, L.E.; Ugbaje, S.U.; Jephcott, T.G.; Paterson, S.E.; Whelan, B.M.; Bishop, T.F. An Approach to Forecast Grain Crop Yield Using Multi-Layered, Multi-Farm Data Sets and Machine Learning. *Precis. Agric.* **2019**, *20*, 1015–1029. [CrossRef]
17. Fabio, O.; Carlo, S.; Tommaso, B.; Luigia, R.; Bruno, R.; Marco, F. Yield Modelling in a Mediterranean Species Utilizing Cause–Effect Relationships between Temperature Forcing and Biological Processes. *Sci. Hortic.* **2010**, *123*, 412–417. [CrossRef]
18. Galán, C.; García-Mozo, H.; Vázquez, L.; Ruiz, L.; De La Guardia, C.D.; Domínguez-Vilches, E. Modeling Olive Crop Yield in Andalusia, Spain. *Agron. J.* **2008**, *100*, 98–104. [CrossRef]
19. García-Mozo, H.; Perez-Badía, R.; Galán, C. Aerobiological and Meteorological Factors' Influence on Olive (*Olea europaea* L.) Crop Yield in Castilla-La Mancha (Central Spain). *Aerobiologia* **2008**, *24*, 13–18. [CrossRef]
20. Ribeiro, H.; Cunha, M.; Abreu, I. Quantitative Forecasting of Olive Yield in Northern Portugal Using a Bioclimatic Model. *Aerobiologia* **2008**, *24*, 141–150. [CrossRef]
21. Galán, C.; Vázquez, L.; Garcıa-Mozo, H.; Domınguez, E. Forecasting Olive (*Olea europaea*) Crop Yield Based on Pollen Emission. *Field Crops Res.* **2004**, *86*, 43–51. [CrossRef]
22. Ribeiro, H.; Cunha, M.; Abreu, I. Improving Early-Season Estimates of Olive Production Using Airborne Pollen Multi-Sampling Sites. *Aerobiologia* **2007**, *23*, 71–78. [CrossRef]
23. Rapoport, H.F.; Hammami, S.B.; Martins, P.; Pérez-Priego, O.; Orgaz, F. Influence of Water Deficits at Different Times during Olive Tree Inflorescence and Flower Development. *Environ. Exp. Bot.* **2012**, *77*, 227–233. [CrossRef]
24. Fornaciari, M.; Pieroni, L.; Orlandi, F.; Romano, B. A New Approach to Consider the Pollen Variable in Forecasting Yield Models. *Econ. Bot.* **2002**, *56*, 66–72. [CrossRef]
25. Oteros, J.; Orlandi, F.; García-Mozo, H.; Aguilera, F.; Dhiab, A.B.; Bonofiglio, T.; Abichou, M.; Ruiz-Valenzuela, L.; del Trigo, M.M.; Díaz de la Guardia, C.; et al. Better Prediction of Mediterranean Olive Production Using Pollen-Based Models. *Agron. Sustain. Dev.* **2014**, *34*, 685–694. [CrossRef]
26. Padilla, F.A.; Valenzuela, L.R. Forecasting Olive Crop Yields Based on Long-Term Aerobiological Data Series and Bioclimatic Conditions for the Southern Iberian Peninsula. *Span. J. Agric. Res.* **2014**, *12*, 215–224.
27. Dhiab, A.B.; Mimoun, M.B.; Oteros, J.; Garcia-Mozo, H.; Domínguez-Vilches, E.; Galán, C.; Abichou, M.; Msallem, M. Modeling Olive-Crop Forecasting in Tunisia. *Theor. Appl. Climatol.* **2017**, *128*, 541–549. [CrossRef]
28. Aguilera, F.; Ruiz-Valenzuela, L. A New Aerobiological Indicator to Optimize the Prediction of the Olive Crop Yield in Intensive Farming Areas of Southern Spain. *Agric. For. Meteorol.* **2019**, *271*, 207–213. [CrossRef]
29. López-Bernal, Á.; Fernandes-Silva, A.A.; Vega, V.A.; Hidalgo, J.C.; León, L.; Testi, L.; Villalobos, F.J. A Fruit Growth Approach to Estimate Oil Content in Olives. *Eur. J. Agron.* **2021**, *123*, 126206. [CrossRef]
30. Ramesh, D.; Vishnu Vardhan, B. Analysis of Crop Yield Prediction Using Data Mining Techniques. *Int. J. Res. Eng. Technol.* **2015**, *4*, 470–473. [CrossRef]
31. Sonnberger, H. Regression Diagnostics: Identifying Influential Data and Sources of Collinearity, by D. A. Belsley, K. Kuh and R. E. Welsch. (John Wiley & Sons, New York, 1980, Pp. Xv + 292, ISBN 0-471-05856-4, Cloth $39.95. *J. Appl. Econom.* **1989**, *4*, 97–99. [CrossRef]
32. Allen, D.M.; Foster, C.B. *Analyzing Experimental Data by Regression*; Wadsworth Pub Co: Belmont, CA, USA, 1982; ISBN 978-0-534-97963-8.
33. Cameron, A.C., Trivedi, P.K. Regression Analysis of Count Data. In *Econometric Society Monographs*; Cambridge University Press: Cambridge, UK; New York, NY, USA, 1998; ISBN 978-0-521-63201-0.
34. Meteorología, A.E.; de Agencia Estatal de Meteorología—AEMET. Gobierno de España. Available online: http://www.aemet.es/es/portada (accessed on 13 October 2020).
35. Dobson, A.J.; Barnett, A.G. *An Introduction to Generalized Linear Models*, 3rd ed.; CRC: Boca Raton, FL, USA, 2008; ISBN 978-1-58488-950-2.
36. Chalapathy, R.; Menon, A.K.; Chawla, S. Anomaly Detection Using One-Class Neural Networks. *arXiv* **2018**, arXiv:1802.06360.
37. Oza, P.; Patel, V.M. One-Class Convolutional Neural Network. *IEEE Signal Process. Lett.* **2019**, *26*, 277–281. [CrossRef]
38. Grünwald, P.D.; Myung, J.I.; Pitt, M.A. *Advances in Minimum Description Length: Theory and Applications*; Neural Information Processing Series; A Bradford Book; MIT Press: Cambridge, MA, USA, 2005; ISBN 978-0-262-07262-5.
39. Bolker, B.M.; Brooks, M.E.; Clark, C.J.; Geange, S.W.; Poulsen, J.R.; Stevens, M.H.H.; White, J.-S. Generalized Linear Mixed Models: A Practical Guide for Ecology and Evolution. *Trends Ecol. Evol.* **2009**, *24*, 127–135. [CrossRef]
40. Dibike, Y.B.; Velickov, S.; Solomatine, D.; Abbott, M.B. Model Induction with Support Vector Machines: Introduction and Applications. *J. Comput. Civ. Eng.* **2001**, *15*, 208–216. [CrossRef]
41. Cristianini, N.; Shawe-Taylor, J. *An Introduction to Support Vector Machines and Other Kernel-Based Learning Methods*; Cambridge University Press: Cambridge, MA, USA, 2000; ISBN 978-0-521-78019-3.
42. Janjanam, D.; Ganesh, B.; Manjunatha, L. Design of an Expert System Architecture: An Overview. *J. Phys. Conf. Ser.* **2021**, *1767*, 012036. [CrossRef]
43. Hardie, W. Oracle Database 19c Introduction and Overview. *White Paper*, 4 February 2019.
44. Rodriguez, J.D.; Perez, A.; Lozano, J.A. Sensitivity Analysis of K-Fold Cross Validation in Prediction Error Estimation. *IEEE Trans. Pattern Anal. Mach. Intell.* **2010**, *32*, 569–575. [CrossRef]

45. Piekutowska, M.; Niedbała, G.; Piskier, T.; Lenartowicz, T.; Pilarski, K.; Wojciechowski, T.; Pilarska, A.A.; Czechowska-Kosacka, A. The Application of Multiple Linear Regression and Artificial Neural Network Models for Yield Prediction of Very Early Potato Cultivars before Harvest. *Agronomy* **2021**, *11*, 885. [CrossRef]
46. Sharifi, A. Yield Prediction with Machine Learning Algorithms and Satellite Images. *J. Sci. Food Agric.* **2021**, *101*, 891–896. [CrossRef] [PubMed]
47. Lodolini, E.M.; Neri, D. *How Growth and Reproduction Cycles Affect Alternate Bearing in Olive*; International Symposium on Olive Growing: San Juan, Argentina, 2008; pp. 191–198.
48. Darpreet, K.; Parshant, B.; Wali, V.K.; Nirmal, S.; Arti, S.; Mudasir, I. Alternate Bearing in Olive. *Int. J. Curr. Microbiol. App. Sci.* **2018**, *7*, 2281–2297. [CrossRef]
49. Rodrigo-Comino, J.; Senciales-González, J.M.; Yu, Y.; Salvati, L.; Giménez-Morera, A.; Cerdà, A. Long-term changes in rainfed olive production, rainfall and farmer's income in Bailén (Jaén, Spain). *Euro-Mediterr. J Environ. Integr.* **2021**, *6*, 58. [CrossRef]
50. Arenas-Castro, S.; Gonçalves, J.F.; Moreno, M.; Villar, R. Projected climate changes are expected to decrease the suitability and production of olive varieties in southern Spain. *Sci. Total Environ.* **2020**, *709*, 136–161. [CrossRef]
51. Mafrica, R.; Piscopo, A.; De Bruno, A.; Poiana, M. Effects of Climate on Fruit Growth and Development on Olive Oil Quality in Cultivar Carolea. *Agriculture* **2021**, *11*, 147. [CrossRef]

agriculture

MDPI

Article

A Novel Lightweight Grape Detection Method

Shuzhi Su [1,2,*], Runbin Chen [1], Xianjin Fang [1,2], Yanmin Zhu [3], Tian Zhang [1] and Zengbao Xu [1]

1 School of Computer Science and Engineering, Anhui University of Science & Technology, Huainan 232001, China
2 Institute of Artificial Intelligence, Hefei Comprehensive National Science Center, Hefei 230031, China
3 School of Computer Mechanical Engineering, Anhui University of Science & Technology, Huainan 232001, China
* Correspondence: szsu@aust.edu.cn

Abstract: This study proposes a novel lightweight grape detection method. First, the backbone network of our method is Uniformer, which captures long-range dependencies and further improves the feature extraction capability. Then, a Bi-directional Path Aggregation Network (BiPANet) is presented to fuse low-resolution feature maps with strong semantic information and high-resolution feature maps with detailed information. BiPANet is constructed by introducing a novel cross-layer feature enhancement strategy into the Path Aggregation Network, which fuses more feature information with a significant reduction in the number of parameters and computational complexity. To improve the localization accuracy of the optimal bounding boxes, a Reposition Non-Maximum Suppression (R-NMS) algorithm is further proposed in post-processing. The algorithm performs repositioning operations on the optimal bounding boxes by using the position information of the bounding boxes around the optimal bounding boxes. Experiments on the WGISD show that our method achieves 87.7% mAP, 88.6% precision, 78.3% recall, 83.1% F1 score, and 46 FPS. Compared with YOLOx, YOLOv4, YOLOv3, Faster R-CNN, SSD, and RetinaNet, the mAP of our method is increased by 0.8%, 1.7%, 3.5%, 21.4%, 2.5%, and 13.3%, respectively, and the FPS of our method is increased by 2, 8, 2, 26, 0, and 10, respectively. Similar conclusions can be obtained on another grape dataset. Encouraging experimental results show that our method can achieve better performance than other recognized detection methods in the grape detection tasks.

Keywords: grape detection; convolutional neural network; self-attention; deep learning

Citation: Su, S.; Chen, R.; Fang, X.; Zhu, Y.; Zhang, T.; Xu, Z. A Novel Lightweight Grape Detection Method. *Agriculture* **2022**, *12*, 1364. https://doi.org/10.3390/agriculture12091364

Academic Editor: Dimitre Dimitrov

Received: 14 July 2022
Accepted: 31 August 2022
Published: 1 September 2022

1. Introduction

The grape has rich nutritional value and good taste, and it is widely popular among people. As an important part of the fruit industry, grape harvesting is labor-intensive and time-consuming [1]. Traditional manual picking is no longer sufficient to meet the needs of the fruit industry since the population is aging and the agricultural labor force is decreasing. It is urgent to develop automated grape-picking machines to harvest grapes in the field. Identifying and locating grapes in real-time is the first step to automating grape harvesting. However, traditional machine learning methods and deep learning-based grape detection methods fall short of practical requirements in speed and accuracy. Hence, developing a rapid and accurate grape detection method has great significance.

Numerous traditional methods have been proposed for grape detection in orchards in recent years. A grape image segmentation method [2] based on different color spaces was proposed and achieved a high recognition rate, but it did not consider the effect of leaf occlusion. Based on k-means clustering, Luo et al. [3] segmented stacked grapes to capture their contours and eventually achieved an 88.89% recognition rate. However, the recognition process required a great deal of time and could not meet the need for real-time grape detection. Pérez-Zavala et al. [4] used a support vector machine to classify information, combining shape and texture information, and achieved high precision and recall, but this method could not detect different varieties of grapes in complex scenes.

With the rapid development of deep learning, various convolutional neural network-based object detection methods have been applied to fruit detection tasks, with good detection results [5,6]. Some researchers investigated the effect of fruit datasets of scales and image resolutions on detection accuracy [7]. Parvathi et al. [8] took ResNet50 as the backbone network of Faster R-CNN and used it for coconut ripeness detection, achieving an accuracy of 89.4%. Fu et al. [9] proposed to remove the background from apple images by depth features before using Faster R-CNN for apple detection, which led to an average detection accuracy of 89.3%, and the detection time was 0.181 s. Faster R-CNN is a two-stage object detection method that has a high detection accuracy in fruit detection tasks [10], but the detection speed cannot meet the requirements. Some researchers proposed using one-stage object detection methods for fruit detection [11,12] to solve the problem. Aguiar et al. [13] used MobileNetV1 and Inception-V2 as networks for SSD and trained them using grape images at the different growth stages, achieving good detection results. Xiong et al. [14] added a residual network to the YOLOv3 model. They used the improved model for citrus recognition at night, resulting in a 2.27% improvement in average detection accuracy and a 26% increase in detection speed. Kateb et al. [15] proposed a modified attention mechanism based on the YOLO architecture to improve the stability of the detection model. In addition, they proposed a blackout regularization to provide better detection capability. Wu et al. [16] added a depth-separable convolution to the YOLOv3 model to improve the real-time detection performance of the model. Li et al. [17] improved the YOLOv4-tiny model by introducing the attention block and Soft-NMS algorithm into the network, increasing the identification accuracy. Meanwhile, the standard convolutions in the YOLOv4-tiny are replaced by depth-separable convolutions, improving the detection speed.

Compared with traditional methods for grape detection, the above detection methods usually utilize CNN as the backbone network to extract features and achieve good performance. However, the limited receptive field of convolutional kernels makes it hard to capture global dependency. Up to now, far too little attention has been paid to Vision Transformer. Vision Transformer models [18–20] take the self-attention to build a connection between pixels, obtaining the complete information of the image. Therefore, using Transformer as a backbone network is an intuitive way to enhance the detection performance. Additionally, feature fusion networks fuse high-level semantic information with low-level detailed information. Adding original features to the fused feature maps is an effective way to enrich the feature information. In post-processing, the bounding box with the highest confidence score is selected as the optimal bounding box. However, the low correlation between the confidence score and positioning accuracy [21] resulted in an optimal bounding box that could not surround the grape well. Thus, proposing a new Non-Maximum Suppression algorithm [22] is to be expected.

The objective of this study is to propose a novel lightweight grape detection method based on YOLOv4 [23] architecture. Inspired by the global information capture ability in Vision Transformer models, a hybrid convolution and transformer network called Uniformer [24] can be used as the backbone network of our method, which combines the advantages of convolution and self-attention to improve the feature extraction capability. To fuse more feature information, a novel feature fusion network based on the Path Aggregation Network (PANet) [25] is desired to be proposed. Compared with PANet, the feature fusion network has fewer parameters and computational complexity. Finally, we expect to propose a Relocation Non-Maximum Suppression (R-NMS) algorithm to improve the localization accuracy of the optimal bounding boxes.

2. Materials and Methods

2.1. Dataset

The Wine Grape Instance Segmentation Dataset (WGISD) [26] was used as the experimental object. The dataset consists of images captured by different camera devices, and it has 5 grape varieties and 300 images, including 240 images with 2048 × 1365 resolution and 60 images with 2048 × 1536 resolution. Some grape images under different scenes are shown in Figure 1. In the experiment, these images were divided into a training set and a test set in the ratio of 4:1 for training and testing of the model: the training set had 240 images, and the test set had 60 images. A total of 4432 annotation boxes were annotated in the WGISD. The specific information of each grape annotation box is shown in Table 1. To further demonstrate the robustness of our method, another grape dataset called wGrapeUNIPD-DL [27] was used. There are 186 images with a resolution of 4288 × 2848, 17 images with a resolution of 4608 × 3456, and 65 images with a resolution of 4032 × 3024. The detailed information about the dataset is shown in Table 2. We divided this dataset in the ratio of 4:1 to obtain 214 images in the training set and 54 images in the test set. As shown in Figure 2, the grapes in this dataset have similar colors to the leaves under occlusion scenes, which brings some difficulty to the grape detection tasks.

Table 1. The detailed information about the WGISD.

Categories	Species of Grapes					Total Number	Number of the Training Set	Number of the Test Set
	Chardonnay	Cabernet Franc	Cabernet Sauvignon	Sauvignon Blanc	Syrah			
Number of images	65	65	57	65	48	300	240	60
Number of labeled grapes	840	1069	643	1317	563	4432	3500	932

Figure 1. Grapes under different light, color, and overlapping scenes. (**a**) Grapes under sunny, cloudy, and rainy. (**b**) Green grapes, purple grapes, and overlapping grapes.

Table 2. The detailed information about the wGrapeUNIPD-DL.

Categories	Total Number	Number of the Training Set	Number of the Test Set
Number of images	268	214	54
Number of labeled grapes	2155	1744	411

Figure 2. Grapes under different light conditions.

2.2. Method

This study aims to design a novel lightweight grape detection method. As shown in Figure 3, our method has three parts. First, Uniformer was used as the backbone network to build the correlation between all pixels to enhance the feature extraction ability. Then, the BiPANet was proposed to fuse the different scale feature maps to enrich the feature information. For the last part, we proposed the R-NMS algorithm to enhance the localization accuracy of the optimal bounding boxes.

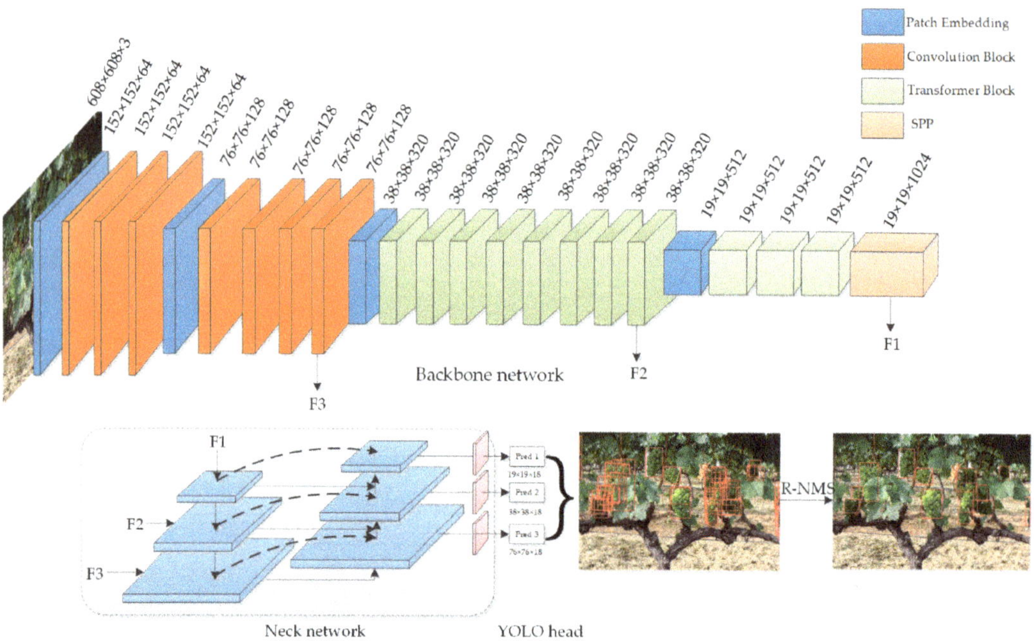

Figure 3. The network structure of our method.

2.2.1. Backbone Network

Generally, CNN is used as the backbone network of various object detection methods, and it builds global perceptual fields by stacking multiple convolutional layers. However, stacking the convolutional layers tends to make the backbone network have a large number of parameters and high computational complexity. To make the backbone network model global information efficient with fewer parameters and low computational complexity, Uniformer was used as the backbone network of our method. As shown in Table 3, Uniformer contains four stages, and each stage includes one patch-embedding layer. In the first and second stages, there are three and four Convolution Blocks, respectively. In the last two stages, there are eight and three Transformer Blocks, respectively. The patch-embedding layer is used to compress the information in the spatial dimension to the channel dimension, reducing the parameters and computational complexity for the next step.

Table 3. Model configurations for the backbone network of our method.

Stage	Input	Operation	Output
Stage 1	608 × 608 × 3	$4 \times 4, 64, stride = 4$ $\begin{bmatrix} [3 \times 3, 64], \begin{bmatrix} 1 \times 1, 64 \\ 5 \times 5, 64 \\ 1 \times 1, 64 \end{bmatrix}, \begin{bmatrix} 1 \times 1, 256 \\ 1 \times 1, 64 \end{bmatrix} \end{bmatrix} \times 3$	152 × 152 × 64
Stage 2	152 × 152 × 64	$2 \times 2, 128, stride = 2$ $\begin{bmatrix} [3 \times 3, 128], \begin{bmatrix} 1 \times 1, 128 \\ 5 \times 5, 128 \\ 1 \times 1, 128 \end{bmatrix}, \begin{bmatrix} 1 \times 1, 512 \\ 1 \times 1, 128 \end{bmatrix} \end{bmatrix} \times 4$	76 × 76 × 128
Stage 3	76 × 76 × 128	$2 \times 2, 320, stride = 2$ $\begin{bmatrix} [3 \times 3, 320], [MHSA, 320], \begin{bmatrix} 320, 1280 \\ 1280, 320 \end{bmatrix} \end{bmatrix} \times 8$	38 × 38 × 320
Stage 4	38 × 38 × 320	$2 \times 2, 512, stride = 2$ $\begin{bmatrix} [3 \times 3, 512], [MHSA, 512], \begin{bmatrix} 512, 2048 \\ 2048, 512 \end{bmatrix} \end{bmatrix} \times 3$	19 × 19 × 512

The network structure of the Convolution Block is shown in Figure 4a. It is composed of the Dynamic Position Encoding (DPE) module, the Local Attention (LA) module, and the Feed Forward Network (FFN) module. To prevent network degradation, a residual skip connection was inserted in these layers. Convolution Block is calculated as follows:

$$\begin{aligned} X &= \mathrm{DPE}(X_{in}) + X_{in} \\ Y &= \mathrm{LA}(\mathrm{Norm}(X)) + X \\ Z &= \mathrm{FFN}(\mathrm{Norm}(Y)) + Y \end{aligned} \tag{1}$$

where $X_{in} \in \mathbb{R}^{C \times H \times W}$ represents the input feature map. The feature map passes through the DPE module to obtain the output feature map $X \in \mathbb{R}^{C \times H \times W}$. The DPE module was used to encode the position of the feature maps, whose convolution kernel size is 3 × 3. Then, the feature map $X \in \mathbb{R}^{C \times H \times W}$ was used as input to the LA module. To speed up the convergence of the model during training, the Norm layer was introduced into the LA module and the FFN module. We used the Norm layer to batch normalize the feature map $X \in \mathbb{R}^{C \times H \times W}$ and then input it to the LA module to obtain the feature map $Y \in \mathbb{R}^{C \times H \times W}$. The LA module was used to extract local features, and consists of two 1 × 1 Point-Wise Convolution (PWConv) layers and a 3 × 3 Depth-Wise Convolution (DWConv) layer. The structure of PWConv-DWConv-PWConv in the LA module comes from MobileNet [28]. Compared with standard convolution, it has fewer parameters. Finally, the feature map $Y \in \mathbb{R}^{C \times H \times W}$ was processed by the FFN module to obtain the final output feature map $Z \in \mathbb{R}^{C \times H \times W}$. The FFN module uses convolution to enhance the expression ability of features, and the convolution kernel sizes are both 1 × 1.

The Convolution Block uses the DPE module, LA module, and FFN module to extract the local information. However, it cannot capture long-range dependencies. To overcome this shortcoming, the Transformer Block was proposed. As shown in Figure 4b, the Transformer Block comprises the DPE module, the Global Attention (GA) module, and the Multi-Layer Perceptron (MLP) module. The Transformer Block is calculated as follows:

$$\begin{aligned} X &= \mathrm{DPE}(X_{in}) + X_{in} \\ Y &= \mathrm{GA}(\mathrm{Norm}(X)) + X \\ Z &= \mathrm{MLP}(\mathrm{Norm}(Y)) + Y \end{aligned} \tag{2}$$

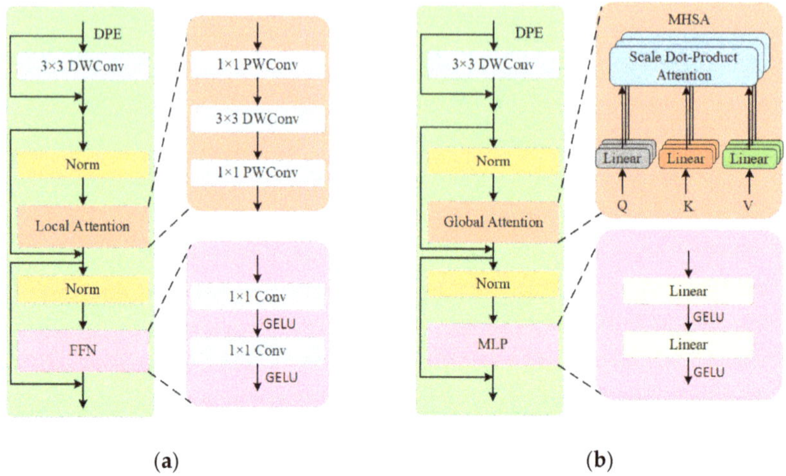

Figure 4. The network structure of the Convolution Block and the Transformer Block. (**a**) Convolution Block and (**b**) Transformer Block.

First, the DPE module encoded the position of the feature map $X_{in} \in \mathbb{R}^{C \times H \times W}$ using a 3×3 convolution kernel to obtain the feature map $X \in \mathbb{R}^{C \times H \times W}$. Then, the GA module and MLP module were used to process the feature maps $X \in \mathbb{R}^{C \times H \times W}$ and $Y \in \mathbb{R}^{C \times H \times W}$, respectively. Unlike the Convolution Block, the Transformer Block uses the GA module and MLP module to extract features and enhance them. The GA module uses self-attention to capture long-range information. Therefore, introducing the Transformer Block into the network can increase the feature extraction ability. The MLP module contains two full connection layers and two GELU functions, which can enhance the expression ability of features. Therefore, introducing the Transformer Block into the network can increase the ability to model global information.

Compared with CNN, which needs to stack multiple convolutional layers to expand the perceptual field for global information, Uniformer can model global information with only a GA module. Ultimately, Uniformer can fully extract global feature information with fewer parameters and computational complexity and provides more useful feature information with the neck network.

2.2.2. Neck Network

As the backbone network of our method, Uniformer takes advantage of self-attention to sufficiently extract feature information. The low-level feature maps contain detailed information, and the high-level feature maps consist of semantic information. To enrich the feature information, PANet was proposed and used for multi-level feature fusion. The network structure of PANet is shown in Figure 5a. There are two pathways in PANet. To obtain more semantic information in low-level feature maps, PANet up-samples the low-resolution feature map in the top-down pathway. Then, the up-sampled feature map is concatenated with the feature map of the next layer. This process is iterated until the highest-resolution map fusion progress is finished. In the bottom-up pathway, PANet down-samples the high-resolution feature map and fuses it with the previous level feature map, enriching the detailed information in the high-level feature map. However, PANet ignores that the fused feature maps contain only a small amount of original feature information.

To fuse more feature information, the effective way is to add the original feature to the fused feature maps. Therefore, we proposed a cross-layer feature enhancement strategy and further constructed BiPANet based on PANet. The network structure of BiPANet is shown in Figure 5b. Compared with PANet, there are some cross-layer feature map fusion pathways in BiPANet. These pathways are used to add the original feature to the fused feature map. As a feature reuse approach, the cross-layer feature enhancement strategy can increase feature information in the feature map with almost no increase in computational cost. Consequently, BiPANet can fuse more information at different scale feature maps, and the fused feature maps contain richly detailed and semantic information.

The structure difference between PANet and BiPANet is described above. The specific construction process of BiPANet in our method is as follows. First, to reduce the number of parameters and computational complexity of the model, we obtained PANet-Lite by reducing the number of convolutional kernels in PANet. Compared with PANet, the smaller number of parameters and the computational complexity in PANet-Lite led to a small decrease in detection performance. Then, we further constructed BiPANet based on PANet-Lite. Reusing features can reduce the effect on network performance as the number of network parameters and computational complexity decrease [29]. Although the number of parameters and computational complexity of BiPANet were reduced compared with PANet, the feature reuse characteristic in the cross-layer feature enhancement strategy improved the feature expression, enhancing the detection performance. Ultimately, compared with PANet, BiPANet has better detection performance with fewer parameters and less computational complexity. To demonstrate the effectiveness of BiPANet, the ablation experiments were performed, as discussed in Section 3.3.

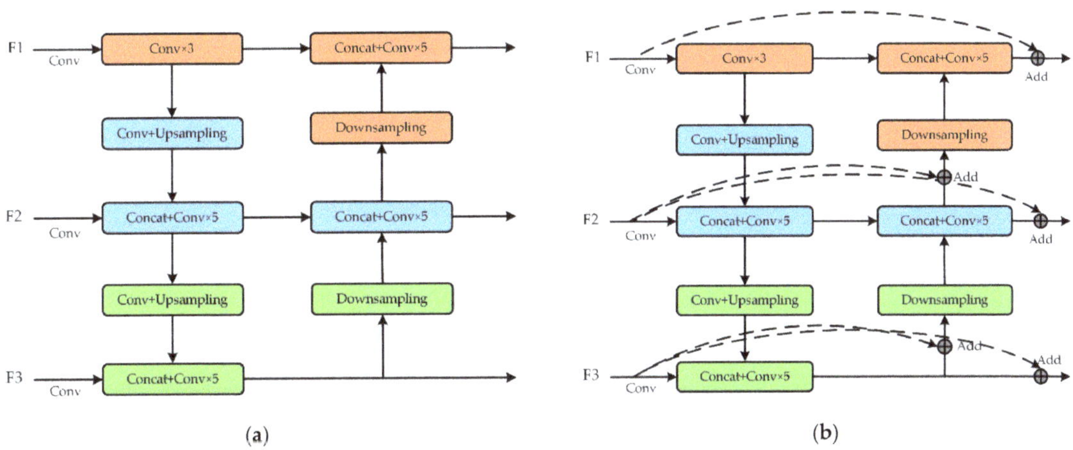

Figure 5. The network structure of PANet and BiPANet. Conv denotes convolution. (**a**) PANet and (**b**) BiPANet. Up-sampling represents reducing the scales of the feature maps to twice their original scales. In contrast to up-sampling, down-sampling represents enlarging the scales of the feature maps to twice the original scales. Add means that the values in two feature maps are summed.

2.2.3. Bounding Box Prediction

The YOLO head uses multi-scale feature maps from BiPANet to predict the position of grapes, and a large number of bounding boxes were obtained. Figure 6 shows these bounding boxes with a significant amount of redundancy. In post-processing, the Non-Maximum Suppression (NMS) algorithm was used to preserve and suppress these bounding boxes. The algorithm selects the bounding box with the highest confidence score as the optimal bounding box. The low correlation between the confidence score and the localization accuracy of bounding boxes results in the optimal bounding box selected by the NMS algorithm, which cannot surround the grapes well.

Figure 6. The predicted bounding boxes.

To improve the localization accuracy of the optimal bounding boxes, we proposed the R-NMS algorithm, which uses the location information of the bounding boxes around the optimal bounding boxes to perform the repositioning operations on the optimal bounding boxes. The new optimal bounding boxes have better localization accuracy than the former optimal bounding boxes. The Manhattan distance measures the proximity between two bounding boxes:

$$\begin{aligned} MH(u,v) &= |y_1 - q_1| + |x_1 - p_1| \\ MH(m,n) &= |y_2 - q_2| + |x_2 - p_2| \\ MH &= MH(u,v) + MH(m,n) \end{aligned} \tag{3}$$

Figure 7 shows the Manhattan distance between two bounding boxes. Since the Manhattan distance cannot accurately measure their overlap degree when their scales are significantly different, we normalized the bounding box coordinates. The process of coordinates' normalization is as follows:

$$\begin{aligned} X &= \{x_1, x_1, p_1, p_2\} \\ Y &= \{y_1, y_1, q_1, q_2\} \\ norm(x_i, y_i) &= \left(\frac{x_i - \min(X)}{\max(X) - \min(X)}, \frac{y_i - \min(Y)}{\max(Y) - \min(Y)}\right) \end{aligned} \tag{4}$$

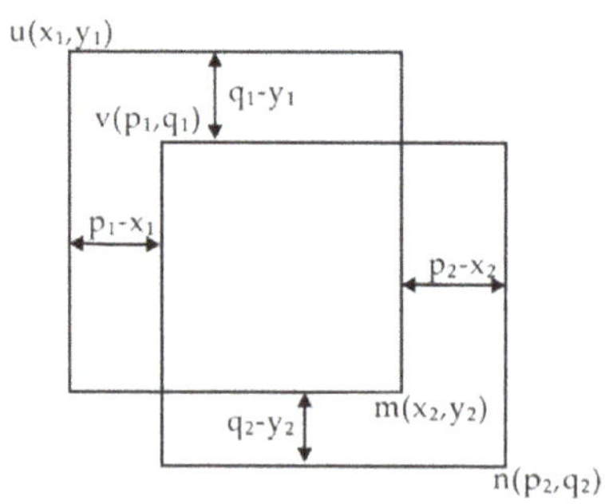

Figure 7. The Manhattan distance between two boxes. u and m are the upper-left and lower-right coordinates of box 1, respectively. v and n are the upper-left and lower-right coordinates of the box 2, respectively.

In Formula (4), the set of horizontal and vertical coordinates are denoted X and Y, respectively, and the function *norm* is used to normalize the coordinates. After normalization, the Manhattan distance between two bounding boxes was computed. Then, we used the position information of the bounding boxes around the optimal bounding to relocate the optimal bounding box. The repositioning operation proceeds as follows:

$$\begin{gathered} MH(B_i, M) \leq P_{tr}, i \in (1, 2, \ldots, n) \\ O = \frac{\sum_{i=1}^{n} |B_i - M|}{n} \\ M_R = M + O \end{gathered} \tag{5}$$

In Formula (5), M means the optimal bounding box, and $B_i, i \in (1, 2, \ldots, n)$ denotes the bounding boxes whose Manhattan distance from the optimal bounding box is lower than the threshold P_{tr}. O represents the average offset between those bounding boxes around the optimal bounding box. The optimal bounding box M is repositioned with the offset to obtain the new optimal bounding box M_R. The experimental results of the R-NMS algorithm on the WGISD are analyzed in Section 3.6. to show how well it performed on the grape detecting task.

3. Results and Discussion

3.1. Implementation Details

In our experiment, we used GPU GeForce RTX 2080Ti to accelerate model training, and the CPU was AMD Ryzen 9 3900X. The operating system was Ubuntu18.04, and the programming language was Python 3.8. The other six models in the experiments were implemented using Pytorch, and their code was obtained from GitHub (Code available at: https://github.com/bubbliiiing/ (accessed on 31 May 2022)). The main code of our model was derived from the YOLOv4 model, and the code and pre-trained model of Uniformer were obtained from the Uniformer research team (code and pre-trained model available at: https://github.com/SenseX/UniFormer (accessed on 31 May 2022)). All models used the same training parameters and were trained by pre-trained models during training, and none of them used image enhancement strategies.

The various parameters set during the model training are shown in Table 4. We used Adam gradient descent to train models. A total of 150 training epochs were set, and we divided them into two phases. The training strategy of freezing the parameters in the backbone network was adopted for the first 75 epochs. In this phase, the learning rate was set to 0.001, the weight decay rate was set to 0.0005, and the batch size was set to 8. In the last 75 epochs, the learning rate was set to 0.0001, the weight decay rate was set to 0.0005, and the batch size was set to 4. In the model test phase, IOU and classification confidence were set to 0.5 and 0.001, respectively, and the threshold, P_{tr}, in the R-NMS algorithm was set to 0.6.

Table 4. The setting of training model parameters.

Parameters	Values
Image resolution	608 × 608
Batch size 1	8
Batch size 2	4
Learning rate 1	0.001
Learning rate 2	0.0001
Weight decay rate 1	0.0005
Weight decay rate 2	0.0005
Optimizer	Adam
Epochs	150

3.2. Evaluation Metrics

We used seven evaluation metrics: precision, recall, F1 score, mean Average Precision (mAP), Frames Per Second (FPS), Params, and Floating-Point Operations (FLOPs), to evaluate the performance of our method and other popular grape detection methods.

The recall represents the proportion of positive samples with correct predictions to all positive samples and is calculated as follows:

$$Recall = \frac{TP}{TP + FN} \tag{6}$$

$$Precision = \frac{TP}{TP + FP} \tag{7}$$

TP indicates that positive samples were predicted as positive samples, and *FN* indicates that positive samples were predicted as negative samples. Precision represents the proportion of true positive samples among all predicted positive samples and is calculated as shown in Equation (7).

FP indicates predicting negative samples as negative samples. The average precision (*AP*) is used as a composite measure of the model performance. The equation of *AP* is as follows:

$$AP = \int_0^1 P(R)dR \tag{8}$$

$$mAP = \frac{\sum_{i=1}^{n} AP_i}{n} \tag{9}$$

there is only one class: the grape class, in the grape datasets, so $mAP = AP$.

*F*1 score is used as an evaluation metric for the combined measure of accuracy and recall, with the following equation:

$$F1 = \frac{2 \times Precision \times Recall}{Precision + Recall} \tag{10}$$

Params mean the number of parameters in the algorithm. Algorithm complexity is measured by GFLOPs. FPS represents the number of images detected by the algorithm per second.

3.3. Ablation Experiments

The effectiveness of Uniformer, BiPANet, and the R-NMS is examined in this section through a series of ablation experiments. Tables 5 and 6 present the experimental results of various methods on the WGISD and wGrapeUNIPD-DL. A represents the baseline YOLOv4, and B was obtained by replacing the backbone network CSPDarknet53 with Uniformer. In post-processing, the NMS algorithm was used in A and B. The evaluation metrics of B on the WGISD and wGrapeUNIPD-DL were better than A. Uniformer takes advantage of Convolution and Transformer to improve the feature extraction ability of the network. As a result, the performance of B was improved. In addition, compared to CSPDarknet53, Uniformer has fewer parameters and low computation complexity, which made the FPS of B higher than that of A. C was improved based on B. C uses PANet-Lite to replace the PANet in B. PANet-Lite was obtained by reducing the number of convolution kernels in PANet. Compared with PANet, PANet-Lite has a smaller number of parameters and lower computational complexity, resulting in the performance of C being poorer than B on the WGISD and wGrapeUNIPD-DL. To solve the problem, we proposed a cross-layer feature enhancement strategy and further constructed BiPANet. D was constructed by replacing PANet-Lite in C with BiPANet. Attributed to the cross-layer feature enhancement strategy, more feature information was fused by BiPANet, so the mAP of D was better than B and C on the WGISD and wGrapeUNIPD-DL. Compared with B, E used the R-NMS algorithm to retain and suppress the candidate bounding boxes in post-processing and achieved good performance. The algorithm used the position information of the bounding boxes around the optimal bounding boxes and performed the repositioning operations on the optimal bounding boxes to improve the localization accuracy. However, the process is time-consuming and leads to decreased FPS. Unlike D, our method used the R-NMS algorithm to retain and suppress the candidate bounding boxes. The R-NMS algorithm improved the localization accuracy of the optimal bounding boxes and reduced the redundant bounding boxes, so the mAP, precision, and recall of our method were increased. Compared with the NMS algorithm, the R-NMS algorithm increased the mAP, and the FPS decreased from 50 to 46. The 46 FPS can also meet the real-time detection requirement. For detection tasks with high real-time requirements, we can utilize tensorRT with the R-NMS algorithm implemented by using GPU to speed up the inference. Our method makes it possible to have high detection accuracy with a fast detection speed.

Table 5. Experimental results of different methods on the WGISD.

Methods	Uniformer	PANet-Lite	BiPANet	R-NMS	Precision	Recall	F1	mAP	Params	FLOPs	FPS
A	✗	✗	✗	✗	87.2%	76.5%	81.5%	86.0%	64.0 M	63.9 G	38
B	✓	✗	✗	✗	87.4%	77.5%	82.2%	87.0%	54.0 M	51.2 G	44
C	✓	✓	✗	✗	87.2%	77.2%	81.9%	86.7%	34.8 M	36.3 G	50
D	✓	✗	✓	✗	87.7%	77.7%	82.4%	87.3%	34.8 M	36.3 G	50
E	✓	✗	✗	✓	87.9%	78.1%	83.1%	87.5%	54.0 M	51.2 G	39
Our method	✓	✗	✓	✓	88.6%	78.3%	83.1%	87.7%	34.8 M	36.3 G	46

Table 6. Experimental results of different methods on the wGrapeUNIPD-DL.

Methods	Uniformer	PANet-Lite	BiPANet	R-NMS	Precision	Recall	F1	mAP	Params	FLOPs	FPS
A	✗	✗	✗	✗	84.3%	60.6%	70.5%	70.4%	64.0 M	63.9 G	38
B	✓	✗	✗	✗	85.0%	61.8%	71.6%	72.0%	54.0 M	51.2 G	44
C	✓	✓	✗	✗	85.5%	60.4%	70.8%	71.7%	34.8 M	36.3 G	50
D	✓	✗	✓	✗	85.2%	62.4%	72.0%	72.4%	34.8 M	36.3 G	50
E	✓	✗	✗	✓	85.4%	62.0%	71.8%	72.2%	54.0 M	51.2 G	39
Our method	✓	✗	✓	✓	85.7%	62.3%	72.2%	72.8%	34.8 M	36.3 G	46

3.4. Experimental Analysis of PANet and BiPANet

In this section, we further compare and analyze the performance of PANet and BiPANet. In Tables 5 and 6, B, C, and D represent the models which use PANet, PANet-Lite, and BiPANet as the feature fusion network, with Uniformer as the backbone network, respectively. PANet-Lite was obtained by reducing the number of convolutional kernels in PANet. Compared with PANet, PANet-Lite has fewer parameters and less computational complexity. Compared with B, the number of parameters and the computational complexity of C were reduced by 19.2 M and 14.9 GFLOPs, respectively. The mAP of B and C on the WGISD dataset were 87.0%, and 86.7%, respectively. Compared with B, the mAP of C on the wGrapeUNIPD-DL was reduced by 0.3%. Experiments on the WGISD and wGrapeUNIPD-DL showed that C had a poorer detection performance than B. To overcome this shortcoming, we proposed a cross-layer feature enhancement strategy and further constructed BiPANet based on PANet-Lite. As a feature reuse approach, the cross-layer feature enhancement strategy can improve the representational power of the network with almost no increase in the number of parameters and computational complexity. Therefore, the detection performance of D was increased. Compared with B and C, the mAP of D on the WGISD was increased by 0.3% and 0.6%, respectively. The mAP of D on the wGrapeUNIPD-DL was 72.4%, which is higher than B and C. Besides, the FPS of D was 50, and the FPS of B was 44. Consequently, selecting BiPANet as the feature fusion network can increase the detection accuracy and the speed.

3.5. Experimental Results and Analysis

In this section, we compare the performance of our method and some popular object detection methods on the WGISD and wGrapeUNIPD-DL. The experimental results are shown in Tables 7 and 8. Our method achieved an 87.7% mAP. Compared with Faster R-CNN, SSD, RetinaNet, YOLOv3, YOLOv4, and YOLOx, the mAP of our method was increased by 21.4%, 2.5%, 13.3%, 3.5%, 1.7%, and 0.8%, respectively. The precision was 72.8%, 85.1%, 78.9%, 83.4%, 87.2%, and 85.6% in Faster R-CNN, SSD, RetinaNet, YOLOv3, YOLOv4, and YOLOx, respectively, which are lower than our method. The number of parameters in our method, Faster R-CNN, SSD, RetinaNet, YOLOv3, YOLOv4, and YOLOx was 34.8 M, 28.3 M, 24.4 M, 36.5, 61.6 M, 64.0 M, and 54.2 M, respectively. On the wGrapeUNIPD-DL, the mAP of our method, Faster R-CNN, SSD, RetinaNet, YOLOv3, YOLOv4, and YOLOx was 72.8%, 43.2%, 56.2%, 45.7%, 65.6%, 70.4%, and 72.6%, respectively. The number of parameters in our method was not the least among all methods, but the precision, recall, F1, and mAP of our method outperformed the other methods. Besides, among all the compared methods, the computational complexity of our method was the smallest, only 36.3 G FLOPs, and the FPS of our method was one of the highest.

The detection results obtained by our method on the WGISD and wGrapeUNIPD-DL are shown in Figures 8 and 9, and our method could accurately identify and locate most grapes. Table 9 shows the detailed detection results of various methods for the images in Figures 8 and 9. There are 75 ground-truth bounding boxes in Figure 8, and our method detected 73 TP bounding boxes. The number of TP bounding boxes detected by Faster R-CNN, SSD, RetinaNet, YOLOv3, YOLOv4, and YOLOx was 68, 69, 66, 68, 70, and 72, respectively. The TN and FN bounding boxes detected by our method were 1 and 3, which is better than other methods. There are 21 ground-truth bounding boxes in Figure 9. Our method and YOLOx detected 39 ground-truth bounding boxes, and no TN and FN bounding boxes were found in the detection results. However, there were FN and PN bounding boxes in the detection results of the methods such as Faster R-CNN. The detection results in Figures 8 and 9 demonstrate that our method achieved good detection performance.

In our method, Uniformer was used as the backbone network. The backbone networks of the other methods are based on CNN, and the poor ability of CNN to model global information prevents them from fully extracting feature information. From the ablation experiment results in Section 3.3, when the backbone network was Uniformer, the mAP was increased by 1%, and the number of parameters and computational complexity were decreased by 10 M and 12.7 GFLOPs, respectively. Uniformer can capture long-range information to improve the feature extraction capability, which led to a significantly good performance in the detection accuracy of our method on the grape detection task. According to the experimental results in Table 5, due to the small number of convolutional kernels in BiPANet and the cross-layer feature enhancement strategy, BiPANet used with Uniformer improved the mAP from 87.0% to 87.3%, and the number of parameters and computational complexity were reduced by 19.2 M and 14.9 GLOPs, respectively. Additionally, the R-NMS algorithm used the position information of the bounding boxes around the optimal bounding boxes to perform repositioning operations on the optimal bounding boxes, which led to an improvement in the localization accuracy of the optimal bounding boxes. As a result, the precision, recall, and mAP of our method were increased. According to the above results, it can be concluded that our method is efficient and lightweight.

Figure 8. The detection results of our method on the WGISD. (**a**) NMS; (**b**) R-NMS.

Table 7. The experimental results on the WGISD using different methods.

Methods	Precision	Recall	F1	mAP	Params	FLOPs	FPS
Faster R-CNN	72.8%	51.6%	60.4%	66.3%	28.3 M	196.5 G	20
SSD	85.1%	75.9%	80.2%	85.2%	24.4 M	124.6 G	46
RetinaNet	78.9%	63.8%	70.1%	74.4%	36.5 M	74.7 G	36
YOLOv3	83.4%	77.7%	80.4%	84.2%	61.6 M	70.0 G	44
YOLOv4	87.2%	76.5%	81.5%	86.0%	64.0 M	63.9 G	38
YOLOx	85.6%	80.4%	82.9%	86.9%	54.2 M	70.1 G	44
Our method	88.6%	78.3%	83.1%	87.7%	34.8 M	36.3 G	46

Figure 9. The detection results of our method on the wGrapeUNIPD-DL.

Table 8. The experimental results on the wGrapeUNIPD-DL using different methods.

Methods	Precision	Recall	F1	mAP	Params	FLOPs	FPS
Faster R-CNN	65.7%	40.2%	49.9%	43.2%	28.3 M	196.5 G	20
SSD	72.8%	51.9%	60.6%	56.2%	24.4 M	124.6 G	46
RetinaNet	67.7%	46.1%	54.9%	45.7%	36.5 M	74.7 G	36
YOLOv3	83.0%	53.6%	65.1%	65.6%	61.6 M	70.0 G	44
YOLOv4	84.3%	60.6%	70.5%	70.4%	64.0 M	63.9 G	38
YOLOx	79.6%	65.4%	71.8%	72.6%	54.2 M	70.1 G	44
Our method	85.7%	62.3%	72.2%	72.8%	34.8 M	36.3 G	46

Table 9. The detection results on the WGISD and wGrapeUNIPD-DL using different methods.

Methods	Grape Detection Results in Figure 8			Grape Detection Results in Figure 9		
	Number of TP Bounding Boxes	Number of FP Bounding Boxes	Number of FN Bounding Boxes	Number of TP Bounding Boxes	Number of FP Bounding Boxes	Number of FN Bounding Boxes
Faster R-CNN	68	22	41	15	11	10
SSD	69	6	10	12	0	0
RetinaNet	66	17	26	14	1	0
YOLOv3	68	8	7	17	0	0
YOLOv4	70	3	5	19	1	2
YOLOx	72	5	3	19	0	0
Our method	73	1	3	19	0	0

3.6. Experimental Analysis of the R-NMS Algorithm

This section analyzes the detection results under the R-NMS algorithm on the WGISD. The performance of the R-NMS algorithm depends on the setting of the threshold, P_{tr}. The experimental results under the R-NMS algorithm with different thresholds, P_{tr}, on the WGISD are shown in Figure 10. When the threshold, P_{tr}, was 0, the R-NMS algorithm degenerated to the NMS algorithm. As the threshold value increased, more valid position information of surrounding bounding boxes was obtained and used for the optimal bounding boxes' repositioning, improving the localization accuracy. The precision, recall, and mAP of our method were improved in this range. When the threshold, P_{tr}, was greater than 0.6, the number of bounding boxes surrounding the optimal bounding box increased. However, the added bounding boxes were far away from the optimal bounding box, and their position information was invalid and had a suppressing effect on the positioning accuracy. Consequently, the localization accuracy of the optimal bounding boxes by performing repositioning operations using this location information was reduced, which led to a decrease in the precision, recall, and mAP. To achieve a better detection performance, when we use the R-NMS algorithm in post-processing, the threshold, P_{tr}, should be set to about 0.5 if possible.

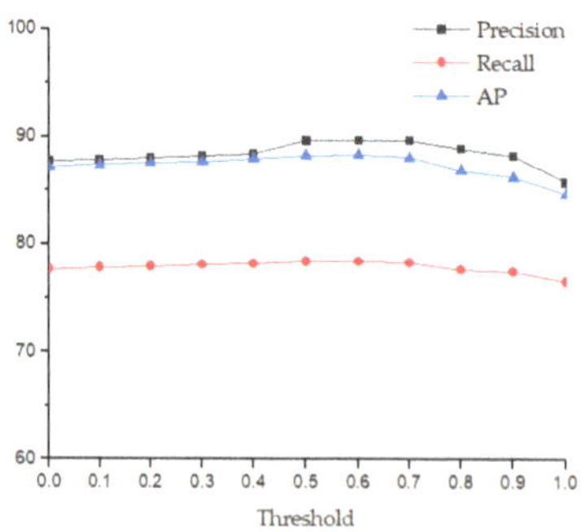

Figure 10. Detection accuracy under different thresholds,P_{tr}.

To obtain good performance, the threshold, P_{tr}, was set to 0.6. Compared with the NMS algorithm, the optimal bounding box obtained by the R-NMS algorithm had a higher localization accuracy and enclosed the grapes well. The result in Figure 11a was obtained by the NMS algorithm, and the result in Figure 11b was obtained by the R-NMS algorithm. The localization accuracy of the bounding box in Figure 11a was significantly better than that in Figure 11b. Figure 12 shows the detection results under the NMS and R-NMS algorithms. There was a redundant bounding box in Figure 12a. In the bounding box suppression phase, the IOU between the redundant bounding box and the optimal bounding box was less than the threshold, P_{tr}. Therefore, the redundant bounding box was not suppressed by the optimal bounding box under the NMS algorithm. In Figure 12b, the R-NMS algorithm performed a repositioning operation on the optimal bounding box by using the position information around the optimal bounding box and obtaining a new optimal bounding box. Since the IOU between the new optimal bounding box and the redundant bounding box was greater than the threshold, P_{tr}, the redundant bounding box was suppressed. Consequently, the R-NMS algorithm can improve the localization accuracy of the optimal bounding boxes and reduce redundant bounding boxes.

(a) (b)

Figure 11. The detection results under different non-maximum suppression algorithms. (**a**) NMS; (**b**) R-NMS.

(a) (b)

Figure 12. The detection results under different non-maximum suppression algorithms. (**a**) NMS; (**b**) R-NMS.

4. Conclusions

In this study, a novel lightweight grape detection method was proposed. The backbone network of our method is a hybrid Convolution and Transformer network called Uniformer, which builds global correlations between all pixels. Compared with CNN, Uniformer fully used the advantages of CNN and Transformer to improve the feature extraction capability with a small number of parameters and less computational complexity. Then, to embed more feature information in fused feature maps, the method provided a novel cross-layer feature enhancement strategy and further constructed BiPANet based on PANet. Due to the low correlation between localization accuracy and confidence, the optimal bounding boxes selected by NMS had low localization accuracy. To solve this problem, the method proposed the R-NMS algorithm. The algorithm used the position information of the bounding boxes around the optimal bounding boxes to perform the repositioning operations on the optimal bounding boxes, obtaining new optimal bounding boxes with high localization accuracy. Special ablation experiments were designed to verify the effectiveness of Uniformer, BiPANet, and the R-NMS algorithm in our method. The experiment results showed that all the algorithms can improve the detection performance. Besides, we analyzed the effect of the R-NMS algorithm from the extensive experimental results. According to those experiments, the R-NMS algorithm can improve the localization accuracy of the bounding boxes. Experimental results on the WGISD demonstrated that our method achieved an 87.7% mAP and 46 FPS, indicating that our method is an effective grape detection method with good accuracy and a fast detection speed.

Author Contributions: Funding acquisition, X.F.; Methodology, R.C.; Software, R.C. and Z.X.; Validation, Y.Z.; Writing—original draft, R.C.; Writing—review & editing, S.S. and T.Z. All authors have read and agreed to the published version of the manuscript.

Funding: This research was funded by the Institute of Energy, Hefei Comprehensive National Science Center (under Grant No. 19KZS203), the National Natural Science Foundation of China (No. 61806006), and the China Postdoctoral Science Foundation (No. 2019M660149).

Institutional Review Board Statement: Not applicable.

Informed Consent Statement: Not applicable.

Data Availability Statement: Publicly available datasets were analyzed in this study. The data can be found here: https://github.com/thsant/wgisd, accessed on 20 April 2022.

Conflicts of Interest: The authors declare no conflict of interest.

References

1. Peng, Y.; Wang, A.; Liu, J.; Faheem, M. A comparative study of semantic segmentation models for identification of grape with different varieties. *Agriculture* **2021**, *11*, 997. [CrossRef]
2. Ma, B.; Jia, Y.; Mei, W.; Gao, G.; Lv, C.; Zhou, Q. Study on the recognition method of grape in different natural environment. *Mod. Food Sci. Technol.* **2015**, *31*, 145–149. [CrossRef]
3. Luo, L.; Zou, X.; Wang, C.; Chen, X.; Yang, Z.; Situ, W. Recognition method for two overlapping and adjacent grape clusters based on image contour analysis. *Trans. Chin. Soc. Agric. Mach.* **2017**, *48*, 15–22. [CrossRef]
4. Pérez-Zavala, R.; Torres-Torriti, M.; Cheein, F.A.; Troni, G. A pattern recognition strategy for visual grape bunch detection in vineyards. *Comput. Electron. Agric.* **2018**, *151*, 136–149. [CrossRef]
5. Yan, B.; Fan, P.; Lei, X.; Liu, Z.; Yang, F. A real-time apple targets detection method for picking robot based on improved YOLOv5. *Remote Sens.* **2021**, *13*, 1619. [CrossRef]
6. Liu, F.; Liu, Y.; Lin, S.; Guo, W.; Xu, F.; Zhang, B. Fast recognition method for tomatoes under complex environments based on improved YOLO. *Trans. Chin. Soc. Agric. Mach.* **2020**, *51*, 229–237. [CrossRef]
7. Wang, X.; Tang, J.; Whitty, M. Data-centric analysis of on-tree fruit detection: Experiments with deep learning. *Comput. Electron. Agric.* **2022**, *194*, 106748. [CrossRef]
8. Parvathi, S.; Selvi, S.T. Detection of maturity stages of coconuts in complex background using Faster R-CNN model. *Biosyst. Eng.* **2021**, *202*, 119–132. [CrossRef]
9. Fu, L.; Majeed, Y.; Zhang, X.; Karkee, M.; Zhang, Q. Faster R–CNN–based apple detection in dense-foliage fruiting-wall trees using RGB and depth features for robotic harvesting. *Biosyst. Eng.* **2020**, *197*, 245–256. [CrossRef]

10. Gao, F.; Fu, L.; Zhang, X.; Majeed, Y.; Li, R.; Karkee, M.; Zhang, Q. Multi-class fruit-on-plant detection for apple in SNAP system using Faster R-CNN. *Comput. Electron. Agric.* **2020**, *176*, 105634. [CrossRef]
11. Peng, H.; Huang, B.; Shao, Y.; Li, Z.; Zhang, C.; Chen, Y.; Xiong, J. General improved SSD model for picking object recognition of multiple fruits in natural environment. *Trans. Chin. Soc. Agric. Eng.* **2018**, *34*, 155–162. [CrossRef]
12. Zhao, D.; Wu, R.; Liu, X.; Zhao, Y. Apple positioning based on YOLO deep convolutional neural network for picking robot in complex background. *Trans. Chin. Soc. Agric. Eng.* **2019**, *35*, 172–181. [CrossRef]
13. Aguiar, A.S.; Magalhães, S.A.; Dos Santos, F.N.; Castro, L.; Pinho, T.; Valente, J.; Martins, R.; Boaventura-Cunha, J. Grape bunch detection at different growth stages using deep learning quantized models. *Agronomy* **2021**, *11*, 1890. [CrossRef]
14. Xiong, J.; Zheng, Z.; Liang, J.E.; Zhong, Z.; Liu, B.; Sun, B. Citrus detection method in night environment based on improved YOLO v3 Network. *Trans. Chin. Soc. Agric. Mach.* **2020**, *51*, 199–206. [CrossRef]
15. Kateb, F.A.; Monowar, M.M.; Hamid, A.; Ohi, A.Q.; Mridha, M.F. FruitDet: Attentive feature aggregation for real-time fruit detection in orchards. *Agronomy* **2021**, *11*, 2440. [CrossRef]
16. Wu, X.; Qi, Z.; Wang, L.; Yang, J.; Xia, X. Apple detection method based on light-YOLOv3 convolutional neural network. *Trans. Chin. Soc. Agric. Mach.* **2020**, *51*, 17–25. [CrossRef]
17. Li, H.; Li, C.; Li, G.; Chen, L. A real-time table grape detection method based on improved YOLOv4-tiny network in complex background. *Biosyst. Eng.* **2021**, *212*, 347–359. [CrossRef]
18. Carion, N.; Massa, F.; Synnaeve, G.; Usunier, N.; Kirillov, A.; Zagoruyko, S. End-to-end object detection with transformers. In Proceedings of the European Conference on Computer Vision, Glasgow, UK, 23–28 August 2020; pp. 213–229.
19. Wang, W.; Xie, E.; Li, X.; Fan, D.-P.; Song, K.; Liang, D.; Lu, T.; Luo, P.; Shao, L. Pyramid vision transformer: A versatile backbone for dense prediction without convolutions. In Proceedings of the IEEE/CVF International Conference on Computer Vision, Virtual, 11–17 October 2021; pp. 568–578.
20. Wu, H.; Xiao, B.; Codella, N.; Liu, M.; Dai, X.; Yuan, L.; Zhang, L. Cvt: Introducing convolutions to vision transformers. In Proceedings of the IEEE/CVF International Conference on Computer Vision, Virtual, 11–17 October 2021; pp. 22–31.
21. Jiang, B.; Luo, R.; Mao, J.; Xiao, T.; Jiang, Y. Acquisition of localization confidence for accurate object detection. In Proceedings of the European Conference on Computer Vision, Munich, Germany, 8–14 September 2018; pp. 784–799.
22. Neubeck, A.; Van Gool, L. Efficient non-maximum suppression. In Proceedings of the 18th International Conference on Pattern Recognition, Hong Kong, China, 20–24 August 2006; pp. 850–855.
23. Bochkovskiy, A.; Wang, C.Y.; Liao, H.Y.M. Yolov4: Optimal speed and accuracy of object detection. *arXiv* **2020**, arXiv:2004.10934.
24. Li, K.; Wang, Y.; Zhang, J.; Gao, P.; Song, G.; Liu, Y.; Li, H.; Qiao, Y. Uniformer: Unifying convolution and self-attention for visual recognition. *arXiv* **2022**, arXiv:2201.09450.
25. Liu, S.; Qi, L.; Qin, H.; Shi, J.; Jia, J. Path aggregation network for instance segmentation. In Proceedings of the IEEE Conference on Computer Vision and Pattern Recognition, Salt Lake City, UT, USA, 18–22 June 2018; pp. 8759–8768.
26. Santos, T.; de Souza, L.; dos Santos, A.; Sandra, A. Embrapa Wine Grape Instance Segmentation Dataset–Embrapa WGISD. Zenodo. 2019. Available online: https://doi.org/10.5281/zenodo.3361736 (accessed on 23 June 2021).
27. Sozzi, M.; Cantalamessa, S.; Cogato, A.; Kayad, A.; Marinello, F. wGrapeUNIPD-DL: An open dataset for white grape bunch detection. *Data Brief.* **2022**, *43*, 108466. [CrossRef]
28. Sandler, M.; Howard, A.; Zhu, M.; Zhmoginov, A.; Chen, L.-C. Mobilenetv2: Inverted residuals and linear bottlenecks. In Proceedings of the IEEE Conference on Computer Vision and Pattern Recognition, Salt Lake City, UT, USA, 18–22 June 2018; pp. 4510–4520.
29. Li, S.; Zhang, G.; Luo, Z.; Liu, J. Dfan: Dual feature aggregation network for lightweight image super-resolution. *Wirel. Commun Mob. Comput.* **2022**, *2022*, 1–13. [CrossRef]

agriculture

Article

Identification of Buffalo Breeds Using Self-Activated-Based Improved Convolutional Neural Networks

Yuanzhi Pan [1,2,3,*], Hua Jin [4], Jiechao Gao [5,6] and Hafiz Tayyab Rauf [7]

1 Faculty of Business and Economics, The University of Hong Kong, Hong Kong 999077, China
2 School of Electronic Information and Electrical Engineering, Shanghai Jiao Tong University, Shanghai 200030, China
3 Artificial Intelligence Lab, Zhenjiang Hongxiang Automation Technology Co., Ltd., Zhenjiang 212050, China
4 School of Computer Science and Communication Engineering, Jiangsu University, Zhenjiang 212013, China
5 Department of Computer Science, University of Virginia, Charlottesville, VA 22903, USA
6 Department of Electrical Engineering, Columbia University, New York City, NY 10027, USA
7 Centre for Smart Systems, AI and Cybersecurity, Staffordshire University, Stoke-on-Trent ST4 2DE, UK
* Correspondence: yzpan@connect.hku.hk or yzpan1984@gmail.com

Abstract: The livestock of Pakistan includes different animal breeds utilized for milk farming and exporting worldwide. Buffalo have a high milk production rate, and Pakistan is the third-largest milk-producing country, and its production is increasing over time. Hence, it is essential to recognize the best Buffalo breed for a high milk- and meat yield to meet the world's demands and breed production. Pakistan has the second-largest number of buffalos among countries worldwide, where the Neli-Ravi breed is the most common. The extensive demand for Neli and Ravi breeds resulted in the new cross-breed "Neli-Ravi" in the 1960s. Identifying and segregating the Neli-Ravi breed from other buffalo breeds is the most crucial concern for Pakistan's dairy-production centers. Therefore, the automatic detection and classification of buffalo breeds are required. In this research, a computer-vision-based recognition framework is proposed to identify and classify the Neli-Ravi breed from other buffalo breeds. The proposed framework employs self-activated-based improved convolutional neural networks (CNN) combined with self-transfer learning. Moreover, feature maps extracted from CNN are further transferred to obtain rich feature vectors. Different machine learning (Ml) classifiers are adopted to classify the feature vectors. The proposed framework is evaluated on two buffalo breeds, namely, Neli-Ravi and Khundi, and one additional target class contains different buffalo breeds collectively called Mix. The proposed research achieves a maximum of 93% accuracy using SVM and more than 85% accuracy employing recent variants.

Keywords: buffalo breeds; Neural Networks; Self Activated CNN; deep learning

Citation: Pan, Y.; Jin, H.; Gao, J.; Rauf, H.T. Identification of Buffalo Breeds Using Self-Activated-Based Improved Convolutional Neural Networks. *Agriculture* **2022**, *12*, 1386. https://doi.org/10.3390/agriculture12091386

Academic Editor: Dimitre Dimitrov

Received: 25 July 2022
Accepted: 29 August 2022
Published: 3 September 2022

1. Introduction

The livestock of Pakistan includes goats, sheep, buffalos, cattle, and their variants. They have been used for milk, meat, wool, and hidden purposes. Rearing animals for milk and meat has always been a desirable task for our farmers [1]. However, lots of aspects of red meat are still to be illuminated, and there is a lot more to do in that sector; on the other hand, the milk sector of Pakistan is prosperous so far, and the need of the hour is to sustain this opulence and success [2]. According to the IFCN (international farms comparison network) report, Pakistan is the third-largest milk-producing country globally, and the figures are expected to mount in upcoming years [3]. Interestingly, Pakistan's buffalos carry much of this burden as they constitute about 65% of total milk production in the entire country. Buffalos are the most widely used animal in South Asia, especially in Pakistan; Buffalo have been bred here for a long time, and Pakistan is fortunately rich in many world-class breeds of buffalo, such as Nili Ravi and Kundi [3].

Before the late nineties, a revolution occurred in Pakistan's milk sector before almost 60% of animals were kept by rural people in small units. The grazing fulfilled the feed requirements and other green fodder [3]. In the last two decades, the dairy sector of Pakistan has witnessed rapid growth; lots of dairy farms have been established, and especially Kundi is a vital player [4]. Worldwide, there is a dearth of buffalos; previously, the developed world has primarily considered cattle as a principal source for getting milk requirements. The USA's milk industry relies chiefly on dairy cattle, but now it is moving forward toward the utilization of Buffalo milk [5]. There are many reasons behind this shift, such as the low percentage of fat in cattle milk (buffalos have twice as much as cattle), its low viscosity (buffalo milk is much more viscous, thick, and creamy than that of cattle), and the inability of it to be rendered into some refining kind of cheese such as Mozzarella cheese and some varieties of Cheddar cheese, etc. To manufacture these from cattle milk, millions of dollars have been invested, but the results have been negative [6].

These buffalo breeds tolerate wide variations in fodder supply, feed quality, and adaptation to low-input monitoring systems. Their home track is in central Punjab's canal-irrigated areas, where they are fed plenty of green fodder [7]. They produce much less milk in their home tract despite their capacity, which may be due to (1) a long calving interval, (2) silent heat, (3) late maturity, and (4) a lack of attention paid to the past to improvement by selection and progeny testing [7].

The classification of native buffalo breeds has traditionally been based on morphological differences, but regional variations are insufficient to distinguish breeds closely related [8]. For their effective and meaningful improvement and conservation, the detailed characterization and evaluation of differences among these breeds using modern technology tools is required.

The technological developments in the field of artificial intelligence (AI) [9], blockchain [10], IoT [11], cloud technologies [12], and data science [13] are going to transform the practice of traditional dairy farming into smart dairy farming. The latest challenge in dairy farming, which is in focus and an essential aspect of dairy farming, is to achieve animal health and welfare at the best possible standards, keeping in mind the high-performance potential of animals [14]. It is necessary for animal sciences to meet these challenges faced by the farmer's introduction of technological advancement.

Artificial neural networks (ANNs) are one of the advanced computing methods used mainly for classification and prediction [15]. ANN is a multi-layer connected neural network that looks like a web. The basic ANN unit consists of an input, a hidden layer, and an output layer. Each layer has neurons or nodes, which are interconnected to each other in the next layer. Each node in the hidden layer has some weight and bias values. ANN is a nonlinear technique that can take any input. Some of the advantages of ANN include its ability to learn and model nonlinear and complex associations between inputs and outputs, as well as its ability to generalize as it can derive unperceived connections after learning from the primary inputs and their correlation with the outputs, making the model capable of generalizing and predicting undiscovered data [16].

Pakistan is the world's second-most buffalo-populated country, accounting for 16.83% of the global buffalo population, with Nili-Ravi constituting the majority [17]. Due to extensive crossbreeding, the Nili and Ravi breeds have been combined into a single Nili-Ravi since the 1960s [18]. Compared to native breeds, this single Nili-Ravi is now considered the best-performing breed for meat and milk production. However, the differentiation between the Nili-Ravi breed and other native breeds is complex, and only experts can identify it. Visual feature-based classification is an exciting research topic with emerging AI-based techniques. The visual features include the shape of the head, horns, texture, and color, and the formation of the udder, foot, and tail [18].

Correct identification of the Nili-Ravi breed will help improve the number of quality animals, ultimately helping to elevate poverty in rural areas of Punjab, where most of the house expenses are met by selling buffalo milk. Using morphological markers and visual features to identify animals remains an expensive and time-consuming manual task by researchers. As a result, we demonstrate that a deep convolutional neural network can detect different breeds.

1.1. Objectives

The objective of this study is to propose a computer-vision-based recognition system for the identification and classification of Neli-Ravi buffalo breeds from other buffalo breeds. The proposed framework used self-activated-based enhanced CNN coupled with self-transfer learning. Information-oriented feature vectors were obtained by transferring feature maps and then classified using machine learning-based classifiers. The proposed method was tested on Neli-Ravi, Khundi, and Mix (a group of several breeds).

1.2. Related Work

As per our knowledge, no work has been done so far to segregate the Neli-Ravi breed from other buffalo breeds using machine learning and deep learning. The morphological features of Nili-Ravi buffalo are given with details in Table 1.

Table 1. Morphological features of Nili-Ravi buffalo.

Sr #	Characters	Description	Image
1	Marking	White markings are often found on hind legs, fore legs, white spots on forehead, muzzle, and white switch on tail (30–40%)	
2	Eye	Eyes are prominent, especially in the females. They are usually walled eyes (73.5%).	
3	Body color	The color is black but brown color is not uncommon (10–15 percent).	
4	Horns	Horns are short, broad at the base, and closely curled behind the base. A few animals (1–2 percent) have loose hanging horns.	
5	Tail	The tail is well set on, broad at the base, and tapering at the ends at the fetlock or just below it in a big tuft of hair, which may hand on the ground with a white switch.	

Table 1. *Cont.*

Sr #	Characters	Description	Image
6	Body	Nili-Ravi buffalos are large size and have deep and low-set frames (massive and wedge shaped).	
7	Head	The head is long, convex in the upper third but with a depression between the orbits. It shows good depth from the angle of the jaw to the base of the horns.	
8	Neck	The neck is long and thin in females, while it is thick and powerful in the male. There is no dewlap, and the umbilical fold is small.	
9	Udder	The teats are long, even squarely placed.	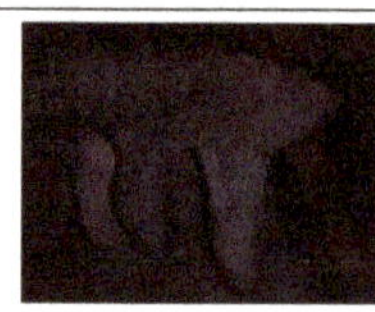

The authors previously contributed in [19], which included establishing an experimental image processing framework on a sheep farm, creating a database containing 1642 sheep images of four breeds recorded on a farm and labeled by an expert with their breed, and training a sheep breed classifier using computer vision and machine learning to obtain an overall precision of 95.8% with 1.7 mean difference. Their method may help sheep farmers distinguish between breeds efficiently, allowing for more accurate meat yield estimation and cost control.

The authors in [20] presented a deep-learning-based method for identifying individual cattle based on their predominant muzzle point image sequence features, which tackles the issue of missing or switched livestock and false insurance claims. They developed a set of muzzle-point images that were not publicly accessible. They also used a deep-learning-based CNN to extract meaningful shape information and visualize a cattle's muzzle point picture. The derived function of muzzle point images is encoded using the stacked denoising auto-encoder technique.

The classification of horse breeds in natural scene images is addressed in [21]. The author collected a data set by exploring the internet. The dataset includes 1693 photos of six different horse breeds. For the classification and identification of horse breeds, deep learning techniques, specifically, CNNs, were used. The proposed dataset is used to train and fine-tune well-known deep CNN architectures pre-trained on the ImageNet dataset. As pre-trained CNN classifiers, VGG architectures with 16 and 19 layers, InceptionV3, ResNet50, and Xception, are used. The average classification accuracy of ResNet50 was 95.90%.

In the other study [22], the authors tried to classify six distinct goat breeds using mixed-breed goat photos. The images of goat breeds were taken at various recorded goat herds in India. Without causing discomfort to the livestock, nearly 2000 digital images of individual goats were collected in limited and unregulated conditions. For the efficient recognition and localization of goat breeds, a pre-trained deep-learning-based object-detection model called Faster R-CNN was fine-tuned using transfer-learning on the acquired images. The fine-tuned model can locate the goat and identify its breed in the image. The model was assessed using the Pascal VOC object-detection assessment metrics.

The authors compare various deep learning models and propose an Android application [23] that uses a smartphone camera to determine a given cat's position and breed. The finalized model's accuracy rate was 81.74%. Another research [24] aims to propose an AI-driven system for fish species identification based on the CNN framework. The proposed CNN architecture has 32 deep layers, each very deep, to extract useful and distinguishing features from the image. The VGGNet architecture is subjected to deep supervision to improve classification efficiency by incorporating various convolutional layers into each level's training. They built the Fish-Pak dataset to test the performance of the proposed 32-layer CNN architecture.

The authors used deep learning models [25] to assess the absolute maximum accuracy in detecting the Canchim breed species, which is virtually identical to the Nelore breed. They also mapped the optimal ground sample distance (GSD) for detecting the Canchim breed. There were 1853 UAV images with 8629 animal samples in the experiments, and 15 different CNN architectures were tested. A total of 900 models were trained (2 datasets, 10 spatial resolutions, K-fold cross-validation, and 15 CNN architectures), allowing for a thorough examination of the factors that influence cattle detection using aerial images captured by UAVs. Several CNN architectures were found to be robust enough to accurately detect animals in aerial images even under less-than-ideal conditions, suggesting the feasibility of using UAVs for cattle monitoring.

Another study [26] suggested using CNN to develop a recognition system for the Pantaneira cattle breed. The researchers examined 51 animals from the Aquidauana Pantaneira cattle center (NUBOPAN). The area is situated in Brazil's Midwest area. Four surveillance cameras were mounted on the walls and took 27,849 photographs of the Pantaneira cattle breed from multiple distances and locations.

2. Materials and Methods

Deep-learning-based identification and recognition approaches are widely used in medical, natural, animal, and many other fields. The deep convolutional network inspires researchers to identify and classify different instances based on the visual features [24,27]. The proposed research offers a self-activated CNN for the identification of buffalos breeds. The proposed CNN contains a 23-layer architecture with 5 blocks of convolution, batch normalization, reLU, and a max-pooling layer. Feature vectors extracted from CNN layers without transfer learning and self-activation reach up to 77.61% validation accuracy. However, applying transfer learning from the fully connected layer of the proposed CNN tends to increase the overall performance. All of the extracted feature vectors are employed for classification using classical ML classifiers with various test and train ratios such as 50–50, 60–40, and 70–30. The K-fold cross-validation techniques with 5 and 10 folds are also used to validate different ML classifiers' performance. The primary steps adopted in the proposed research are shown in Figure 1.

Figure 1 determines that the input data are augmented with rotation, clipping, and the pixel range. The augmented data are further transmitted to the proposed CNN to classify buffalos breeds. To extract the rich feature vectors, we applied transfer learning to the 21st layer of CNN, which splits into various ratios to classify the breeds.

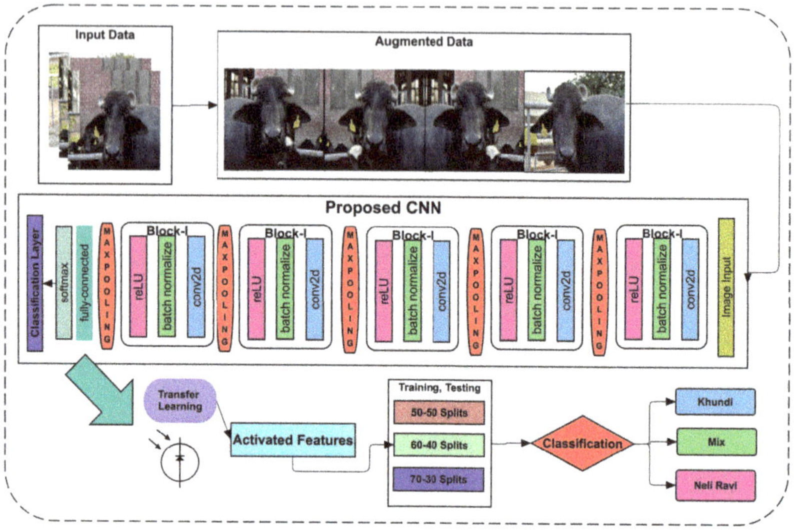

Figure 1. Proposed framework used to identify and segregate Neli-Ravi breed from all other breeds.

2.1. Data-Set Description

We collected data sets from Buffalo research center Pakistan. The acquired data-set is publically available at the data-set repository called Mendeley [28]. We normalized and preprocessed the data, including all class labels for self-activated CNN training. The size of the data set and details are shown in Table 2.

Table 2. Data-set description.

Category	Frequency
Khundi	150
Mix	154
Neli-Ravi	146
Total	450
Format	JPG
Dimension	$256 \times 256 \times 3$

Table 2 has shown three categories of used data-set in the proposed research. The three classes include Khundi, Mix, and Neli-Ravi. All images' dimensions and formats are kept the same as given, $256 \times 256 \times 3$, and JPG, respectively. However, data are primarily augmented before training and testing to avoid over-fitting and other deep learning issues.

2.2. Data Augmentation

The data augmentation is performed using various augmentation techniques. The bounding points of images are being transformed using reflection in the X and Y direction at random locations. The X and Y transformation specifies that the image has left, right, up, and down reflections. The random X, Y translation is set to $[-4, 4]$. It defines the random translation in X and Y dimensions with a specific range of pixel intensities on the image. The used pixel range is -4 to 4. Moreover, random cropping is applied, which performs a specific bitwise hide and seek operation, including cut-out and random erasing on the given image. The data samples are shown in Figure 2.

Figure 2. Number of data samples increased for training through augmentation.

2.3. Proposed CNN

The self-activated CNN architecture principally depends upon 23 layers containing five convolutional blocks with a different number of filters and sizes. The first layer is an image ($img[r,c]$) input layer having 256 × 256 × 3 input size of each given image instance. The 2nd layer is a convolutional layer with half the image size as several filters with a 3 × 3 kernel size. The padding and stride taken as one mean the sliding over the pixel to pixel for a given image are one so that each 3 × 3 kernel evolves on the whole image. The 2D convolutional computations are calculated as shown in Equations (2) and (3) by getting input from Equation (1) [29].

$$imput\ image = I = img[r,c] \tag{1}$$

$$kernel = \begin{pmatrix} kernel(1,1) & kernel(1,2) & kernel(1,3) \\ kernel(2,1) & kernel(2,2) & kernel(2,3) \\ kernel(3,1) & kernel(3,2) & kernel(3,3) \end{pmatrix} \tag{2}$$

$$y(i,j) = \sum_{i=-1}^{1} \sum_{j=-1}^{1} \mathrm{I}(x+i, y+j) * kernel(2+i, 2+j) \tag{3}$$

The Equation (2) represents the general kernel mask of 3 × 3, which later multiplies with the input image, where i and j represent the rows and columns; at the end of convolve operation, the output matrix is calculating in $y(i,j)$.

Each batch normalization layer is used in all convolutional blocks, where batch normalization is used for input normalization by calculating the variance ${\sigma_B}^2$ and mean μ_B for each colour channel.

$$g_{x_i} = (x_i - \mu_B) \ / \ \sqrt{{\sigma_B}^2 + \epsilon} \tag{4}$$

g_{x_i} in Equation (4) is calculating the activated normalization, which is anticipated by using the division of difference between input x_i and mean μ_B with the square root of the sum of squared variance ${\sigma_B}^2$ and constant value. The constant value invariance is added to stable the minimal variance value. The re-consideration of this activation is taken as g_{y_i}; the zero mean and variance with the unit value are essential to scaling. The scaling is shown in Equation (5).

$$g_{y_i} = \gamma g_{x_i} + \beta \tag{5}$$

In Equation (5), the learnable factors γ and β add to scale the activated output. The multiplication will shift the previous activated output into a new value with learnable parameters.

The max-pooling layer is down-sampling the given upper inputs into higher intensity values. The max-pooling for each convolve block is computed in Figure 3.

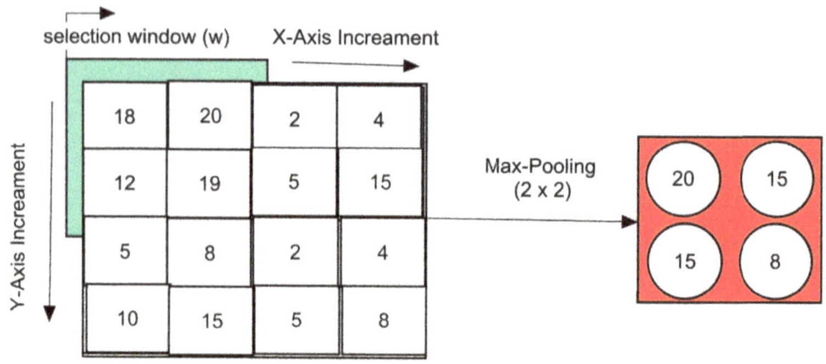

Figure 3. Max-pooling operation employed for proposed CNN.

For iteration over the 2D matrix of an image, the input image selected the window, as explained in Equation (6).

$$w = I_{selected_{2x2}} \quad (6)$$

The max-values from each incremented window 'w' are taken firstly from the left to right of the image, and then the next row is started; in this way, the y-direction increment starts, and lastly, a max-pooled downloaded max-window is gathered for each activated input of upper layer. The max-pooling is shown in Equation (7).

$$Max - Pooling = \max(w_{(x,y)}) \quad (7)$$

After getting the convolving data, the data are normalized using the batch normalization layer. In this way, the input batch data variations are reduced. After normalizing the data, the rectified linear unit (ReLU)-based activations are performed. The ReLU function works like a threshold value that specifies the negative value to zero and upper as it is for the same intensity values as calculated in Equation (8).

$$ReLU = \begin{cases} g_{y_i} & g_{y_i} \geq 0 \\ 0, & g_{y_i} < 0 \end{cases} \quad (8)$$

The normalized feature vectors are then passed to the max-pooling layer with a kernel size of 2, and the stride is also kept as 2 to slide over 2 points to take as the next input pool. All these layers' parameters are shown in Table 3.

Table 3. Convolution layers' description of proposed CNN.

Layer Names	Number of Kernels	Kernel Sizes
Input layer	256 × 256 × 3	-
Convolve-1	128	3 × 3
Convolve-2	64	3 × 3
Convolve-3	32	3 × 3
Convolve-4	16	3 × 3
Convolve-5	8	3 × 3
Max-pooling	-	2 × 2

Table 3 has shown that only the convolutional layer changes the number of filter variations throughout the network where the batch normalization and ReLU activations are performed in each convolutional block. However, the architecture with each layer's weights have been shown in CNN architect in Figure 4. Each layer's weights are activated using the trained network to see the pooling effect in each architecture. Therefore, in the proposed architecture diagram, the weights of convolutions with batch normalization and reLU activations are shown as a stack. The convolve showing the colorful inputs is taken where the ongoing following batch normalization shows how the colors are being activated and changes for following layer input. Finally, in the fifth block, the convolve weights and batch normalization highlight more decision-making intensities. The final activated weights are passed on to the fully connected layer, taking it as the three classes and applying the soft-max operation-based activations.

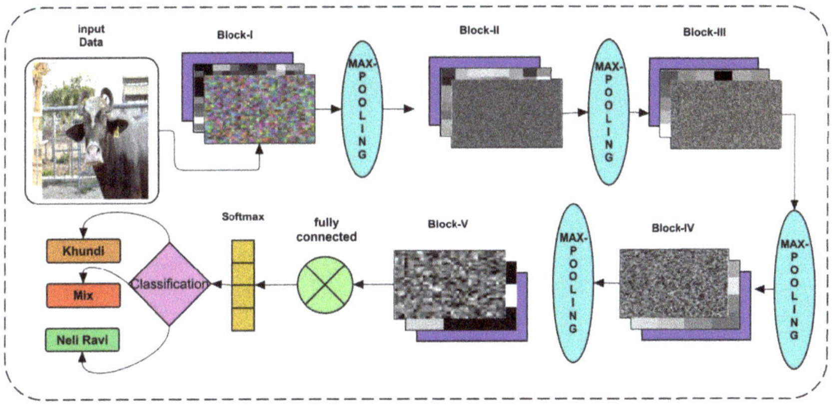

Figure 4. Block diagram of proposed CNN architecture.

Figure 4 has shown the input data layer as augmented data input with all the layers' flow towards final classification.

2.4. *Self-Transfer Learning Features*

Transfer learning is used in many recent studies [30], and it motivated us to use it for better classification results in terms of accuracy. However, many studies have employed state-of-the-art pre-trained ImageNet variants for transfer learning to get the numeric features. The proposed study does not get any promising results using the proposed CNN, but the transfer learning of data-specific CNN makes it more efficient. The proposed CNN's use of self-transfer learning is also suitable for getting more accurate precision values.

2.5. *Classification*

Firstly, the proposed study self-activated CNN for classification purposes and then used seven classical ML classification methods. CNN is not performed well even on a higher number of iterations. It observes that a greater number of iterations tends to decrease the validation accuracy. The optimal number of parameters and their values are discussed in the results section. However, the seven used classifiers include 3 KNN variants fine-KNN, medium-KNN, and Coarse-KNN. The ensemble and boosting algorithms named total boost, LP-boost, and bag-ensemble are likewise applied. In order to make more certainty in the results, different data splits were applied using random instances selection.

3. Results and Discussion

Each image's background has a static context since a static camera captures the various buffalo breeds' images. Each pixel has its intensity, and the background of each buffalo image is generated by merging all pixels. Statistical concepts underpin the majority of context models. The context model can be divided into two groups based on various background descriptions: one is a reference model resulting in each pixel's feature details. The other model is one that was generated by extracting the region function. Since image division requires high-level visual prior information, context models of this kind are typically complicated. Parameters setting to train proposed CNN on the acquired data set are described in Table 4. These parameters were found as the most optimal features via performance measures.

Table 4. Parameters setting considered during the training and testing of proposed CNN.

Parameters	Values
Epochs	100
Activation function	Stochastic gradient decent (sgdm)
Training time	227 min and 22 s
Iterations	1900
Initial Learning Rate	0.004
Batch Size	8

Table 4 indicates a proposed CNN's optimal parameters with the most accurate found CNN as 77.62% validation accuracy. The training time with different values is validated to improve the accuracy. Therefore, the performance-based selection of CNN is used with the given parameters of CNN. The training and validation graph is shown in Figure 5.

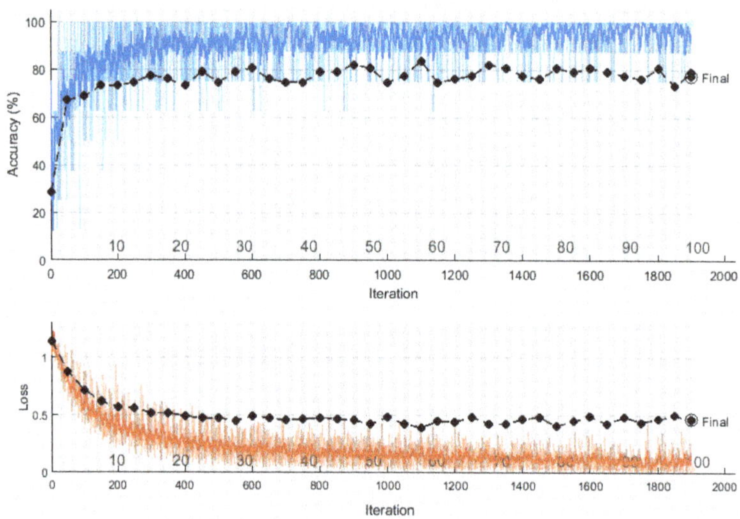

Figure 5. Proposed CNN training and validation plots.

Figure 5 designates the absolute accuracy by finding the 77.62% as validation accuracy where the loss is up to 0.5 with a bit of variation. The data-specific learning of a given data set is more favorable as it covers all aspects of features, such as location, geometry, or other local features. At this stage, the feature vector is taken as transfer learning to get the numeric features from the CNN fully connected layer encodings. The activated features are then later used as a classification feature vector.

3.1. Data Splitting

The activated features are of size n × 3, where all image features are split up with randomization to remove the data's biases. The used data split ratios are 50–50% (training, testing), 60–40% (training-testing), and 70–30% (training, testing). All data divisions employ seven different classifiers.

3.2. Experiment 1 (50–50)

In the first experimental phase, we split the features vectors as 50% for training and 50% for testing. The features matrix displays 225 × 3 and 225 × 3 for training and testing vectors. The trained classifiers-based predictions are then evaluated in terms of accuracy, precision, sensitivity, and F1 score. Moreover, a statistical evaluation metric such as Kappa has also been computed to evaluate the results. The prediction results are given in Table 5.

Table 5. Results obtained using different ML classifiers on 50–50 Split.

Method	Accuracy (%)	Precision (%)	Sensitivity (%)	F1-Score (%)	Kappa
Fine-KNN	83.11	82.88	83.11	83.03	0.62
Medium-KNN	84.44	84.44	84.64	84.53	0.65
Coarse-KNN	85.33	86.36	85.52	85.57	0.67
LP-Boost	82.22	82.21	82.44	82.32	0.60
Total-boost	85.33	85.63	85.46	85.30	0.67
Bag-Ensemble	78.67	78.65	78.93	78.78	0.52
SVM	84.00	84.09	84.16	83.97	0.64

Table 5 has bestowed that the maximum prediction accuracy is achieved by two classifiers, including 85.33% by coarse-knn and total-boost. The additional algorithms also performed well, with the accuracy range nearer to the two best-achieved algorithms. Their accuracy remains in the range of 80%+, where the accuracy obtained using the bag-ensemble method is not promising. Prediction plots obtained using different ML classifiers on 50–50 split are given in Figure 6.

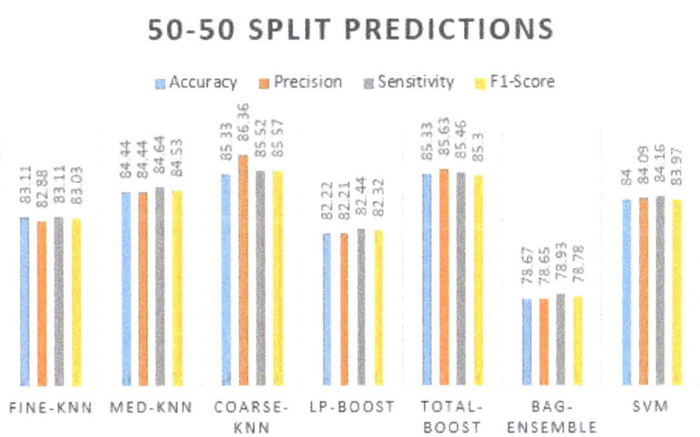

Figure 6. Prediction plot obtained using different ML classifiers on 50–50 split.

We can observe that other performance measures of classification predictions are also higher as the precision value is 86.36% by coarse-knn, which is more reliable than the precision obtained using the total-boost method. Hence, in terms of more true optimistic predictions over the summation of true positives and false positives, = the sensitivity and F1s-core are also improved by the coarse-knn compared to other classifiers. The confusion matrix obtained using Coarse-KNN is presented in Table 6.

Table 6. Confusion matrix obtained using Coarse-KNN.

Classes	Khundi	Mix	Neli-Ravi
Khundi	66	8	1
Mix	20	57	0
Neli-Ravi	3	1	69

The confusion matrix of coarse-knn indicated that the accurately predicted instances do not satisfy the Mix class as 20 instances are wrongly predicted in the khundi target class that belongs to the Mix class. However, the correctly predicted number of instances is 192 in total.

3.3. *Experiment 2 (60–40)*

Experiment 2 uses features vectors obtained from the last layer of self-activated CNN, splitting into 60–40% with 60% for training and 40% for testing. The overall results were almost identical to the 50–50 split strategy. Results obtained using different ML classifiers on 60–40 Split are presented in Table 7.

Table 7. Results obtained using different ML classifiers on 60–40 split.

Method	Accuracy (%)	Precision (%)	Sensitivity (%)	F1-Score (%)	Kappa
Fine-KNN	80.74	80.83	81.03	80.91	0.56
Medium-KNN	84.07	86.05	84.13	83.71	0.64
Coarse-KNN	84.44	85.22	84.66	84.76	0.65
LP-Boost	81.48	81.51	81.79	81.62	0.58
Total-boost	78.89	78.98	79.21	79.08	0.52
Bag-Ensemble	80.74	80.83	81.03	80.91	0.56
SVM	84.07	84.54	84.24	84.03	0.64

In Table 7, we can observe that the coarse-knn got the highest accuracy (84.44%) with slightly less accuracy precision, sensitivity, F1-score, and kappa values. Moreover, the other classifiers also perform in the range of 80%+. We can see in the 2nd row; the medium-knn achieved 84.07% accuracy. Similarly, SVM achieved the same accuracy value but with a lower precision rate of 84.54%. Fine-knn and bag-ensemble obtained the same accuracy value of 80.74% with all the same other evaluation measures. A prediction plot obtained using different ML classifiers on 60–40 split is displayed in Figure 7.

The first and second phases of experimental chunks obtained almost the same results, which were more promising as different classifiers of different categories providing almost identical accuracy values—the most accurate classifier's prediction-based confusion matrix is shown in Table 8. The primarily used conventional 70–30% split ratio is also utilized for more confident results.

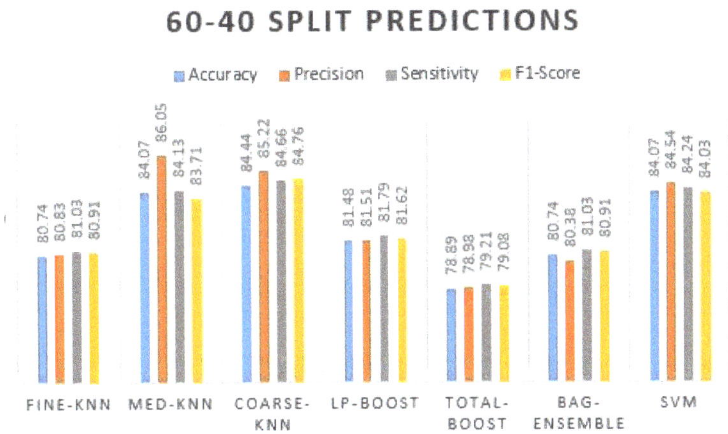

Figure 7. Prediction plot obtained using different ML classifiers on 60-40 split.

Table 8. Confusion matrix obtained using SVM.

Classes	Khundi	Mix	Neli-Ravi
Khundi	76	13	1
Mix	23	70	0
Neli-Ravi	3	2	82

The confusion matrix identified that the khundi class predictions are more accurate, whereas the 'Mix' class needs more improvement. The 76, 70, and 82 are the accurate predictions, whereas the wrongly predicted numbers are 13,1 and 23,3 and 2. Those wrongly predicted instances are on the upper and lower diagonal of the table.

3.4. Experiment 3 (70–30)

In the third experiment phase, we performed experiments using standard 70% random features splitting with 30% data testing. The evaluation measures are the same as in experimental phases one and two. The obtained results show gradual improvements as compared to the previous ones. Results obtained using different ML classifiers on 70–30 split are exhibited in Table 9.

Table 9. Results obtained using different ML classifiers on 70–30 Split.

Method	Accuracy (%)	Precision (%)	Sensitivity (%)	F1-Score (%)	Kappa
Fine-KNN	88.15	88.55	88.28	88.33	0.73
Medium-KNN	88.89	89.53	89.07	88.96	0.75
Coarse-KNN	89.63	90.09	89.71	89.82	0.76
LP-Boost	88.89	89.10	89.03	89.01	0.75
Total-boost	87.41	87.79	87.58	87.52	0.71
Bag-Ensemble	88.15	88.94	88.34	88.21	0.73
SVM	93.33	93.51	93.43	93.40	0.85

In Table 9, the maximum accuracy recorded using SVM is 93.33%, with the most accurate precision value. The given accuracy of this phase is justified and reportable. The other classifiers also display more than 87% accuracy. A prediction plot obtained using different ML classifiers on a 70–30 split is manifested in Figure 8.

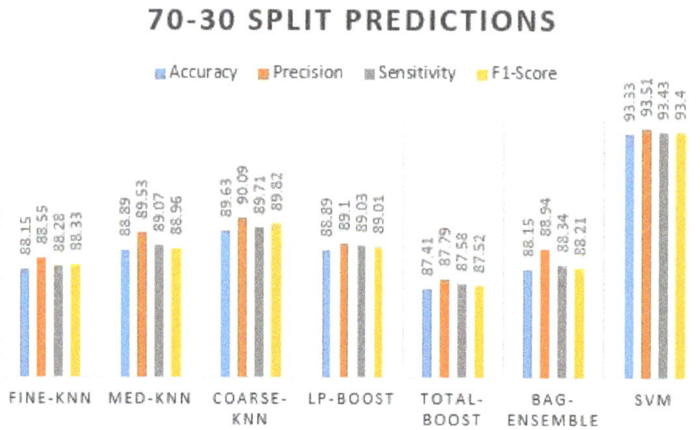

Figure 8. Prediction plot obtained using different ML classifiers on 70-30 split.

It is observed that in all experiments, the performance measure does not vary. The confusion matrix for the best-achieved accuracy algorithm is also shown in Table 10.

Table 10. Confusion matrix obtained using SVM on 70–30 split.

Classes	Khundi	Mix	Neli-Ravi
Khundi	42	3	0
Mix	6	40	0
Neli-Ravi	0	0	44

The confusion matrix of the most accurate SVM method shows that the Mix class still has more wrong predictions. The overall precision rate is improved to 93.51% over the summation of true positives and false positives. There are fewer wrong predictions in the other two classes; the F1-score provides both precision and a sensitivity response, which is also improved to the extent of 93.40%.

3.5. *Experiment 4 (Five-Fold Cross Validation)*

After the holdout validation techniques, the results are further evaluated on K-fold cross-validation. The primarily used values for K are 5 and 10 folds on the extracted feature vectors. The results, in this case, were also promising; the results were obtained using different ML classifiers on five-fold cross-validation in Table 11.

Table 11 confirms that coarse-knn and SVM produce the best accuracy and other effects. The accuracy value reaches up to 86.22% with 86.52% precision, 86.39% sensitivity, and 86.41% of the F1-score. By examining the precision value, the true positive value seems to be truer where the F1-score cross-validated it by taking sensitivity and precision over it and again received 86.41%. The prediction plot obtained using different ML classifiers on five-fold cross validation is manifested in Figure 9.

Table 11. Results obtained using different ML classifiers on five-fold cross validation.

Method	Accuracy (%)	Precision (%)	Sensitivity (%)	F1-Score (%)	Kappa
Fine-KNN	82.89	83.08	83.07	83.07	0.61
Medium-KNN	82.89	83.10	83.07	83.02	0.61
Coarse-KNN	85.56	86.08	85.69	85.83	0.67
LP-Boost	84.22	84.48	84.41	84.43	0.65
Total-boost	83.56	83.93	83.73	83.82	0.63
Bag-Ensemble	83.11	83.28	83.35	83.27	0.62
SVM	86.22	86.52	86.39	86.41	0.69

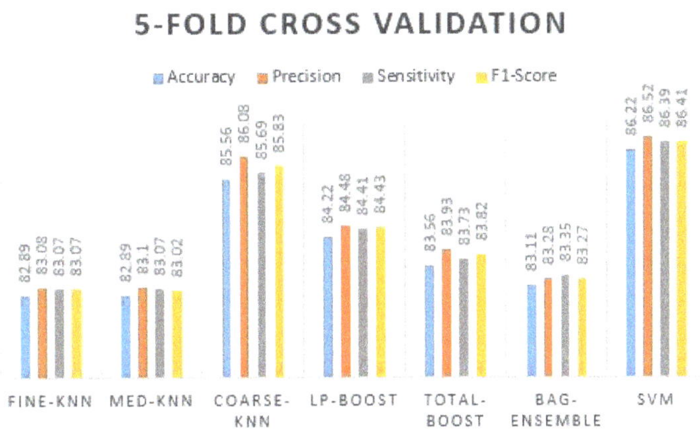

Figure 9. Prediction plot obtained using different ML classifiers on 5-Fold Cross Validation

We can see from the graph that all five other classifiers are above 80% to 84% of the accuracy values, where coarse-knn and SVM have accurate results. Furthermore, we evaluated the proposed framework on 10-fold validation to ensure the extracted feature vectors' integrity.

3.6. Experiment 5 (10-Fold Cross Validation)

All extracted features are given to 10-fold validation, making each split during the training and validation tenfold. All the results have appeared approximately similar to the five-fold cross-validation; no significant degradation seems in any instance of performance. Results obtained using different ML classifiers on 10-fold cross-validation are presented in Table 12.

In Table 12, again coarse-knn and SVM received the highest accuracy results, with more than 85% accuracy and more than 85% of precision, sensitivity, and F1-score. The prediction plot obtained using different ML classifiers on 10-fold cross validation is displayed in Figure 10.

It is observed that the performance of overall classifiers is improved. This indicates that higher training folds can increase the performance of all the methods used. The SVM confusion matrix is shown in Tables 13 and 14.

Table 12. Results obtained using different ML classifiers on 10-fold cross validation.

Method	Accuracy (%)	Precision (%)	Sensitivity (%)	F1-Score (%)	Kappa
Fine-KNN	83.56	83.79	83.73	83.76	0.63
Medium-KNN	84.00	84.12	84.21	84.15	0.64
Coarse-KNN	85.11	85.76	85.29	85.38	0.67
LP-Boost	83.78	83.96	83.96	83.96	0.64
Total-boost	83.78	84.10	83.99	84.00	0.64
Bag-Ensemble	84.22	84.36	84.42	84.39	0.65
SVM	85.56	85.74	85.75	85.71	0.68

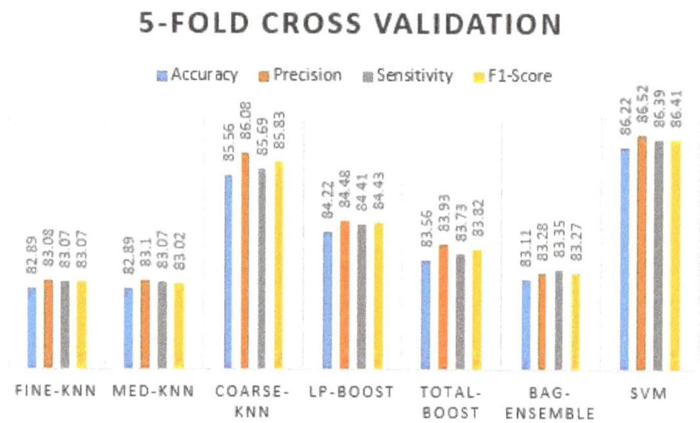

Figure 10. Prediction plot obtained using different ML classifiers on 10-fold cross validation.

Table 13. Confusion matrix obtained using SVM on five-fold cross validation split.

Classes	Khundi	Mix	Neli-Ravi
Khundi	124	24	2
Mix	33	121	0
Neli-Ravi	3	0	143

Table 14. Confusion matrix obtained using SVM on 10-fold cross validation split.

Classes	Khundi	Mix	Neli-Ravi
Khundi	121	27	2
Mix	35	119	0
Neli-Ravi	1	0	145

It seems that the 'Mix' class is devising inaccuracy, which suggests it may contain both types of features of Neli-Ravi and Khundi. However, the Neli-Ravi and Khundi were classified more accurately.

The evaluation metrics used for the experiments are given in the Equations (9)–(12).

$$Precision = 100 \times \left(\frac{T_p}{T_p + F_p}\right) \tag{9}$$

$$Sensitivity = 100 \times (\frac{T_p}{T_p + F_N}) \tag{10}$$

$$F1 - Score = 100 \times (2 \times (\frac{Precision \times Sensitivity}{Precision + Sensitivity})) \tag{11}$$

$$Accuracy = 100 \times (\frac{T_p + T_N}{T_p + T_N + F_p + F_N}) \tag{12}$$

In evaluation measures, one statistical coefficient is used to validate the results. The value of kappa shows the kappa agreement on predicted values compared to the actual values. The kappa value remains from 0 to 1, showing confidence in the given rules. It provides no agreement on the 0–0.20 value and shows minimal agreement on 0.21 to 0.39. Additionally, it shows a weak agreement from 0.40 to 0.59. The substantial ranges start from 0.60 to 0.90 or above. This shows confident agreement upon a 0.90 to 1 value. However, the calculation is performed using a confusion matrix by obtaining predicted and actual class numbers [31].

$$cm_n = Margin_{col} \tag{13}$$

$$rm_n = Margin_{row} \tag{14}$$

The margin values of the calculated confusion matrix are used for the calculation of Pr (*e*), which is actually the major nominator value for Equation (16).

$$\Pr(e) = (cm_1 \times rm_{1)} \ / \ n) + (cm_2 \ \times \ rm_2 \ )/ \ n) \ / \ n \tag{15}$$

$$Cohen - Kappa = \frac{\Pr(a) \ - \Pr(e)}{1 - \Pr(e)} \tag{16}$$

Equation (15) shows the actual calculation of Pr(*e*). The cm_1 shows the confusion matrix margin values of all classes, whereas rm_1 shows the row margin values of the confusion matrix. However, the cohen-kappa shows confidence in its 0.8 to 0.9 strong agreement. We can see from the table that the SVM has a 0.85 value of kappa coefficient, which is a far more substantial value to rely on data content or features.

4. Conclusions

Pakistan's dairy production centers face issues in identifying and separating the Neli-Ravi breed from other buffalo breeds. As a result, automatic buffalo breed identification and classification are needed. The proposed study utilized the Pakistani buffalos data set for computer-vision-based recognition and Neli-Ravi classification from other buffalo breeds. The proposed deep CNN architecture deduced the spatial information from the buffalo images and assembled the rich feature vectors. Furthermore, a learning-based approach transfers rich features into activated layers for better classification. Several ML-based classifiers are adopted to classify the instances into relevant target classes. The confidence in the data set and features are validated using different data splits, which do not abruptly vary. All results are normalized and reliable, making the proposed research an accurate recognition method for different buffalo breeds. The proposed framework achieves a maximum accuracy of 93%, and using recent versions, it achieves more than 85% accuracy.

In the future, we intend to employ optimized deep-learning-based architectures using optimization algorithms such as genetic algorithm (G.A), bee-swarm optimization (BSO), and the differential evolution algorithm (D.E) to enhance the accuracy.

Author Contributions: Conceptualization, Y.P., H.J., J.G. and H.T.R.; methodology, Y.P. and H.J.; software, Y.P., H.J., J.G. and H.T.R.; visualization, Y.P., H.J., J.G. and H.T.R.; writing—original draft, Y.P. and H.J.; and supervision, H.J. All authors have read and agreed to the published version of the manuscript.

Funding: This research was funded by the National Natural Science Foundation of China [61902156], Zhenjiang Science and Technology Bureau; Zhenjiang Finance Bureau high-tech enterprise storage cultivation funds; Jiangsu Science and Technology Department, Jiangsu Provincial Finance Department high-tech enterprise storage cultivation funds; and Zhenjiang Hongxiang Automation Financial support from Technology Co., Ltd.

Institutional Review Board Statement: Not applicable.

Informed Consent Statement: Not applicable.

Data Availability Statement: Dataset is available here https://data.mendeley.com/datasets/vdgnxsm692/2. (accessed on 1 August 2022).

Conflicts of Interest: The authors declare no conflict of interest.

References

1. Sikandar, B.A.; Shi, X.; Gou, X.; Zhaobing, G.; Qing, L.; Jamal, M.A.; Khederzadeh, S.; Talpur, M.Z.; Ming, M.H. Genetic relationship and diversity analysis in four buffalo breeds of Pakistan. *Pak. J. Agric. Sci.* **2020**, *57*.
2. Bilal, M.; Suleman, M.; Raziq, A. Buffalo: Black gold of Pakistan. *Livest. Res. Rural Dev.* **2006**, *18*, 140–151.
3. Shah, S.K. *Buffaloes of Pakistan*; Pakistan Agricultural Research Council (PARC): Istanbul, Turkey, 1991.
4. Luo, X.; Zhou, Y.; Zhang, B.; Zhang, Y.; Wang, X.; Feng, T.; Li, Z.; Cui, K.; Wang, Z.; Luo, C.; et al. Understanding divergent domestication traits from the whole-genome sequencing of swamp-and river-buffalo populations. *Natl. Sci. Rev.* **2020**, *7*, 686–701. [CrossRef]
5. Pasha, T. Comparison between bovine and buffalo milk yield in Pakistan. *Ital. J. Anim. Sci.* **2007**, *6*, 58–66. [CrossRef]
6. Murtaza, M.A.; Pandya, A.J.; Khan, M.M.H. Buffalo milk. In *Handbook of Milk of Non-Bovine Mammals*, 2nd ed.; Park, Y.W., Haenlein, G.F.W., Wendorff, W.I., Eds.; John and Wiley and Sons: Hoboken, NJ, USA, 2017, pp. 261–368.
7. Aujla, K.M.; Hussain, A. Economics of milk production of major dairy buffalo breeds by agro-ecological zones in Pakistan. *Pak. J. Agric. Res.* **2015**, *28*.
8. Sajid, I.; Babar, M.; Javed, K. Genetic diversity of Nili-Ravi from Nili and Ravi buffalo breeds of Pakistan. *Ital. J. Anim. Sci.* **2007**, *6*, 314–317. [CrossRef]
9. Murase, H. Artificial intelligence in agriculture. *Comput. Electron. Agric.* **2000**, *29*, 4377–4383. [CrossRef]
10. Queiroz, M.M.; Telles, R.; Bonilla, S.H. Blockchain and supply chain management integration: A systematic review of the literature. *Supply Chain. Manag. Int. J.* **2020**, *25*, 241–254. [CrossRef]
11. Hossain, M.S.; Rahman, M.H.; Rahman, M.S.; Hosen, A.S.; Seo, C.; Cho, G.H. Intellectual Property Theft Protection in IoT Based Precision Agriculture Using SDN. *Electronics* **2021**, *10*, 1987. [CrossRef]
12. Gao, J.; Wang, H.; Shen, H. Machine Learning Based Workload Prediction in Cloud Computing. In Proceedings of the 2020 29th International Conference on Computer Communications and Networks (ICCCN), Honolulu, HI, USA, 3–6 August 2020; IEEE: New York, NY, USA, 2020; pp. 1–9.
13. Lokhorst, C.; De Mol, R.; Kamphuis, C. Invited review: Big Data in precision dairy farming. *Animal* **2019**, *13*, 1519–1528. [CrossRef]
14. Faye, B.; Konuspayeva, G. The sustainability challenge to the dairy sector–The growing importance of non-cattle milk production worldwide. *Int. Dairy J.* **2012**, *24*, 50–56. [CrossRef]
15. Yegnanarayana, B. *Artificial Neural Networks*; PHI Learning Pvt. Ltd.: New Delhi, India, 2009.
16. Mijwel, M.M. Artificial Neural Networks Advantages and Disadvantages. 2018. Available online: https://www.linkedin.com/pulse/artificial-neural-networks-advantages-disadvantages-maad-m-mijwel (accessed on 1 August 2022).
17. Islam, S.; Reddy, U.K.; Natarajan, P.; Abburi, V.L.; Bajwa, A.A.; Imran, M.; Zahoor, M.Y.; Abdullah, M.; Bukhari, A.M.; Iqbal, S.; et al. Population demographic history and population structure for Pakistani Nili-Ravi breeding bulls based on SNP genotyping to identify genomic regions associated with male effects for milk yield and body weight. *PLoS ONE* **2020**, *15*, e0242500. [CrossRef]
18. Mirza, R.; Waheed, A.; Akhtar, M.; Khan, M.; Dilshad, S.; Faraz, A. Correlation of linear type traits with milk yield in Nili Ravi buffaloes of Pakistan. *JAPS J. Anim. Plant Sci.* **2020**, *30*, 780–785.
19. Jwade, S.A.; Guzzomi, A.; Mian, A. On farm automatic sheep breed classification using deep learning. *Comput. Electron. Agric.* **2019**, *167*, 105055. [CrossRef]
20. Kumar, S.; Pandey, A.; Satwik, K.S.R.; Kumar, S.; Singh, S.K.; Singh, A.K.; Mohan, A. Deep learning framework for recognition of cattle using muzzle point image pattern. *Measurement* **2018**, *116*, 1–17. [CrossRef]
21. Atabay, H.A. Deep Learning for Horse Breed Recognition. *CSI J. Comput. Sci. Eng.* **2017**, *15*, 45–51.
22. Ghosh, P.; Mustafi, S.; Mandal, S.N. Image-Based Goat Breed Identification and Localization Using Deep Learning. *Int. J. Comput. Vis. Image Process.* **2020**, *10*, 74–96. [CrossRef]
23. Zhang, X.; Yang, L.; Sinnott, R. A Mobile Application for Cat Detection and Breed Recognition Based on Deep Learning. In Proceedings of the 2019 IEEE 1st International Workshop on Artificial Intelligence for Mobile (AI4Mobile), Hangzhou, China, 24 February 2019; IEEE: New York, NY, USA, 2019. [CrossRef]

24. Rauf, H.T.; Lali, M.I.U.; Zahoor, S.; Shah, S.Z.H.; Rehman, A.U.; Bukhari, S.A.C. Visual features based automated identification of fish species using deep convolutional neural networks. *Comput. Electron. Agric.* **2019**, *167*, 105075. [CrossRef]
25. Barbedo, J.G.A.; Koenigkan, L.V.; Santos, T.T.; Santos, P.M. A Study on the Detection of Cattle in UAV Images Using Deep Learning. *Sensors* **2019**, *19*, 5436. [CrossRef]
26. De Lima Weber, F.; de Moraes Weber, V.A.; Menezes, G.V.; da Silva Oliveira Junior, A.; Alves, D.A.; de Oliveira, M.V.M.; Matsubara, E.T.; Pistori, H.; de Abreu, U.G.P. Recognition of Pantaneira cattle breed using computer vision and convolutional neural networks. *Comput. Electron. Agric.* **2020**, *175*, 105548. [CrossRef]
27. Meraj, T.; Rauf, H.T.; Zahoor, S.; Hassan, A.; Lali, M.I.; Ali, L.; Bukhari, S.A.C.; Shoaib, U. Lung nodules detection using semantic segmentation and classification with optimal features. *Neural Comput. Appl.* **2019**, *33*, 10737–10750. [CrossRef]
28. Rauf, H.T. Buffalo-Pak: Buffalo Breed Dataset from Pakistan for Visual Features Based Classification. Mendeley Data. 2021. Available online: https://data.mendeley.com/datasets/vdgnxsm692/2 (accessed on 1 August 2022).
29. Gilg, M. 2 - Representation of Networks of Wireless Sensors with a Grayscale Image: Application to Routing. In *Building Wireless Sensor Networks*; Femmam, S., Ed.; Elsevier: Amsterdam, The Netherlands, 2017; pp. 31–66. doi: doi: 10.1016/B978-1-78548-274-8.50002-7. [CrossRef]
30. Tan, C.; Sun, F.; Kong, T.; Zhang, W.; Yang, C.; Liu, C. A survey on deep transfer learning. In Proceedings of the International Conference on Artificial Neural Networks, Rhodes, Greece, 4–7 October 2018; Springer: Berlin/Heidelberg, Germany, 2018; pp. 270–279.
31. McHugh, M.L. Interrater reliability: The kappa statistic. *Biochem. Medica* **2012**, *22*, 276–282. [CrossRef]

agriculture

MDPI

Article

Tea Sprout Picking Point Identification Based on Improved DeepLabV3+

Chunyu Yan [1,2], Zhonghui Chen [1], Zhilin Li [1], Ruixin Liu [1], Yuxin Li [1,2], Hui Xiao [1], Ping Lu [3] and Benliang Xie [1,2,*]

1 College of Big Data and Information Engineering, Guizhou University, Guiyang 550025, China
2 Power Semiconductor Device Reliability Engineering Center of the Ministry of Education, Guiyang 550025, China
3 State Key Laboratory Breeding Base of Green Pesticide and Agricultural Bioengineering, Key Laboratory of Green Pesticide and Agricultural Bioengineering, Ministry of Education, Guizhou University, Guiyang 550025, China
* Correspondence: blxie@gzu.edu.cn

Abstract: Tea sprout segmentation and picking point localization via machine vision are the core technologies of automatic tea picking. This study proposes a method of tea segmentation and picking point location based on a lightweight convolutional neural network named MC-DM (Multi-Class DeepLabV3+ MobileNetV2 (Mobile Networks Vision 2)) to solve the problem of tea shoot picking point in a natural environment. In the MC-DM architecture, an optimized MobileNetV2 is used to reduce the number of parameters and calculations. Then, the densely connected atrous spatial pyramid pooling module is introduced into the MC-DM to obtain denser pixel sampling and a larger receptive field. Finally, an image dataset of high-quality tea sprout picking points is established to train and test the MC-DM network. Experimental results show that the MIoU of MC-DM reached 91.85%, which is improved by 8.35% compared with those of several state-of-the-art methods. The optimal improvements of model parameters and detection speed were 89.19% and 16.05 f/s, respectively. After the segmentation results of the MC-DM were applied to the picking point identification, the accuracy of picking point identification reached 82.52%, 90.07%, and 84.78% for single bud, one bud with one leaf, and one bud with two leaves, respectively. This research provides a theoretical reference for fast segmentation and visual localization of automatically picked tea sprouts.

Keywords: DeepLabv3+; deep learning; semantic segmentation; picking point identification

Citation: Yan, C.; Chen, Z.; Li, Z.; Liu, R.; Li, Y.; Xiao, H.; Lu, P.; Xie, B. Tea Sprout Picking Point Identification Based on Improved DeepLabV3+. *Agriculture* **2022**, *12*, 1594. https://doi.org/10.3390/agriculture12101594

Academic Editor: Dimitre Dimitrov

Received: 9 August 2022
Accepted: 28 September 2022
Published: 2 October 2022

1. Introduction

Premium tea is a general term for high-quality tea. The picking method of tea sprouts is a crucial factor in determining their economic value [1]. In general, the single bud, one bud with one leaf, and one bud with two leaves are the primary tender materials of premium tea [2]. Mechanical picking can automatically pick tea leaves, but they cannot select a specific picking point. Machine picking will also destroy the integrity of the sprouts, which will reduce the economic value of the premium tea [3]. Premium tea sprouts are picked nearly by manual means. However, this picking method is inefficient and costly and cannot meet the immense demand for premium tea in the market.

In recent years, computer vision-based robots have been applied in the automatic picking of premium tea sprouts, which is expected to improve picking efficiency and reduce picking costs [4]. Many scholars have researched this topic and obtained many excellent results. Zhao et al. performed RGB analysis of tea bud images and obtained distinct tea bud characteristics using HSI color conversion, and they used HSV spatial transformation for tea bud segmentation [5]. Qian et al. applied the jump connection in U-Net and contrastive-center loss function to the SegNet model and achieved good results in tea shoot segmentation [6]. Qi et al. applied the Otsu to the traditional watershed algorithm. They found that the algorithm showed a better ability to identify tea buds in complex backgrounds [7]. Hu et al. proposed a discriminative pyramid network for semantic segmentation of tea geometries in natural fields, and the MIoU and PA values of the method were 84.46% and 94.48%, respectively [8]. The tea garden environment has complexities, such as uncontrolled light conditions and high similarity between sprouts and old leaves [9]. The aforementioned methods can separate tea sprouts from the old leaves under certain conditions, but they cannot accurately identify and locate the sprout picking point. This incapability limits the automatic picking of tea sprouts.

The development of computer vision technology has improved the performance of convolutional neural networks (CNNs) in locating objects in images with complex backgrounds. CNNs are perceptrons composed of millions of neurons. After they are trained, they can localize objects in images with nearly no processing. DeepLabV3+ network is a CNN that can be used for pixel-level object detection. The method has been widely applied in the field of agriculture. Peng et al. applied DeepLabV3+ with Xception to segment lychees, and the MIoU of this method reached 0.765 [10]. Song et al. utilized DeepLabV3+ with ResNet-101 to classify calyces, branches, and wires in orchards, which obtained MIoU of 0.694 for uniform weights and 0.480 for median frequency weights on the homemade kiwifruit canopy image dataset [11]. Ayhan and Kwan applied DeepLabV3+ with Xception to segment the forest, grassland, and shrubland in the Slovenia dataset, which achieved an IoU of 0.9086 for the forest, 0.7648 for grassland, and 0.1437 for shrubland [12]. Zhang et al. adopted DeepLabV3+ with ResNet-18 to segment trunk, leaves, and apple in a commercial "Fuji" apple orchard [13]. The method achieved 94.8%, 97.5%, and 94.5% pixel classification accuracy for trunk, apple, and leaf (background), respectively. These studies have shown that DeepLabV3+ can segment tea sprouts from the background. It can also separate single buds, one leaf, and two leaves. The segmentation results of DeepLabV3+ can be used to identify the coordinates of tea sprout picking points.

In this study, we propose a lightweight tea sprout segmentation network, termed MC-DM, to segment tea sprouts. The recognition of tea sprout picking points was achieved on the basis of the segmentation results of the MC-DM. The architecture of the MC-DM was improved from the DeeplabV3+. First, the detection efficiency is essential in tea garden picking tasks. MC-DM used the modified MobilenetV2 as the backbone network to reduce the calculation while fully extracting image features. Second, the densely connected atrous spatial pyramid pool (DASPP) module was introduced into the MC-DM to improve the network feature extraction capability. Finally, RGB color separation and the Shi–Tomasi algorithm were used to detect partial corner points of single bud, one leaf, and two leaves. The corner point with the lowest value of the vertical coordinate was identified as the picking point corresponding to a single bud, one bud with one leaf, and one bud with two leaves. The experimental results show that the segmentation accuracy of MC-DM is nearly the same as that of DeepLabV3+. However, the number of parameters MC-DM has is greatly reduced. After the segmentation results of MC-DM are applied to picking point recognition, the recognition accuracy also meets the requirements of actual premium tea picking. We provide a new method for realizing real-time detection under the circumstance of photographing in outdoor tea fields.

2. Materials and Methods

2.1. *Dataset*

The collection dates of the tea sprout images used in the experiment were from mid-March 2021 to mid-April 2021, the period of premium tea picking. The collection location was the tea garden in Guiyang City, China. We used Redmi Note 7pro mobile phone (20 MP + 5 MP + 2 MP rear triple camera) to capture tea sprout images, and the size of the original image was 4000 × 3000 pixels. Weather conditions included cloudy, overcast, and clear skies. The picking times were from 09:00 to 17:00. The sampled tea sprout data have wide background variation, which was convenient for strengthening the robustness and generalization ability of the segmentation network. A total of 1110 images of tea sprouts were collected, which satisfied the data requirements for the pixel-based semantic segmentation.

Data augmentation, as a method of data preprocessing, plays an important role in deep learning. In general, effective data augmentation methods can avoid over-fitting and improve the robustness of the model [14]. The general methods of data augmentation are changing brightness, flipping, and adding noise. The experiments were conducted using manual data augmentation given the enormous labor cost of fully supervised training. As shown in Figure 1, the number of datasets increased from 1110 to 5550 by flipping horizontally, changing brightness, and adding Gaussian noise. All images were scaled to 640 × 480 pixels by the linear interpolation algorithm to improve the efficiency of model training. The expanded samples were divided into the training set, validation set, and test set according to the ratio of 6:2:2.

Figure 1. Selected samples of data enhancement. (**a**) Original image; (**b**) enhanced brightness; (**c**) reduced brightness; (**d**) horizontal flip; (**e**) noise addition.

2.2. Image Annotation

We manually labeled the tea sprout images with the LabelMe labeling tool into 4-pixel categories: background (cls0, black), single bud (cls1, red), one leaf (cls2, green), and two leaves (cls3, yellow). The original image is a 24-bit RGB image, and its corresponding visualization label is shown in Figure 2.

Figure 2. Tea sprout image and its corresponding visualization label. (**a**) Tea sprout image; (**b**) visualization label.

2.3. Description of Tea Sprout Picking Point

According to the different grades of premium tea and the frying process, the picking point of tea sprouts can be divided into single bud picking (A), one bud with one leaf picking (B), and one bud with two leaves picking (C). Among them, one bud with one leaf picking is the most extensive. Figure 3 shows the schematic of the corresponding standard picking point. The picking point is located at the center of the tea stalk connected with the tea sprout.

Figure 3. Picking point of tea sprout.

2.4. Models

2.4.1. MC-DM Architecture

Picking point recognition is a core technology of tea sprout picking systems [15]. The segmentation of single buds, one leaf, and two leaves are the prerequisite for identifying the picking point. DeepLabV3+ is a representative algorithm in the semantic segmentation field [16]. It includes two parts: an encoding module and a decoding module. In the encoding stage, the input image is first passed through the Xception backbone network to obtain a feature map of 16-time down-sampling. Then, the feature map of the 16-time down-sampling is placed into the atrous spatial pyramid pool (ASPP) module. The ASPP module consists of a 1×1 convolution, an average pooling layer with global information, and three 3×3 atrous convolutions with dilated rates of 6, 12, and 18. Finally, the feature maps obtained from the ASPP module are spliced, and the number of channels is compressed to 256 through a 1×1 convolution. In the decoding part, the feature map output from the encoding part is upsampled four times by a bilinear interpolation algorithm. Then, it is concatenated with the same-resolution feature maps extracted by the Xception backbone network. Finally, the tandem features are refined by 3×3 convolution, and the segmentation result is obtained by upsampling for four times. The DeepLabV3+ architecture is shown in Figure 4.

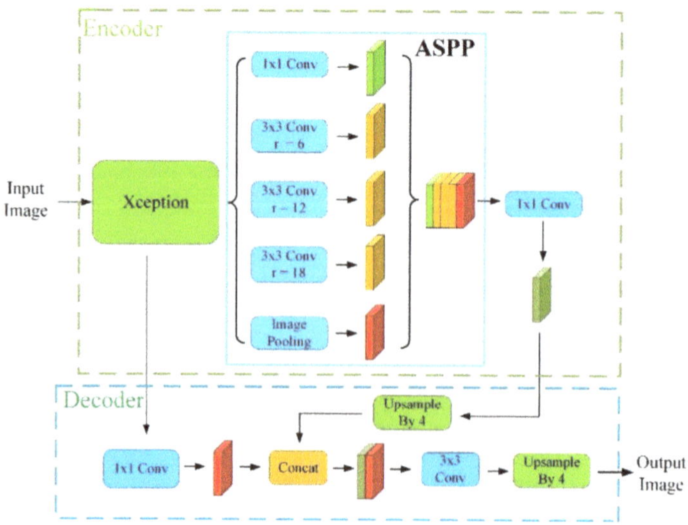

Figure 4. DeepLabv3+ network architecture.

MC-DM retains the excellent encoding–decoding structure of DeepLabV3+ to better identify tea sprouts. However, some improvements have been implemented. Figure 5 shows the network architecture of the MC-DM, and the improved parts are marked in red.

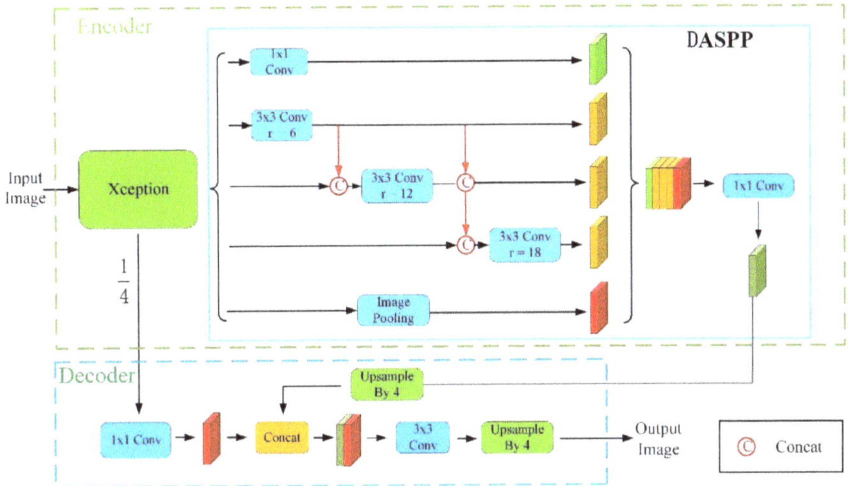

Figure 5. MC-DM network architecture.

As shown in Figure 5, MC-DM is an end-to-end network. The network input is a tea sprout image, and the output is a segmentation result of tea sprouts. In the encoding phase, the backbone network Xception is first replaced by a lightweight network. MobileNetV2 reduces the computational effort and mines the tea sprout features. The MobileNetV2 can extract feature maps of down-sampling for 2, 4, 8, and 16 times. Then, we place the feature map output of down-sampling for 16 times from the MobileNteV2 network into the DASPP. Compared with the atrous convolutions in the ASPP, those in the DASPP are cascaded by the series structure. The output of the atrous convolution with a smaller dilatation rate is contacted with the output of the backbone network. Then, they are sent to the atrous convolution with a larger dilatation rate to achieve a better feature extraction effect. In the decoding part, MC-DM retains the decoding network architecture of DeepLabV3+. The feature maps are restored to the original size of the input image through continuous upsampling.

2.4.2. Lightweight Backbone Network

MobileNetV2 is a lightweight network that can be embedded in mobile devices [17]. Its core contribution is to replace the standard convolution with deep separable convolution, which reduces the computational effort and model parameters. The deep separable convolution decomposes the standard convolution into two parts: deep convolution and point convolution. For the first time, the input features are first decomposed into multiple single-channel ones and convolved in each channel with a 3 × 3 convolution kernel to extract features, which is called deep convolution. Then, point convolution convolves the result of depth convolution with a 1 × 1 convolution to assemble the output features. Compared with standard convolution, the 3 × 3 deep separable convolution reduces the computation by 90% with only a slight reduction in inaccuracy [18]. On the basis of the deep separable convolution, the mobileNetV2 introduces an inverse residual structure, which further improves the network performance. In the inverse residual module, the input feature channels are first expanded by a 1 × 1 convolution. Then, the expended features are convolved with the 3 × 3 depth separable convolution to extract feature information. Finally, a 1 × 1 convolution is used to restore the number of feature channels. Compared with the residual architecture, the inverse residual structure avoids the drawback that the network only has good extraction of low-dimensional features and loses information on high-dimensional features when the number of input feature channels is small [19]. The network architecture of the MobileNetV2 is shown in Table 1.

Table 1. MobileNetV2 network architecture.

Input	Operator	*t*	*c*	*n*	*s*
224 × 224 × 3	Conv2d	-	32	1	2
112 × 112 × 32	bottleneck	1	16	1	1
112 × 112 × 16	bottleneck	6	24	2	2
56 × 56 × 24	bottleneck	6	32	3	2
28 × 28 × 32	bottleneck	6	64	4	2
14 × 14 × 64	bottleneck	6	96	3	1
14 × 14 × 96	bottleneck	6	160	3	2
7 × 7 × 160	bottleneck	6	320	1	2
7 × 7 × 320	Conv2d 1 × 1	-	1280	1	1
7 × 7 × 1280	7 × 7 Avgpooling	-		1	-
1 × 1 × 1280	1 × 1	-	k	-	-

The input represents the size of the input feature map of the current layer. The operator represents the operations performed by MobileNetV2, including the normal convolution layer, the inverse residual structure, and the average pooling layer. t is the expansion multiple of the channel in the inverse residual structure, n denotes the number of repetitions for the current layer, c is the number of output channels, and s is the stride of the convolution in the current layer.

In this study, the modified MobileNetV2 is used as the MC-DM backbone network, and its architecture is shown in Table 2. First, the average pooling layer and the fully connected layer of MobileNetV2 are removed given that the semantic segmentation task needs to preserve the location information of image pixels. Second, only the first eight convolutional layers of MobileNetV2 are retained to extract features for reducing the computation and memory consumption of MC-DM. The original MobileNetV2 is aimed at the image classification task, and the size of the output feature map in the eighth layer is 1/32 of the original map. The stride size of convolution in this layer is changed to 1 to retain more feature information of tea sprouts. Furthermore, the size of the output feature map in this layer is increased to 1/16 of the original map.

Table 2. Backbone network architecture.

Input	Operator	*t*	*c*	*n*	*s*
640 × 480 × 3	Conv2d	-	32	1	2
320 × 240 × 32	bottleneck	1	16	1	1
320 × 240 × 16	bottleneck	6	24	2	2
160 × 120 × 24	bottleneck	6	32	3	2
80 × 690 × 32	bottleneck	6	64	4	2
40 × 30 × 64	bottleneck	6	96	3	1
40 × 30 × 96	bottleneck	6	160	3	2
40 × 30 × 160	bottleneck	6	320	1	1

We do not randomly initialize the weights of the MC-DM model but instantiate the pre-trained weight of ImageNet in the modified MobilenetV2 network. ImageNet's pre-trained weight has shown extraordinary achievements in the image analysis field. It contains over 14 million images of different scenes. During optimization, migrating the pre-training weights of ImageNet can fit the pre-training model to a specific region of interest, which accelerates the model convergence and optimizes the model weights.

2.4.3. DASPP

DeepLabV3+ uses the ASPP module that contains atrous convolution to extract the multi-scale features of the input image for retaining the detailed features of image convolution and increasing the receptive field. When the size of the convolution kernel is k with a dilation rate of r, the receptive field of the atrous convolution in the ASPP module is

$$R = (r - 1) \times (k - 1) + k \tag{1}$$

As shown in Equation (1), the receptive field of the atrous convolution is proportional to its dilation rate. The atrous convolution expands the convolution receptive field by filling zeros in the convolution kernel and outputs the convolution results of nonzero sampling points. As the dilation rate increases, the nonzero pixel sampling of the atrous convolution becomes sparser. The information obtained by atrous convolution is seriously lost under the same computing conditions. This situation is detrimental to the learning and training of the model.

The ASPP module of the DeepLabV3+ network is improved through dense connection by referring to the DenseNet network to solve the problem of information loss due to the enlarged receptive field of the ASPP module. The network architecture of DASPP is shown in the yellow dashed box in Figure 5. Compared with those of the ASPP in Figure 4, the three atrous convolutions of DASPP are cascaded by the series structure. The output of the atrous convolution with less dilation rate is contacted with the resulting output from the backbone network. Then, it is sent to the atrous convolution with more dilation rate to achieve a better feature extraction effect. The process is illustrated by an atrous convolution with a dilation rate of 12. Figure 6a shows the pixel sampling of the atrous convolution with a dilation rate of 12 in the ASPP. The receptive field of the atrous convolution is 25, and the number of elements involved in the computation is only 9. Figure 6b shows the pixel sampling of the atrous convolution with a dilation rate of 12 in DASPP. The number of pixels involved in the computation is increased from 3 to 7. The receptive field is raised from 25 to 49.

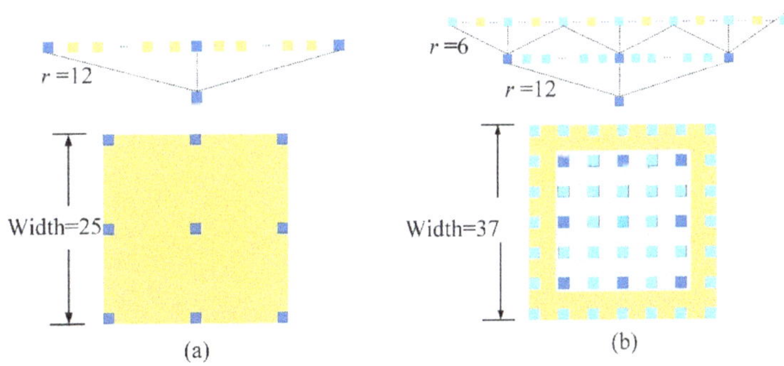

Figure 6. Distribution of atrous convolution sampling before and after dense connection. (**a**) Before dense connection; (**b**) After dense connection.

For the computation of the receptive field, DASPP can provide a larger receptive field. $R_{k,r}$ denotes the receptive field provided by the atrous convolution with the kernel size k and dilation rate r. The atrous convolutions in the ASPP module work in parallel and do not share any information. Therefore, the maximum receptive field of ASPP is the maximum of the receptive fields provided by each atrous convolution. From Equation (2), the maximum receptive field of the ASPP module is

$$R = \max(R_{3,6}, R_{3,12}, R_{3,18}) = 37 \tag{2}$$

A larger receptive field can be obtained by cascading two atrous convolutions. The calculation of the cascaded receptive field is shown in Equation (3), where R_1 and R_2 are two different receptive fields of atrous convolution, respectively.

$$R = R_1 + R_2 - 1 \tag{3}$$

According to Equation (3), the maximum receptive field of DASPP is

$$R = \max(R_{3,6} + R_{3,12} + R_{3,18}) = 73 \tag{4}$$

We define information utilization β to measure the relationship between receptive field and information utilization β. The calculation of β is shown in Equation (5), where $X1$ is the number of elements in the receptive field involved in computation and $X2$ is the total number of elements in the receptive field. Table 3 shows the performance of the atrous convolution with different connections in the feature map.

$$\beta = \frac{X1}{X2} \tag{5}$$

Table 3. Effect of ASPP connection method on the atrous convolution.

ASPP Connection	Dilation	Receptive Field	Effective Elements	β (%)
ASPP	6, 12	25	9	1.44
	6, 12, 18	37	9	0.66
	6, 12, 18, 24	49	9	0.37
DASPP	6, 12	37	49	2.18
	6, 12, 18	73	255	4.79
	6, 12, 18, 24	122	961	6.46

As shown in Table 3, the atrous convolution in the ASPP module works in parallel, which ignores the correlation between the atrous convolution with different dilution rates. As the number of atrous convolutions increases, the pixel utilization decreases. By contrast, DASPP achieves information sharing between different atrous convolution branches, which increases the range of receptive fields and significantly improves pixel utilization.

The DASPP module increases the receptive field and reduces the information loss, but it increases the model parameters and decreases the model computation speed. A 1×1 convolution is added before the atrous convolution of DASPP to reduce the channel dimension of the input features, which decreases the model parameters, for solving the abovementioned problem. In this study, the modified MobilenetV2 is used as the backbone network. The input feature channel of the ASPP module is 320, and the output feature channel dimension is 256. Then, the parameter number of the ASPP module is calculated, as shown in Equation (6).

$$N_1 = 320 \times 256 \times 9 \times 3 = 2,211,840 \tag{6}$$

The number of input feature channels for the atrous convolution in the DASPP module is reduced from 320 to 256, and the output feature channel dimension is 256. The parameter number of the DASPP module is calculated as shown in Equation (7).

$$N_2 = (320 + 576 + 832) \times 256 + 256 \times 256 \times 9 \times 3 = 2,211,840 \tag{7}$$

As shown in Equations (6) and (7), the proposed DASPP maintains the same number of parameters as the ASPP.

2.4.4. Picking Point Positioning Method

The quality of premium teas requires ensuring the integrity of the sprouts and limiting the picking area to the sprout stems. The identification process of tea sprout picking points is shown in Figure 7. First, three binary images of the same size as the original image need to be obtained. The reason is that the color difference among the background, single bud, one leaf, and two leaves of the divided tea buds is obvious and the internal color components are the same. Therefore, the binary operation for a single bud, one leaf, and two leaves can be completed according to the color range of the pixel points in the tea bud image. Then, the pixel points within the discrete area of the binary image are counted. This way determines the maximum connected area for single bud, one leaf, and two leaves, which removes the interference of background information on picking point recognition. Finally, the corner points of the maximum connected area are calculated using the Shi–Tomasi algorithm. The corner point with the lowest vertical coordinate is identified as the corresponding pick point. Table 4 shows the RGB color segmentation thresholds for a single bud, one leaf, and two leaves.

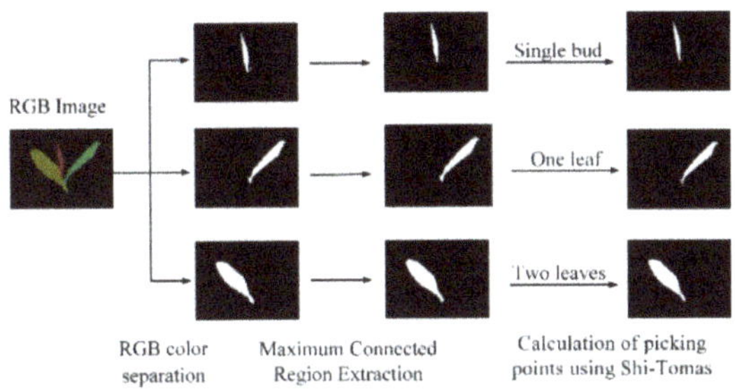

Figure 7. Picking point identification process.

Table 4. Segmentation thresholds of tea buds.

	Bud	One Leaf	Two Leaves
R	255	0	255
G	0	255	255
B	0	0	0

2.5. Evaluation Metrics

The experiment uses multiple levels of control parameter variables for the evaluation to more effectively measure the performance of each model. The main evaluation indexes include the accuracy of the model prediction, the prediction speed of the model, and the size of the model parameters. Many criteria are used to measure the accuracy of image segmentation. In general, MIoU is the most representative evaluation metric in the fields of image segmentation. IoU refers to the intersection between the predicted set of values and the actual set of values for a single class of pixels. MIoU is the average of IoUs of all categories. It reflects the ability of the model to segment the image pixels. The mathematical expression of MIoU is shown in Equation (8).

$$MIoU = \frac{1}{k+1}\sum_{i=0}^{k}\frac{p_{ii}}{\sum_{j=0}^{k}p_{ij}+\sum_{j=0}^{k}p_{ji}-p_{ii}} \tag{8}$$

where $k+1$ is the number of total classes, including a background class; p_{ii} is the number of pixels predicted to be correct. p_{ij} is the number of pixels belonging to class i but predicted to be class j. p_{ji} is the number of pixels belonging to class j but is predicted to be class j.

3. Results and Discussion

3.1. Implementation Details

We conducted experiments and validated our proposed method based on the PyTorch platform on the same workstation, which has the following configuration and installed software: Intel Core i7-11700@ 3.4 GHZ×8 threads, NVidia GeForce GTX 3080 GPU with RAM of 12 GB, 64-bit Windows 10 operating system, Python 3.8, OpenCV 3.4.1 and PyTorch1.4.1. During the model training process, tea sprout images were used as the input, and the stochastic gradient descent was used as the optimization function. All networks were trained for 800 iterations with a batch size of 16. The initial learning rate was set to 0.04 and multiplied by 0.5 every five training iterations. The momentum parameter is 0.9.

3.2. Backbone Network Validation

Considering the effectiveness of the Deeplabv3+ network used for semantic image segmentation, we apply it to the semantic segmentation of tea sprout images in this study. Tea picking robots have high requirements on the prediction speed of embedded models and the number of model parameters. Thus, Deeplabv3+ with a lightweight Mobilenetv2 backbone is designed as the base network. This section compares Deeplabv3+-Mobilenetv2 with the original Deeplabv3+, which is trained and tested under the same experimental conditions, to verify the effectiveness of the design choice. Their performance metrics are shown in Table 5.

Table 5. Comparison of the influence of different backbone networks.

Method	Backbone	MIoU/%	Parameters/MB	Speed (f/s)
Original Deeplabv3+	Xception	91.33	104.61	25.10
Deeplabv3+-Mobilenetv2	MobilenetV2	90.33	11.18	41.14

Table 5 shows that DeepLabV3+ with different backbone networks has different detection results for the tea sprout dataset. The MIoU of the Deeplabv3+-Mobilenetv2 is 90.33%, the model prediction speed is 41.14 f/s, and the number of model parameters is 11.18 MB. The MIoU of the DeepLabV3+-MobilenetV2 model is only slightly reduced compared with that of the original DeepLabV3+ model. However, the DeepLabV3+-MobilenetV2 model has a 16.04 f/s improvement in model segmentation speed and an 86.13% reduction in the number of model parameters. By using MobilenetV2 instead of Xception as the backbone of the DeepLabV3+ model, the real-time performance and lightweight property of the algorithm have been significantly improved. This optimization is extremely beneficial for future deployment on mobile tea-picking devices. The design choice of MobileNetV2 as the backbone network is verified.

3.3. Analysis of DASPP and Dilation Rate Combinations

This section of the experiment is to verify the effect of the ASPP connection method and different dilation rates on the experimental results. Table 6 shows the effect of varying ASPP connection methods and dilation rate groups on MC-DM network performance when the backbone network is MobilenetV2. For the same expansion rate group, the segmentation speed and the number of parameters are essentially the same for the DASPP and ASPP models. However, the MIoUs of the DASPP model are higher than those of the ASPP model. The maximum improvement of the MIoU reaches 2.13%. These results show that the DASPP model with a larger receptive field can segment the tea sprout images more accurately, which validates the effectiveness of the DASPP connection.

Table 6. Comparison of the influence of different ASPP connection methods and dilation rates.

ASPP Connection	Dilation Rates	Receptive Field	MIoU	Parameters/MB	Speed (f/s)
ASPP	6, 12	25	85.47	6.75	45.05
	6, 12, 18	37	90.33	7.56	24.77
	6, 12, 18, 24	49	90.50	8.36	37.06
DASPP	6, 12	37	87.60	6.76	44.78
	6, 12, 18	73	91.85	7.63	40.82
	6, 12, 18, 24	122	92.02	8.56	36.89

In terms of dilation rate selection, the segmentation accuracy of the dilation rate DASPP(6,12) model is the lowest, and it has an MIoU of only 87.60%. The MIoU for the dilation rate DASPP(6,12,18) model and the dilation rate DASPP(6,12,18,24) model are 91.85% and 92.02%, respectively. The difference in MIoU values between them is 0.17. Thus, they can be considered to have the same segmentation accuracy. However, compared with the dilation rate DASPP(6,12,18) model, the model parameters of the dilation rate DASPP(6,12,18,24) model is increased by 0.93 MB, and the segmentation speed is decreased by 3.93 f/s. Therefore, we chose DASPP(6,12,18) as the combination of the atrous convolution dilation rate for MC-DM.

3.4. Comparison with Other Segmentation Models

We conducted comparative experiments between MC-DM and other segmentation models to further verify the effectiveness and feasibility of the MC-DM. Table 7 shows the evaluation results of different models for tea sprout segmentation.

Table 7. Evaluation results of different methods.

Model	MIoU/%	Parameters/MB	Speed (f/s)
PSPNet	83.23	29.93	40.85
SegNet	85.79	29.45	24.77
U-Net	86.29	13.40	25.79
DeepLabV3+	91.33	54.52	25.10
MC-DM	91.85	7.63	40.82

As shown in Table 7, the MIoU of MC-DM reaches 91.85%, which is 8.62%, 6.06%, 5.56%, and 0.52% higher than those of PSPNet [20], SegNet [21], U-Net [22], and the DeepLabV3+, respectively. The MC-DM model has the highest segmentation accuracy. In terms of model lightweight property, the model parameter of MC-DM is only 7.63 MB, which is 74.51%, 74.09%, 43.06%, and 86.01% lower than those of PSPNet, SegNet, U-Net, and the DeepLabV3+, respectively. The MC-DM model performs optimally in terms of model lightweight property. In addition, the segmentation speed of the MC-DM model reaches 40.82 f/s, which is 16.05 f/s, 15.03 f/s, and 15.72 f/s better than those of SegNet, U-Net, and DeepLabV3+, respectively. The segmentation speed of the PSPNet model is the same as that of the MC-DM model. After the above-mentioned comparison, the MC-DM model achieves excellent performance in three metrics: segmentation accuracy, model lightweight, and segmentation speed.

Figure 8 shows the prediction results of different segmentation models for tea sprout images. The first and fifth columns are the original image, annotated image, PSPNet segmentation results, SegNet segmentation results, U-Net segmentation results, DeepLabV3+ segmentation results, and MC-DM segmentation results, respectively. The segmentation results of PSPNet are the worst. For images where the sprout color is similar to the background color, PSPNet nearly loses its segmentation ability. SegNet and U-Net segmentation results are better than PSPNet segmentation results. However, many pixel segmentation errors are found in their segmentation results. MC-DM and DeepLabV3+ segmentation results are close to the annotated images. However, the segmentation results of the MC-DM model for tea sprouts are more refined.

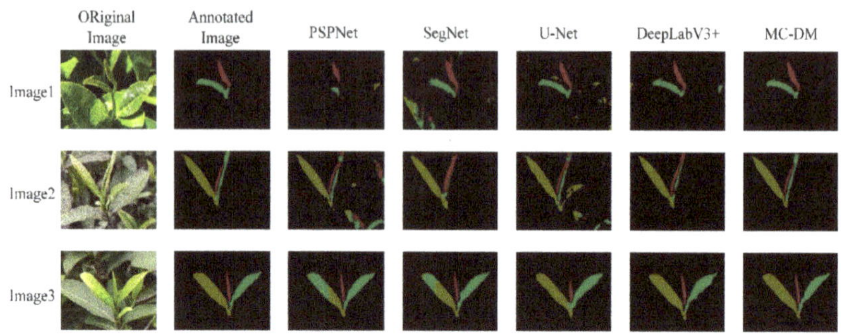

Figure 8. Prediction results of different segmentation models.

The comparison results in Table 7 and Figure 8 also show that the MC-DM model can achieve better tea sprout segmentation with lower computational resources. It can effectively segment the background, single bud, one leaf, and two leaves in the tea sprout image.

3.5. Picking Point Marking Validation

The two-dimensional coordinates of the tea sprout picking point are calculated using the tea images segmented with the trained MC-DM model and the coordinates are marked on the tea sprout pictures to further evaluate the segmentation effect of MC-DM and verify the effectiveness of the pick point localization method. When the picking point is located on the corresponding tea stalk, it is recorded as successful marking; otherwise, it is recorded as failed marking. Figure 9 shows the marking results of some picking points. Notably, the blue points are picking points for single bud, the green points are picking points for one bud with one leaf, and the red points are picking points for one bud with two leaves. Failed picking points are marked with a red frame. Our method has excellent results in locating the tea sprout picking points in the natural background. Most of the tea sprout picking points under different illumination are correctly marked.

Figure 9. Results of picking point marking. (**a**–**d**) Correctly marked picking points; (**e**,**f**) Incorrectly marked picking points.

Tables 8–10 show the picking point identification statistics of 1100 tea sprouts. The identification accuracy of all three types of picking points is over 80%. The recognition accuracy of one bud with one leaf picking point reaches 90.07%. Considering that premium tea is mainly picked from one bud with one leaf, the proposed picking point identification method meets the requirements of the picking robot for the picking point identification accuracy.

Table 8. Identification results of picking point for single bud.

Sample	1	2	3	4	5	Total
Number of picking points	222	222	222	222	222	1110
Number of successful identifications	187	177	182	186	184	916
Number of failed identifications	35	45	40	36	38	194
recognition accuracy/%	84.23	79.73	81.98	83.78	82.88	82.52

Table 9. Identification results of picking point for one bud with one leaf.

Sample	1	2	3	4	5	Total
Number of picking points	221	221	222	222	222	1108
Number of successful identifications	201	199	193	201	204	998
Number of failed identifications	20	22	29	21	18	110
recognition accuracy/%	90.95	90.00	86.94	90.54	91.89	90.07

Table 10. Identification results of picking point for one bud with two leaves.

Sample	1	2	3	4	5	Total
Number of picking points	162	167	161	172	166	828
Number of successful identifications	132	149	136	145	140	702
Number of failed identifications	30	18	25	27	26	126
recognition accuracy/%	81.48	89.22	84.47	84.30	84.34	84.78

The main reasons for the failure of picking point identification are as follows. First, the MC-DM model has segmentation errors for sprout stems in some tea images due to the small target size of spout stems compared with the overall tea sprout. This condition leads to the failure in marking the corresponding picking points. Second, some of the single bud stems are wrapped by young leaves (Figure 9e), which leads to the incorrect identification of the picking point. It is the main reason for poor single bud picking point identification. The stems of one bud with two leaves are also located at the bottom of the tea sprout. It is easily obscured by old leaves (Figure 9f), which results in failure in picking point identification.

4. Conclusions

This study proposed a lightweight segmentation network based on DeeplabV3+, named MC-DM, to segment tea sprouts and identify the picking points. First, the lightweight MobileNetV2 replaced the original backbone network Xception to reduce the number of model parameters and improve the speed of model segmentation. Then, a densely connected spatial pyramidal pooling module was introduced into the MC-DM network, which enabled the network to obtain denser pixel sampling and a large receptive field. Finally, the tea sprout picking points were identified using the Shi–Tomasi algorithm based on the segmentation results of the MC-DM network. The experimental results show that MC-DM achieves 91.58% of MIoU for tea sprouts. MC-DM has higher segmentation accuracy and fewer network parameters than other segmentation methods. In addition, the recognition accuracy rates of picking points for single bud, one bud with one leaf, and one bud with two leaves are 82.52%, 90.07%, and 84.78%, respectively. Its performance meets the requirements of the tea picking robot for picking point recognition accuracy.

Author Contributions: Conceptualization, C.Y. and Z.C.; methodology, C.Y.; software, C.Y. and Z.C.; validation, Z.C. and Z.L.; formal analysis, C.Y. and Z.L.; investigation, C.Y., Z.C. and Y.L.; data curation, R.L. and Y.L.; writing—original draft preparation, C.Y.; writing—review and editing, C.Y., Z.C.; visualization, R.L. and H.X.; supervision, P.L. and B.X.; project administration, P.L. and B.X.; funding acquisition, P.L. and B.X. All authors have read and agreed to the published version of the manuscript.

Funding: This research was funded by the National Natural Science Foundation of China (grant number 61562009), National Key Research and Development Program (grant number 2016YFD0201305-07), Open Fund Project in Semiconductor Power Device Reliability Engineering Center of Ministry of Education (grant number ERCMEKFJJ2019-06), and the Guizhou University Introduced Talent Research Project (grant number 2015-29).

Institutional Review Board Statement: Not applicable.

Informed Consent Statement: Not applicable.

Data Availability Statement: Not applicable.

Conflicts of Interest: The authors declare no conflict of interest.

References

1. Zhang, L.; Zou, L.; Wu, C.; Chen, J.; Chen, H. Locating Famous Tea's Picking Point Based on Shi-Tomasi Algorithm. *Comput. Mater. Contin.* **2021**, *69*, 1109–1122. [CrossRef]
2. Zhu, Y.; Wu, C.; Tong, J.; Chen, J.; He, L.; Wang, R.; Jia, J. Deviation Tolerance Performance Evaluation and Experiment of Picking End Effector for Famous Tea. *Agriculture* **2021**, *11*, 128. [CrossRef]
3. Chen, Y.; Chen, S. Localizing plucking points of tea leaves using deep convolutional neural networks. *Comput. Electron. Agric.* **2020**, *171*, 105298. [CrossRef]
4. Zhang, L.; Zou, L.; Wu, C.; Jia, J.; Chen, J. Method of famous tea sprout identification and segmentation based on improved watershed algorithm. *Comput. Electron. Agric.* **2021**, *184*, 106108. [CrossRef]
5. Zhao, B.; Wei, D.; Sun, W.; Liu, Y.; Wei, K. Research on tea bud identification technology based on HSI/HSV color transformation. In Proceedings of the 2019 6th International Conference on Information Science and Control Engineering (ICISCE), Shanghai, China, 20–22 December 2019; pp. 511–515. [CrossRef]
6. Qian, C.; Li, M.; Ren, Y. Tea sprouts segmentation via improved deep convolutional encoder-decoder network. *IEICE Trans. Inf. Syst.* **2020**, *E103-D*, 476–479. [CrossRef]
7. Qi, F.; Xie, Z.; Tang, Z.; Chen, H. Related study based on Otsu Watershed Algorithm and New Squeeze-and-Excitation Networks for Segmentation and Level Classification of Tea Buds. *Neural Process. Lett.* **2021**, *53*, 2261–2275. [CrossRef]
8. Hu, G.; Li, S.; Wan, M.; Bao, W. Semantic segmentation of tea geometrid in natural scene images using discriminative pyramid network. *Appl. Soft Comput.* **2021**, *113*, 107984. [CrossRef]
9. Yang, H.; Chen, L.; Chen, M.; Ma, Z.; Deng, F.; Li, M.; Li, X. Tender Tea Shoots Recognition and Positioning for Picking Robot Using Improved YOLO-V3 Model. *IEEE Access* **2019**, *7*, 80998–1810111. [CrossRef]
10. Peng, H.; Xue, C.; Shao, Y.; Chen, K.; Xiong, J.; Xie, Z.; Zhang, L. Semantic segmentation of litchi branches using DeepLabV3+ model. *IEEE Access* **2020**, *8*, 164546–164555. [CrossRef]
11. Song, Z.; Zhou, Z.; Wang, W.; Gao, F.; Fu, L.; Li, R.; Cui, Y. Canopy segmentation and wire reconstruction for kiwifruit robotic harvesting. *Comput. Electron. Agric.* **2021**, *181*, 105933. [CrossRef]
12. Ayhan, B.; Kwan, C. Tree, Shrub, and Grass Classification Using Only RGB Images. *Remote Sens.* **2020**, *12*, 1333. [CrossRef]
13. Zhang, X.; Fu, L.; Karkee, M.; Whiting, M.D.; Zhang, Q. Canopy Segmentation Using ResNet for Mechanical Harvesting of Apples. *IFAC-PapersOnLine* **2019**, *52*, 300–305. [CrossRef]
14. Liu, C.; Zhao, C.; Wu, H.; Han, X.; Li, S. ADDLight: An Energy-Saving Adder Neural Network for Cucumber Disease Classification. *Agriculture* **2022**, *12*, 452. [CrossRef]
15. Xiong, J.; Lin, R.; Liu, Z.; He, Z.; Tang, L.; Yang, Z.; Zou, X. The recognition of litchi clusters and the calculation of picking point in a nocturnal natural environment. *Biosyst. Eng.* **2018**, *166*, 44–57. [CrossRef]
16. Peng, Y.; Wang, A.; Liu, J.; Faheem, M. A Comparative Study of Semantic Segmentation Models for Identification of Grape with Different Varieties. *Agriculture* **2021**, *11*, 997. [CrossRef]

17. Ma, D.; Li, P.; Huang, X.; Zhang, Q.; Yang, X. Efficient semantic segmentation based on improved DeepLabV3+. *Comput. Eng. Sci.* **2022**, *44*, 737–745. [CrossRef]
18. Yao, X.; Guo, Q.; Li, A. Light-Weight Cloud Detection Network for Optical Remote Sensing Images with Attention-Based DeeplabV3+ Architecture. *Remote Sens.* **2021**, *13*, 3617. [CrossRef]
19. Sandler, M.; Howard, A.; Zhu, M.; Zhmoginov, A.; Chen, L. Mobilenetv2: Inverted residuals and linear bottlenecks. In Proceedings of the IEEE Conference on Computer Vision and Pattern Recognition, Salt Lake City, UT, USA, 17–19 June 2018; pp. 4510–4520. [CrossRef]
20. Zhao, H.; Shi, J.; Qi, X.; Wang, X.; Jia, J. Pyramid Scene Parsing Network. In Proceedings of the IEEE Conference on Computer Vision and Pattern Recognition, Honolulu, HI, USA, 21–26 July 2017; pp. 2881–2890. [CrossRef]
21. Badrinarayanan, V.; Kendall, A.; Cipolla, R. SegNet: A Deep Convolutional Encoder-Decoder Architecture for Image Segmentation. *IEEE Trans. Pattern Anal. Mach. Intell.* **2017**, *3*, 2481–2495. [CrossRef] [PubMed]
22. Ronneberger, O.; Fischer, P.; Brox, T. U-net: Convolutional networks for biomedical image segmentation. *IEEE Trans. Pattern Anal. Mach. Intell.* **2015**, *39*, 234–241. [CrossRef]

agriculture

MDPI

Article

Influences of Government Policies and Farmers' Cognition on Farmers' Participation Willingness and Behaviors in E-Commerce Interest Linkage Mechanisms during Farmer–Enterprise Games

Xiaolu Wei and Junhu Ruan *

College of Economics and Management, Northwest A&F University, Xianyang 712100, China
* Correspondence: rjh@nwsuaf.edu.cn

Abstract: E-commerce interest linkage mechanisms serve as an effective solution to the problems of farmer–market cooperation, agricultural supply-side reforms, and farmers' income growth. This study, guided by the theory of planned behavior, undertook an evolutionary game analysis of farmer–enterprise cooperation with government interventions with farmers. Based on data from 554 questionnaires administered in Mei County, Shaanxi Province, China, this study found a difference between the realistic and optimal choices of farmers. In addition, this study used a structural equation model to investigate the influence of government policies and farmers' cognition on the participation willingness and behaviors of farmers in e-commerce interest-linkage mechanisms. The results showed that the optimal choice for farmers in a farmer–enterprise cooperative game is participation in e-commerce, and government policies can be used to improve farmer–enterprise e-commerce interest-linkage mechanisms. Farmers' basic characteristics and experiences impacted their cognition of e-commerce, which, in turn, had a significant positive effect on their e-commerce participation willingness and behaviors. Government policies had a positive effect on farmers' experiences, cognition of e-commerce, and participation behaviors, but no direct positive impact on farmers' willingness to participate. Government policies and farmers' basic characteristics interacted and acted together on the participation willingness and behavior of farmers.

Keywords: e-commerce interest linkage; participation willingness and behaviors; government policies; farmers' cognition; evolutionary game model; structural equation model

Citation: Wei, X.; Ruan, J. Influences of Government Policies and Farmers' Cognition on Farmers' Participation Willingness and Behaviors in E-Commerce Interest Linkage Mechanisms during Farmer–Enterprise Games. *Agriculture* **2022**, *12*, 1625. https://doi.org/10.3390/agriculture12101625

Academic Editor: Dimitre Dimitrov

Received: 24 August 2022
Accepted: 3 October 2022
Published: 6 October 2022

1. Introduction

The rapid growth of the economy and the fast rise of internet enterprises in China in recent years have led to tremendous changes in domestic business models. E-commerce is gradually developing in rural areas that are otherwise dominated by traditional sales methods. In 2005, the Chinese government proposed e-commerce business models, and in 2022, it provided clear, specific, and long-term development paths for rural e-commerce and digital rural areas: continue to promote the integrated development of primary, secondary, and tertiary industries in rural areas; encourage various regions to expand various agricultural functions; explore the diversified values of rural areas; and focus on the development of rural e-commerce. E-commerce is now accelerating its penetration into rural areas, with China's rural online retail sales reaching CNY 2.05 trillion in 2021, up 11.3 percent year on year. Various types of e-commerce interest linkage (EIL) mechanisms were established in the Yangtze Triangle area, Greater Bay area, and Chinese central and western areas, and rural e-commerce is booming. Rural e-commerce not only changes the models for the sale of agricultural products but also facilitates employment, income, and other aspects in rural areas. However, most farmers currently continue to sell agricultural products through traditional methods [1]. Their willingness to participate and rural e-commerce behaviors are strongly impacted by their cognition of e-commerce and government policies. For this reason, research is needed on the factors that influence government policies and the impact of farmers' cognition on farmers' participation willingness and behaviors regarding EIL mechanisms [2,3]. This research, which used a farmer–enterprise cooperative game, can play a vital role in the implementation of agricultural supply-side reforms and the development of the rural economy.

Different from the agricultural food market electronic trading platforms [4] with industry chain-type, intermediary, professional, and alliance-type operation models in the United States and Europe, multiple types of interest linkage models have taken shape regarding lands, funds, labor, technologies, and sites as core elements in China; among these, the "farmer + enterprise" model is relatively common and effective [5].

Most studies on e-commerce participation explore factors that significantly influence the operation of these mechanisms from the perspective of enterprises [6]. The factors explored include the logistics service quality [7] and soundness of product supply chains, [8] development of scale economy and market internet coverage rate, [9] internal and external environment assessment of enterprises, logistics target setting, and strategic supply of supply chains [10]. In their research from the perspective of farmers, Cui et al. found that farmers' cognitive dimensions [11], social innovation [12], endowment, and regional environment [13] have a notable impact on their willingness to participate in e-commerce. Zhang found that joining a cooperative can decrease farmers' willingness to participate in e-commerce [14]. Luo et al. found that farmers' age, level of education [15], family income [16], personal characteristics [17], transaction costs [18], and other factors exert effects on farmers' participation behaviors regarding e-commerce. In terms of research methods, most scholars used a multivariate logistic model, structural equation model, and structural equation model, while a few scholars studied the related problems of supply chain coordination and put forward coordination schemes through evolutionary game theory [19–21].

As noted, most existing research on farmers' participation in e-commerce mechanisms takes the position of enterprises as primary. There is, by contrast, little game theory research on the participation behaviors of farmers and enterprises regarding EIL mechanisms. Although there is current research on the factors that influence their willingness to participate and their engagement in e-commerce, the perspectives are singular, and the farmers' willingness and behaviors are not combined for analysis. The multivariate logistic, interpretative structural, and structural equation models are the methods that are most used. Concrete reasons for variations in farmers' participation willingness and behaviors in e-commerce and reality have not been studied in any depth using the evolutionary game model.

Therefore, compared with the relevant literature, the novelty of this study lay in the use of the theory of planned behavior, which was applied to conduct an evolutionary game analysis on farmer–enterprise interest linkage mechanisms with government interventions and farmers as the subject. The game analysis results, along with the results of previous empirical studies of farmers in fruit planting (such as kiwi and strawberry) in Mei County, Shaanxi Province, revealed a difference between farmers' realistic and optimal choices. The structural equation model was then used to estimate and examine the causal relationship between farmers' characteristics (F1), experiences (F2), cognition (F3), behaviors (F4), willingness (F5), and government policies (F6). A corresponding analysis was carried out to put forward future directions.

2. Materials and Methods

2.1. Theoretical Hypotheses

Ajzen's theory of planned behavior predicts that all factors influencing behavior do so indirectly via behavioral intention. According to this theory, indicators such as behavioral attitude, external subjective norms, and perceived behavioral control can be used to measure the extent of behavioral intention to participate. For this reason, it was assumed in this study that with EIL mechanisms, farmers' willingness to participate and their e-commerce behaviors are influenced by internal and external factors, including farmers' basic characteristics and cognition, as well as government policies; various influencing factors were thus analyzed in this case.

Farmers' willingness to participate in EIL mechanisms is highly correlated with their behavior; this relationship was previously fully considered in theoretical and empirical analysis and will thus not be considered here. The theoretical game analysis concluded that farmers' optimal choices are inconsistent with reality. Hence, farmers' cognition, government policies, and other factors influencing farmers' participation willingness and behaviors in relation to EIL mechanisms were further considered.

2.2. Research Hypotheses

2.2.1. Farmers' Characteristics

Farmer characteristics can be thought of as their intrinsic resources. Farmers with a greater resource endowment have a comparative advantage and are able to enhance their competitiveness in market transactions. Income, level of education, social relations, and other characteristics have a significant impact on farmers' participation in cooperative economic organizations [22–24]. On this basis, the research hypothesis H1 was proposed.

2.2.2. Experiences

Experiences represent farmers' valuable knowledge about relevant technologies and markets accumulated in past agricultural production and operation activities they were occupied in, including transferring the external training into self-ability, long-term labor accumulation, and other activities. These previous experiences help farmers to know about e-commerce and accumulate knowledge, which then influences their participation willingness and behaviors regarding e-commerce [25,26]. On this basis, the research hypothesis H2 was proposed.

2.2.3. Government Policies

Government policies are crucial external factors to guarantee farmers' participation in EIL mechanisms. Governments have established agricultural cooperatives and other village organizations [27,28], promoted farmer–enterprise cooperation [29,30], and other policies to improve the level of farmers' knowledge of e-commerce mechanisms, thereby facilitating their participation in them [17,31–33]. On this basis, the research hypotheses H5–H8 were proposed.

2.2.4. Farmers' Cognition

Farmers' cognition concerns farmers' self-perception of the level of difficulty, marketing effects, and development anticipation of participating in EIL mechanisms. As more and more farmers start to learn about online shopping, their attitude toward the Internet and information is gradually changing [34]; problems of distrust, security concerns, and their sense of loyalty have improved; and their cognitive attitudes toward participation are more positive, making them more likely to take part in e-commerce mechanisms [35–38]. On this basis, the research hypotheses H3 and H4 were proposed.

2.2.5. Interaction

Under the theory of planned behavior, three variables, namely, behavioral attitudes, external subjective norms, and perceived behavioral control, interact and act together on farmers' participation willingness and behaviors regarding e-commerce. On this basis, the research hypothesis H9 was proposed. Based on the abovementioned hypotheses, several research hypotheses were formulated and are shown in Table 1.

Table 1. Research hypotheses.

Hypothesis Path	Hypothesis Content	Supporting References
H1	Farmers' basic characteristics positively influence their cognition of e-commerce.	[22–24]
H2	Farmers' experiences positively influence their cognition of e-commerce.	[25,26]
H3	Farmers' cognition of e-commerce positively impacts their participation willingness.	[34–38]
H4	Farmers' cognition of e-commerce positively impacts their participation behaviors regarding e-commerce.	[34–38]
H5	Government policies positively affect farmers' experiences.	[17,31–33]
H6	Government policies positively affect farmers' cognition of e-commerce.	[17,27–33]
H7	Government policies exert a positive influence on farmers' participation willingness.	[26,39]
H8	Government policies exert a positive influence on farmers' participation behaviors regarding e-commerce.	[17,31–33]
H9	Government policies interact with farmers' basic characteristics and together act on farmers' participation willingness and behaviors regarding e-commerce.	The theory of planned behavior

2.3. Theoretical Models

In line with the aforesaid theoretical and research hypotheses, a model of farmers' participation willingness and behaviors regarding EIL mechanisms was constructed and is depicted in Figure 1.

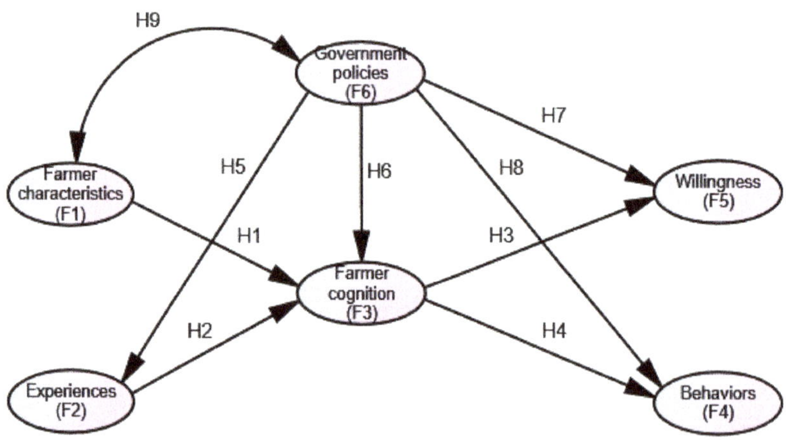

Figure 1. Model of farmers' participation willingness and behaviors regarding EIL mechanisms.

2.4. Variable Descriptions

Shaanxi Province, a major province of traditional agriculture in China, actively responds to the national call to encourage the construction and development of digital agricultural and rural systems and advances the establishment of rural EIL mechanisms. This research group planned to analyze influencing factors in the regions where e-commerce systems had sound development but were less adopted by farmers and put forward targeted suggestions to provide experience for the development in other regions. Therefore, empirical research was carried out on 56 villages in one street and seven towns in Mei County, Baoji City, Shaanxi Province, China, from March to May 2022, employing questionnaires, interviews, and other survey sampling methods. A total of 604 questionnaires were distributed, and 554 valid questionnaires were recovered, with an effectiveness rate of

91.7%. The proportion of agricultural products examined by the sample group that were subject to e-commerce was 66%.

IBM SPSS 26.0 was employed to conduct the descriptive analysis (Tables 2 and 3) using six variables and 19 items in the model of farmers' participation willingness and behaviors in EIL mechanisms. As Table 2 shows, the indicator of participation behaviors took values of 0 or 1, while the remaining items had more than three options on average.

Table 2. Sample basic information statistics.

Variable Name	Classification	Sample Number	Frequency (%)
Gender	Male	389	70.1
	Female	166	29.9
Age	Under 30 years old	216	38.9
	30–40 years old	238	42.9
	40–50 years old	68	12.3
	50–60 years old	29	5.2
	Over 60 years old	4	0.7
Level of education	Zero	5	0.9
	Elementary school	17	3.1
	Junior high school	124	22.3
	Senior high school	224	40.4
	College or above	185	33.3
Planting year	Within 3 years	29	5.2
	3–5 years	43	7.7
	5–10 years	256	46.1
	10–20 years	106	19.1
	Over 20 years	121	21.8

Table 3. Variable definitions and descriptive statistics.

Variable Name	Item	Variable Definition and Assignment	Average	Standard Deviation
Participation behaviors	Participation in EIL mechanisms or not	1 = participation, 0 = non-participation	0.66	0.475
Willingness to participate	Willingness to participate in EIL mechanisms	1 = absolutely not, 2 = partially not, 3 = normally, 4 = partially willing, 5 = totally willing	3.70	1.119
Willingness to encourage surrounding people to participate	Willingness to encourage acquaintances to participate in EIL mechanisms	1 = absolutely not, 2 = partially not, 3 = normally, 4 = partially willing, 5 = totally willing	3.66	1.093
Daily online time	Daily online time	1 = 0–2 h, 2 = 2–4 h, 3 = 4–6 h, 4 = 6–8 h, 5 = over 8 h	4.18	0.818
Social relations	Association with relatives and friends who participate in e-commerce	1 = never, 2 = sometimes, 3 = normally, 4 = often, 5 = always	4.00	0.926
Level of education	Level of education	1 = zero, 2 = elementary school, 3 = junior high school, 4 = senior high school, 5 = college or above	4.02	0.873
Age	Age interval	1 = over 60 years old, 2 = 50–60 years old, 3 = 40–50 years old, 4 = 30–40 years old, 5 = under 30 years old	4.14	0.876
Planting year	Agricultural planting year	1 = within 3 years, 2 = 3–5 years, 3 = 5–10 years, 4 = 10–20 years, 5 = over 20 years	3.45	1.074
E-commerce training	Frequency of participation in e-commerce training	1 = never, 2 = sometimes, 3 = normally, 4 = often, 5 = always	3.46	1.173
Technical training	Frequency of participation in planting technical training	1 = never, 2 = sometimes, 3 = normally, 4 = often, 5 = always	3.42	1.189
Cognition of EIL mechanisms	Cognition of EIL mechanisms	1 = absolutely not, 2 = partially not, 3 = normally, 4 = partially, 5 = totally	3.32	1.194
Mastery of and proficiency in e-commerce operation	Mastery of and proficiency in e-commerce operation	1 = absolutely not, 2 = partially not, 3 = normally, 4 = partially, 5 = totally	3.63	1.240

Table 3. *Cont.*

Variable Name	Item	Variable Definition and Assignment	Average	Standard Deviation
Cognition of marketing effects of e-commerce platforms	Marketing effects of e-commerce platforms	1 = very poor, 2 = partially poor, 3 = normal, 4 = partially good, 5 = very good	3.49	1.123
E-commerce future development anticipation	E-commerce future development anticipation	1 = very poor, 2 = partially poor, 3 = normally, 4 = partially good, 5 = very good	4.06	0.836
Understanding of government e-commerce policies	Understanding of government policies on e-commerce	1 = absolutely not, 2 = partially not, 3 = normally, 4 = partially, 5 = totally	3.58	1.183
Publicity	Government publicity for policies on e-commerce	1 = very small, 2 = partially small, 3 = normally, 4 = partially large, 5 = very large	3.38	1.217
Subsidies	Government subsidy for e-commerce participation	1 = never heard of nor accepted, 2 = heard of but not accepted, 3 = heard of and partially accepted, 4 = heard of and totally accepted	3.08	1.348
Training	E-commerce and technical training organized by the government	1 = never heard of nor accepted, 2 = heard of but not accepted, 3 = heard of and partially accepted, 4 = heard of and totally accepted	3.08	1.334
Supervision	Government supervision of e-commerce policies	1 = very small, 2 = partially small, 3 = normally, 4 = partially large, 5 = very large	3.24	1.229

2.5. Model Construction

2.5.1. Evolutionary Game Model

Based on bounded rationality, evolutionary game theory goes against the assumption of the perfect rationality of economic actors in traditional game theory. The choice of strategy of game subjects is continuously adapting and tends to be locally stable in the end. In this study, game subjects are farmers and enterprises and, according to evolutionary game theory, they act under bounded rationality. In addition, government interventions are included in the game, and farmers who participate in EIL mechanisms are subject to only positive external influences, such as technologies and publicity. The strategy space of farmers is {participation in mechanisms, non-participation in mechanisms}, while that of enterprises is {providing platforms, not providing platforms}. Their strategy selection is shown in Table 4.

Table 4. Selectable strategies of farmers and enterprises.

Participant	Selectable Strategy	
Farmers	Participation	Non-participation
Enterprises	Providing platforms	Not providing platforms

The probability of farmers' participation in EIL mechanisms is expressed as x and $(1 - x)$ refers to the probability of their non-participation. The probability of enterprises providing platforms is denoted by y, and $(1 - y)$ denotes the probability of not providing platforms; specifically, $(0 \leq x,\ y \leq 1)$. R_1 signifies the available revenue of farmers in the case of participation in EIL mechanisms, R_2 signifies the available revenue of farmers in the case of non-participation in EIL mechanisms, R_3 signifies the available revenue of enterprises in the case of providing platforms, and R_4 signifies the available revenue of enterprises in the case of not providing platforms. C_1 represents the costs of farmers in the case of participation in EIL mechanisms, C_2 represents the costs of farmers in the case of non-participation in EIL mechanisms, C_3 represents the costs of enterprises in the case of providing platforms, and C_4 represents the costs of enterprises in the case of not providing platforms. S represents positive government influences on farmers' participation in EIL mechanisms and P is the resource wasting caused by enterprises' providing platforms and farmers' not participating.

2.5.2. Structural Equation Model

In this study, the structural equation model was used. This overcomes the limitations of the general linear regression method and is applicable to the research of multiple variables with joint action. A total of six variables, namely, farmers' characteristics (F1), experiences (F2), cognition (F3), behaviors (F4), willingness (F5), and government policies (F6), were assumed as latent variables. In addition, 19 items in the questionnaires were selected as observed variables reflecting these latent variables. Path analysis was conducted by constructing the structural equation model to test the causal relationships between the latent variables. The general form of the model is stated as Equation (1):

$$Y = \alpha Y + \beta X + \varepsilon \quad (1)$$

where Y expresses the endogenous variable vectors, X expresses the exogenous variable vectors, α is a structural coefficient matrix that represents the relationships between endogenous variables, β is a structural coefficient matrix that represents the influences of exogenous variables on endogenous variables, and ε is a residual term that represents the unexplained parts.

3. Results and Discussion

3.1. Evolutionary Game Model

First, the evolutionary game model of the farmer–enterprise interest linkage mechanisms under government interventions was analyzed.

3.1.1. Model Analysis

Following the method proposed by Friedman, a local stability analysis was undertaken using the interest payoff matrix (shown in Table 5) and the Jacobian matrix to explore the evolutionarily stable strategy (ESS) formed by both sides of the games.

Table 5. Game payoff matrix for the strategy selection of farmers and enterprises with government intervention.

Farmer	Enterprise	
	Providing platforms	Not providing platforms
Participation in the linkage mechanisms	$R_1 - C_1 + S, R_3 - C_3$	$R_2 + S, R_4 - C_4$
Non-participation in the linkage mechanisms	$R_2 - C_2, R_3 - C_3 - P$	$R_2 - C_2, R_4 - C_4$

Following the correlation theory of evolutionary games, the calculation properties of expected revenues, the expected revenues of farmers in the case of participation U_1 and in the case of non-participation U_2 could be expressed as follows:

$$U_1 = y(R_1 - C_1 + S) + (1 - y)(R_2 + S) = y(R_1 - C_1 - R_2) + R_2 + S \quad (2)$$

$$U_2 = y(R_2 - C_2) + (1 - y)(R_2 - C_2) = R_2 - C_2 \quad (3)$$

The average expected revenue of the farmers was

$$U_A = xU_1 + (1 - x)U_2 \quad (4)$$

Based on the Malthusian dynamic equation, the replicated dynamic equation (t represents time) of the probability x of farmers choosing the "cooperation" strategy was

$$F(x) = \frac{dx}{dt} = x(U_1 - U_A) = x(1 - x)[y(R_1 - C_1 - R_2) + S + C_2] \quad (5)$$

Similarly, the expected revenue of enterprises in the case of providing platforms U_3 and not providing platforms U_4 was expressed as follows:

$$U_3 = x(R_3 - C_3) + (1 - x)(R_3 - C_3 - P) = xP + R_3 - C_3 \quad (6)$$

$$U_4 = x(R_4 - C_4) + (1 - x)(R_4 - C_4) = R_4 - C_4 \tag{7}$$

The average expected revenues of the enterprises was

$$\mathrm{U_B = yU_3 + (1 - y)U_4} \tag{8}$$

The replicated dynamic equation of the probability y of enterprises choosing the "cooperation" strategy was expressed as follows:

$$F(y) = \frac{dy}{dt} = y(U_3 - U_B) = y(1 - y)(xP + R_3 - C_3 - R_4 + C_4) \tag{9}$$

The replicated dynamic equation group consisting of Equations (5) and (9) was expressed as follows:

$$\begin{cases} F(x) = \frac{dx}{dt} = x(U_1 - U_A) = x(1 - x)[y(R_1 - C_1 - R_2) + S + C_2] \\ F(y) = \frac{dy}{dt} = y(U_3 - U_B) = y(1 - y)(xP + R_3 - C_3 - R_4 + C_4) \end{cases}$$

With the replicated dynamic equation group set as $F(x) = \frac{dx}{dt} = 0$ and $F(y) = \frac{dy}{dt} = 0$, the following five local equilibrium points were obtained: $A(0,0)$, $B(0,1)$, $C(1,0)$, $D(1,1)$, and $E(x^*, y^*)$.

$$(x^* = \frac{R_4 - C_4 + C_3 - R_3}{P}, \; y^* = \frac{S + C_2}{R_2 + C_1 - R_1}, \text{ and } (0 \le x, y \le 1))$$

$$(x^* = \frac{R_4 - C_4 + C_3 - R_3}{P}, \; y^* = \frac{S + C_2}{R_2 + C_1 - R_1}, \text{ and } (0 \le x, y \le 1))$$

The Jacobian matrix obtained via the replicated dynamic equation group (Equations (5) and (9)) was expressed as

$$J = \begin{bmatrix} (1 - 2x)[y(R_1 - C_1 - R_2) + S + C_2] & x(1 - x)(R_1 - C_1 - R_2) \\ y(1 - y)P & (1 - 2y)(xP + R_3 - C_3 - R_4 + C_4) \end{bmatrix} \tag{10}$$

$$\begin{aligned} DetJ = {} & (1 - 2x)[y(R_1 - C_1 - R_2) + S + C_2] \times (1 - 2y)(xP + R_3 - C_3 - R_4 + C_4) \\ & - x(1 - x)(R_1 - C_1 - R_2) \times y(1 - y)P \end{aligned} \tag{11}$$

$$TrJ = (1 - 2x)[y(R_1 - C_1 - R_2) + S + C_2] + (1 - 2y)(xP + R_3 - C_3 - R_4 + C_4) \tag{12}$$

With $DetJ > 0$ and $TrJ < 0$, the equilibrium point of the replicated dynamic equations was an ESS, which was obtained by considering the symbols of the determinants and the trace values of the Jacobian matrix at five local equilibrium points based on the assumed conditions and $(0 \le x, y \le 1)$.

According to the analysis results set out in Table 6, among the five local equilibrium points, only the point $D(1,1)$ was an ESS, indicating that the cooperation strategy was chosen by both farmers and enterprises. There were also three unstable points—$A(0,0)$, $B(0,1)$, and $C(1,0)$—indicating that non-cooperation or other strategies were chosen by farmers and enterprises. Additionally, there was a saddle point at $E(x^*, y^*)$.

Table 6. Local stability analysis of the evolutionary game system with government intervention.

Equilibrium Point	*DetJ*	Symbol	*TrJ*	Symbol	Stability
$(0,0)$	$(S+C_2)(R_3-C_3-R_4+C_4)$	$-$	$(S+C_2)+(R_3-C_3-R_4+C_4)$	$+/-$	Unstable
$(0,1)$	$(R_1+S+C_2-C_1-R_2)(R_4+C_3-C_4-R_3)$	$+$	$(R_1+S+C_2-C_1-R_2)+(R_4+C_3-C_4-R_3)$	$+/-$	Unstable
$(1,0)$	$-(S+C_2)(P+R_3-C_3-R_4+C_4)$	$-$	$-(S+C_2)+(P+R_3-C_3-R_4+C_4)$	$+/-$	Unstable
$(1,1)$	$(R_1+S+C_2-C_1-R_2)(P+R_3-C_3-R_4+C_4)$	$+$	$-(R_1+S+C_2-C_1-R_2)-(P+R_3-C_3-R_4+C_4)$	$-$	ESS
(x^*,y^*)	E	$+$	0		Saddle point

Note: $E=\frac{(C_3+R_4-R_3-C_4)(S+C_2)(P-C_3-R_4+R_3+C_4)(R_2+C_1-R_1-S-C_2)}{P(R_2+C_1-R_1)}$

3.1.2. Evolutionary Simulation

The dynamic evolutionary simulation of the game was carried out using MATLAB to allow for a visual assessment of the game between farmers and enterprises. The simulation cycles were set to 10, along with the following variables: $R_1=0.5$, $R_2=0.3$, $R_3=6$, $R_4=5$, $C_1=0.1$, $C_2=0.1$, $C_3=1$, $C_4=0.5$, $S=0.5$, and $P=0.1$. The initial values (x,y) of the numerical simulation were set as $(0.3,0.5)$, $(0.5,0.5)$, and $(0.5,0.8)$. The dynamic evolution process of the strategy selection of farmers and enterprises changing over time is displayed in Figure 2.

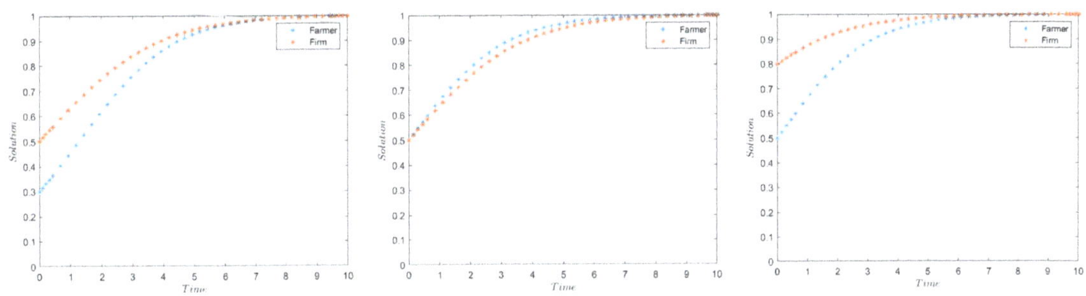

Figure 2. Dynamic evolution diagram of the strategy selection of farmers and enterprises.

As indicated in Figure 2, when the initial probability value was $(0.3,0.5)$, that is, when the probability of farmer participation was less than the probability of enterprise participation, the final dynamic evolution strategy was both parties participating. When the initial probability value was $(0.5,0.5)$, that is, when the probability of farmer participation was equal to the probability of enterprise participation, the final dynamic evolution strategy was both parties participating. When the initial probability value was $(0.5,0.8)$, that is, when the probability of farmer participation was greater than the probability of enterprise participation, the final dynamic evolution strategy was both parties participating. Thus, the various initial probability values for the strategy selection of farmers and enterprises, i.e., (x,y), resulted in final game evolution results that converged to the point $D(1,1)$, i.e., the ESS, meaning that the cooperation strategy was chosen by both farmers and enterprises.

From the government's point of view, the final evolution results for both sides of the games were also influenced by differences in the equation parameter S, as shown in Figure 3. With the other conditions unchanged and a change in the positive government influence S (set as 0.5, 1, and 2), regarding the farmers' participation in the linkage mechanisms, the evolutionary strategy of game subjects exhibited the trend depicted in Figure 3 with the adjustment of S. As depicted in Figure 3, as S increased, the change in the probability of farmers tending to cooperation became increasingly rapid. This indicated that the positive government influence S had a positive impact on the farmers' participation in the linkage mechanisms.

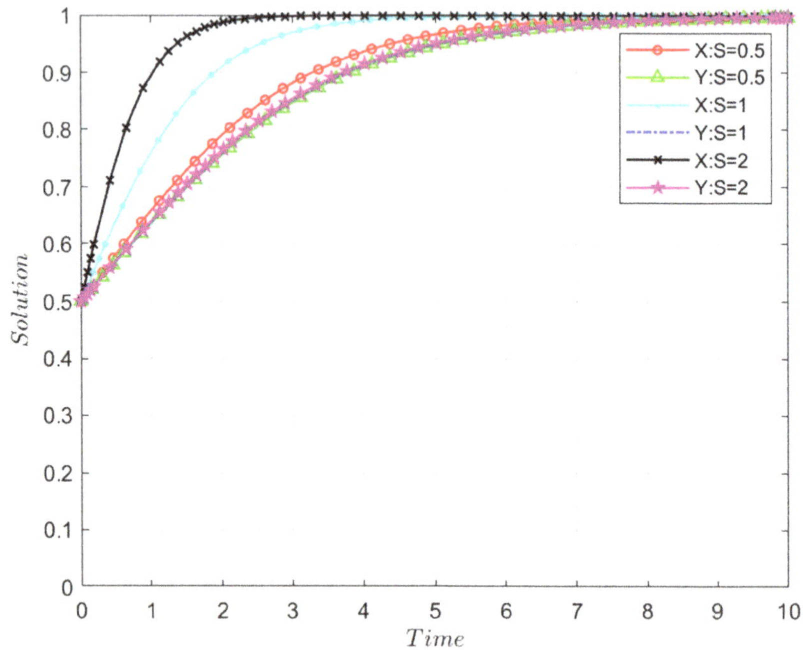

Figure 3. Trend of the farmer–enterprise evolutionary game strategy under government interventions.

In summary, the replicated dynamic equations $F(x) = \frac{dx}{dt} = 0$ and $F(y) = \frac{dy}{dt} = 0$ were solved through the construction of the evolutionary game model to obtain equilibrium points, and a local stability analysis was conducted to identify the ESS. Furthermore, MATLAB software was used for the numerical simulation. It was found that the optimal choice for farmers was to participate in the EIL mechanisms, and the implementation of government policies and systems could promote the improvement process of the EIL mechanisms for farmers and enterprises. This was contradictory to the reality that farmers' participation in the EIL mechanisms was not high. For this reason, a structural equation was constructed in terms of the farmers' basic characteristics and cognition and government policies to analyze the factors that affected farmers' participation willingness and behaviors.

3.2. Structural Equation Analysis

3.2.1. Reliability and Validity Tests

Reliability and validity analysis of the data collected via the questionnaires were conducted using the software SPSS 26.0 and Amos 24.0, with the participation behaviors (PB) in e-commerce not involved since they had only one observed variable. The test results of the other five variables are displayed in Table 7. The Cronbach's alphas of behavioral willingness, farmers' characteristics, experiences, farmers' cognition, and government policies were greater than or close to 0.8, with good measurement reliability. As for the validity test, the load capacities of the standard factors, as well as the KMO values, were all greater than 0.6, and the convergent validity AVE was greater than or close to 0.5, revealing good validity.

Table 7. Reliability and validity analysis.

Variable	Variable Setting	Dimensionality	Reliability	Validity		
			Cronbach's α	Load Capacity of Factors	KMO	AVE
F1	FCS1	Daily online time	0.893	0.766	0.794	0.689
	FCS2	Social relations		0.932		
	FCS3	Level of education		0.94		
	FCS4	Age		0.645		
F2	EXP1	Planting year	0.798	0.614	0.662	0.589
	EXP2	E-commerce training		0.88		
	EXP3	Technical training		0.785		
F3	COG1	Cognition of EIL mechanisms	0.838	0.619	0.825	0.528
	COG2	Mastery of and proficiency in e-commerce operation		0.83		
	COG3	Cognition of marketing effects of e-commerce platforms		0.658		
	COG4	E-commerce development anticipation		0.87		
	COG5	Understanding of government policies on e-commerce		0.614		
F5	WTP1	Willingness to participate	0.856	0.813	0.600	0.755
	WTP2	Willingness to encourage acquaintances to participate		0.921		
F6	GP1	Publicity	0.784	0.693	0.771	0.477
	GP2	Subsidy		0.617		
	GP3	Training		0.808		
	GP4	Supervision		0.628		

3.2.2. Model Fitting

The model fitness was judged using Amos 24.0. It was found that $\chi^2/df = 2.616$, meeting the standard value (smaller than 3), while $GFI = 0.935$ and $AGFI = 0.913$, both of which were greater than the ideal value (0.9). In addition, $RMSEA = 0.054$, which was smaller than 0.8 and met the ideal standard value, thus indicating an ideal absolute fit index. Furthermore, $CFI = 0.959$ and $TLI = 0.951$, both of which were greater than the ideal standard value (0.9), thus indicating an ideal value-added fit index. Therefore, the model had an ideal overall fitness, and the structural equation model was effective. The diagram of the final standardized path coefficients of the model is depicted in Figure 4.

3.2.3. Hypothesis Tests

According to Table 8, the path coefficient of government policies –> farmers' participation willingness in e-commerce (H7) failed to pass the significance test, with the absolute value of the statistic C.R. also being smaller than 2, suggesting that the relationship between the variables in the proposed H7 had no significant influence. The path coefficients of the other variables all passed the significance test, revealing that the relationships between the corresponding latent variables are significant.

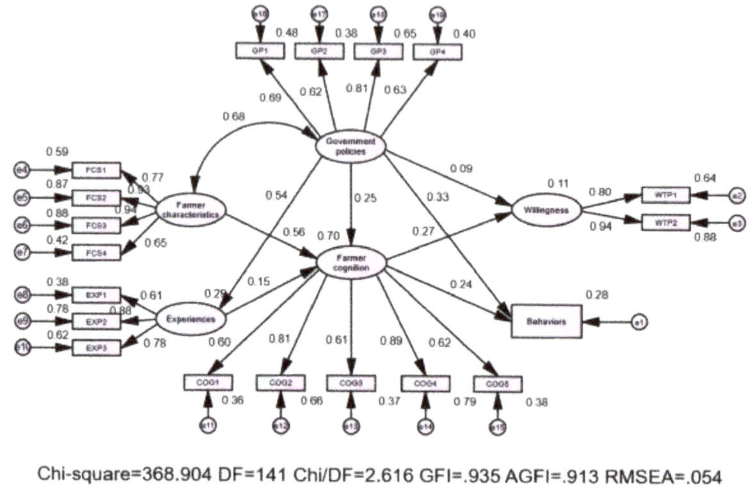

Figure 4. Diagram of the standardized path coefficients of the model.

Table 8. Path coefficients and their significance tests.

Path	Non-Standardization	S.E.	C.R.	P	Standard Path Coefficient	Corresponding Hypothesis
F1 –> F3	0.720	0.079	9.063	***	0.559	H1
F2 –> F3	0.121	0.033	3.688	***	0.155	H2
F3 –> F5	0.325	0.094	3.437	***	0.266	H3
F3 –> F4	0.156	0.043	3.642	***	0.241	H4
F6 –> F2	0.598	0.064	9.336	***	0.542	H5
F6 –> F3	0.220	0.051	4.354	***	0.254	H6
F6 –> F5	0.094	0.080	1.178	0.239	0.089	H7
F6 –> F4	0.185	0.039	4.719	***	0.329	H8
F1 <–> F6	0.325	0.034	9.500	***	0.679	H9

Note: F1—farmers' characteristics, F2—experiences, F3—cognition, F4—behaviors, F5—willingness, F6—government policies. The *** indicates that the statistical test has reached a 1% significance level.

3.2.4. Results Analysis

According to the analysis of the model results:

(1) Hypotheses H1 and H2 held, which was in agreement with the aforementioned hypothesis [22–26]. In other words, farmers' basic characteristics and experiences had a significant positive effect on their e-commerce cognition, with path coefficients of 0.559 and 0.155, respectively. This implied that farmers' characteristics had a significant effect on the improvement of their cognition. The standardized path coefficients of the four latent farmer characteristic variables FCS1–4 were 0.766, 0.932, 0.940, and 0.645, respectively, showing that the farmers' level of education had the greatest positive effect, followed by social relations, daily hours online, and age. The standardized path coefficients of the latent variables of experiences EXP1–3 were 0.614, 0.880, and 0.785, respectively, indicating that e-commerce training exhibited the greatest positive effects on farmers' experiences, followed by technical training and planting year.

(2) Hypotheses H3 and H4 held, which was in agreement with the aforementioned hypothesis [34–38]. That is, farmers' e-commerce cognition positively impacted their participation willingness and behaviors in e-commerce, with path coefficients of 0.266 and 0.241, respectively, implying that farmers' participation willingness and behaviors

in e-commerce were greatly affected by their cognition, which had a greater influence on their participation willingness. The standardized path coefficients of the latent variables of farmers' e-commerce cognition COG1–5 were 0.619, 0.830, 0.658, 0.870, and 0.614, respectively, signifying that e-commerce future development anticipation had the greatest positive effect on farmers' e-commerce cognition, followed by mastery of and proficiency in e-commerce operation, cognition of marketing effects of e-commerce platforms, cognition of EIL mechanisms, and understanding of government policies on e-commerce.

(3) Hypotheses H5, H6, and H8 held, which was in agreement with the aforementioned hypothesis [17,27–33]. Government policies positively affected farmers' experiences, cognition of e-commerce, and participation behaviors, with path coefficients of 0.542, 0.254, and 0.329, respectively, implying that farmers' e-commerce cognition and participation behaviors were greatly affected by the policies with the greatest influence on their experiences. However, hypothesis H7 did not hold, meaning that government policies had no direct positive influence on farmers' participation willingness, which was contrary to the extant literature [26,39]. They could, to some extent, indirectly affect farmers' participation willingness by affecting their basic characteristics, experiences, and cognition due to the lagging and weakening of the transmission mechanisms. As a result, farmers' participation willingness and behaviors in real life deviated from reality. The standardized path coefficients of the latent government-policy variables were 0.693, 0.617, 0.808, and 0.628, respectively, implying that government training had the greatest positive effect on government policies, followed by publicity, supervision, and subsidy.

(4) Hypothesis H9 held, which fits with the theory of planned behavior. In other words, government policies interacted with farmers' basic characteristics with a path coefficient of 0.679, implying that farmers' participation willingness and behaviors in e-commerce were jointly affected by government policies and their basic characteristics, in line with the theory of planned behavior.

3.3. Structural Equation Analysis

With the rapid development of the Internet, the cross-border integration of traditional agriculture and the Internet is inevitable. Therefore, in the EIL mechanisms, it is very important to determine the strategic choices of all parties. Based on the findings, the following conclusions were reached.

First, farmers' participation in the EIL mechanisms was the optimal choice in the farmer–enterprise cooperation games with government interventions, and the implementation of relevant government policies and systems could promote the improvement of the EIL mechanisms for farmers and enterprises.

Second, farmers' cognition of e-commerce was affected by their basic characteristics and experiences, and farmers' basic characteristics positively influenced the improvement of farmers' cognition in a more significant manner.

Third, farmers' cognition of e-commerce had a positive influence on their participation willingness and behaviors in e-commerce and exhibited a greater influence on their participation willingness.

Fourth, government policies had a significant positive effect on farmers' experiences, cognition of e-commerce, and participation behaviors, without any direct positive effect on their participation willingness.

Fifth, government policies and farmers' basic characteristics interacted and jointly acted on farmers' participation willingness and behaviors in e-commerce.

To some extent, this study made a more comprehensive demonstration of the previous relevant studies and combined the willingness and behavior from the perspective of farmers. A combination of an evolutionary game and structural equation model was used. Moreover, hypothesis H7 contradicted the existing literature [26,39] by showing that government policies had no direct impact on farmers' willingness to participate. However, at the same time, government policies can indirectly influence farmers' participation willingness in e-commerce by affecting their basic characteristics, experiences, and cognition. This explained the difference between farmers' realistic and optimal choices and showed that government policies were an external influencing factor of great importance.

4. Conclusions

Based on the theory of planned behavior, this study employed an evolutionary game analysis of farmer–enterprise cooperation with government intervention from the perspective of farmers. There was a difference between the realistic and optimal choices of farmers according to the evolutionary game analysis results combined with empirical research results. In addition, structural equation modeling was used to explore the impacts of government policies and farmers' cognition on farmers' participation willingness and behaviors regarding the EIL mechanisms. The following suggestions are proposed to increase farmers' participation willingness and behaviors regarding the EIL mechanisms.

Considering the great importance of government policies in encouraging farmers to participate in the EIL mechanisms, the relevant policies should be optimized, adjusted, and implemented. The government should attach great importance to e-commerce and technical training for farmers, increase publicity concerning e-commerce policies, grant further subsidies to e-commerce enterprises and participating farmers, promote the e-commerce knowledge and literacy of farmers, and improve farmers' satisfaction with e-commerce. In this way, farmers' participation willingness will be boosted, and the deviation between their participation willingness and behaviors will be reduced. Additionally, government departments should also strengthen everyday communication and interaction with farmers, understand the actual situation, and regulate and supervise with a view to securing effective protection of the interests of farmers.

Moreover, the government should constantly improve the construction of e-commerce infrastructure in rural areas and help to reduce the participation cost faced by farmers in e-commerce. According to the results here, farmers' cognition of e-commerce had a significant positive influence on their participation willingness and behaviors regarding e-commerce. The results revealed that farmers' self-perception of the level of difficulty, marketing effects, and development anticipation of participating in the EIL mechanisms were important factors that restricted the farmers' participation willingness and behaviors. Therefore, a diversity of policies should be adopted by governments at all levels in keeping with the local context to strengthen financial support for rural e-commerce; improve network infrastructures in rural areas, local logistics network design, and local logistics resource allocation; and reduce the cost to farmers of participation in e-commerce. In this way, the real interests of farmers will be ensured, and farmers will recognize that participation in e-commerce can increase their income. As a result, they will develop a positive attitude toward participating in e-commerce and grow a considerable interest in the EIL mechanisms, thus facilitating their participation in e-commerce mechanisms.

All forces should be fully mobilized to allow for participation in the EIL mechanisms to boost the high-quality development of e-commerce in rural areas. It requires the radiation effects of big farmers and e-commerce enterprises and the leadership example of village officials for participation in e-commerce mechanisms. This will allow for an increase in farmers' e-commerce participation and strengthen their interest linkage with e-commerce enterprises. It also helps to link small farmers to big markets, promotes agricultural supply-side reform, and boosts rural revitalization.

This study can be extended in several directions. This study presents an evolutionary game analysis of agricultural enterprise cooperation under government intervention, assuming the impact of government policies without practical quantification. Therefore, it would be interesting to study the impact of government-specific policy data on farmers' evolutionary stabilization strategies. In this study, government policy was considered to be deterministic. However, in other cases, government policy may be uncertain. Therefore, the introduction of a three-way evolutionary game will enrich the research. In this study, through the structural equation model, we studied what factors promoted farmers' willingness and behavior and found that government policies had no direct impact on farmers' willingness to participate, which is inconsistent with the expected hypothesis. Next, we can expand the model, reverse study which factors hinder farmers' participation, and conduct a comparative analysis.

Author Contributions: Conceptualization, X.W.; methodology, X.W.; software, X.W.; validation, X.W.; formal analysis, X.W.; investigation, J.R.; resources, X.W. and J.R.; data curation, X.W. and J.R.; writing—original draft preparation, X.W.; writing—review and editing, X.W.; visualization, X.W.; supervision, J.R.; project administration, J.R.; funding acquisition, J.R. All authors have read and agreed to the published version of the manuscript.

Funding: This research was funded by the National Nature Science Foundation of China (71973106, 72271202); Shaanxi Science Fund for Distinguished Young Scholars under grant 2021JC-21; Key Scientific Research Projects of Shaanxi Provincial Department of Education under grant 21JT043; SCO Institute of Modern Agricultural Development Program under grant 4, Northwest A&F University; and Tang Scholar of Northwest A&F University.

Institutional Review Board Statement: Not applicable.

Data Availability Statement: Not applicable.

Acknowledgments: I am very grateful for the support of Northwest A&F University.

Conflicts of Interest: The authors declare no conflict of interest.

References

1. Ackermann, S.; Adams, I.; Gindele, N.; Doluschitz, R. The role of e-commerce in the purchase of agricultural input materials. *Landtechnik* **2018**, *73*, 10–19.
2. Li, L.; Zeng, Y.; Ye, Z.; Guo, H. E-commerce Development and Urban-rural Income Gap: Evidence from Zhejiang Province, China. *Pap. Reg. Sci.* **2021**, *100*, 475–494. [CrossRef]
3. Karine, H. E-commerce development in rural and remote areas of BRICS countries. *J. Integr. Agric.* **2021**, *20*, 979–997.
4. Fritz, M.; Hausen, T.; Schiefer, G. Developments and development directions of electronic trade platforms in US and European agri-food markets: Impact on sector organization. *Int. Food Agribus. Manag. Rev.* **2004**, *7*, 1–21.
5. Pavlou, P.A.; Liang, H.; Xue, Y. Understanding and mitigating uncertainty in online exchange relationships: A principal-agent perspective. *MIS Quart.* **2007**, *31*, 105–136. [CrossRef]
6. Negrão, C.S.V. Impact of E-Commerce on Agricultural Business Success. In *Improving Business Performance Through Effective Managerial Training Initiatives*; IGI Global: Hershey, PA, USA, 2018; pp. 223–253.
7. Xing, Y.; Grant, D.B. Developing a framework for measuring physical distribution service quality of multi-channel and "pure player" internet retailers. *Int. J. Retail Distrib. Manag.* **2006**, *34*, 78–289. [CrossRef]
8. Bodini, A.; Zanoli, R. Competitive factors of the agro-food e-commerce. *J. Food Prod. Market.* **2011**, *17*, 241–260. [CrossRef]
9. Jiong, M.; Xu, L.; Huang, Q.; Li, C. Research on the e-commerce of agricultural products in Sichuan Province. *J. Digit. Inform. Manag.* **2013**, *11*, 97.
10. Dello Stritto, G.; Schiraldi, M.M. A strategy oriented framework for food and beverage e-supply chain management. *Int. J. Eng. Bus. Manag.* **2013**, *5*, 5–50. [CrossRef]
11. Schwering, D.S.; Sonntag, W.I.; Kühl, S. Agricultural E-commerce: Attitude segmentation of farmers. *Comput. Electron. Agric.* **2022**, *197*, 106942.
12. Cui, L.L.; Wang, L.J.; Wang, J.Q. An empirical analysis of social innovation factors' facilitation effects on the development of e-commerce in 'Taobao village'-taking Lishui, Zhejiang as an example. *Chin. Rural Econ.* **2014**, *12*, 50–60.
13. Ma, Z.B. Farmer's endowment, regional environment and their participation willingness in poverty alleviation through e-commerce–based on a questionnaire survey of 630 farmers in the ethnic border areas. *China Circul. Econ.* **2017**, *31*, 47–54.
14. Zhang, Y.F. Fresh fruit sales through e-commerce, farmers' participation willingness and the cooperative involvement-a survey data of farmers in Yantai cherry-producing area. *J. Nanjing Agric. Univ.* **2016**, *49–58*, 163–164.
15. Luo, W.C. Analysis of farmers' selection of the sales channels for agricultural products and influencing factors. *World Surv. Res.* **2013**, *35–37*, 52.
16. Bai, Y.W.; Ji, T.; Wang, J. Analysis of smallholder farmers' selection of e-commerce channels and influencing factors-an empirical investigation based on the Yantai cherry-producing area. *Rural Econ. Technol.* **2016**, *11*, 71–75.
17. Fecke, W.; Danne, M.; Musshoff, O. E-Commerce in Agriculture—The Case of Crop Protection Product Purchases in a Discrete Choice Experiment. *Comput. Electron. Agric.* **2018**, *151*, 126–135. [CrossRef]
18. Chen, H.W.; Mu, Y.Y. Social networks, transaction costs, and farmers' market participation behavior. *Econ. Surv.* **2020**, *37*, 45–53.

19. Johari, M.; Hosseini-Motlagh, S.M. Evolutionary behaviors regarding pricing and payment-convenience strategies with uncertain risk. *Eur. J. Oper. Res.* **2022**, *297*, 600–614. [CrossRef]
20. Hosseini-Motlagh, S.M.; Johari, M.; Nematollahi, M.; Pazari, P. Reverse supply chain management with dual channel and collection disruptions: Supply chain coordination and game theory approaches. *Ann. Oper. Res.* **2022**, 1–34. [CrossRef]
21. Hosseini-Motlagh, S.M.; Johari, M.; Pazari, P. Coordinating pricing, warranty replacement and sales service decisions in a competitive dual-channel retailing system. *Comput. Ind. Eng.* **2022**, *163*, 107862. [CrossRef]
22. Zhang, Q.W.; Zhou, H.P.; Lyu, S.J.; Hu, N.P. Analysis of influencing factors of farmers' willingness to participate in cooperatives-taking Liaodian Township, Acheng City, Heilongjiang Province as an example. *Agric. Technol. Econ.* **2013**, *3*, 98–104.
23. Wen, T.; Wang, X.H.; Yang, D.; Zhu, J. Behavioral characteristics, benefit mechanisms and decision-making effects of farmers' participation in cooperative economic organizations under the new situation. *Manag. World* **2015**, *7*, 82–97.
24. Briggeman, B.C.; Whitacre, B.E. Farming and the internet: Factors affecting input purchases online and reasons for non-adoption. In Proceedings of the 2008 Annual Meeting, Dallas, TX, USA, 2–6 February 2008.
25. Zeng, Y.W.; Qiu, D.M.; Shen, Y.T.; Guo, H.D. Study on the formation of Taobao Villages: Taking Dongfeng Village and Junpu village as examples. *Econ. Geogr.* **2015**, *35*, 90–97.
26. Li, X.; Sarkar, A.; Xia, X.; Memon, W.H. Village Environment, Capital Endowment, and Farmers' Participation in E-Commerce Sales Behavior: A Demand Observable Bivariate Probit Model Approach. *Agriculture* **2021**, *11*, 868. [CrossRef]
27. Mishra, A.K.; Williams, R.P.; Detre, J.D. Internet access and internet purchasing patterns of farm households. *Agric. Resour. Econ. Rev.* **2009**, *38*, 240–257. [CrossRef]
28. Hazell, P.; Poulton, C.; Wiggins, S.; Dorward, A. The future of small farms: Trajectories and policy priorities. *World Dev.* **2010**, *38*, 1349–1361. [CrossRef]
29. Markelova, H.; Meinzen-Dick, R.; Hellin, J.; Dohrn, S. Collective action for smallholder market access. *Food Policy* **2009**, *34*, 1–7. [CrossRef]
30. Key, N.; Runsten, D. Contract farming, smallholders, and rural development in Latin America: The organization of agroprocessing firms and the scale of outgrower production. *World Dev.* **1999**, *27*, 381–401. [CrossRef]
31. Baourakis, G.; Kourgiantakis, M.; Migdalas, A. The Impact of E-commerce on Agro-food Marketing: The Case of Agricultural Cooperatives, Firms and Consumers in Crete. *Br. Food J.* **2002**, *104*, 580–590. [CrossRef]
32. Banerjee, T.; Mishra, M.; Debnath, N.C.; Choudhury, P. Implementing E-Commerce Model for Agricultural Produce: A Research Roadmap. *Period. Eng. Nat. Sci.* **2019**, *7*, 302–310. [CrossRef]
33. Beckman, J.; Ivanic, M.; Jelliffe, J. Market Impacts of Farm to Fork: Reducing Agricultural Input Usage. *Appl. Econ. Perspect. Policy* **2021**. [CrossRef]
34. Abebe, G.K.; Bijman, J.; Kemp, R.; Omta, O.; Tsegaye, A. Contract farming configuration: Smallholders' preferences for contract design attributes. *Food Policy* **2013**, *40*, 14–24. [CrossRef]
35. Hennessy, T.; Läpple, D.; Moran, B. The digital divide in farming: A problem of access or engagement? *Appl. Econ. Perspect. Policy* **2016**, *38*, 474–491. [CrossRef]
36. Ardrey, J.; Denis, N.; Magnin, C.; Revellat, J. Unlocking the Online Retail Opportunity with European Farmers. *McKinsey Co.*. 2020. Available online: https://www.mckinsey.com/industries/agriculture/our-insights/unlocking-the-online-retail-opportunity-with-european-farmers (accessed on 20 July 2022).
37. Dongsheng, L.I.; Yulian, Y. Research on Farmers' Adoption Intention to E-Commerce of Agricultural Products Based on UTAUT Model. *Converter* **2021**, *2021*, 947–957.
38. Hasan, B. Exploring gender differences in online shopping attitude. *Comput. Hum. Behav.* **2010**, *26*, 597–601. [CrossRef]
39. KC, S.K.; Timalsina, A.K. Challenges for Adopting E-Commerce in Agriculture in Nepalese Context—A Case Study of Kathmandu Valley. In Proceedings of the IOE Graduate Conference. 2016, pp. 305–312. Available online: http://conference.ioe.edu.np/publications/ioegc2016/IOEGC-2016-40.pdf (accessed on 20 July 2022).

agriculture

MDPI

Article

An Attention Mechanism-Improved YOLOv7 Object Detection Algorithm for Hemp Duck Count Estimation

Kailin Jiang [1,†], Tianyu Xie [2], Rui Yan [2], Xi Wen [2], Danyang Li [2,†], Hongbo Jiang [2], Ning Jiang [3], Ling Feng [2], Xuliang Duan [2] and Jianjun Wang [1,*]

[1] College of Science, Sichuan Agricultural University, Ya'an 625000, China
[2] College of Information Engineering, Sichuan Agricultural University, Ya'an 625000, China
[3] College of Electrical Engineering, Anhui Polytechnic University, Wuhu 241000, China
* Correspondence: jianjunw@sicau.edu.cn; Tel.: +86-18-0901-00891
† These authors contributed equally to this work.

Abstract: Stocking density presents a key factor affecting livestock and poultry production on a large scale as well as animal welfare. However, the current manual counting method used in the hemp duck breeding industry is inefficient, costly in labor, less accurate, and prone to double counting and omission. In this regard, this paper uses deep learning algorithms to achieve real-time monitoring of the number of dense hemp duck flocks and to promote the development of the intelligent farming industry. We constructed a new large-scale hemp duck object detection image dataset, which contains 1500 hemp duck object detection full-body frame labeling and head-only frame labeling. In addition, this paper proposes an improved attention mechanism YOLOv7 algorithm, CBAM-YOLOv7, adding three CBAM modules to the backbone network of YOLOv7 to improve the network's ability to extract features and introducing SE-YOLOv7 and ECA-YOLOv7 for comparison experiments. The experimental results show that CBAM-YOLOv7 had higher precision, and the recall, mAP@0.5, and mAP@0.5:0.95 were slightly improved. The evaluation index value of CBAM-YOLOv7 improved more than those of SE-YOLOv7 and ECA-YOLOv7. In addition, we also conducted a comparison test between the two labeling methods and found that the head-only labeling method led to the loss of a high volume of feature information, and the full-body frame labeling method demonstrated a better detection effect. The results of the algorithm performance evaluation show that the intelligent hemp duck counting method proposed in this paper is feasible and can promote the development of smart reliable automated duck counting.

Keywords: object detection; YOLOv7; attention mechanism; deep learning; hemp duck count

Citation: Jiang, K.; Xie, T.; Yan, R.; Wen, X.; Li, D.; Jiang, H.; Jiang, N.; Feng, L.; Duan, X.; Wang, J. An Attention Mechanism-Improved YOLOv7 Object Detection Algorithm for Hemp Duck Count Estimation. *Agriculture* **2022**, *12*, 1659. https://doi.org/10.3390/agriculture12101659

Academic Editor: Dimitre Dimitrov

Received: 18 August 2022
Accepted: 3 October 2022
Published: 10 October 2022

Publisher's Note: MDPI stays neutral with regard to jurisdictional claims in published maps and institutional affiliations.

1. Introduction

With the continuous development of modern society and the economy, the global consumption level continues to rise. People's demand for poultry meat, eggs, and other poultry-related products is increasing, and the livestock and poultry farming industry bear a wide scope for development. Such a large-scale demand for livestock and poultry products will inevitably lead to a continuous expansion in the scale of the farming industry. However, in the context of tight feed grain supplies, soil resources needed for breeding, and scarce water resources, the farming industry needs to continuously improve the quality and efficiency of production. Inefficient farming methods will increasingly worsen farming pollution, leading to an increased environmental burden and deviating from the concept of environmental protection.

Sparrow ducks, commonly known as "hemp ducks", are the main species of domestic ducks and one of the most abundant, widely distributed, and diverse species of domestic ducks in the world. Occupying about 70% or more of the total waterfowl breeding, duck breeding is roughly divided into three types: meat, egg, and both meat and egg, which have high economic value. Large-scale hemp duck farming can meet the huge market demand for poultry meat and eggs, but at the same time, it faces pressure and challenges in many aspects [1]. As countries around the world pay attention to the ecological environment, the development of waterfowl farming has been subject to certain restrictions and regulations. Many areas have been prohibited and restricted, and the spatial range suitable for farming hemp ducks continues to shrink [2]. At present, farming is developing in the direction of intensification and ecology. Large-scale farming and higher rearing density will have a greater impact on the temperature, humidity, ventilation, harmful gases, dust, and microbial content of poultry houses. It indirectly has a series of adverse effects on the intake, growth performance, and animal welfare of birds. For example, unreasonable duck flock density rates will lead to poor living conditions, causing physiological diseases, such as body abrasions, skin damage, and fractures. Considering animal behavior, such as pecking and fighting among a species, the unreasonable density will bear a negative impact on the efficiency and economy of the livestock and poultry industry [3,4].

From the above information, it can be concluded that rearing density is one of the key factors affecting livestock and poultry production on a large scale, as well as animal welfare, and the key to solving the problem of improving breeding efficiency lies in the real-time monitoring of breeding density and the reasonable scheduling of the spatial quantity of the flock: our work is focused on the former. At present, in the hemp duck farming industry, much of the counting is carried out manually or by artificial machinery, which are both very laborious. When hemp duck flocks are in motion, it further increases the difficulty of manual counting, thus affecting the breeding efficiency. In essence, the density of the hemp duck flock depends on the size of the effective activity space as well as the population size, and given the constant limitations of the current breeding area, the main factor affecting the problem is therefore the number of hemp ducks. For this reason, we focused on the hemp duck flock count.

With the development of technology, monitoring equipment plays a huge role in farming. There are various methods to monitor the behavior of individual animals, such as the insertion of chips to record physiological data, the use of wearable sensors, and (thermal) imaging techniques. Some methods employ wearable sensors attached to the feet of birds to measure their activity, but this may have an additional impact on the monitored animals [5–7]. In particular, in commercial settings, technical limitations and high costs lead to the low feasibility of such methods. Therefore, video assessment based on optical flow would be an ideal method to monitor poultry behavior and physiology [4]. Initially, many surveillance videos were manually observed, inefficient, and relied on the staff's empirical judgment without standards [8]. However, in recent years, due to the advent of the era of big data and the rapid development of computer graphics cards, the computing power of computers has been increasing, accelerating the development of artificial intelligence. Research related to artificial intelligence is increasing, and computer vision is becoming more and more widely used in animal detection.

For example, the R-CNN proposed by Girshick et al., in 2014 introduced a two-stage detection method for the first time. This method uses deep convolutional networks to obtain excellent target detection accuracy, but its many redundant operations greatly increase space and time costs, and is difficult to deploy in actual duck farms [9,10]. Law et al., proposed a single-stage detection method, CornerNet, and a new pooling method: corner pool. However, the method, based on key points, often encounters a large number of incorrect object bounding boxes, which limits its performance and cannot meet the high performance requirements of the duck breeding model [11]. Duan et al., constructed the CenterNet framework on the basis of CornerNet to improve the accuracy and recall and designed two custom modules with stronger robustness to feature-level noise, but the anchor-free method is a process with key point combinations of the first two, and due to the simple network structure, time-consuming processing, low rate, and unstable measurement results, it cannot meet the requirements of high performance and high accuracy rate needed in the industrial farming of hemp ducks [12].

Our work uses a single-stage object detection algorithm, which only needs to extract features once to achieve object detection, and its performance is higher compared to the multi-stage algorithm. At present, the mainstream single-stage target detection algorithms mainly include the YOLO series, Single Shot MultiBox Detector (SSD), RetinaNet, etc. In this paper, we transfer and apply the idea of crowd counting based on CNN to the problem of counting ducks [13,14]. Along with the output of the detection results, we embedded an object counting module to respond to industrialization needs. Object counting is also a common task in the computer vision community. Object counting can be divided into multi-category object counting and single-category object counting; this work employed single-category counting of a flock of hemp ducks [15–18].

The objectives that this paper hopes to achieve are:

(1) We built a new large-scale dataset of drake images and named it the "Hemp Duck Dataset". The Hemp Duck Dataset contains 1500 labels for the whole body frame and head frame for duck target detection. The Hemp Duck Dataset was released for the first time by the team. We have made it public and provide the access method at the end of the article.
(2) This study constructed a comprehensive working baseline, including hemp duck identification, hemp duck object detection, and hemp duck image counting, to realize the intelligent breeding of hemp ducks.
(3) This project model introduced the CBAM module to build the CBAM-YOLOv7 algorithm.

2. Materials and Methods

2.1. Acquisition of Materials

The hemp duck is one of the most abundant, widely distributed, and diverse species of domestic ducks in China, with the characteristics of small size, feed saving, and high egg production efficiency, which is of great research value. We used the DJI Pocket 2, an extremely adaptable and flexible miniature gimbal camera, to capture the image and video datasets used in this study. Data were collected from the original waterfowl farm in Ya'an, Sichuan Province, China, founded by Professor Lin-Quan Wang, a renowned waterfowl breeder from Sichuan Agricultural University.

In the process of preparing the dataset, we first collected data from 10 different hemp duck houses by changing the image shooting angle and distance several times. Then, we manually screened and discarded some data with high repetition and some redundant data that were not captured due to the obstruction of the hemp ducks' house. In the end, our dataset contained a total of 1500 images, including 1300 images in the training set and 200 images in the test set. Figure 1 shows the analysis of the challenges posed by non-maximum suppression for the hemp duck detection, identification, and counting tasks. Figure 2 shows an example of a dataset labeling effort.

In the prediction phase of the object detection work, the network output multiple candidate anchor boxes, but many of them were overlapping around the same object, as shown in Figure 1b. Non-maximum suppression was able to retain the best one among this group of candidate anchor boxes, as shown in Figure 1c. We named two different ducks hemp duck A and hemp duck B. When hemp duck A and hemp duck B are too close, the prediction box of hemp duck A may be eliminated due to the screening of non-maximum intrusion. Therefore, it is a challenge to accurately estimate the number of dense Hemp Duck Datasets with inclusion.

Since labeling the whole hemp duck body resulted in many overlapping labeling boxes, which affected the accuracy of individual hemp duck counting, we chose the method of labeling only the hemp duck head and conducted a comparison experiment between the two.

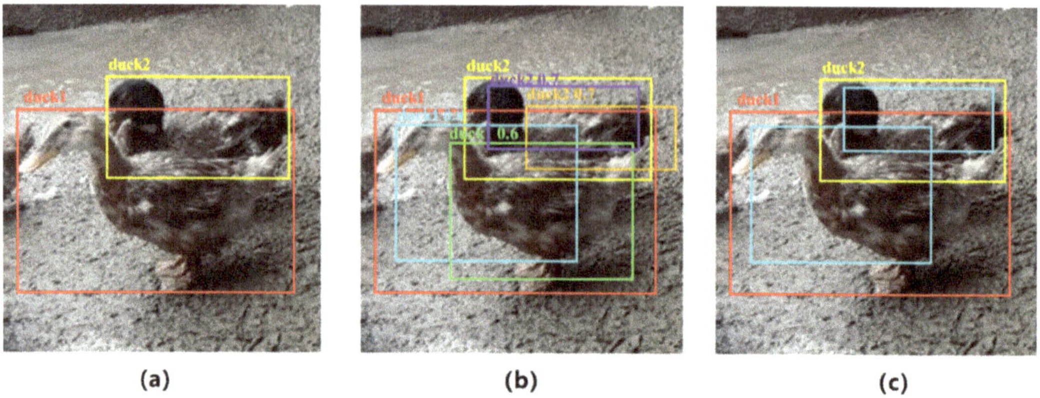

Figure 1. (**a**) Two ground truth boxes of hemp ducks; (**b**) output prediction boxes of the simulated network for the two hemp ducks; (**c**) effect of removing the redundant prediction boxes after non-maximum suppression.

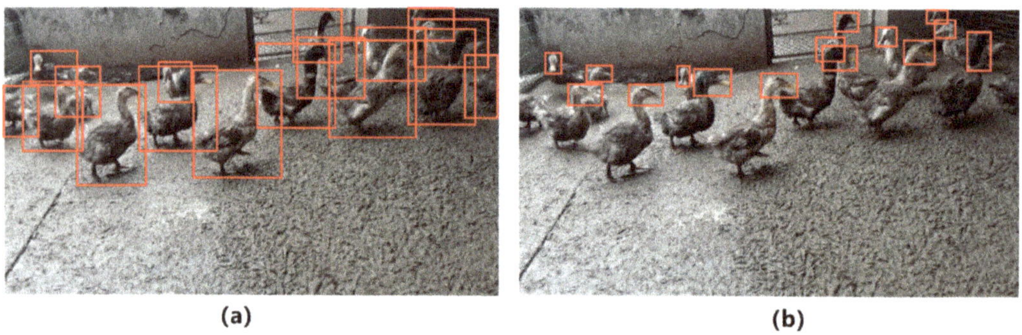

Figure 2. (**a**) Example of data annotation of the whole body of hemp ducks; (**b**) example of data annotation of hemp ducks with only the head annotated.

2.2. Data Pre-Processing

2.2.1. Mixup Data Augmentation

Mixup is an unconventional data enhancement method based on a simple data-independent data enhancement principle that uses linear interpolation to construct new training samples and labels [19]. The formula for processing the data labels is as follows:

$$\widetilde{x} = \lambda x_i + (1 - \lambda)x_j \tag{1}$$

$$\widetilde{y} = \lambda y_i + (1 - \lambda)y_j \tag{2}$$

Among it, the two data pairs (x_i, y_i) and (x_j, y_j) are the training sample pairs in the original dataset (the training sample and its corresponding label); λ is a parameter that follows the distribution of β; $\widetilde{x}$ is the training sample of the mixup after the data enhancement operation; $\widetilde{y}$ is the label of $\widetilde{x}$. Figure 3 shows the data results of the hemp ducks after the mixup data enhancement process with different fusion proportions.

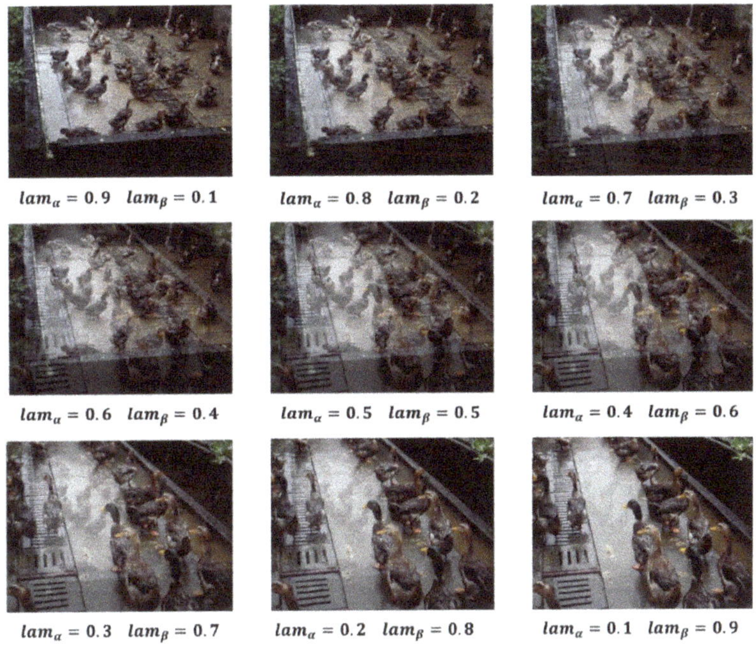

Figure 3. The results of the hemp duck data enhanced by mixup data with different fusion proportions, where lam_{α} and lam_{β} are the fusion proportions of the images and $lam_{\alpha} + lam_{\beta} = 1$.

2.2.2. Mosaic Data Augmentation

The YOLOv4 network uses Mosaic data augmentation, the idea of which is to randomly cut four images and combine them into one image as newly generated training data, greatly enriching the detection dataset, making the network more robust, and reducing the GPU video memory occupation [14]. Figure 4 represents the workflow of Mosaic's data augmentation operation.

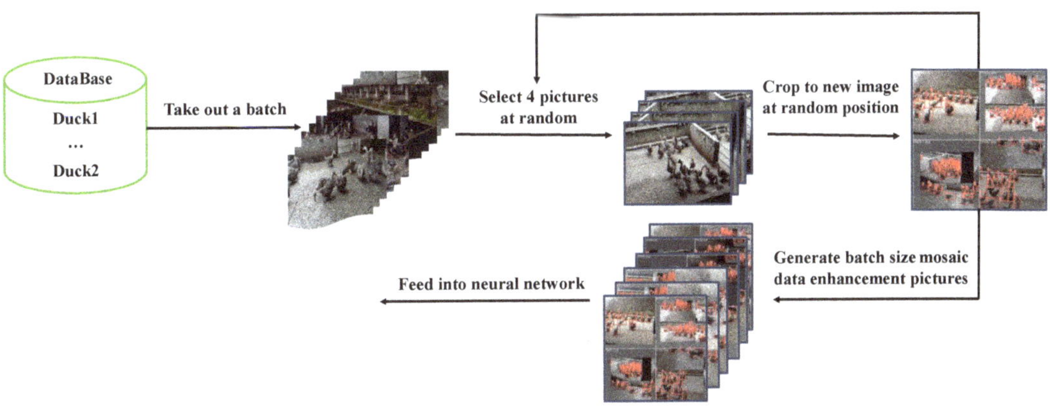

Figure 4. Mosaic data augmentation. Firstly, a batch of image data was randomly extracted from the dataset of mallard ducks. Then, four images were randomly selected, randomly scaled, randomly distributed, and spliced into new images, and the above operations were repeated for batch size times. Finally, the Mosaic data augmentation data were fed into the neural network for training.

2.3. Experimental Environment

The GPU of the project team computer was an NVIDIA GeForce RTX3090, with 3840 CUDA stream processors and 24 GB video memory. The CPU included 14-core Intel (R) Xeon (R) gold 6330, 2.00 GHz, and 60 GB memory. The operating system was Windows 10 and PyTorch version 1.8.1, Python version 3.8, and CUDA version 11 were used.

2.4. Training Parameters

The training parameters of the training process used in the experiment are shown in Table 1.

Table 1. Training parameters.

Parameter	Value	Parameter	Value
Learning Rate	0.01	Weight Decay	0.0005
Batch Size	16	Momentum	0.937
Image Size	640 × 640	Epochs	300

2.5. Evaluation Metrics

In order to evaluate the performance of the algorithm, the evaluation indices used in this study were precision (P), recall (R), mean average precision (mAP), F1 score, and frames per second (FPS).

Precision represents the proportion of positive samples in the samples with positive prediction results. The calculation formula is as follows:

$$\text{Precision} = \frac{TP}{TP + FP} \tag{3}$$

Recall represents the prediction result as the proportion of the actual positive samples in the positive samples to the positive samples in the whole sample. The calculation formula is as follows:

$$\text{Recall} = \frac{TP}{TP + FN} \tag{4}$$

The F1 score is the weighted average of precision and recall, calculated as follows:

$$F1 = \left(\frac{2}{\text{Recall}^{-1} + \text{Precision}^{-1}}\right) = 2 \cdot \frac{\text{Precision} \cdot \text{Recall}}{\text{Precision} + \text{Recall}} \tag{5}$$

Precision reflects the model's ability to distinguish negative samples. The higher the precision, the stronger the model's ability to distinguish negative samples. Recall reflects the model's ability to identify positive samples. The higher the recall, the stronger the model's ability to identify positive samples. The F1 score is a combination of the two. The higher the F1 score, the more robust the model.

The average precision (AP) is the average value of the highest precision under different recall conditions (generally, the AP of each category is calculated separately). The calculation formula is as follows:

$$AP = \frac{1}{11} \sum_{0,0.1\ldots1.0} P_{smooth}(i) \tag{6}$$

In Pascal VOC 2008 [20], the threshold value of the IOU is set to 0.5. If one object is repeatedly detected, the one with the highest confidence is the positive sample and the other is the negative sample. On the smoothed PR curve, the precision value of 10 bisectors (including 11 breakpoints) was obtained on the horizontal axis 0–1, and the average value was calculated as the final AP value.

The mean average precision (mAP) is the mean value of the average precision and the mean AP value of each category. The calculation formula is as follows:

$$mAP = \frac{\sum_{j=1}^{S} AP(j)}{S} \quad (7)$$

where S is the number of all categories, and the denominator is the sum of the APs of all categories. The object detection object in this study was only one type of hemp duck, therefore, AP = mAP.

2.6. Related Network

In this section, the YOLOv7 algorithm is first introduced, and then the improvement proposed in this paper of adding an attention mechanism to YOLOv7 is introduced in detail [13].

2.6.1. YOLOv7

In this paper, a recognition and detection algorithm based on computer vision is proposed for object detection and population statistics of farm ducks. By using this algorithm, breeders can obtain the quantity and behavior dynamics of mallard ducks in real time so as to realize the rapid management and strategy formulation of farms, optimize the reproduction rate and growth of ducks, and help to maximize the economic benefits.

In view of the small density of individuals in the duck population and the real-time requirement of population statistics, we chose the latest Yolov7 model. You Only Look Once (Yolov7) is a single-stage object detection algorithm. Figure 5 shows the network structure diagram of Yolov7 [13]. The Yolov7 model preprocessing method is integrated with Yolov5, and the use of Mosaic data augmentation is suitable for small object detection [13,14,21]. In terms of architecture, extended ELAN (E-ELAN) based on ELAN is proposed. Expand, shuffle, and merge cardinality are used to continuously enhance the learning ability of the network without destroying the original gradient path. Group convolution is used to expand the channel and cardinality of the computing block in the architecture of the computing block. Different groups of computational blocks are guided to learn more diverse features [13].

Then, it focuses on some optimization modules and methods known as trainable "bag-of-freebies" [13], including the following:

1. RepConv without identity connection is used to design the architecture of planned reparametrized convolution, which provides more gradient diversity for different feature maps [22].
2. The auxiliary detection head is introduced, and the soft labels generated by the optimization process are used for lead head and auxiliary head learning. Therefore, the soft labels generated from it should better represent the distribution and correlation between source data and object and obtain more accurate results [23].
    (1) The batch normalization layer is directly connected to the convolution layer so that the normalized mean and variance of the batch are integrated into the deviation and weight of the convolution layer in the inference stage.
    (2) By using the addition and multiplication method of implicit knowledge in YOLOR combined with the convolution feature map, it can be simplified into vectors by precomputation in the inference stage so as to combine with the deviation and weight of the previous or subsequent convolution layer [24].
    (3) The EMA model is used purely as the final inference model. Finally, real-time object detection can greatly improve the detection accuracy without increasing the reasoning cost so that the speed and accuracy in the range of 5–160 FPS exceed all known object detectors, and fast response and accurate prediction of object detection can be achieved [25].

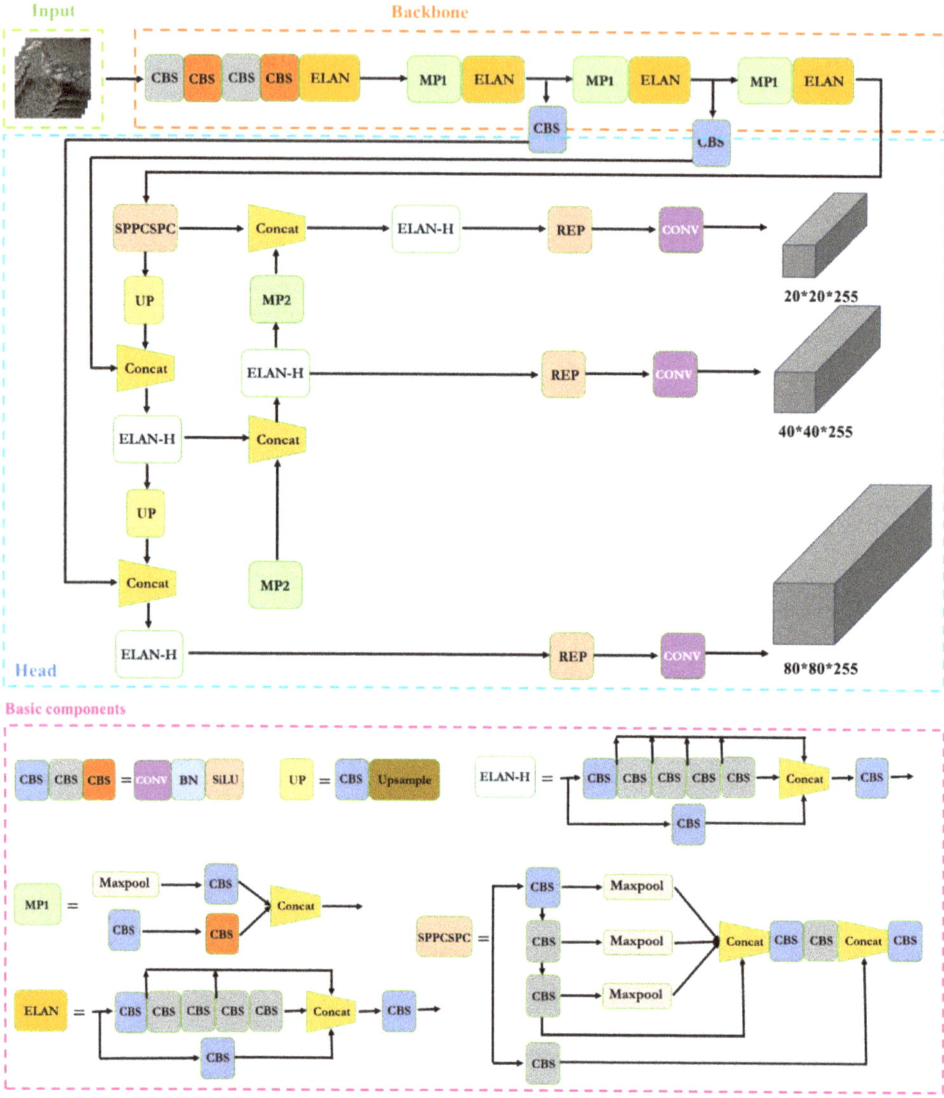

Figure 5. The network architecture diagram of Yolov7 contains four general modules: input terminal, backbone, head, and prediction, and five basic components: CBS, MP, ELAN, ELAN-H, and SPPCSPC.

2.6.2. Improved YOLOv7 with Attention Mechanism

The attention mechanism is a common data processing method that is widely used in machine learning tasks in various fields [26]. The core idea of the attention mechanism of computer vision is to find the correlation between the original data, and then highlight the important features, such as channel attention, pixel attention, multi-order attention, and so on.

The CBAM mainly includes a channel attention module and a spatial attention module [10]. The module structure is shown in Figure 6.

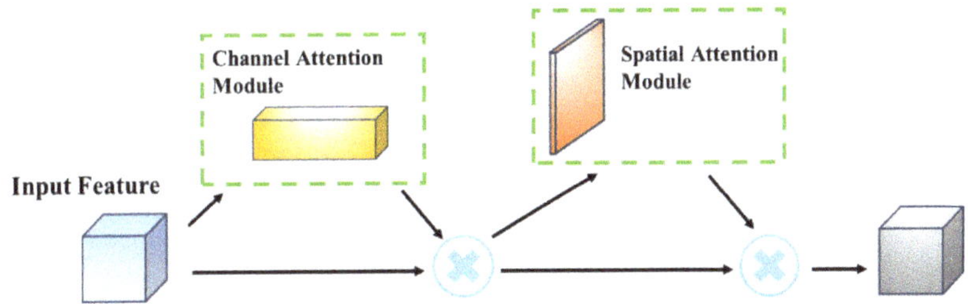

Figure 6. CBAM module structure diagram.

CBAM [27] is a lightweight attention module, which can perform attention operations in the channel and spatial dimensions. It is composed of a channel attention module (CAM) and a spatial attention module (SAM). The CAM can make the network pay more attention to the foreground of the image and the meaningful area, while the SAM can make the network pay more attention to a position rich in contextual information of the whole picture [28,29].

2.6.3. YOLOv7 Introduces the CBAM Attention Mechanism

The CBAM attention mechanism was added to the YOLOv7 network structure [13,27], and the network structure is shown in Figure 7. The function of this module is to further improve the feature extraction ability of the feature extraction network. Once we added the attention mechanism to the backbone network, the attention mechanism module destroyed some of the original weights of the backbone network. This led to errors in the prediction results of the network. In this regard, we chose to add the attention mechanism to the part of enhancing feature network extraction without destroying the original features of network extraction.

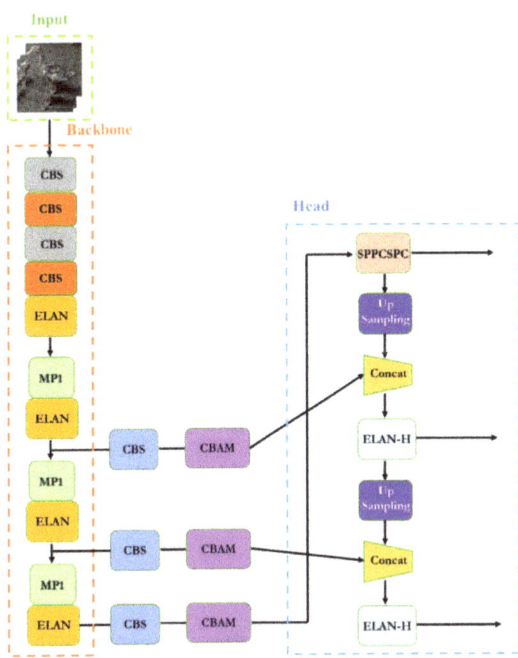

Figure 7. Addition of the YOLOv7 network structure to the CBAM.

The CBAM attention mechanism works as follows:

In the channel attention module, the input feature map of H × W × C is subjected to a global max pooling (GMP) and a global average pooling (GAP), and two feature maps with a size of 1 × 1 × C are obtained. The two feature maps are sent to a two-layer multilayer perceptron. The number of neurons in the first layer of the MLP is C/r (r is the reduction rate), and the activation function is ReLU. The number of neurons in the second layer is C, and the weights of these two layers of neural networks are shared. Then, the output features are added based on element-wise computation, and the final channel attachment feature is generated through sigmoid activation. Finally, the channel attention feature is multiplied by the original input feature map to obtain the input feature of the spatial attention module [10].

In the spatial attention module, the feature map in the previous step is used as the input.

After GMP and GAP, two feature maps with a size of H × W × 1 are obtained. Then the Concat operation is performed. After the dimensionality reduction of the feature map, the spatial attention feature is generated by sigmoid activation. Finally, the spatial attention feature is multiplied by the input feature map to obtain the final feature map [27].

3. Experiment Results

In order to evaluate the effect of the CBAM-YOLOv7 algorithm, we borrowed SE and ECA modules to replace the CBAM modules for ablation experiments. SE mainly includes squeeze and excitation operations [30]. The module structure is shown in Figure 8.

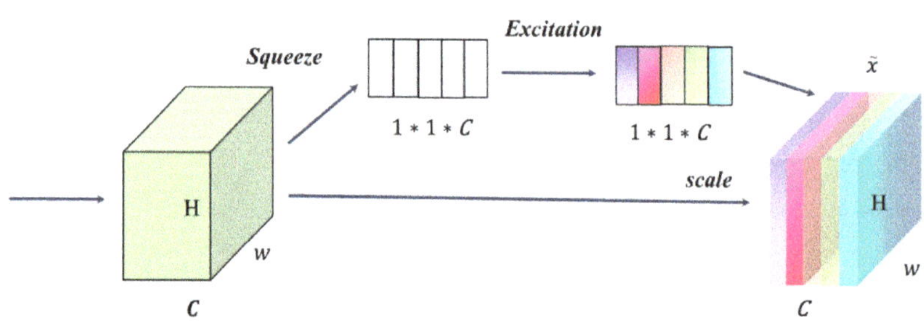

Figure 8. Structure of the SE.

The SE mechanism can flexibly capture the connection between global information and local information, allowing the model to obtain the object area that needs to be focused on and assign it more weight, highlighting significant useful features and suppressing and ignoring irrelevant features, thereby improving accuracy.

The ECA module proposes a local cross-channel interaction strategy without dimensionality reduction, which can effectively avoid the influence of dimensionality reduction on the learning effect of channel attention. The ECA module consists of a one-dimensional convolution determined by nonlinear adaptation, which captures local cross-channel interaction information by considering each channel and its k neighbors. Since only a few parameters are involved, it is a very lightweight plug-and-play block, but with significant effect gain [31]. The structure of ECA is shown in Figure 9.

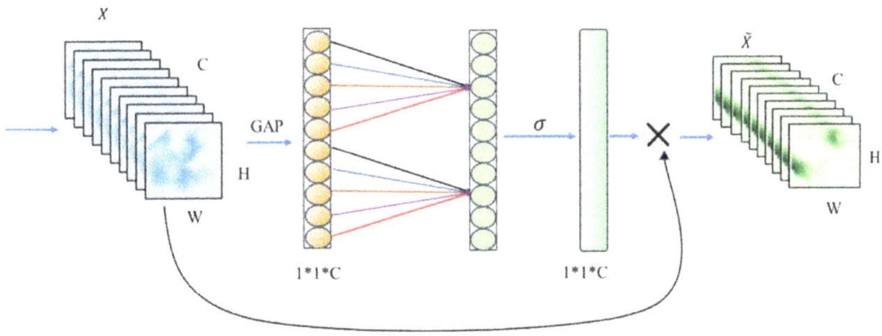

Figure 9. Structure of ECA.

3.1. Object Detection Network Comparison Experiment Results

In the process of selecting the object detection model of the hemp ducks, we applied the existing popular object detection model to the Hemp Duck Dataset for training and testing, and compared the precision, recall, F1 score, mAP@0.5, and other evaluation indicators. Finally, YOLOv7 was selected as the object detection model in this study for subsequent experiments. Table 1 shows the comparison of the evaluation indexes of each object detection model on the Hemp Duck Dataset. The experimental results are shown in Table 2.

Table 2. Comparison of object detection algorithms.

Method	P	R	F1	mAP@0.5	mAP@0.5:0.95	FPS
CenterNet	92.16%	95.12%	0.94	95.41%	62.80%	33
SSD	86.03%	82.40%	0.84	89.03%	45.90%	39
EfficientDet	87.66%	92.98%	0.90	95.91%	60.40%	26
RetinaNet	88.00%	89.17%	0.89	94.04%	56.40%	13
YOLOv4s	92.26%	78.04%	0.85	89.82%	44.10%	22
YOLOv5s	95.50%	88.70%	0.92	94.90%	66.70%	62
YOLOv7	95.80%	93.64%	0.95	97.57%	65.50%	60

As can be seen in Table 2, YOLOv7 performed overall better than the other detection algorithms tested, leading in terms of the precision, F1 score, and mAP@0.5 and a close second in terms of the recall, mAP@0.5:0:0.95, and detection speed. For example, the recall rate of the YOLOv7 algorithm was 15.6% higher than that of YOLOV4. The remaining indicators are basically superior to the other target detection algorithms. Finally, we chose YOLOv7 as the target detection algorithm used in the experiment.

3.2. Contrast Experiment Results of Introducing Attention Mechanism

In order to verify the effectiveness of the improved algorithm, this study used CBAM as the attention mechanism and added it to the YOLOv7 object detection algorithm to conduct experiments on the Hemp Duck Dataset. The experimental results are shown in Table 3, and the recall rate, mAP@0.5, and mAP@0.5:0.95 were used as the measures.

Table 3. Comparative experiments.

Attention Mechanism			P	R	F1	mAP@ 0.5	mAP@ 0.5:0.95	FLOPS (G)
SE	CBAM	ECA						
×	×	×	95.80%	93.64%	0.95	97.57%	65.50%	106.47
√	×	×	95.36%	93.53%	0.94	97.48%	65.10%	106.49
×	√	×	96.84%	94.57%	0.95	98.72%	66.10%	106.49
×	×	√	95.55%	93.75%	0.95	97.41%	65.20%	106.49

As can be seen in Table 3, compared to the original YOLOv7 algorithm, the accuracy rate of the SE-YoloV7 algorithm decreased by 0.44%, the recall rate decreased by 0.11%, the mAP decreased by 0.09%, and the FLOPS increased by 0.02G. The accuracy rate of the ECA-YOLOv7 algorithm decreased by 0.25%, the recall rate increased by 0.11%, the mAP also decreased, and the FLOPS increased by 0.02G. The results in Table 3 show that the SE-Yolov7 and ECA-YOLOV7 algorithms not only had a lower effect than the original YOLOv7, but also increased the model parameters and the computational pressure. Compared to the original YOLOv7 algorithm, the accuracy of the CBAM-YOLOV7 algorithm increased by 1.04%, the recall increased by 0.93%, the mAP@0.5 by 1.15%, and the mAP@0.5:0.95 by 0.60%. In addition, the value of the FLOPS parameter of the CBAM-Yolov7 model is equal to that of the SE-Yolov7 and ECA–YOLOv7 models. By comparing and analyzing the experimental results, it can be concluded that the algorithm in this paper demonstrated better performance than both the original algorithm and the algorithm with the SE and ECA modules. Compared to the SE-YOLOv7 and ECA-YOLOv7, the CBAM module not only improved the channel attention module, but also added a spatial attention module, analyzed it from two dimensions, and determined the order from the channel to the space.

Figure 10 shows the detection effect of the CBAM-YOLOv7 algorithm on the Hemp Duck Dataset.

Figure 10. CBAM-YOLOv7 network prediction result graph.

3.3. Comparison of Experiment Results with Different Data Annotation Methods

In a previous article, we took into account the data sheet annotation section and compared the data annotation of the whole body with the data annotation of the hemp duck with only the head. For this, we used the improved CBAM-YOLOv7 algorithm on two different annotation methods. The experimental results are shown in Table 4.

Table 4. Comparison of two different annotation experiments.

Annotation Method		P	R	F1	mAP@ 0.5	mAP@ 0.5:0.95	FPS
Head Annotation	Whole Body Annotation						
√	×	95.06%	91.90%	0.93	94.01%	47.20%	46
×	√	95.80%	93.64%	0.95	97.57%	65.50%	60

As can be seen in Table 4, the experimental data of the head annotation are far inferior to that of the whole body annotation. We discuss the reasons for this. Usually, the model extracts the features contained in the pixels of the original image through convolution, and the receptive field reflects the correspondence between the information of a single high-level feature and the original pixels, which is determined by the network convolution kernel. In this case, as the number of network layers increases, a single high-level feature reflects the larger range of pixels in the original image, and the field of view is wider and the high-level information can better reflect the macro outline of the object in the original image. Then, as the number of network layers increases and the receptive field becomes larger, the microscopic information is lost, so that the information about the small object will be aggregated to a point, and the small object originally contains fewer pixels. If it increases, there will be fewer features after aggregation. For example, a small object of 10 × 10 pixels may have only 1 × 1 features after convolution, or even multiple small objects of 10 × 10 pixels. After multiple convolution operations, only one feature may be generated, which leads to the failure of pixel recognition. Therefore, the head annotation method is unsuitable for the task of count estimation on the Hemp Duck Dataset.

Figure 11 shows the detection results of two different labeling methods based on the YOLOv7 algorithm on the Hemp Duck Dataset.

In the graph of the results, it can be seen that the two labeling methods obtained different results in estimating the number of hemp ducks. Based on the comparison, the experimental results of labeling the whole body method are more accurate

3.4. Results of Ablation Experiment

Figure 12 shows the prediction chart of Yolov7 algorithm without adding training skills. Based on the original YOLOv7 algorithm, we tried some training techniques in the ablation experiment, using different tricks to process the model, such as mosaic processing and image fusion. Through the experiments, we verified the experimental data obtained when using the above treatments. The experimental results are shown in Table 5.

Table 5. Ablation experiments.

Group	Mosaic	MixUp	P	R	F1	mAP@ 0.5	mAP@ 0.5:0.95	FPS
1	×	×	95.30%	93.64%	0.94	97.64%	65.10%	56
2	√	×	95.55%	93.64%	0.95	97.26%	65.40%	57
3	×	√	95.45%	93.75%	0.95	97.65%	64.80%	56
4	√	√	95.80%	93.64%	0.95	97.57%	65.50%	60

Figure 11. Prediction results of two annotation methods.

Figure 12. Yolov7 algorithm prediction chart (without tricks).

In the ablation experiments of YOLOv7, each group of experiments corresponded to a set of training skills and evaluation indicators. Among them, "√" indicates that this training technique was used, and "×" indicates that this training technique was not applicable.

As can be seen in Table 5, overall, the simultaneous use of Mosaic and mixup fared better than using just one method or neither. Figure 13 shows the comparative effect of the results of the four groups of ablation experiments on Precision, Recall and mAP@0.5 indicators.

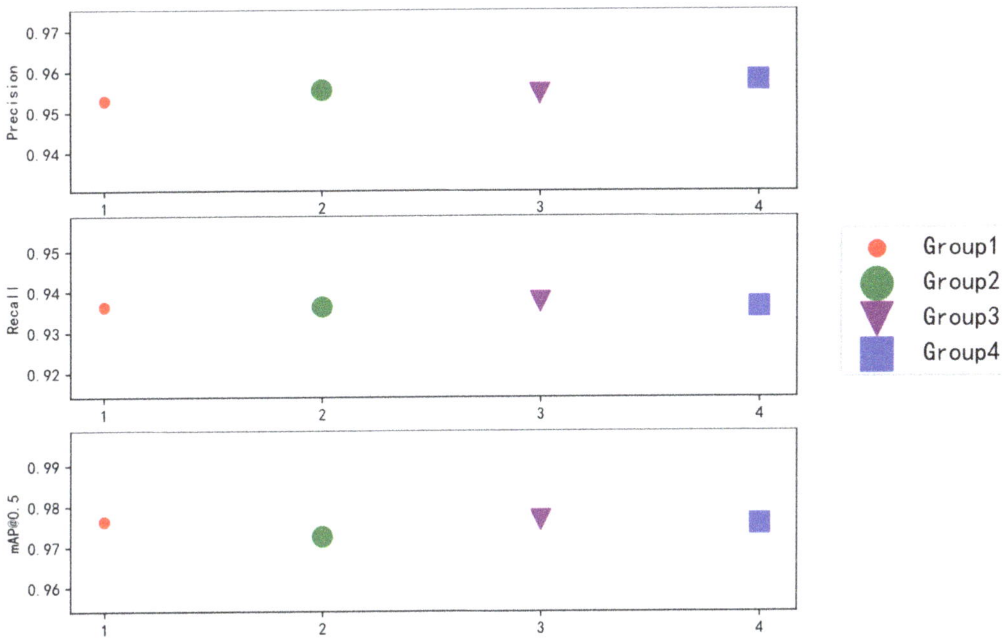

Figure 13. Ablation experiment. Each point in the figure corresponds to a set of training skills and is distinguished by different colors and shapes of the points. The comparison of the experimental results shows that when Mosaic and mixup were used, the precision, recall, F1, map, FPS, and other data improved. The results show that Group 4 worked best.

4. Discussion

4.1. Contribution to Animal Target Detection and Counting

Poultry farming has a very large market size; however, the number of experiments for target testing and counting of poultry is low. Therefore, conducting reasonable data annotation of the original poultry pictures and improving the algorithmic network structure to obtain good feature information had a very significant impact on the experiments achieving good results. There are few methods for estimating intelligent poultry target detection counts, and this study fills this gap to some extent. The experiments in this project present a reflection on the implementation of automated poultry counting and provide an experimental basis for the target detection counting of poultry.

4.2. Contribution to Intelligent Farming of Sisal Ducks

The manual counting method is inefficient and the results are not always accurate, both because of the constant movement of the ducks and the high labor costs. As a result, it is difficult and costly to avoid duplication of effort. In this study, we used an improved YOLOv7 algorithm to obtain an efficient and accurate stocking density and to control the stocking density to a suitable value. The accuracy rate of the algorithm increased by 1.04%, the recall rate increased by 0.93%, the mAP@0.5 increased by 1.15%, and the mAP@0.5:0.95

increased by 0.60%. The use of minimal manual labor costs will improve farming efficiency and reduce problems such as irrational disease and pecking and fighting among species. At the same time, making hemp duck farming intelligent by using deep neural network models provides an efficient approach to duck target detection and breeding density management.

This study was based on the YOLOv7 model, migrating the learning method to the intelligent farming of hemp ducks while improving the original YOLOv7 so that the model has higher target detection, improves the robustness of the model, and is more adaptable to realistic scenarios. Moreover, we have filled the gap of few automated estimation algorithms for agricultural poultry farming numbers.

4.3. Comparison of Methods

Two-stage: First stage: focus on finding the location of the target object and obtaining a suggestion frame, ensuring sufficient accuracy and recall. Second stage: focus on classifying the suggestion frame and finding a more accurate location. Two-stage structure sampling is used to deal with category imbalance, but at the same time, it is slow and generally seen in competition scenarios.

One-stage: Instead of obtaining a suggestion frame stage, the class probability and position coordinates of the object are generated directly, and the final detection result is obtained directly after a single detection, which is faster than the multi-stage algorithm and generally has a slightly lower accuracy. The YOLOv7 and improved algorithms used in this study are single-stage detection, but the collection of some methods as well as the module re-referencing and dynamic tag assignment strategies make it much faster and more accurate. For large-scale, high-density breeding of sisal ducks, the single-stage algorithm can better focus on the real-time changes in the number of sisal ducks and obtain better results.

4.4. Limits and Future Work

It should not be overlooked that there are still limitations to this study. Firstly, the study did not consider disturbances caused by the external environment, such as birds in the farm field environment, and most of the images in the dataset were based on high light and average weather conditions, which may not be sufficiently representative. In the future, we will collect more duck datasets under different conditions.

Secondly, there are false detection and missing detection data in the detection results. For error detection, we adopted two methods to find a more appropriate confidence level, namely, adding more negative samples for training, and using the image dataset generated by GAN [32] for data enhancement. For missed detection, we used two measures. One was to filter out the samples with a large loss value in each training iteration and add them to the training set of the next iteration so that the detection model could pay more attention to the samples that were easily missed. The second was to use the method proposed by Soft NMS [33] to combine multiple weighted frames according to the confidence level of the IoU, optimize the final generated frame, and improve the detection performance of dense small target objects.

To sum up, this study aimed to explore different algorithms to determine the best algorithm under a wide range of environmental conditions with duck count uncertainty and deploy the network model to hardware devices for practical application in farms.

5. Conclusions

In this study, a large-scale dataset for estimating the count of hemp ducks was constructed, including 1500 pictures of hemp ducks, which can be collected by individuals to provide data support for visual research in the field of poultry. In this study, three CBAM modules were added to the backbone network of the YOLOv7 algorithm to optimize the YOLOv7 network structure. An improved YOLOv7 algorithm with an attention mechanism was proposed, and SE-YOLOv7 and ECA-YOLOv7 were introduced for comparative experiments. In comparison, the precision rate, recall rate, and mAP all improved, and the

FLOPS only increased by 0.02 G with no increase in computational pressure. In addition, this study considered the overlapping problem of hemp duck labeling frames, proposed two labeling methods, whole body labeling and head-only labeling, and conducted comparative experiments. The feature information was lost, and the dense counting task of hemp ducks could not be realized. The algorithm in this paper achieved good detection results on the task of counting dense hemp duck groups. The CBAM-YOLOv7 algorithm was proposed to improve the detection accuracy, and the advantages and disadvantages of the two labeling methods were discussed. Future research will continue to optimize the network structure of the proposed algorithm and deploy the network to the hardware environment used in field farming.

Author Contributions: Conceptualization, K.J.; methodology, K.J., T.X., R.Y. and D.L.; software, K.J., T.X., R.Y., L.F., X.W., H.J. and D.L.; formal analysis, K.J. and H.J.; investigation, K.J., T.X. and N.J.; resources, K.J., T.X., D.L. and X.D.; data curation, K.J., T.X., L.F. and N.J.; writing—original draft, K.J., T.X., R.Y., X.W. and H.J.; writing—review and editing, K.J., T.X. and D.L.; visualization, K.J. and T.X.; supervision, J.W. and X.D. All authors have read and agreed to the published version of the manuscript.

Funding: Subsidy for University Student Innovation Training Program (No. 202210626024).

Institutional Review Board Statement: The animal study protocol was approved by the Institutional Animal Care and Use Committee of Sichuan Agricultural University (protocol code 20200052, 23 May 2020).

Data Availability Statement: The data are available online at: https://pan.baidu.com/s/1Hg6nyhsp4WYLY-2qb24Sug (Extraction code: anpf. accessed on 24 September 2022).

Conflicts of Interest: The authors declare no conflict of interest.

References

1. Zhang, Y.; Wang, L.; Bian, Y.; Wang, Z.; Xu, Q.; Chang, G.; Chen, G. Marginal diversity analysis of conservation of Chinese domestic duck breeds. *Sci. Rep.* **2019**, *9*, 13141. [CrossRef] [PubMed]
2. Wu, Z. The current situation and development countermeasures of the edemon industry. *Vet. Guide* **2021**, *15*, 96–97.
3. Fu, M.; Wang, J.; Wu, W.; Li, Y.; Jiang, D.; Huang, Y.; Tian, Y.; Zhang, X. Influence of high-breeding density on the feeding environment and growth performance of miassing duck. *Guangdong Agric. Sci.* **2021**, *48*, 126–136.
4. Neethirajan, S. ChickTrack—A quantitative tracking tool for measuring chicken activity. *Measurement* **2022**, *191*, 110819. [CrossRef]
5. Yang, X.; Zhao, Y.; Street, G.M.; Huang, Y.; Filip To, S.D.; Purswell, J.L. Classification of broiler behaviours using triaxial accelerometer and machine learning. *Animal* **2021**, *15*, 100269. [CrossRef] [PubMed]
6. Abdoli, A.; Alaee, S.; Imani, S.; Murillo, A. Fitbit for chickens? Time series data mining can increase the productivity of poultry farms. In Proceedings of the 26th ACM SIGKDD International Conference on Knowledge Discovery & Data Mining, Virtual Event, CA, USA, 6–10 July 2020; pp. 3328–3336.
7. Van der Sluis, M.; de Haas, Y.; de Klerk, B.; Rodenburg, T.B.; Ellen, E.D. Assessing the activity of individual group-housed broilers throughout life using a passive radio frequency identification system—A validation study. *Sensors* **2020**, *20*, 3612. [CrossRef] [PubMed]
8. Gu, Y.; Wang, S.; Yan, Y.; Tang, S.; Zhao, S. Identification and Analysis of Emergency Behavior of Cage-Reared Laying Ducks Based on YoloV5. *Agriculture* **2022**, *12*, 485. [CrossRef]
9. Girshick, R.; Donahue, J.; Darrell, T.; Malik, J. Rich feature hierarchies for accurate object detection and semantic segmentation. In Proceedings of the IEEE Conference on Computer Vision and Pattern Recognition, Columbus, OH, USA, 23–28 June 2014; pp. 580–587.
10. He, K.; Zhang, X.; Ren, S.; Sun, J. Spatial pyramid pooling in deep convolutional networks for visual recognition. *IEEE Trans. Pattern Anal. Mach. Intell.* **2015**, *37*, 1904–1916. [CrossRef] [PubMed]
11. Law, H.; Deng, J. Cornernet: Detecting objects as paired keypoints. In Proceedings of the European Conference on Computer Vision (ECCV), Munich, Germany, 8–14 September 2018; pp. 734–750.
12. Duan, K.; Bai, S.; Xie, L.; Qi, H.; Huang, Q.; Tian, Q. Centernet: Object detection with keypoint triplets. *arXiv* **2019**, arXiv:1904.08189.
13. Wang, C.Y.; Bochkovskiy, A.; Liao, H.Y.M. YOLOv7: Trainable bag-of-freebies sets new state-of-the-art for real-time object detectors. *arXiv* **2022**, arXiv:2207.02696.
14. Bochkovskiy, A.; Wang, C.Y.; Liao, H.Y.M. Yolov4: Optimal speed and accuracy of object detection. *arXiv* **2020**, arXiv:2004.10934.
15. Hsieh, M.R.; Lin, Y.L.; Hsu, W.H. Drone-based object counting by spatially regularized regional proposal network. In Proceedings of the IEEE International Conference on Computer Vision, Venice, Italy, 22–29 October 2017; pp. 4145–4153.

16. Aich, S.; Stavness, I. Improving object counting with heatmap regulation. *arXiv* **2018**, arXiv:1803.05494.
17. Arteta, C.; Lempitsky, V.; Noble, J.A.; Zisserman, A. Interactive object counting. In Proceedings of the European Conference on Computer Vision, Zurich, Switzerland, 6–12 September 2014; Springer: Cham, Switzerland, 2014; pp. 504–518.
18. Cholakkal, H.; Sun, G.; Khan, F.S.; Shao, L. Object counting and instance segmentation with image-level supervision. In Proceedings of the IEEE/CVF Conference on Computer Vision and Pattern Recognition, Long Beach, CA, USA, 16–17 June 2019; pp. 12397–12405.
19. Zhang, H.; Cisse, M.; Dauphin, Y.N.; Lopez-Paz, D. mixup: Beyond empirical risk minimization. *arXiv* **2017**, arXiv:1710.09412.
20. Hoiem, D.; Divvala, S.K.; Hays, J.H. Pascal VOC 2008 challenge. *World Lit. Today* **2009**, *24*, 2.
21. Available online: https://github.com/ultralytics/yolov5 (accessed on 15 August 2022).
22. Ding, X.; Zhang, X.; Ma, N.; Han, J.; Ding, G.; Sun, J. Repvgg: Making vgg-style convnets great again. In Proceedings of the IEEE/CVF Conference on Computer Vision and Pattern Recognition, Nashville, TN, USA, 20–25 June 2021; pp. 13733–13742.
23. Available online: https://github.com/RangiLyu/nanodet (accessed on 15 August 2022).
24. Wang, C.Y.; Yeh, I.H.; Liao, H.Y.M. You only learn one representation: Unified network for multiple tasks. *arXiv* **2021**, arXiv:2105.04206.
25. Tarvainen, A.; Valpola, H. Mean teachers are better role models: Weight-averaged consistency objects improve semi-supervised deep learning results. *Adv. Neural Inf. Process. Syst.* **2017**, *30*, 1–10.
26. Niu, Z.; Zhong, G.; Yu, H. A review on the attention mechanism of deep learning. *Neurocomputing* **2021**, *452*, 48–62. [CrossRef]
27. Woo, S.; Park, J.; Lee, J.Y.; Kweon, I.S. Cbam: Convolutional block attention module. In Proceedings of the European Conference on Computer Vision (ECCV), Munich, Germany, 8–14 September 2018; pp. 3–19.
28. Muhammad, M.B.; Yeasin, M. Eigen-cam: Class activation map using principal components. In Proceedings of the 2020 International Joint Conference on Neural Networks (IJCNN), IEEE, Glasgow, UK, 19–24 July 2020; pp. 1–7.
29. Ying, X.; Wang, Y.; Wang, L.; Sheng, W.; An, W.; Guo, Y. A stereo attention module for stereo image super-resolution. *IEEE Signal Process. Lett.* **2020**, *27*, 496–500. [CrossRef]
30. Hu, J.; Shen, L.; Sun, G. Squeeze-and-excitation networks. In Proceedings of the IEEE Conference on Computer Vision and Pattern Recognition, Salt Lake City, UT, USA, 18–22 June 2018; pp. 7132–7141.
31. Wang, Q.; Wu, B.; Zhu, P.; Li, P.; Zuo, W.; Hu, Q. Supplementary material for "ECA-Net: Efficient channel attention for deep convolutional neural networks". In Proceedings of the 2020 IEEE/CVF Conference on Computer Vision and Pattern Recognition, IEEE, Seattle, WA, USA, 13–19 June 2020.
32. Lee, D.D.; Pham, P.; Largman, Y.; Ng, A. Advances in neural information processing systems 22. *Tech. Rep.* **2009**, *13*, 10.
33. Bodla, N.; Singh, B.; Chellappa, R.; Davis, L.S. Soft-NMS—Improving object detection with one line of code. In Proceedings of the IEEE International Conference on Computer Vision, Venice, Italy, 22–29 October 2017; pp. 5561–5569.

agriculture

MDPI

Article

A LoRaWAN IoT System for Smart Agriculture for Vine Water Status Determination

Antonio Valente [1,2,*], Carlos Costa [1,3], Leonor Pereira [4], Bruno Soares [4], José Lima [2,5] and Salviano Soares [1,6]

1 Engineering Department, School of Sciences and Technology, UTAD, 5000-801 Vila Real, Portugal
2 INESC TEC—INESC Technology and Science, 4200-465 Porto, Portugal
3 CISeD—Research Centre in Digital Services, Polytechnic of Viseu, 3504-510 Viseu, Portugal
4 CoLAB Vines&Wines—National Collaborative Laboratory for the Portuguese Wine Sector, Associação para o Desenvolvimento da Viticultura Duriense (ADVID), Edifício Centro de Excelência da Vinha e do Vinho, Régia Douro Park, 5000-033 Vila Real, Portugal
5 Research Centre in Digitalization and Intelligent Robotics (CeDRI) and Laboratório para a Sustentabilidade e Tecnologia em Regiões de Montanha (SusTEC), Instituto Politécnico de Bragança, 5300-253 Bragança, Portugal
6 IEETA—Institute of Electronics and Informatics Engineering of Aveiro, 3810-193 Aveiro, Portugal
* Correspondence: avalente@utad.pt; Tel.: +351-917885934

Abstract: In view of the actual climate change scenario felt across the globe, resource management is crucial, especially with regard to water. In this sense, continuous monitoring of plant water status is essential to optimise not only crop management but also water resources. Currently, monitoring of vine water status is done through expensive and time-consuming methods that do not allow continuous monitoring, which is especially inconvenient in places with difficult access. The aim of the developed work was to install three groups of sensors (Environmental, Plant and Soil) in a vineyard and connect them through LoRaWAN protocol for data transmission. The results demonstrate that the implemented system is capable of continuous data communication without data loss. The reduced cost and superior range of LoRaWAN compared to WiFi or Bluetooth is especially important for applications in remote areas where cellular networks have little coverage. Altogether, this methodology provides a remote, continuous and more effective method to monitor plant water status and is capable of supporting producers in more efficient management of their farms and water resources.

Keywords: smart agriculture; IoT; LoRaWAN; WSN; water status

Citation: Valente, A.; Costa, C.; Pereira, L.; Soares, B.; Lima, J.; Soares, S. A LoRaWAN IoT System for Smart Agriculture for Vine Water Status Determination. *Agriculture* **2022**, *12*, 1695. https://doi.org/10.3390/agriculture12101695

Academic Editor: Dimitre Dimitrov

Received: 26 August 2022
Accepted: 3 October 2022
Published: 14 October 2022

Publisher's Note: MDPI stays neutral with regard to jurisdictional claims in published maps and institutional affiliations.

1. Introduction

Agricultural production consumes large amounts of freshwater worldwide. According to estimates by the Food and Agricultural Organization (FAO), irrigation in agriculture accounts for 70% of freshwater consumption [1]. Although grapes for wine production are grown under water deficits or with no irrigation, monitoring of vine water status is extremely important, as it allows assertive water management in irrigated vines, thus contributing to optimizing the use of water resources. In addition, water management becomes crucial in the actual climate change scenario [2] caused by global warming, where weather patterns are more difficult to predict and natural resources such as water availability become uncertain.

Monitoring the water status of vines essentially depends on crossing climatic data and measurements made by operators in the vineyard based on a method developed in 1965 by Schölander [3]. This method is time-consuming and expensive [4] and does not allow monitoring with a sufficient degree of detail for more efficient management of water in vineyards. This is due to the need for specific equipment with little mobility and the short window of opportunity for measurements, which does not allow extensive measurements to be carried out over large areas. Thus, at present, this monitoring is an arduous task for which the execution entails the allocation of many resources, making it a practice accessible only to large companies. Specifically, the need for technicians specially trained to transport and handle the Schölander chamber and the gas bottle—in addition to the price of this equipment—combined with the fact that measurements have to be carried out before sunrise, makes this kind of monitoring not possible for general use.

On the other hand, García-Tejera et al. [5] show that evaporative demand, the hydraulic architecture of the plant, and the texture and depth of the soil play key roles in the final water potential observed. They also state that to establish irrigation programs based on water potential without considering the environmental and plant factors that influence it can create the paradox of having a plant suffer greater water stress even when high irrigation volumes are applied.

These studies highlight the importance of the soil–plant–atmosphere continuum (SPAC) model, both when considering that "water moves from the soil, through a plant, out into the surrounding atmosphere." [6] or with "water moving the 'wrong way' through the plant: from the atmosphere, through a plant, towards the soil" [7].

This model has been frequently used to estimate water status, even with the use of sensory fusion [8]. However, this always entails centralized weather stations and GSM communications, which in agricultural areas normally have different coverage problems or have few sensors with low spatial distribution [9].

In recent years, the Internet of Things (IoT) concept has become popular, and agriculture is no exception. The number of published papers evidences the increasing discussion and relevancy of IoT applied to smart farming [10].

Despite this, in our opinion, this is the first work that uses an IoT infrastructure (in this case based on the LoRaWAN protocol) to communicate with different sensors (sensors for the soil, plant and atmosphere) distributed in order to monitor the water potential using the SPAC model.

As previously mentioned, IoT solutions have been gaining importance globally. The benefit of using integrated soil–plant–atmosphere sensor systems coupled to an online platform will allow the continuous monitoring of water status in real-time and is economically more accessible to a wider range of producers. In addition, water management in agricultural activities is of major importance to achieve an environmentally sustainable sector in line with United Nations Sustainable Development Goals (SDG), namely, Goals 12 ("sustainable production and consumption") and 13 ("climate action") through the development of new innovative solutions in accordance with Goal 9 ("industry, innovative and infrastructures") [11].

Therefore, in light of the above issues, the objective of this work is to make the following contributions:

- Develop an integrated network of sensors for soil–plant–atmosphere with wireless communication through the LoRaWAN protocol;
- Create a platform for receiving data in real time obtained through sensors positioned in places with difficult access and connectivity;
- Enable continuous, remote and general-time monitoring of the water status of vines so that in the near future it is possible to optimize the management of vineyards and water resources.

Related Work

The assessment of the water status in a vineyard is usually obtained through sensors in the soil (soil matrix potential), sensors in the vine (stomatal conductance and leaf water potential) and data from meteorological stations [12]. In this work, the soil matrix potential is obtained through Watermark® granular matrix sensors (Irrometer Company, Inc., Riverside, CA, USA) that are read using a hand-held soil moisture meter (Watermark®, model 30KTCD [13]). Leaf water potential measurements are made with a pressure chamber, and stomatal conductance is evaluated using a portable, open-system, gas exchange analyser (LI-6400; LI-COR, Lincoln, NE, USA [14]). The whole process is laborious and requires several trips to the field (at least twice a day every two weeks). To overcome these limits, some authors assess plant or soil water status from modelling using data from weather stations [15]. Some authors also measure sap flow through thermal dissipation ('Granier' method) as a complement to leaf water potential measurements and soil water content as indicators of water status [16].

The Internet of Things (IoT) is a rapidly evolving paradigm that integrates smart electronic devices (such as sensors, actuators, and controllers) and computers throughout the internet to facilitate how human beings live and to help optimize present and future resource consumption and eventually fundamentals. This rapid evolution brings several new research problems that need solutions from multidisciplinary fields [17].

Extensive research has been done globally potentiating needed IT- and IoT-based transformations. IoT provides its benefits to several application domains such as connected industry, smart traffic, security and surveillance, smart agriculture and automation, healthcare and medicine, smart cities and homes, energy consumption, environment and pollution, etc. A detailed discussion of major IoT applications from both technological and social perspectives can be found in [18]. It has been stated, and we agree, that "Agriculture is one of the important domain around the world".

Environmental aspects are relevant to agriculture. Talavera et al. [19], in their survey study, explore fundamental efforts to use IoT applications for agro-industrial and environmental aspects. They were driven by the need to identify application areas, trends, architectures, and challenges that are open in these fields. They followed a systematic literature review published with peer-review from 2006 to 2016. From an initial pool of 3578 papers, 2652 were selected, 720 were eligible, and 72 met the inclusion and quality criteria; these were clustered into four application domains corresponding to: monitoring, control, logistics and prediction. The results from the review were compiled into an IoT architecture roughly common for the found solutions. The selected studies came from worldwide sources. Most research still addresses monitoring applications (62%), 25% also focusses on control, and the rest (l13%) are preliminary solutions in logistics and prediction. The temperature and humidity of the air, as well as soil moisture and solar radiation, were recognized as universally measured variables. Similarly, actuators such as valves, pumps, motors, sprinklers, humidifiers and lamps were widely used in irrigation, fertilization, pesticide management and illumination. They also observed that cloud storage has not been widely adopted, and communication technologies used were Wireless Personal Area Network (WPAN) protocols such as Bluetooth and ZigBee, followed by Wireless Metropolitan Area Networks (WMANs) supported by cellular technologies (GPRS/GSM/3G/4G).

Recently, low-power WAN (LPWAN) technologies such as LoRa and NB-IoT are becoming commonplace in IoT applications due to their low power requirements, wide coverage range and low cost compared to other long-distance technologies. In one survey [20], the authors concluded that LoRa is the best option for smart agriculture applications. They affirm that for LPWAN, narrowband (NB)-IoT and long range (LoRa) are the two leading technologies. Thus, they provide a comprehensive survey of NB-IoT and LoRa as efficient solutions for connecting smart devices. They demonstrate that unlicensed LoRa has advantages in terms of battery life, capacity and cost. On the other hand, licensed NB-IoT offers benefits in terms of Quality of Service (QoS), latency, reliability and range. These technologies are appropriate for IoT applications that need to communicate tiny amounts of information over a long range. IoT solutions based on cellular technology can provide large coverage, but they consume a great amount of power. After comparing and describing the technical differences (physical features, network architecture and MAC protocol) and IoT factors (QoS, battery life, latency, network coverage, range, deployment model and cost) between LoRA and NB-IoT, the authors defined suitable application domains for each one. For Smart Agriculture, they selected LoRa based on device cost, battery life and coverage. The communication protocol and the system architecture are designated a LoRaWAN network, while LoRa defines the physical layer. Standardized technical development and advancement with technical solutions are the main goal of the LoRa Alliance, which was established in 2015. The LoRa Alliance® (Fremont, CA, USA) is an open, non-profit association with the mission to support and promote the global adoption of the LoRaWAN® standard [21].

There are actual studies exploring this new era of precision agriculture. The application of IoT technologies with the integration of unmanned aerial vehicles (UAVs) for sensing and automation of agricultural fields is a current and future trend. One study [22] conducted a survey of recent research in IoT and UAV technology applied to smart agriculture. Smart sensors, network protocols and solutions for smart farming were described. Further, the fundamental role of UAV technology in smart agriculture was presented by analysing its application in various scenarios such as irrigation, fertilization, pesticide spraying, weed treatment, plant growth monitoring, etc. Moreover, the use of UAV systems in complex agricultural environments was also analysed. In addition to the increasing use of UAV technology, the authors also stated that the growing trend is to use LoRaWAN and Wireless Sensor Networks (WSN). Most of the reported applications used a single group of parameters (e.g., sensor-based irrigation systems, nutrient portion definition based on soil sensors, monitoring of various soil characteristics, automatic irrigation and water quality by moisture estimation based on acquired image processing) or were targeted to the implementation of smart greenhouses.

2. Materials and Methods

As previously mentioned, there are several works with IoT systems for precision agriculture. However, in most cases, these systems are not suitable for multiple groups of parameters (atmosphere, plant and soil), and when they are, they are centred around applications in greenhouses. The developed system aims to determine the water stress of a vineyard, but it allows for other factors. In order to determine water stress, it is necessary to understand the exchange of water (in its liquid or gaseous state) or, in other words, evapotranspiration. Figure 1 is a representation of these exchanges, which consist of transpiration of the vegetation, evaporation of daily water (irrigation or precipitation) and the humidity of the leaves and the soil.

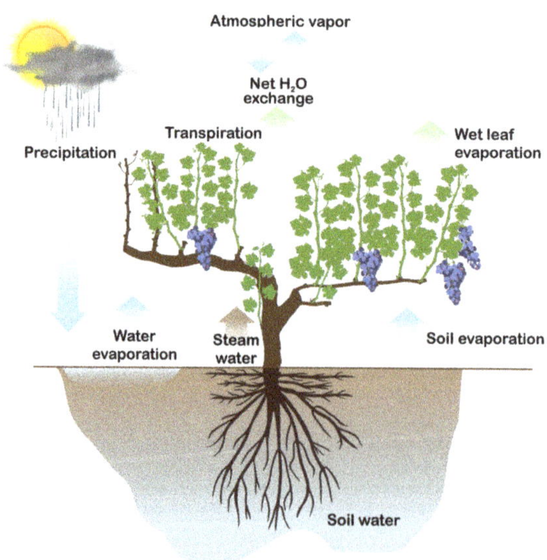

Figure 1. Land evapotranspiration.

Taking this model into account, our system was developed using sensors for the three groups of parameters: atmosphere, plant and soil. Table 1 discriminates between the parameters measured by the different sensors and organizes them into distinct groups ("Atmosphere", "Plant" and "Soil"). This choice also took into account energy consumption and the existence of a digital output by allowing more than one sensor to be connected to a module.

Table 1. Groups of sensors tested in field experiments along with parameters measured.

Group	Parameters	Sensor	Output	Manufacturer
Atmosphere monitoring	Air temperature, air humidity, solar radiation, precipitation, number of lightning strikes, lightning strike distances, wind speed, wind direction, wind gust speed, vapour pressure, atmospheric pressure,relative humidity, humidity sensor temperature	ATMOS 41 [23]	Digital (SDI-12)	METER Group, Pullman, WA, USA
	Air temperature, air humidity, barometric pressure	BME680 [24]	Digital (I2C)	Robert Bosch GmbH, Gerlingen-Schillerhöhe, Germany
Plant monitoring	Leaf wetness	PHYTOS 31 [25]	Analog	METER Group, Pullman, WA, USA
	Body temperature and object temperature	SIL-411 [26]	Digital (SDI-12)	Apogee Instruments, Inc., Logan, UT, USA
	Steam water potential	FloraPulse [27]	Digital (SDI-12)	FloraPulse Co., Davis, CA, USA
Soil monitoring	Soil water content, electrical conductivity, soil temperature	TEROS 12 [28]	Digital (SDI-12)	METER Group, Pullman, WA, USA
	Soil water potential and soil temperature	TEROS 21 [29]	Digital (SDI-12)	METER Group, Pullman, WA, USA
	Soil water content	SoilWatch10 [30]	Analog	Pino-Tech, Stargard, Poland

2.1. The Implemented System

The implemented system (Figure 2) consists of sensor modules with wireless transmission using the LoRaWAN protocol (class A) communicating every 15 min, a gateway connected to The Things Network [31] through a GSM/LTE connection, and a server with a time-series database in InfluxDB [32] and Grafana [33] as an observability platform.

2.2. LoRaWAN Modules

The requirements for IoT modules based on the LoRaWAN protocol are: low cost, low consumption, small dimensions, fast prototyping and easy programming (compatible with the Arduino environment). In view of these requirements, we chose the modular system from RAK Wireless (RAK) [34]. The developed module is depicted in Figure 3.

As the outputs required for connection to the chosen sensors (see Table 1) are SDI-12, I2C and analogue, the modules for the RAK modular system were: the base module RAK5005-O (already supplied with connections for a 3.6 V lithium battery and solar charging—max 6 V); the core module RAK4631, based on Nordic nRF52840, with LoRa (SX1262)—the LoRaWAN protocol is implemented through a library; and the RAK5802 module, based on the 3PEAK TP8485E, which is designed to interface with the RS485 protocol. So that this last module could serve as an interface with the SDI-12 protocol, a dedicated library was developed. An example of a complete module connected to a sensor with SDI-12 communication (the TEROS 12) can be seen in Figure 4.

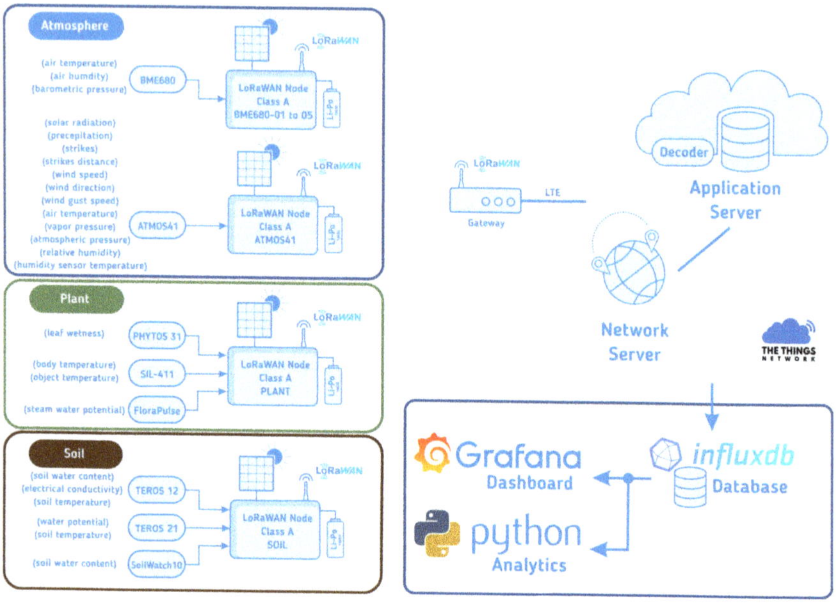

Figure 2. Graphical scheme of the implemented system.

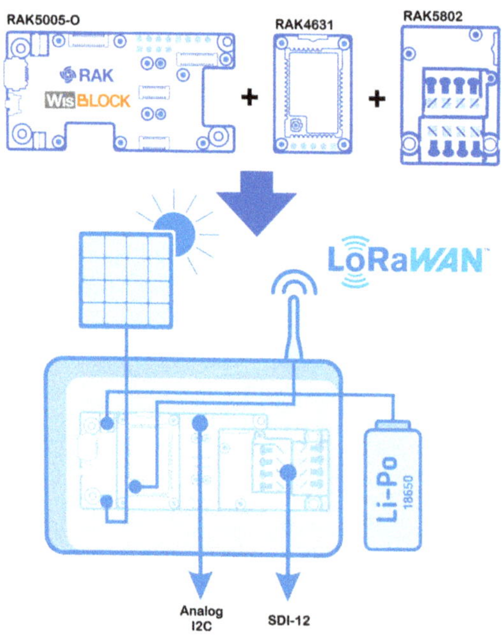

Figure 3. Module used in the developed system, based on RAK modular system WisBlock.

Figure 4. LoRaWAN module with TEROS 12 sensor from Meter Group.

All modules with the BME680 sensor and ATMOS41 have 2600 mAh lithium-ion batteries, and the remaining modules have batteries that provide a total of 4600 mAh of capacity. This is due to the fact that both the 'PLANT' module and the 'SOIL' module have three sensors connected to each module.

2.3. *Implementation at Quinta dos Aciprestes in the Douro UNESCO Region*

The entire system (the gateway and modules) was placed in a vineyard at Quinta dos Aciprestes (Real Companhia Velha, SA, Douro, Portugal) [35]. Figure 5 shows the location of the installed modules and gateway. The 'ATMOSPHERE' sensors were placed in the vineyard and included five modules with BME680 (01 to 05) and one module with the ATMOS41 All-in-One weather station. A module for the 'PLANT' sensor group with stem water potential sensor (FloraPulse), leaf wetness sensor (PHYTOS 31) and the infrared radiometer sensor (SIL-411) was placed on a vine. Soil water content (TEROS 12 and SoilWatch10) and soil water potential (TEROS 21) were also placed in soil next to a vine and were connected to the module for the 'SOIL' group.

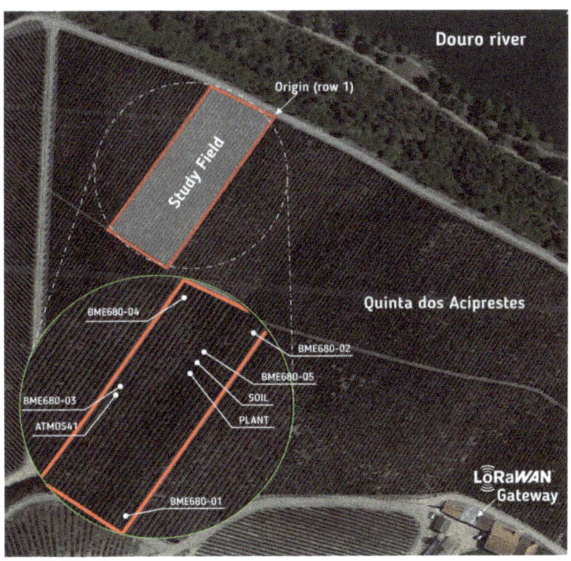

Figure 5. Location of the study field, delimited in red. Inset shows a zoom of the field of study, in which it is possible to see the location of the eight LoRaWAN modules. The gateway is situated about 300 m from the study field. Satellite imagery courtesy of Google Maps™.

The installation of the gateway and all sensors and modules can be seen in Figures 6 and 7.

(**a**) (**b**) (**c**)

Figure 6. Placement of the (**a**) LoRaWAN gateway, (**b**) soil sensors and (**c**) vine sensors (FloraPulse sensor on a vine trunk) at Quinta dos Aciprestes.

(**a**) (**b**) (**c**)

Figure 7. Images of the modules (**a**) for the 'SOIL' group, (**b**) for the 'PLANT' group and (**c**) for the 'ATMOSPHERE' group (ATMOS 41) in loco at Quinta dos Aciprestes.

3. Results

The system was implemented on-site (Quinta dos Aciprestes), and data have been recorded in the database since July 2022. For better visibility, only one week's worth of data are presented: from August 8, 2022 until August 15, 2022. The presentation is divided into the groups previously described, i.e. 'ATMOSPHERE', 'PLANT' and 'SOIL'. Finally, visualization of the data on the Grafana platform is shown, along with a module power consumption analysis.

3.1. 'ATMOSPHERE' *Results*

As previously mentioned, the 'ATMOSPHERE' group comprises sensor modules with the BME680 sensor (five modules from 01 to 05) and the ATMOS 41 module, which is an all-in-one weather station.

Figure 8 shows the air temperature curves of all 'ATMOSPHERE' sensors (both BME680 (BME680-01 to 05) and ATMOS41). This is just an example, because the BME680 sensors also send values for air humidity, barometric pressure and battery voltage.

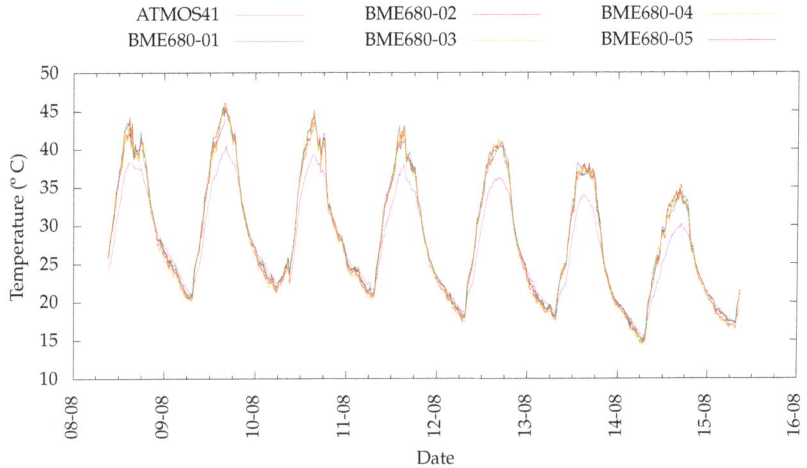

Figure 8. Temperature data from all the 'ATMOSPHERE' devices: ATMOS41 and the five devices with BME680 Bosch sensors.

The ATMOS41 sensor sends, in addition to the previous data, solar radiation, rainfall (Figure 9 shows solar radiation and precipitation) wind speed, gusts and direction (Figure 10 shows wind data) and lightning count and distance.

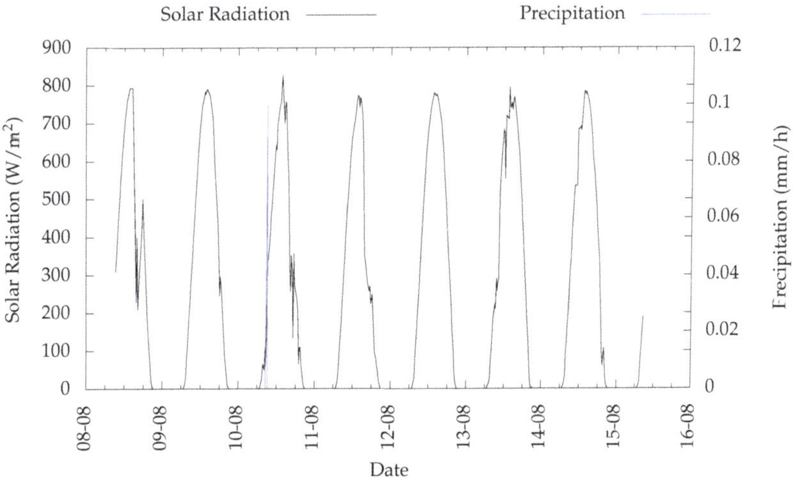

Figure 9. Solar radiation data from ATMOS41.

3.2. 'PLANT' *Results*

The 'PLANT' plant group results incorporate data from leaf wetness (PYTHOS 31, Figure 11), steam water potential (FloraPulse, Figure 12) and canopy temperature (SIL-411, Figure 11) sensors. These sensors are linked to the 'PLANT' module.

3.3. 'SOIL' *Results*

The 'SOIL' module contains the sensors of the 'SOIL' group, to which soil water content sensors (TEROS 12 and SoilWatch10) and a soil water potential sensor (TEROS 21) are connected. In Figure 13, the curves of the soil water tension value and the raw value of the soil water content based on TEROS 12 data are represented.

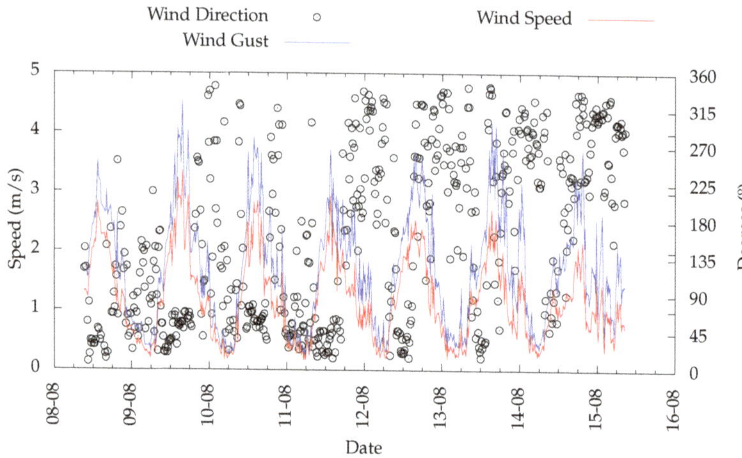

Figure 10. Wind data from ATMOS41.

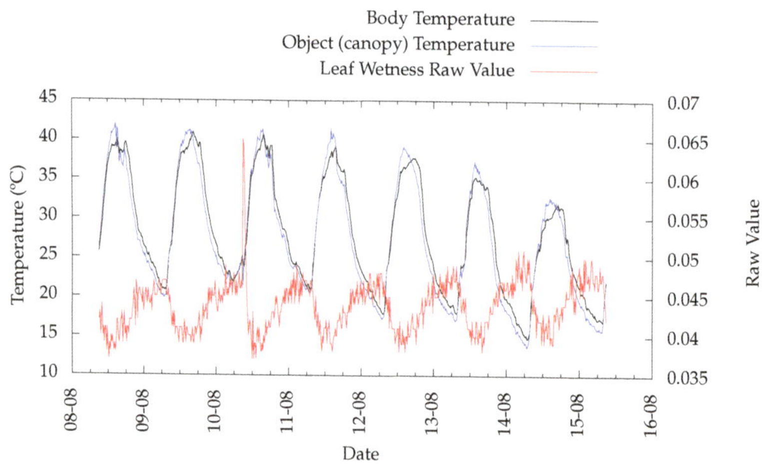

Figure 11. IR data from SIL-411 and leaf wetness data from PYTHOS31.

3.4. *Grafana Dashboard*

The data presented above are indicative of the data present in the database. Through the time-series database, it is also possible to work the data through scripts in Python, as the Listing 1 shows. In this way, it is possible to use artificial intelligence on the collected data to determine and predict, among other factors, water stress. However, for better visualization and understanding of all the data being collected, a dashboard was created on the Grafana platform (Figure 14). All system data, together with data collected on-site with a Schölander camera, will feed training data to a machine-learning system.

In addition to the data collected by the system, weather forecast data obtained through the Pirate Weather API [36] was also added to the dashboard and uses, among other sources, data from the Global Forecast System (GFS) [37]. Thus, the dashboard is divided into: current ATMOS41 data and temperature histogram for the last 7 days (Figure 14A); data for the 'ATMOSPHERE' group (Figure 14B), which includes air temperature and humidity, wind speed, gust and direction, barometric pressure, solar radiation, precipitation and lightning count; weather forecast data (Figure 14C) with daily forecasts for up to four days and hourly for up to 48 h for air temperature and humidity, wind speed and direction, and precipitation; and 'PLANT' group data (Figure 14D) with vine steam water potential, canopy temperature and leaf wetness; and data from the 'SOIL' group (Figure 14E), which includes data on soil water content and potential.

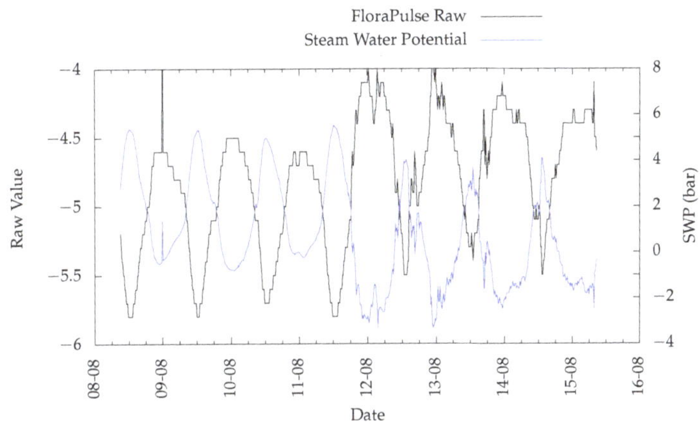

Figure 12. Steam data from FloraPulse sensor.

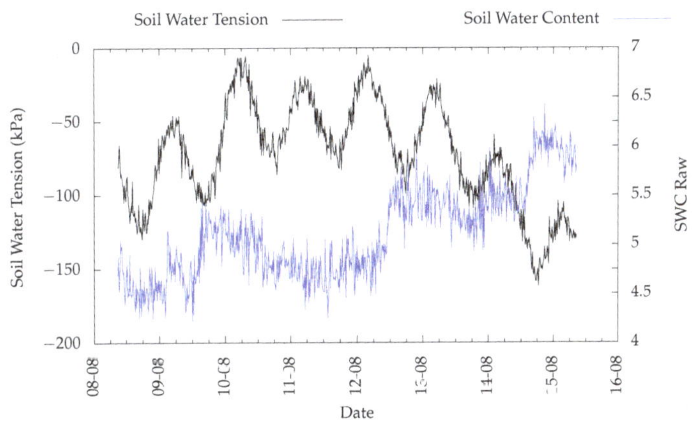

Figure 13. Soil moisture from TEROS21, water tension and TEROS12, volumetric water content.

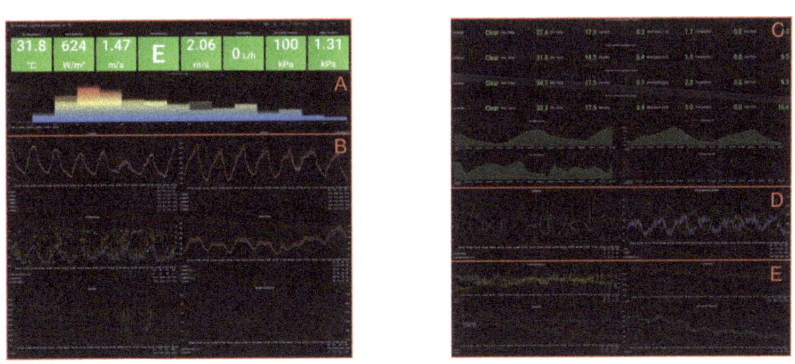

Figure 14. Images of the Grafana dashboard: (**A**) actual ATMOS41 values and a 7-day air temperature histogram; (**B**) atmospheric data; (**C**) weather forecast data; (**D**) plant data; and (**E**) soil data.

Listing 1: Python script example to access InfluxDB.

```
from influxdb_client import InfluxDBClient, Point
from influxdb_client.client.write_api import SYNCHRONOUS
import matplotlib.pyplot as plt
import numpy as np
import csv
import pandas
import datetime as dt

bucket = "IOT_ACIPRESTES"

client = InfluxDBClient(url="http://xxx.xxx.xxx.xxx:8086", token="********************************", org="UTAD_IOT")

#write_api = client.write_api(write_options=SYNCHRONOUS)
query_api = client.query_api()

tables = query_api.query('''
from(bucket: "IOT_ACIPRESTES")
  |> range(start: -4d)
  |> filter(fn: (r) => r["_measurement"] == "mqtt_consumer")
  |> filter(fn: (r) => r["_field"] == "uplink_message_decoded_payload_AirTemperature")
  |> filter(fn: (r) => r["host"] == "avalente01")
  |> filter(fn: (r) => r["topic"] == "v3/iot-aciprestes@ttn/devices/advid-atmos41-01/up" or r["topic"] == "v3/iot-aciprestes@ttn/
       devices/advid-bme680-01/up"or r["topic"] == "v3/iot-aciprestes@ttn/devices/advid-bme680-03/up" or r["topic"] == "v3/iot-
       aciprestes@ttn/devices/advid-bme680-03/up" or r["topic"] == "v3/iot-aciprestes@ttn/devices/advid-bme680-04/up" or r["topic
       "] == "v3/iot-aciprestes@ttn/devices/advid-bme680-05/up" )
  |> aggregateWindow(every: 15m, fn: mean, createEmpty: false)
''')
values = []
time = []
for table in tables:
    #print(table.records)
    for row in table.records:
      values = np.append(values, row.values['_value'])
      time = np.append(time, row.values['_time'])

plt.plot(time, values)
plt.xticks(rotation = 90)
plt.show()
```

3.5. Module Power Consumption

The power consumption of the four types of modules (BME680, ATMOS41, Plant and Soil) was measured using a Nordic Semiconductor Power Profiler Kit II and is summarized in Table 2, where *current* is the average current consumed by the module (takes into account the different operating times during the 15 min sampling period: 2 s sampling time, 2 s transmission time, 3 s reception time and 893 s sleep), *battery* is the capacity of the lithium-ion battery used in the module and *days* is the number of days the module has been operating without solar charging. The ATMOS 41 module must always be powered to obtain wind gust and precipitation values, even though the microcontroller is in sleep mode.

Table 2. Module power consumption.

Model Type	Current	Battery	Days
BME680	3.5 mA	2600 mA h	31
ATMOS41	4.4 mA	2600 mA h	24
'PLANT'	10.1 mA	4600 mA h	19
'SOIL'	7.3 mA	4600 mA h	26

4. Discussion

The implemented system, in terms of data communication, has been operating without losses. All eight modules have their batteries with voltage values higher than 4 V, which demonstrates that the battery–solar panel set is well-dimensioned for all modules.

Regarding the data collected, it should be noted that there is a difference between the temperature values of the ATMOS41 and the BME680 sensors (Figure 8). This may be due to the difference in shields, because in terms of accuracy, the two systems are nearly identical (±0.6 °C). However, the BME680 sensors use a 3D-printed PLA shield [38] (Figure 15), for which studies indicate that the error in the measurement of air temperature is not greater than 1.5 °C [39]. As in the implemented system, the difference when solar radiation is high is greater (about 4 °C) than when it is low (about 2 °C); more studies will have to be carried out to determine the origin of this difference.

Figure 15. Example of the 3D-printed BME680 sensor shield.

In relation to the remaining data, these are within the expected values. It should be noted that during the period presented in this study, there was only one episode of rainfall that can be observed, both in Figure 9 on the precipitation curve, and in Figure 11 on the leaf wetness curve.

A similar and, eventually, more generic study was developed in [40]. On it, a low-cost, modular, Long-Range Wide-Area Network (LoRaWAN)-based IoT platform, called "LoRaWAN-based Smart Farming Modular IoT Architecture" (LoRaFarM) was proposed and aimed to improve the management of generic farms in a highly customizable way. The authors stated that the platform, built around a middleware core, is easily extensible with ad-hoc low-level modules (feeding the middleware with data coming from the sensors deployed in the farm) or high-level modules (providing advanced functionalities to the farmer). The proposed platform was evaluated on a farm in Italy, where it collected environmental data (air/soil temperature and humidity) related to the growth of farm products such as grapevines and greenhouse vegetables over a period of three months from July to September 2019. It should be noted that in their work, for soil moisture, air humidity sensors were used in water-resistant casings, which does not give the water content in the soil as is necessary and is provided by the system implemented and presented here. A web-based visualization tool for the collected data is also presented to validate the LoRaFarM architecture. In general, the LoRaFarM platform inherits its topological structure from the LoRaWAN architecture, as low-level communication patterns are built around the LoRaWAN technology. Specifically, data obtained from farm-level modules are collected by LoRaWAN-oriented End Nodes (ENs) and forwarded to a Network Server (NS) by a LoRaWAN Gateway (GW). In their case, the NS was built on The Things Network [31], and the core middleware was developed to retrieve the data collected from the NS to feed high-layer modules (i.e., the Application Server (AS)) and to be available to end users. The results and discussion of the vineyard scenario reported are in concordance with the results we obtained and the discussion presented for the same environmental data. The actual study goes further, with data collected from plant groups, and it can be extended to other data for which the sensors are already implemented in the modules.

There are other studies that approach obtaining water stress without a wireless sensor network. As an alternative, timely optical remote sensing and non-invasive evaluation of plant water stress based on unmanned aerial vehicles (UAVs) has become common [41]. In this study, remote and proximal sensing measurements were compared with plant physiological variables to test innovative services and support systems to farmers for optimizing irrigation practices and scheduling. The experiment was conducted in two vineyards located in Sardinia, Italy. The indicators of crop water status (crop water stress index and linear thermal index) were calculated from UAV images and ground infrared thermal images and then related to physiological measurements. Remote and proximal sensing images acquired with high-resolution thermal cameras mounted at ground level or on unmanned aerial vehicles (UAV) have spatial resolutions of a few centimetres. They can provide information accurate enough for both assessing plant water status in the field and implementing appropriate irrigation management strategies. The crop water stress index (CWSI), a thermally derived indicator of water deficit based on leaf/canopy temperature measurements, has been used to assess the water status of crops in several plants, such as grapevines, French beans, wheat, rice, maize and cotton. Many studies of plant water stress have analysed the relationships between air temperature, remote sensing indices, and physiological parameters such as stomatal conductance (Gs) and stem water potential (SWP). However, any image acquisition is costly, even when using low-cost UAV solutions. The technique applied in this study built on the use of the CWSI, which has been tested in several studies using ground and satellite data. The use of CWSI maps gives the main advantage of managing irrigation at a large scale by considering the spatial variability of vine water status and developing an approach for providing precision irrigation recommendations.

Another study was based on low-resolution thermal infrared imaging [42]. The goal of this work was to demonstrate the capability of VineScout, a ground robot designed to assess and map vineyard water status using thermal infrared radiometry. Trials were carried out in Douro Superior (Portugal) under different irrigation treatments during the 2019 and 2020 seasons. Grapevines were non-invasively monitored at different times of the day using leaf water potential as reference indicators of plant water status. Grapevine canopy temperatures, recorded with an infrared radiometer, as well as environmental data acquired with a multispectral sensor were saved on the robot controller's computer. The authors state that the promising outcomes gathered with VineScout using different sensors based on thermography, multispectral imaging and environmental data disclose the need for further studies considering new variables related to plant water status, and more grapevine cultivars, seasons and locations to improve the accuracy, robustness and reliability of the predictive models in the context of precision and sustainable viticulture. Leaf water potential was used as a reference indicator of the plant water status (ground truth), and its measurement was taken simultaneously with vineyard monitoring by the robot by a Schölander pressure bomb. One of the main advantages of the VineScout approach to assess plant water status is that vineyard water status variability can be mapped, expanding the concept and application of precision viticulture—in this case, precision irrigation or variable-rate irrigation to optimize water usage and efficiency. The data collected were extracted by pen drive after the map was completed.

These approaches have some advantages, but for a region such as the Demarcated Region of the Douro, with vineyards on steep slopes and a quite heterogeneous environment where conditions on one level may be very different from those on a neighbouring level, they are not the most suitable. Approaches based on wireless sensor networks are the most suitable for this region, and due to the poor GSM network coverage, LoRaWAN technology is the most suitable.

This work shows that through the combination of different technologies, it is possible, even in remote areas, to monitor atmospheric, plant and soil status remotely and in real time, overcoming the challenges of traditional methods (Schölander method) used for water status determination.

5. Conclusions

Regarding climate change, the effects of high temperatures and water scarcity are increasingly significant across the globe. For the success of agriculture, especially for vineyards, the assessment of plants' water status is essential in order to act in a timely and conscientious manner towards efficient management of the culture and water resources. In this sense, this study leads to lower cost and a more effective way of continuously monitoring crop water status remotely and in real time, overcoming the challenges of the Schölander method. This is particularly important in regions where access to parcels and their management is difficult. Furthermore, installation of the LoRaWAN module adds value due to its reduced costs and superior range compared to WiFi or Bluetooth, which is especially valuable for applications in remote areas where cellular networks have little coverage. Altogether, this will support producers in efficient management of their farms, allowing increased quality while contributing to environmental and economic sustainability.

The developed system aims to monitor water stress of the vineyard; however, it allows for other parameters. Water stress arises as the relation between several biotic and abiotic factors. Following that, it is necessary to understand water flux in the atmosphere, plant and soil, as considered in the development of this sensor network.

The system was implemented in a Douro vineyard (Quinta dos Aciprestes) that shares the connection problems of remote areas. Through the implementation of a wireless transmission system based on LoRaWAN protocol (class A) and an online platform (Grafana) for data observation, the system has been operating without communication losses. The installed batteries present the correct voltage, demonstrating that the battery–solar panel set is well-dimensioned for all modules. Regarding the data collected, it should be noted that there is a difference between the temperature values between the 'ATMOSPHERE' group sensors, and more studies will have to be carried out to determine the origin of this difference. In relation to the remaining data, they are within the expected values.

As future work, all system data, together with data collected on-site with a Schölander camera and meteorological data, will eventually become training data to feed a machine learning system. This will allow more accurate estimation of the water stress of the vineyard and can be the base of an information-support decision system with one or more systems such as smart harvest, smart irrigation, etc.

Author Contributions: Funding acquisition, S.S.; investigation, A.V.; methodology, A.V. and C.C.; hardware, A.V. and J.L.; software A.V. and J.L.; supervision, B.S., L.P., J.L. and S.S; visualization, A.V.; writing—original draft preparation, A.V., C.C., L.P. and B.S.; writing—review and editing, A.V., C.C., L.P., B.S. and L.P. All authors have read and agreed to the published version of the manuscript.

Funding: This work was financed by National Funds through the Portuguese funding agency, FCT—Fundação para a Ciência e a Tecnologia, within project UIDB/50014/2020.

Institutional Review Board Statement: Not applicable.

Data Availability Statement: The data presented in this study are available on request from the corresponding author. The data are not publicly available as it are private data of Quinta dos Aciprestes.

Acknowledgments: Special thanks to ADVID (Association for the Development of Viticulture in the Douro Region) for supporting this work through CoLAB VINES&WINES, and to Real Companhia Velha (RCV), for allowing the development of this project in Quinta dos Aciprestes, and to all the staff who supported the implementation of the trial in the vineyards. Rui Soares and Vitória Rodrigues (RCV) are also acknowledged, as well as Nelson Machado (CoLAB VINES&WINES), and Igor Gonçalves and Luís Marcos, from the Technical Services Department of ADVID, which supported the development of this work.

Conflicts of Interest: The authors declare no conflict of interest.

Abbreviations

The following abbreviations are used in this manuscript:

LoRaWAN	Long-Range Wide-Area Network
SPAC	Soil–Plant–Atmosphere Continuum
GSM	Global System for Mobile Communications
LTE	Long-Term Evolution, fourth-generation (4G) wireless standard

References

1. FAO. *Water at a Glance: The Relationship between Water, Agriculture, Food Security and Poverty*; Technical Report; FAO: Rome, Italy, 2010 .
2. Malhi, G.S.; Kaur, M.; Kaushik, P. Impact of Climate Change on Agriculture and Its Mitigation Strategies: A Review. *Sustainability* **2021**, *13*, 1318. [CrossRef]
3. Scholander, P.F.; Bradstreet, E.D.; Hemmingsen, E.A.; Hammel, H.T. Sap Pressure in Vascular Plants: Negative hydrostatic pressure can be measured in plants. *Science* **1965**, *148*, 339–346. [CrossRef] [PubMed]
4. Kirkham, M. Pressure Chambers. In *Principles of Soil and Plant Water Relations*; Kirkham, M.B., Ed.; Academic Press: Boston, MA, USA, 2014 ; pp. 333–345. [CrossRef]
5. García-Tejera, O.; López-Bernal, Á.; Orgaz, F.; Testi, L.; Villalobos, F.J. The pitfalls of water potential for irrigation scheduling. *Agric. Water Manag.* **2021**, *243*, 106522. [CrossRef]
6. Nobel, P.S. Bioenergetics. In *Physicochemical and Environmental Plant Physiology*; Nobel, P.S., Ed.; Academic Press: Boston, MA, USA, 2009; pp. 276–317. [CrossRef]
7. Eller, C.B.; Lima, A.L.; Oliveira, R.S. Foliar uptake of fog water and transport belowground alleviates drought effects in the cloud forest tree species, *Drimys brasiliensis* (Winteraceae). *New Phytol.* **2013**, *199*, 151–162. [CrossRef] [PubMed]
8. Ohana-Levi, N.; Zachs, I.; Hagag, N.; Shemesh, L.; Netzer, Y. Grapevine stem water potential estimation based on sensor fusion. *Comput. Electron. Agric.* **2022**, *198*, 107016. [CrossRef]
9. Zinkernagel, J.; Maestre-Valero, J.F.; Seresti, S.Y.; Intrigliolo, D.S. New technologies and practical approaches to improve irrigation management of open field vegetable crops. *Agric. Water Manag.* **2020**, *242*, 106404. [CrossRef]
10. Navarro, E.; Costa, N.; Pereira, A. A Systematic Review of IoT Solutions for Smart Farming. *Sensors* **2020**, *20*, 4231. [CrossRef] [PubMed]
11. Nations, U. Sustainable Development Goal (SDG). Available online: https://sdgs.un.org/#goal_section (accessed on 22 August 2022).
12. Centeno, A.; Baeza, P.; Lissarrague, J.R. Relationship between Soil and Plant Water Status in Wine Grapes under Various Water Deficit Regimes. *HortTechnology* **2010**, *20*, 585–593. [CrossRef]
13. Watermark Digital Meter. Available online: https://www.irrometer.com/pdf/sensors/407%20WATERMARK%20Meter-WEB.pdf (accessed on 22 August 2022).
14. LI-6400XT Portable Photosynthesis System—LI-COR Environmental. Available online: https://www.licor.com/env/products/photosynthesis/LI-6400XT/ (accessed on 22 August 2022).
15. Gaudin, R.; Kansou, K.; Payan, J.C.; Pellegrino, A.; Gary, C. A water stress index based on water balance modelling for discrimination of grapevine quality and yield. *OENO One* **2014**, *48*, 1–9. [CrossRef]
16. Malheiro, A.C.; Gonçalves, I.N.; Fernandes-Silva, A.A.; Silvestre, J.C.; Conceição, N.S.; Paço, T.A.; Ferreira, M.I. Relationships between relative transpiration of grapevines and plant and soil water status in Portugal's Douro wine region. *Acta Hortic.* **2011**, *922*, 261–268. [CrossRef]
17. Stankovic, J.A. Research directions for the internet of things. *IEEE Internet Things J.* **2014**, *1*, 3–9. 2312291. [CrossRef]
18. Kumar, S.; Tiwari, P.; Zymbler, M. Internet of Things is a revolutionary approach for future technology enhancement: A review. *J. Big Data* **2019**, *6*, 1–21. [CrossRef]
19. Talavera, J.M.; Tobón, L.E.; Gómez, J.A.; Culman, M.A.; Aranda, J.M.; Parra, D.T.; Quiroz, L.A.; Hoyos, A.; Garreta, L.E. Review of IoT applications in agro-industrial and environmental fields. *Comput Electron Agric* **2017**, *142*, 283–297. [CrossRef]
20. Sinha, R.S.; Wei, Y.; Hwang, S.H. A survey on LPWA technology: LoRa and NB-IoT. *ICT Express* **2017**, *3*, 14–21. [CrossRef]
21. Homepage-LoRa Alliance®. Available online: https://lora-alliance.org/ (accessed on 22 August 2022).
22. Boursianis, A.D.; Papadopoulou, M.S.; Diamantoulakis, P.; Liopa-Tsakalidi, A.; Barouchas, P.; Salahas, G.; Karagiannidis, G.; Wan, S.; Goudos, S.K. Internet of Things (IoT) and Agricultural Unmanned Aerial Vehicles (UAVs) in smart farming: A comprehensive review. *Internet Things* **2022**, *18*, 100187. [CrossRef]
23. Weather Station for Research | ATMOS 41 All-in-One Weather Station. https://www.metergroup.com/en/meter-environment/products/atmos-41-weather-station (accessed on 22 August 2022).
24. Gas Sensor BME680 | Bosch Sensortec. Available online: https://www.bosch-sensortec.com/products/environmental-sensors/gas-sensors/bme680/ (accessed on 22 August 2022).
25. PHYTOS 31 | Leaf Wetness Sensor | METER Environment. Available online: https://www.metergroup.com/en/meter-environment/products/phytos-31-leaf-wetness-sensor (accessed on 22 August 2022).

26. SIL-411 SDI-12 Digital Output Standard Field of View Commercial-Grade Infrared Radiometer Sensor | Apogee Instruments. Available online: https://www.apogeeinstruments.com/sil-411-commercial-grade-sdi-12-digital-output-standard-field-of-view-infrared-radiometer-sensor/ (accessed on 22 August 2022).
27. Stem Water Potential Sensors | FloraPulse. Available online: https://www.florapulse.com/ (accessed on 22 August 2022).
28. TEROS 12 | Soil Moisture Sensor | METER Group. Available online: https://www.metergroup.com/en/meter-environment/products/teros-12-soil-moisture-sensor (accessed on 22 August 2022).
29. Soil Water Potential Sensor | TEROS 21 | METER Environment. Available online: https://www.metergroup.com/en/meter-environment/products/teros-21-soil-water-potential-sensor (accessed on 22 August 2022).
30. SoilWatch 10—Soil Moisture Sensor—PINO-TECH. Available online: https://pino-tech.eu/soilwatch10/ (accessed on 22 August 2022).
31. The Things Network. Available online: https://www.thethingsnetwork.org/ (accessed on 22 August 2022).
32. InfluxDB: Open Source Time Series Database | InfluxData. Available online: https://www.influxdata.com/ (accessed on 22 August 2022).
33. Grafana: The Open Observability Platform | Grafana Labs. Available online: https://grafana.com/ (accessed on 22 August 2022).
34. The Official Website of RAKwireless—Where IoT Is Made Easy—RAKwireless-IoT Made Easy. Available online: https://www.rakwireless.com/en-us (accessed on 22 August 2022).
35. Quinta dos Aciprestes—Real Companhia Velha. Available online: https://realcompanhiavelha.pt/en/quintas/quinta-dos-aciprestes/ (accessed on 22 August 2022).
36. PirateWeather Developer Portal. Available online: https://pirateweather.net/ (accessed on 22 August 2022).
37. Global Forecast System (GFS). Available online: http://www.ncei.noaa.gov/products/weather-climate-models/global-forecast (accessed on 22 August 2022).
38. Solar Radiation Shield for Weather Station | 3D CAD Model Library | GrabCAD. Available online: https://grabcad.com/library/solar-radiation-shield-for-weather-station-1 (accessed on 19 August 2022).
39. Botero-Valencia, J.S.; Mejia-Herrera, M.; Pearce, J.M. Design and implementation of 3-D printed radiation shields for environmental sensors. *HardwareX* **2022**, *11*, e00267. [CrossRef] [PubMed]
40. Codeluppi, G.; Cilfone, A.; Davoli, L.; Ferrari, G. LoRaFarM: A LoRaWAN-Based Smart Farming Modular IoT Architecture. *Sensors* **2020**, *20*, 2028. [CrossRef] [PubMed]
41. Matese, A.; Baraldi, R.; Berton, A.; Cesaraccio, C.; Di Gennaro, S.F.; Duce, P.; Facini, O.; Mameli, M.G.; Piga, A.; Zaldei, A. Estimation of Water Stress in Grapevines Using Proximal and Remote Sensing Methods. *Remote Sens.* **2018**, *10*, 114. [CrossRef]
42. Fernández-Novales, J.; Saiz-Rubio, V.; Barrio, I.; Rovira-Más, F.; Cuenca-Cuenca, A.; Santos Alves, F.; Valente, J.; Tardaguila, J.; Diago, M.P. Monitoring and Mapping Vineyard Water Status Using Non-Invasive Technologies by a Ground Robot. *Remote Sens.* **2021**, *13*, 2830. [CrossRef]

agriculture

MDPI

Article

Decision Support in Horticultural Supply Chains: A Planning Problem Framework for Small and Medium-Sized Enterprises

Marius Drechsler and Andreas Holzapfel *

Department of Logistics Management, Hochschule Geisenheim University, Von-Lade-Str. 1, D-65366 Geisenheim, Germany
* Correspondence: andreas.holzapfel@hs-gm.de

Abstract: This paper investigates and systematizes planning problems along the supply chain of small and medium-sized companies in the horticultural market of ornamental plants, perennials, and cut flowers. The sector faces considerable challenges such as multiple planning uncertainties, product perishability, and considerable lead times. However, decisions in practice are often based on rules of thumb. Data-driven decision support is thus necessary to professionalize supply chain, logistics, and operations planning in the sector. We explore the practical planning problems with the help of expert interviews with people in charge of typical companies active in the market. We structure the planning problems along the supply chain according to their time horizon and highlight the critical elements of the planning tasks and horticultural specifics. We examine the status quo of research on decision support for these planning tasks with the help of a structured literature review, highlight research gaps, and outline promising future research directions. We find that the tactical planning domains of material/product requirement, production, and demand planning are especially critical in practice, and that there is a great need for research to develop practically relevant decision support systems. Such systems are currently available only to a limited extent in literature and are not fully compatible with requirements in the ornamental horticultural sector. By structuring and detailing the relevant decision problems, we contribute to an understanding of planning problems and decision-making in horticultural supply chains, and we provide a first comprehensive overview of planning problems, aligned literature, and research gaps for the horticultural business.

Keywords: supply chain; horticulture; logistics; operations; planning framework; decision support

Citation: Drechsler, M.; Holzapfel, A. Decision Support in Horticultural Supply Chains: A Planning Problem Framework for Small and Medium-Sized Enterprises. *Agriculture* **2022**, *12*, 1922. https://doi.org/10.3390/agriculture12111922

Academic Editor: Dimitre Dimitrov

Received: 27 October 2022
Accepted: 11 November 2022
Published: 15 November 2022

1. Introduction

One of the major parts of the horticultural sector is the market for ornamental plants, i.e., flowers, ornamental plants, and perennials. Ornamental horticulture for example makes EUR 9.4 billion in sales per year, accounting for 19.8% of the overall horticultural market in Germany, which is one of the largest producing markets in Europe [1,2]. Small- and medium-sized companies have a particularly large role in this market with about 5000 producers and 16,500 specialized trade and retail companies in Germany [3]. The companies can be divided into producers, wholesalers, and retailers, with most of the companies undertaking multiple functions [4]. Producers mostly specialize in propagating and cultivating plants into (semi-)finished products. This usually requires long lead times and good storage conditions. Retailers are all enterprises selling horticultural products to final customers, which are florists, specialty retailers, market stands, garden centers, DIY markets, and supermarkets. They typically offer a big assortment, no or short lead times, and can only operate marginal stockpiling to guarantee freshness. Wholesalers are all traders in between the other two roles, such as ex- or importers, auctions, and cash and carry markets, which can cover urgent demand by offering short lead times and often also handle planning and execution of transportation. Given their heterogeneous functions, all these market players face specific supply chain planning problems.

Systematized supply chain and operations planning is important for horticultural companies as the sector exhibits several specific challenges such as product perishability, as well as demand and supply volatility and uncertainties [5]. Product surpluses lead to spoilage and waste and have to be avoided. It is equally important to prevent product deficits and thus stockouts. This is complicated by external and sometimes unpredictable factors such as product decay, pests, traffic, and weather conditions [2,6]. The weather influences the production of plants, as well as the demand of final customers, and it is difficult to predict seasonal trends prior to the first sales data of a season. Lead times of sometimes several months and more also have to be factored in, and reliable information is rare.

However, there has been only limited research on supply chain and operations planning problems in the horticultural context. Existing contributions on supply chain management barely touch the prevalent planning problems or analyze selected planning problems in very specific contexts (see Section 2). In consequence, there is no general and holistic overview of planning problems in the horticultural business [7]. Such an overview could however help to systematize and understand the interaction of the multiple decisions. Additionally, guidance towards the research gaps for practically relevant decision support in ornamental horticulture is missing so far. Such research could further support industry practice to take decisions in a more structured, informed manner based on reliable data. Our study has thus the following objectives: Identifying and systematizing relevant supply planning problems of small and medium-sized companies in ornamental horticulture supply chains considering the different agents introduced above and investigating pertinent literature on the different supply chain planning areas to depict the status quo and to derive the research gaps.

We thereby contribute to horticultural supply chain literature by developing a structured supply chain planning matrix providing an overview of planning problems and representing the interrelations of decisions and their time horizon. By detailing the relevant decision problems, we contribute to an understanding of planning problems and decision-making in horticultural supply chains. Thus, we guide horticultural decision-makers in a structured way through the large bandwidth of planning problems. By reviewing existing scientific contributions in the various planning domains, we further pinpoint under-investigated decision problems in literature, and thus guide researchers in filling the existing gaps. Thereby, we also support the development of data-driven decision support systems that are tailored to the requirements of the horticultural sector. This could enhance existing Enterprise Resource Planning (ERP) systems used in horticulture, as we find that these systems currently miss interfaces for using the available data for decision support in an automated manner.

All results presented are based on exploratory expert interviews with managers involved in supply chain and operations decision-making in the business and on structured literature reviews to identify the research gaps.

The paper is organized as follows. We motivate our research objectives in detail and formulate our research questions based on initial literature research on supply chain planning in horticulture in Section 2. Afterward, we explain our methodological approach for identifying and structuring the relevant planning problems in detail in Section 3. We further introduce a general supply chain planning framework that will be adapted to the situation in ornamental horticulture in the following section. Section 4 identifies and systematizes the practical decisions and reflects on the respective gaps for decision support in the pertinent literature. We summarize and reflect our findings in a discussion in Section 5, and finally conclude our paper with an outlook and further research directions in Section 6.

2. Literature Overview and Research Questions

Overall, literature on supply chain planning problems in ornamental horticulture is relatively scarce. Existing contributions on supply chain management in horticulture often focus on describing characteristics, developments, and needs of very specific market segments, but barely touch the prevalent planning problems [6,8].

Nevertheless, there are some dedicated publications that analyze specific planning problems and provide solution approaches for specific ornamental horticultural planning areas. Examples are the contributions of van der Vorst et al. [9] who present innovative logistics concepts in the floricultural sector, Ossevoort et al. [10] who develop an approach for improving strategic distribution planning in Dutch horticulture, or Tromp et al. [11] who present a model for predicting remaining vase life of cut roses. Therefore, it is worthwhile structuring the existing body of contributions by planning area and time horizon to provide a systematic overview of existing suggestions in the literature, but also highlighting the areas that are still under-researched. Besides this, a major proportion of the still small body of publications that explicitly deal with sector-specific supply chain planning problems focuses on the Dutch horticultural market which is driven by auctions and large companies. These findings however can not always be transferred to non-auction-driven markets dominated by small and medium-sized companies, like Germany. Thus, it is also relevant to analyze which approaches are generic enough to provide guidance for different agents in the heterogeneous horticultural market and which contributions require supplementation of further research.

Based on the results of our literature research, so far only Mir and Padma [7] presents an overview of different decision support systems in a horticultural context. They review multiple plant-specific contributions for general purposes in horticulture and applications for integrated plant protection, nutrient management, and land use. Therewith, Mir and Padma [7] provide a first overview of relevant decision support systems for horticulture, but their findings are limited to the aforementioned topics which exclude the areas of distribution and sales planning as well as some procurement planning problems. These are, however, relevant to investigate when aiming to provide a complete overview of supply chain planning problems and contributions. Thus—to the best of our knowledge—there is no general and holistic overview of planning problems and aligned literature for the horticultural business so far. Such an overview could however help to systematize and understand the interaction of the multiple decisions beyond direct plant treatment. This could further support industry practice in taking decisions in a more structured and informed manner based on reliable data.

Our study thus aims at identifying and systematizing the practically relevant planning problems and corresponding decisions along the supply chain of small and medium-sized companies in ornamental horticulture. We further target to identify and systematize existing contributions within the different planning domains and highlight the gaps for future research. We posit the following three research questions to be answered:

1. Which specific decision problems arise in the value chains of ornamental plants, perennials, and cut flowers and how are they correlated?
2. What differences in planning problems can be identified between different positions in the respective supply chain?
3. To what extent have these planning problems already been studied, and what research gaps exist?

3. Methodology

As there is currently no structured overview of supply chain and operations decisions in horticulture, we applied a three-step research approach to identify, systematize and analyze the corresponding planning problems and research gaps. First, we conducted exploratory expert interviews (Section 3.1). Second, we systematized the planning problems identified with the help of a supply chain planning matrix (Section 3.2). Finally, we analyzed pertinent literature in the respective contexts to identify the status and gaps of horticultural supply chain and operations research (Section 3.3).

3.1. Exploratory Expert Interviews

As a first step, we chose an exploratory study design for answering research questions one and two aimed at identifying and portraying the practically relevant planning problems along the supply chain. An exploratory qualitative research study is appropriate for uncovering such unstudied research areas as data from practice enrich the theoretical findings [12]. Exploratory interviews are thereby recommended for use in logistics and supply chain management research [13]. Developing theoretical insights out of empirical data, we followed an approach inspired by grounded theory [14,15].

3.1.1. Data Collection

Our empirical data collection followed the ground theory concept leading to generalizable results [15] and providing a practical perspective on the relevant planning problems in ornamental horticulture. We conducted semi-structured exploratory expert interviews with people in charge of horticultural companies that proved to have a holistic view of all planning activities and processes in their enterprises as recommended in pertinent literature [13,16]. These were (co-)owners or executive managers. We chose an unbiased selected sample of representative average small and medium-sized companies in the German horticultural market of ornamentals, perennials, and cut flowers, following the classification of the European Commission [17]. We focus on the German market to create a unified context and because Germany is one of the largest producing horticultural markets in Europe [1,2]. Furthermore, we limited the study to ten enterprises to keep the investigation to a reasonable scale. The companies are specified in Table 1.

Table 1. Specification of case companies

Company	Size	Assortment	Description
Com 1	Medium	Bed and balcony plants, cut flowers	Cash & carry market wholesaler with their own large production and fleet
Com 2	Small	Pot plants, perennials, cut flowers	Producer with their own retail store, wholesale activities and a limited own fleet
Com 3	Medium	Cut flowers	Online retailer for bouquets, also acting as mediator and wholesaler for cooperating retailers
Com 4	Small	Mediterranean perennials	Specialized importing wholesaler with their own fleet
Com 5	Small	Pot plants, perennials, house plants	Retailer relying on close cooperation with a producer at the same location
Com 6	Small	Herbs, pot plants, bulb plants	Specialized producer expanding their operations by establishing a further modern production facility
Com 7	Medium	Grasses, perennials	Specialized producer with their own fleet serving large customers
Com 8	Small	Cut flowers, pot plants	Specialized producer with their own retail store, also serving other retailers or wholesalers
Com 9	Small	Perennials, blooming plants, herbs	Specialized producer mostly serving retailers with an own fleet and wholesale activities
Com 10	Small	Bed and balcony plants, perennials, house plants	Retailer with a small production operation of their own and mainly wholesaler sourcing

The participating companies cover all major roles in the horticultural supply chain, with some companies taking on multiple functions. Out of the ten companies, we classify seven as producers, six as wholesalers, and five as retailers. We developed an interview guideline and marginally adapted it after conducting the first interview, which also served as a pre-test [14]. The questions targeted the different planning problems in horticulture, their correlations, and circumstances in the sector. A brief overview of the questions' topics was sent to all interview partners prior to the interviews. We used this to make sure that we got insights into current horticultural proceedings and different supply chain roles [12]. All interviews were conducted via video chat by the same two interviewers, who took field notes during the talks and compared and merged them afterward to achieve reliability [18]. The interviews lasted 85 min on average. After conducting the ten interviews there were no important new findings and we stopped the data collection, as we could assume that we had reached theoretical saturation.

3.1.2. Data Analysis

Following the approach of Mayring [19], we coded the interviews freely before merging the codes and structuring them into up to four levels of subcodes. We performed the analysis and evaluation of the interviews with the help of the MAXQDA software and all results of this paper have been validated and confirmed by the interviewees. Thereby we followed the guidelines of Gioia et al. [20] to ensure the trustworthiness of our data analysis. More than 550 codes were recorded and structured for investigating the supply chain and operational decision problems.

3.2. Supply Chain Planning Framework

To structure the planning problems identified, we used the systematization proposed by Fleischmann et al. [21]. They developed a general supply chain planning problem matrix by distinguishing the planning problems according to their time horizon (long-, medium- and short-term) and their position in the supply chain process (procurement, production, distribution, sales) (see Figure 1).

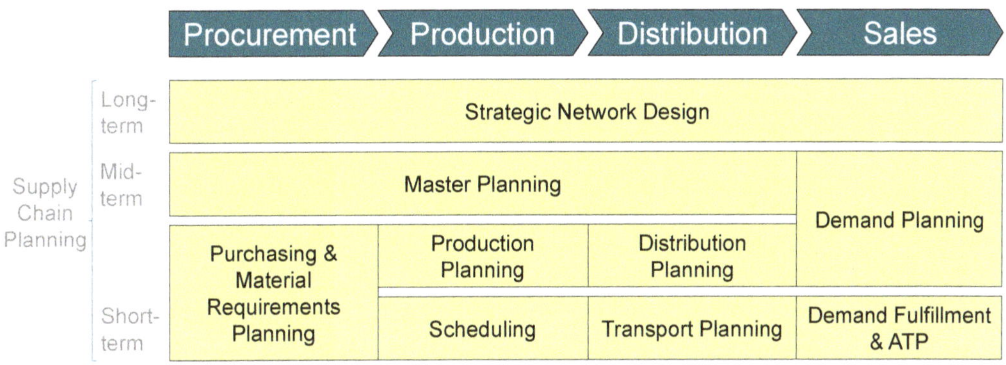

Figure 1. Supply chain planning matrix, own representation based on Fleischmann et al. [21].

Strategic network design covers all long-term planning sections determining the basic supply chain design and the elementary material and product flow between suppliers, producers, wholesalers, retailers, and customers. Master planning coordinates procurement, production, and distribution, which often have to be planned simultaneously, on a tactical level. Demand planning covers mid- to short-term demand estimates. The planning of ordering components and finished products takes place on a mid- and short-term level in the procurement section. Production planning especially covers lot-sizing problems, providing the basis for short-term production scheduling. Product flows are coordinated in distribution planning and subsequent transportation planning, i.e., short-term vehicle routing. The tasks of demand fulfillment and available to promise (ATP) planning cover short-term sales planning [21].

Note that these general supply chain processes and planning tasks usually differ concerning their relevance and context depending on which supply chain functions a company covers. For instance, while horticultural producers naturally face a high variety of production planning tasks, distribution planning is not their main focus. On the other hand, pure wholesalers and retailers have no classic production. Nevertheless, there are similar tasks for these types of companies from a planning perspective in terms of warehousing, which typically substitutes the production step at trading companies [22].

We adapted the general framework proposed by Fleischmann et al. [21] to the planning situation of small and medium-sized horticultural companies by varying the individual contents, highlighting the context specifics and interdependencies that can be found in the horticultural sector. In doing this, we followed a similar procedure to that already taken in literature for other types of business, e.g., for grocery retailing and shipyards [22,23].

3.3. Structured Literature Research

Based on the systematization of planning problems achieved, we performed a structured literature search to investigate the third research question following the approach of Seuring et al. [24]. For each planning problem relevant to horticulture and identified within our study, we conducted a separate literature search focusing on planning problem descriptions and solution approaches to evaluate how intensive research has already been conducted for the practically relevant planning tasks. To do this we first combined keywords regarding the horticultural planning tasks with those of classic decision support system nomenclature in the respective area. In case there were no or only very few publications referring to the context of ornamental horticulture, we extended the perspective to a broader horticultural context, to other industry sectors with similar characteristics or circumstances, or, as a final measure, to generic supply chain and operations planning literature. After this identification step, we completed a qualitative content analysis for portraying the respective contributions [19] and analyzed the gaps comparing the existing literature with the planning problems and necessities identified in the exploratory interviews.

4. Typology of Planning Problems in Horticulture

This section discusses the practical planning along the supply chain of small and medium-sized companies in ornamental horticulture, divided by their respective time horizon. The systematization of planning areas follows the nomenclature of Fleischmann et al. [21] with slight adaptations where necessary due to horticulture specifics. For each planning area, we first detail the most relevant practical planning problems in horticulture, refer to potential supply chain role specifics, and discuss planning interrelations. We then reflect on the current status of pertinent literature regarding the respective planning problems and delineate research gaps and needs for decision support from a practical perspective.

4.1. Strategic Planning

In the long-term strategic planning domain, we can make distinctions according to the supply chain areas between procurement, production, (physical) distribution, and sales planning. We discuss the corresponding strategic planning problems in the following paragraphs.

4.1.1. Strategic Procurement Planning

Planning Problems Identified

Supplier selection and continuous supplier relationship management are key strategic factors for all supply chain roles in the horticultural sector. As the sector is characterized by various uncertainties (e.g., regarding supply and production due to weather extremes or pests) and high perishability of the products, the companies in the sector typically build on long-term cooperations with suppliers to ensure reliable supply and product quality. This means that collaborations over a time span of several decades are common. This highlights the importance of strategic procurement planning decisions. Although long-term relationships and close collaboration are key in ornamental horticulture, this does not mean that the horticultural companies focus exclusively on one supplier per product. Instead, the sector's uncertainties also compel the need to consider suitable alternative suppliers that may also react short term. Screening and contracting suppliers (often wholesalers) that have short-term compensating potential is also very relevant because the horticultural market is characterized by considerable short-term fluctuations in demand due to trends and weather conditions. Overall, supplier selection defines general conditions for product ordering potential, as well as the depth and breadth of the assortment, and lays the framework for make-or-buy decisions.

Pertinent Literature and Gaps

Contributions in literature also highlight cooperation as an important factor for gaining a competitive advantage in horticulture [25,26]. Joint planning as well as risk and reward sharing are postulated as the main drivers for close relationships and information sharing between enterprises [27]. Geerling-Eff et al. [26] additionally outline the importance and potential issues of selecting partners to cooperate with. Matopoulos et al. [27] underline that supplier selection and cooperation are multi-criteria decision problems in horticulture with many factors that are hard to quantify such as reliability, quality, and experience. One of the few quantitative approaches available in the domain of strategic procurement planning in horticulture is the work of de Keizer et al. [28], who model and optimize strategic network decisions for bouquet-making, sourcing cut flowers from auctions. Their approach also takes quality decay into consideration. Outside the domain of ornamental horticulture, there are further models that could be also applicable in the sector. The quality decay of perishable products is for example considered in a source selection model for strawberries [29], and Yazdani et al. [30] present a model for food supplier selection under uncertainty. A holistic supplier selection model reflecting the various specifics of ornamental horticulture has not however been developed so far. This observation is also supported by contributions in horticultural literature that emphasize that especially the consideration of the impact of quality decay and lead times on supply chains of perishables is a research gap in supply chain design and procurement planning [28,29].

4.1.2. Strategic Production Planning

Planning Problems Identified

The strategic production design in horticulture consists of planning acreage and greenhouses as well as the corresponding production systems. These planning tasks are key for producers. Trading companies could also have similar planning issues when deciding about the design of their warehouses [22]. For small and medium-sized horticultural trading companies (wholesalers, retailers), warehouse design, however, plays a minor role due to the fact that the rotation of plants is usually relatively quick. Besides this, wholesalers and retailers often combine their business with the production of their own and thus use the production facilities as warehouses. We, therefore, focus on the planning tasks of producers in the following. The environmental conditions of the production location and the cultivatable land limit potential production and high-impact decisions, on which, plants can be produced. The production system then defines the setup and layout of a production site, which can consist of open crop land, greenhouses, foil tunnels, and roofed indoor areas. Typically, many small and medium-sized horticultural companies grow organically over decades. This often results in fragmented expansion and unfavorable structuring of production sites. A typical strategic production design problem of existing businesses is thus limited by the acreage of multiple in- or outdoor production sites with different levels of automation, conditions, and sizes. The companies in the sector would thus highly benefit from decision support for the suitable reorganization of given production facilities.

Pertinent Literature and Gaps

In literature there are only a few strategic production design models with a dedicated horticultural context. Rath [31] present a model for planning the energy supply of greenhouses for growers, and Vanthoor, et al. [32] develop a model for greenhouse design optimization. Annual financial results are maximized by making optimal choices from alternatives for structure types, covering material, shade screening, whitewash, thermal screens, heating and cooling systems, and CO_2 enrichment. Further research on this topic can be found in the efficient land use planning section presented by Mir and Padma [7]. As shown in our exploratory study, strategic production design requires the consideration of a vast number of additional factors in practice. Implementing a comprehensive optimization model does not appear easy as a result, but there is certainly potential for research on developing decision support systems evaluating alternative production designs that may potentially recur depending on the given infrastructural conditions. This is also supported by Vanthoor et al. [32] who see their optimization model as a first component of a more complex planning system that is still to be developed. Thus, there is considerable improvement potential for the internal organization in horticultural companies, as de Waal and Meingast [33] point out.

4.1.3. Physical Distribution Planning

Planning Problems Identified

As we focus on small and medium-sized enterprises, most of the producers and retailers have no extensive distribution network and often do not even conduct explicit distribution planning. Exceptions are the companies that assume a wholesale function. These often have a wider distribution network and usually assume the responsibility for transportation planning, relying on a fleet of their own. In doing this, they strengthen their market position, especially for cooperation with small producers that are pleased to have the possibility to outsource distribution issues. Nevertheless, in practice, most companies own at least a small truck to handle internal and urgent transportation. However, distribution is often more efficiently operated by larger or specialized partners due to economies of scale. These partners have to make decisions on fleet dimensioning and the definition of transportation links that determine the constraints for tactical distribution planning in terms of transport, packing, storing, and processing capacities. The distribution structure also needs to be set up for fast transportation of fragile and perishable products. Note that direct-to-customer deliveries or customer pick-ups, therefore, dominate in the ornamental horticulture market.

Pertinent Literature and Gaps

There are several contributions in the relevant literature focusing on strategic distribution design in horticulture, especially considering virtual trading networks. In this domain, a trading network for the Dutch horticultural market has been developed that highlights the advantages of direct producer-to-retailer flows focusing not just on cost reduction, but also on the end user´s satisfaction [10]. The concept enables better connections between consumers and producers and does not require a physical presence of the products anymore at intermediate steps. The contributions of van der Vorst et al. [34] and van der Vorst et al. [9] support these benefits of virtual networks that also allow the introduction of virtual auctions for perishable products [35]. Besides this, de Keizer et al. [36] present a mathematical model for logistics network design for perishables, specifically cut flowers, considering quality decay as well as its heterogeneity. Contrary to other planning areas there are thus several suitable contributions to decision support relating to ornamental horticulture's strategic distribution planning providing guidance for the traders involved in transportation. However, models reflecting the distribution conditions and possible particularities outside the Netherlands are missing so far.

4.1.4. Strategic Sales Planning

Planning Problems Identified

In the sales domain, sales channels and customer relationship management as well as the product program have to be defined on a strategic level. Strategic sales planning highly impacts the upstream logistics planning areas, and the general conditions regarding lead times, ordering, and production are predefined. Lead times for example can differ from several months for large grocery retailers to immediate supply for final customers. The product program and the respective plants determine specific requirements regarding temperature, humidity, treatment, and more which affects the production planning and materials program. It is typically small and medium-sized producers that continuously specialize to remain competitive in the market in terms of the product program, but for retailers, a certain level of generalization is important to attract final customers. On a wholesaler level, both the generalization and specialization approaches can be promising, and respective examples can be found on the market. While specialization can make sense when importing specific plants from a dedicated region (e.g., Mediterranean plants), small and medium-sized retailers in particular value a broad assortment of wholesalers as this means they do not have to contract too many different suppliers. Wholesalers act per definition in the B2B market, and retailers in the B2C market. Producers may define their preferred sales channel(s) as they have both options. Rising e-commerce in the horticultural domain is thereby an option to widen the customer base without a physical presence. Close cooperations with customers also play an important role in horticulture, especially for producers and wholesalers. Joint sales planning and the organization of advertising and sales campaigns result in valuable information and planning security on a mid-term basis in a generally uncertain market.

Pertinent Literature and Gaps

Explicit strategic sales planning literature with a focus on horticultural products are rare. In terms of strategic marketing planning, White and Uva [37] analyzes multiple cases and discusses different strategies to provide guidance for developing marketing plans in horticulture. Apart from this, the horticultural market is typically not referred to in strategic sales planning literature, and Engelke [38] outlines a research gap in marketing performance and strategic service differentiation in horticultural retail. Nevertheless, there are some contributions in the literature that exhibit elements that could be transferred to a horticultural setting because of their consideration of characteristics like product perishability and the varying product development stages by different customers. Retailers for example demand plants that are at the beginning of their blooming time to be able to provide attractive plants for sale while still ensuring a long blooming time for final customers. Wholesalers in comparison require plants at an earlier developmental level. In literature, there are for example multiple general approaches for optimizing the product portfolio balancing external variety and internal complexity due to product differentiation [39], or considering product quality functions focusing on customer requirements [40]. Besides this Hübner and Kuhn [41] investigate assortment planning models for perishables and non-perishables with limited shelf space, which could be helpful for horticultural retailers. However, most of the models referred to assume modular products and there is no adaptation of assortment planning models explicitly for horticultural products and their specific requirements.

4.2. *Tactical Planning*

On a mid-term basis, horticultural companies face the planning problems of material or product requirements, production, distribution, and demand. We portray these planning problems in the following.

4.2.1. Material/Product Requirement Planning

Planning Problems Identified

The inherent decision of making or buying makes material and product requirement planning a tactical planning problem for horticultural producers. This task can be seen as more of an operational issue for wholesalers and retailers. However, the typically long lead times in the sector that have to be considered for ensuring the basic availability of products also require mid-term product requirement planning for trading companies.

Producers first have to decide whether to grow plants from seeds or seedlings or to use the cultivation of young plants. These young plants are in turn available at different stages. Some plants such as rose stocks have to be cultivated for years before becoming marketable, which explains the possible long-term aspect of the product requirement decision for producers. Besides the production time, the lead times of seeds and seedlings also have to be considered (which may amount to up to one year). If producers have a B2C channel of their own, they may further decide on whether to grow all the plants or to source additional finished products to supplement their assortment. Generally, self-production may offer higher margins for the producers as well as independence regarding the supply market. However, certain products may be cheaper to source externally from other producers that specialize in these products. Additionally, production of their own may also suffer from uncertainties regarding output. Besides the price, quality aspects are very important in the make-or-buy decision. Self-produced plants offer the possibility of strict quality control in all growing stages. If assortment and order quantity flexibility is prioritized, short-term sourcing via wholesalers is the third basic option for producers to set up their sourcing. Stockouts of seasonal products are just as common as short-term discount offers of overflow products in the sector.

Besides the decision of making or buying products, order decisions also have to be made on a mid-term basis. The earlier horticultural products are ordered or reserved, the higher the available quantities and qualities are. However, production restrictions such as seasonal growing cycles and weather conditions combined with volatile demand and trends can nonetheless result in availability problems in horticulture. The majority of procurement processes in the horticultural sector can therefore be divided into pre- and re-orders or daily spot market sales [42]. A company's main products are mostly sourced in large part via pre-ordered quantities from producing companies. These quantities should cover the basic demand that is forecasted and thus ensure planning security, availability, and quality of the core products. Short-term re-orders can only cover smaller quantities, but provide flexibility to react to fluctuations in demand or unforeseeable supply or production shortages. Wholesalers are typically the main source for re-orders, but producers may also have free quotas for short-term sales. Many companies emphasize that they target coverage of the majority of their demand by pre-orders of large quantities, but that they also want to cover a certain share by re-orders from the same producers or other wholesalers such as cash and carry markets to ensure responsiveness during the high season.

Pertinent Literature and Gaps

While there is literature available on make-or-buy decisions beyond the horticultural context (see e.g., McIvor and Humphreys [43] or van de Water and van Peet [44]), there is so far no specific application in the horticultural domain. Besides this, the prevalent purchasing structure with pre- and re-orders has not so far been covered in horticultural research. This marks a significant opportunity for developing a corresponding decision support system relevant in practice due to the importance of this planning problem emphasized by all participating companies in our exploratory study. Overall, sector-specific expertise is needed for material and product requirement planning to evaluate different materials and sourcing alternatives. This also requires an understanding of the drivers of and barriers to using enterprise resource planning in horticulture. Verdouw et al. [45] examine these to handle the perishability aspects and uncertainties potentially better in horticulture. They outline the importance of improvisation and ad-hoc communication and emphasize the lack of well-structured administrative organization in the sector, which is especially the case for small and medium-sized companies. This further highlights the need for systematic decision support.

Although there is this gap in the literature, specific aspects of tactical material and product requirement planning are covered in horticultural literature. In a recent contribution, Faraudo Pijuan [46] presents an economic order quantity model that considers growing models and dynamic holding costs. The quality aspect also plays a major role in procurement decisions in the supply chain of perishable goods and therefore quality decay in this decision is an important area of research to possibly expand shelf/vase life for the consumers [47,48]. This aspect has not so far been covered in tactical procurement decisions for ornamental horticultural products, but Rijpkema et al. [29] consider costs for quality loss in a source selection decision based on a case study of Egyptian strawberries. Moreover, quality considerations are largely applied in time-temperature distribution approaches for horticultural products (see e.g., Rosset et al. [49], Rijgersberg et al. [50], or Tromp et al. [11]). These approaches could be a starting point when searching for options to include quality aspects in a decision support system for horticultural purchasing.

4.2.2. Production Planning

Planning Problems Identified

Having a production site of one´s own is common in the sector of small and medium-sized horticultural companies. It is therefore not surprising that production planning is—together with the aforementioned product requirement planning—highlighted as a central planning task in our study that impacts all the other planning areas. This involves developing a cultivation plan that makes the most efficient use of available acreage and other capacities. Due to the close cooperation that is typical in the sector, these plans are often developed in close contact with key customers. The producers have to decide which plants to grow in which position at their production sites, coordinating the growing times of multiple products weeks or months ahead. Thus, planning tasks that are usually assumed to be more operational, as for example lot-sizing decisions [21], have to be considered on the tactical level, and an accumulated production plan is not enough. This means the occupancy planning of the acreage has to be performed in a detailed manner. There has to always be enough free acreage when a new product has to be planted for finishing production on time and for fulfilling demand. This can result in tight occupation plans that require a high rotation speed of the plants. High rotation is also needed due to the perishability of the plants and because the plants might otherwise wither or become too big to be marketable. As a result, inventory holding barely plays a role when determining mid-term production volumes. Ultimately, the specific characteristics of the cultivation and occupancy plan greatly depend on the individual plants a producer cultivates.

Pertinent Literature and Gaps

While horticultural production planning has received some attention in operations research and decision support systems literature some decades ago, there is a lack of current follow-up studies that close the research gaps that still exist. In 1989 Annevelink [51] presented a production planning model to optimize the batch size and production start of different glasshouse floricultural crops to maximize profit constrained by production and labor capacity as well as maximum sales. Based on that production plan, he worked out a spacing plan in a follow-up study determining the crops' position using a heuristic approach. He thereby considered the lot sizes and space requirements, which change over the production process [52]. In 2000 Darby-Dowman et al. [53] then presented a stochastic programming model resulting in a planting and harvesting plan in a horticultural context. However, there are several aspects that have still not yet been investigated. One example is that demand is often seen as a maximum of possible sales in the literature available, but fulfillment or service level constraints do not play a role. Generally, the assumption of deterministic demand and processing times is a major limitation in production planning models that merits further research efforts [54]. Besides this, the production and ordering structure that we have identified as typical in the ornamental horticultural context with mid-term production and short-term ordering as a compensating option has also not so far been reflected in the pertinent literature. It might thus be worth investigating models dedicated to general perishable products such as the one of Pauls-Worm et al. [55] and trying to adapt and transfer them into an ornamental horticultural context. Apart from this, there are further opportunities to incorporate dynamic space requirements, seasonality aspects, and sector-specific production requirements into general production planning and lot-sizing concepts (using the models presented in Jans and Degraeve [54] or Guzman et al. [56] as a basis, for example).

4.2.3. Distribution Planning

Planning Problems Identified

Demand fulfillment and customer satisfaction are important goals for companies in the horticultural sector, but providing freshness and preventing spoilage does not allow for large safety stocks, long transport distances, or low delivery frequencies [9]. Distribution planning is therefore highly restricted in horticultural companies. As small and medium-sized companies in the ornamental horticultural sector usually only have one central company facility, distribution planning on a tactical level mostly means defining regular base tours and delivery frequencies and weekdays for key customers. It is also important to consider the previously mentioned constraints and the usually very limited transport capacity. The resulting base tours can, however, only be used as a rough framework for operational transport planning due to the several uncertainties in the market, and significant rearrangements are commonly required at short notice. The question of how much inventory should be held is also aligned with distribution planning. An explicit safety stock level planning is mostly only held by trading companies, especially wholesalers. They apply a safety stock calculation, especially for the supply of large retailers that have high priority. Producers are usually bound to contractually fixed amounts of product surplus when serving large customers.

Pertinent Literature and Gaps

Besides the rich literature on general distribution planning, there are also a few contributions in literature focusing on a horticulture-specific context which thus show ways to master this challenging planning task. De Keizer et al. [57] develop a hub network and Ossevoort et al. [58] a metro model for the horticultural sector. They thereby aim to reduce shipment costs and distances by using consolidation effects. De Keizer et al. [5] also present a floricultural distribution network for the Netherlands. All these contributions, however, focus on the Dutch market, which is highly dominated by the auction trade and its specifics. More research in this domain is thus required in non-auction-driven markets. In a broader context, models for the distribution of perishable products show several similarities with the planning conditions in ornamental horticulture. Jiang et al. [59] for example integrate distribution planning with harvesting decisions for perishable agricultural products, and Gaggero and Tonelli [60] present a two-step optimization model for determining relevant figures for the distribution network such as safety stocks, replenishment cycles, and volumes. Focusing on grocery products, Holzapfel et al. [61] present an approach for determining delivery patterns for customer deliveries considering interdependencies along the supply chain, and Frank et al. [62] extend this approach by considering different temperature requirements of products delivered together. These contributions could, however, be extended by a stream of literature focusing on further horticultural distribution specifics. De Keizer et al. [5] and Jiang et al. [59], for example, see research opportunities in integrating horticulture-specific aspects, such as quality control, packaging or bouquet-making, as well as uncertainties.

4.2.4. Demand Planning

Planning Problems Identified

As demand is very volatile and most horticultural products are perishable, one of the main challenges in mid-term planning is to forecast the demand as precisely as possible. The companies investigated to see the weather, holiday effects, and seasonal trends as the most crucial obstacles for generating reliable forecasts. The forecast for key products like cut flowers is difficult, particularly for horticulturally relevant holidays like Valentine's Day. Most small and medium-sized companies estimate forecasts on a product or product group level based on the previous year's demand, known pre-orders, market information, experience, and gut feeling. Quantitative models are, however, not applied. Corresponding with colleagues and partners, for example at fairs or wholesale markets, is therefore key for small and medium-sized companies to collect the necessary market information. Besides the forecast itself, finding an economic balance between demand fulfillment and product overflow that leads to plant wastage is challenging for companies in the horticultural sector. It is particularly the case that short-term demand in the horticultural market has to be fulfilled via plants produced make-to-stock that need to be planned on a mid-term basis, so demand planning has a huge impact on all other tactical planning tasks.

Pertinent Literature and Gaps

In the pertinent literature, multiple external, horticulture-specific factors in demand forecasting that we have identified within our exploratory study are still not sufficiently factored in. The combination of the effects of weather conditions, last-minute assortment changes, product quality development, and promotions makes horticultural demand forecasting a challenging field of research. However, Haselbeck et al. [63] have recently developed a Gaussian process regression for seasonal data that enables event-triggered augmented refitting. The model was successfully tested on real-world horticulture cashier data and proved the ability to handle plants' seasonal demand and several external impacts. Apart from horticulture there are demand management models like that of Dellino et al. [64] that could also help in a horticultural context. They present a modular decision support system for forecasting and order planning for food supply chains. Horticultural supply chains, however, are even more nontransparent compared to the setting assumed and more driven by short-term adaptations. To consider such sudden changes, augmentations are used in literature with time-related functions, such as a forgetting factor or a moving window [65]. Overall, there is still extensive leeway for providing data-driven decision support in demand forecasting in the horticultural market given the current manual practices that we have found in the business and the status of the literature within this domain.

4.3. *Operational Planning*

Building on the framework set by tactical planning, there have to be solutions for short-term purchasing, production scheduling, transport planning, and demand fulfillment.

4.3.1. Purchasing

Planning Problems Identified

Purchasing is closely connected to tactical material requirement planning and impacts the whole scope of operational transport planning and demand fulfillment. While wholesalers and retailers reserve basic product contingents on a tactical level, these contingents have to be retrieved short-term, but re-orders also have to be placed. Additionally, some product categories are typically sourced entirely on a short-term horizon, such as cut flowers. In this case, wholesalers are the typical short-term source for retailers. This is also true for re-orders that are necessary due to forecasting errors, unforeseeable fluctuations in demand, or because of the flexibility reserve, the companies have calculated on a mid-term basis. Product availability and/or quality can be a problem for these short-term orders as both typically fluctuate in a market of natural products. To complicate things, the majority of products in ornamental horticulture are seasonal and specific plants can be purchased only within a limited time frame of several weeks. Besides wholesalers, producers can also be a short-term source for specific product categories as some companies produce free-for-sale quotas at their own risk. These quotas and sales fluctuations may lead to a product surplus on the part of the producer or wholesaler and result in attractive offers reaching the retailers at short notice. It is usually difficult to anticipate these in mid-term planning, but they may change the short-term assortment and sales planning significantly.

Pertinent Literature and Gaps

To the best of our knowledge, a dedicated optimization model for short-term ordering in horticulture does not yet exist. The planning situation, however, is similar to that assumed in a newsvendor setting, and various newsvendor approaches might therefore be adaptable for purchasing in ornamental horticulture, such as that presented by Matsuyama [66]. We refer to Khouja [67] and Qin et al. [68] for comprehensive literature reviews of newsvendor models. Apart from a newsvendor setting, other models that are dedicated to the ordering and inventory management of perishables might also suit the horticultural context. We, therefore, refer to Goyal and Giri [69] who present a literature review on perishable inventory systems and Broekmeulen and van Donselaar [70] who outline the advantages of considering age distribution for the replenishment of perishables.

4.3.2. Production Scheduling

Planning Problems Identified

Short-term production planning comprises scheduling the activities of planting, plant care, and shop floor control. The tactical cultivation and occupancy plan has to be executed and new production has to be started. Production scheduling also means handling delays or reacting to unexpected circumstances, which are common in horticulture due to the dependence on weather and other uncertainties. Plant care comprises all activities to optimize the plant's growing conditions. These can be watering, adding fertilizer or substrate, plant treatments like cuttings or removing older leaves, re-potting, pest control, and much more. These are tasks that are not exclusive to producers, but basic plant care at the very least is also required at the wholesale and retail stage. The extent of and workload attributed to these tasks, however, differs between the different supply chain roles. Producers usually have a comparatively small assortment, which makes it possible to standardize plant care and limits the heterogeneity of tasks. However, these tasks are then executed intensively. Wholesalers and retailers mostly do not have the capacity for intensive plant care. Re-potting for example is too much effort for most retailers, who usually prefer to take discounts instead of providing additional treatment to the plants. The larger assortments of wholesalers and retailers which would cause an enormous heterogeneity of different treatments enhance the effects of avoiding treatment whenever possible. This of course excludes necessary basic treatments like watering to maintain a plant's quality and to prevent withering. An important task for all supply chain roles is also the continuous evaluation of the plants in production or stock. In practice, companies check their plants frequently to detect deviations from the expected production/storage time and quality early. This allows for possible adjustments in sales, procurement or production planning, or the scheduling of additional treatments. All activities mentioned are usually labor intensive and therefore also require suitable workforce management, and thus are highly interconnected with tactical and operational personnel planning.

Pertinent Literature and Gaps

Some decision support contributions regarding selected aspects of short-term production scheduling can be identified in the relevant literature. Chalabi et al. [71] present a model for optimizing greenhouse heating, minimizing energy consumption while satisfying temperature constraints for the requirements of tomato crops. Besides this, Magarey et al. [72] develop a decision support system for pest management, and Damos and Karabatakis [73] model the population dynamics of orchards and the dissemination of pests. However, there are still numerous necessary plant treatments that have not yet been investigated that exhibit cost and quality trade-offs in ornamental horticulture. Beyond the ornamental horticultural market, there are models for irrigation decisions in agriculture, such as for maize [74] or avocados [75], as well as for harvesting apples [76], oranges [77], sugar [78], and grapes [79]. Elia and Conversa [80] also presents a model for managing the fertigation of open-field vegetable crops as well as optimizing the water and nitrogen supply for plant development. Further research on these topics can be found in the integrated plant protection and nutrient management sections presented by Mir and Padma [7]. These studies are focused on individual crops and support the hypothesis that decision support for horticultural production scheduling offers numerous research and practical optimization opportunities. However, several aspects of ornamental horticulture are still unexplored, and a comprehensive approach that combines the heterogeneous requirements of the individual plants reflecting the characteristic uncertainties is lacking so far. Additionally, pertinent literature claims for more elaborate methods for crop estimation, which implies also further plant-specific research [76].

4.3.3. Transport Planning

Planning Problems Identified

While small and medium-sized retailers in the horticultural sector have typically outsourced their transportation requirements, short-term transport planning is a relevant task for producers with a fleet of their own and wholesalers. While the producers mostly have to serve a limited set of customers, transport planning can be complex, especially for wholesalers who provide transportation for a wide range of customers. Delivery tour planning at horticultural companies is characterized by high volatility, demand fluctuation, and supply insecurity. The operative transport planning has to be very flexible as a result. Fast turnover, short storage times, and cross-dock-like processing are required to achieve operational excellence because of the plants' quality loss and perishability.

Pertinent Literature and Gaps

Vehicle routing problems have been widely studied in the literature (see Vidal et al. [81] and Braekers et al. [82] for an overview) and thus there are numerous extensions and adaptations that may fit the circumstances of delivery tour planning for ornamental horticultural products. For example, Gong and Fu [83] present a model for perishable food products that considers delivery time windows, while Buelvas Padilla et al. [84] focus on the quality loss aspect, and Wu and Wu [85] present a green vehicle routing model within this domain. Some models have already been applied in a horticultural domain. Ginantaka [86] present a route selection model for horticultural transportation that takes into account road and traffic impacts, and Soysal et al. [87] consider the perishability of tomatoes in a similar problem scenario. Given the ambitious freshness requirements for deliveries, an interesting path of research in a horticultural domain might also be the integration of production scheduling and transport planning. We refer to Kuhn et al. [88] for a general overview of models that consider the corresponding interdependencies. In the horticultural domain, Widodo et al. [89] present a model to maximize customer satisfaction by jointly optimizing harvesting patterns and deliveries for fresh products. Further contributions with an explicit application in horticulture are lacking, however. This shows the potential for research on integrative production and transport planning considering the specifics of ornamental horticulture, such as a many-to-many supply chain structure [87].

4.3.4. Demand Fulfillment

Planning Problems Identified

The main objective in short-term demand fulfillment is to match current supply and demand to avoid stock-outs and product surplus. The significance of stock-outs, however, varies across horticultural products, companies, and market situations. Generally, the possibilities for substitution are comparatively high for many horticultural products. A stock-out of one specific product can often be compensated by similar products with different colors, or closely related plants with a similar look or usage. This helps in terms of demand fulfillment especially in the retail sector as it is very hard to guarantee availability on a product level due to the highly fluctuating demand, limited shelf/vase life, limited production and sales seasons, and possibly unavailable short-term supply options. The lack of extensive storage options and the perishability of the plants further lead to limited inventory holding and thus lower on-shelf availability. Freshness and a large welcoming assortment with multiple options for product substitution are typically prioritized instead by horticultural retailers. While stock-outs are a minor issue, active demand control is necessary to prevent product surplus and waste at the end of the sales season. Typical measures applied by retailers are marketing campaigns, promotions, discounts, and active store layout adaptations. High-stock items are therefore positioned prominently in the store, advertised in leaflets, or significantly discounted. The latter measure is especially applied at the end of the sales season. Granting discounts is also the dominant demand control element for wholesalers, who can use this to overcome product overflow and quality deficits. Producers typically use so-called availability lists to steer customer demand. These lists are regularly published and contain the current marketable products. This enables the producers to manage their seasonality in production and avoid information deficits and supply uncertainty among their customers. The producers aim to sell out most of their products during the regular sales season because of higher profit margins. The remaining product surplus is then typically sold at short notice to wholesalers or auction houses at reduced prices. This highlights the interconnections between short-term demand fulfillment and tactical sales, production, and procurement planning, while having to consider the notable consequences of planned procurement, production, and sales quantities on the profits than can be achieved.

Pertinent Literature and Gaps

The practical planning situation in demand fulfillment, which we have found in our exploratory study, highlights its stochastic components and the need for dynamic decision support systems. The contribution of Ludwig [90] reflects this, presenting a stochastic model with a Gaussian distributed demand for potted plants that can incorporate decision makers' risk attitudes. Apart from this, distinct models for demand fulfillment in a horticultural context are lacking. In the agricultural domain, Widodo et al. [89] integrate demand fulfillment into a model for harvesting decisions that highlight possible future research directions in horticultural research, combining short-term demand fulfillment with production scheduling issues. Further contributions focus on demand management via price setting. Matos et al. [91] present a computational decision support system for determining the optimal price for specific fruits during their shelf life that considers quantity and quality deterioration, which could be adapted to ornamental plants as well. Pina et al. [92] develop a similar decision support system for vegetables based on microbiological growth models used for estimating the products´ remaining shelf life. Supporting decisions in terms of what, where, and what quantity of plants to position in a retail outlet, the various assortment and shelf space allocation models reviewed by Hübner and Kuhn [41] can be a starting point for versions that are adapted to the ornamental horticulture setting.

5. Discussion

Our systematization based on the supply chain planning framework of Fleischmann et al. [21] offers guidance regarding the most relevant planning tasks and systematizes the planning problems according to the supply chain area (procurement, production, distribution, sales) and the time horizon (strategic, tactical, operational). We find that most planning problems in horticultural supply chains can be mapped within this matrix analogous to the generic proposal of Fleischmann et al. [21]. However, some planning tasks that are generally seen as more relevant on an operational level such as purchasing and lot-sizing decisions have to be considered on a mid-term basis in ornamental horticulture due to the long production time of most of the plants and the need to establish at least a certain level of planning security in an uncertain market environment. These planning tasks together with tactical demand planning are also identified as the most critical ones for small and medium-sized companies, with a considerable impact on their short-term profitability and operational capacity to act.

In contrast, our structured literature review shows that contributions for decision support within a horticultural domain have so far often focused on strategic decisions and distribution issues. Both aspects, however, play a minor role for small and medium-sized companies in the ornamental horticulture sector. The strategic framework is often predetermined for these companies or does not change for a very long time horizon as the companies have only limited investment resources and supplier and customer relationships often last for decades. Distribution planning is identified as an important topic for wholesalers with a larger set of suppliers and customers, while retailers and producers often outsource distribution tasks or have to serve only a limited set of customers or a single outlet. Overall, although interconnections with generic planning models can be identified and models developed for other sectors, there is still a huge gap in the literature regarding decision support systems that reflect the planning characteristics and specifics in ornamental horticulture. This is especially true for the most relevant previously mentioned planning tasks in practice and underlines the need for future research efforts in the supply chain, logistics, and operations planning in horticulture. Tables 2–4 summarize the relevant contributions in literature for each planning area and highlight the main gaps in the strategic (Table 2), tactical (Table 3), and operational (Table 4) supply chain planning horizon.

Table 2. Literature on strategic supply chain planning in horticulture or aligned sectors.

Planning Area	Literature (Q:Qualitative; M:Model-Based)	Focus and Methodology	Gaps
Strategic Procurement Planning	van der Broek and Smulders [25] (Q)	Cross-border barriers for innovation on the example of international cooperations in horticulture (case study/ interviews)	Development of generic supplier selection models reflecting specifics in horticulture
	Geerling-Eff et al. [26] (Q)	National cooperation between different agents in Dutch horticulture (secondary analysis of publications)	
	Matopoulos et al. [27] (Q)	Collaboration in agri-food grower-processor supply chains of small and medium-sized companies (case study/interviews)	
	de Keizer et al. [28] (M)	Logistics network design considering the quality of perishable (horticultural) products (modeling and optimization)	
	Rijpkema et al. [29] (M)	Sourcing strategies for international perishable product supply chains considering shelf life decay of perishables (modeling and simulation)	
	Yazdani et al. [30] (M)	Multi-tier supplier selection for food supply chains considering uncertainty (modeling)	
Strategic Production Planning	Rath [31] (M)	Energy supply for greenhouse production (modeling)	Consideration of additional relevant factors for horticultural practice
	Vanthoor et al. [32] (M)	Greenhouse design (modeling and optimization)	
Physical Distribution Planning	van der Vorst et al. [34] (Q)	Dutch horticultural network, sector developments, bottlenecks and improvement potentials (qualitative study)	Investigation of distribution conditions and particularities outside the Netherlands
	van der Vorst et al. [9] (Q)	Dutch floricultural sector developments, bottlenecks, and opportunities (qualitative study)	
	Ossevoort et al. [10] (M)	Logistics hub network for Dutch floricultural logistics (scenario analysis)	
	Cheng et al. [35] (M)	Cloud-based auction tower for trading perishable products and information sharing in Dutch horticulture (platform development)	
	de Keizer et al. [36] (M)	Network design considering heterogeneous quality decay of perishables and application to the horticultural sector (modeling and optimization)	
Strategic Sales Planning	White and Uva [37] (Q)	Marketing plan development for horticultural companies (guideline development)	Adaptation of generic assortment planning models for horticultural products and their requirements

Table 3. Literature on tactical supply chain planning in horticulture or aligned sectors

Planning Area	Literature (Q:Qualitative; M:Model-Based)	Focus and Methodology	Gaps
Material/ Product Requirement Planning	Verdouw et al. [45] (Q)	Drivers and barriers of ERP systems in Dutch horticulture (case study/interviews)	Horticultural focus; consideration of quality decay and pre- and re-order distinction
	Faraudo Pijuan [46] (M)	Economic order quantity and price determination for agricultural and livestock industries (modeling and optimization)	
	van der Vorst et al. [47] (Q)	Quality-controlled fresh food distribution and inventory management (qualitative framework)	
	Verdouw et al. [48] (Q)	Information distribution in virtual logistic networks in Dutch horticulture (case study)	
	Rijpkema et al. [29] (M)	Sourcing strategies for international perishable product supply chains considering shelf life decay (modeling and simulation)	
	Tromp et al. [11] (M)	Prediction of remaining vase life of cut roses based on time-temperature sum (modeling)	
Production Planning	Annevelink [51] (M)	Production planning for glasshouse floriculture (modeling and optimization)	Consideration of stochastic demand, service levels, and horticulture-specific production and ordering structure
	Annevelink [52] (M)	Spacing and allocation plan with dynamic space requirements for floricultural glasshouse production (modeling and optimization)	
	Darby-Dowman et al. [53] (M)	Planting and harvest planning considering risk aversion in agriculture (modeling and optimization)	
Distribution Planning	de Keizer et al. [57] (M)	Hub network in a Dutch potted plant supply chain considering logistics costs, working times, and emissions (modeling and simulation)	Investigation of non-auction-dominated markets, considering horticulture specifics
	Ossevoort et al. [58] (M)	Distribution planning for Dutch horticulture using consolidation effects (modeling and simulation)	
	de Keizer et al. [5] (Q)	Control and design of floricultural supply chain networks and market developments (case study/ interviews and literature review)	
	Jiang et al. [59] (M)	Distribution scheduling with integrated harvest planning of perishable products and time windows in an agricultural context (modeling and optimization)	
	Gaggero and Tonelli [60] (M)	Distribution planning for perishable products considering replenishment cycles, safety stocks, and product volumes (modeling, optimization, and simulation)	
Demand Planning	Haselbeck et al. [63] (M)	Demand forecasting using machine learning for horticulture (modeling and forecasting)	Considering horticulture-specific factors in demand forecasting
	Dellino et al. [64] (M)	Sales forecasting for perishable products and order plan selection (modeling and forecasting)	

Table 4. Literature on operational supply chain planning in horticulture or aligned sectors

Planning Area	Literature (Q:Qualitative; M:Model-Based)	Focus and Methodology	Gaps
Purchasing	Matsuyama [66] (M)	Multi-period newsvendor problem for ordering perishables (modeling and optimization)	Horticultural focus and specifics
	Broekmeulen and van Donselaar [70] (M)	Ordering perishables considering age distributions (modeling and optimization)	
Production Scheduling	Chalabi et al. [71] (M)	Control of greenhouse heating considering weather data, forecasts, and set points (modeling and optimization)	Comprehensive approach combining the heterogeneous requirements of different plants; consideration of further plant treatments and uncertainty
	Magarey et al. [72] (Q)	Management of plant diseases with decision support systems in an agricultural context (case study)	
	Damos and Karabatakis [73] (M)	Web-based integrated pest management in an agricultural context (modeling and forecasting)	
	Bergez et al. [74] (M)	Effective water use in irrigated agriculture (modeling, simulation, and optimization)	
	Gurovich et al. [75] (M)	Irrigation scheduling strategies based on phytomonitoring techniques (modeling)	
	Gonzalez-Araya et al. [76] (M)	Harvest planning and labor distribution for apple orchards (modeling and optimization)	
	Caixeta-Filho [77] (M)	Harvest planning of oranges considering chemical, biological, and logistical restrictions (modeling and optimization)	
	Higgins and Laredo [78] (M)	Harvesting and transport planning for sugar supply chains (modeling and optimization)	
	Ferrer et al. [79] (M)	Scheduling wine grape harvest considering cost and quality objectives (modeling and optimization)	
	Elia and Conversa [80] (M)	Real-time irrigation and nitrogen fertilization management (modeling, optimization, and simulation)	
Transport Planning	Gong and Fu [83] (M)	Vehicle routing problem with time windows considering the quality loss of perishable products (modeling and optimization)	Horticultural focus; integrated production and transportation planning considering horticultural specifics
	Buelvas Padilla et al. [84] (M)	Vehicle routing problem considering perishable food damage caused by road conditions (modeling and optimization)	
	Wu and Wu [85] (M)	Vehicle routing problem considering time-dependent split deliveries, time windows, and customer satisfaction (modeling and optimization)	
	Ginantaka [86] (M)	Vehicle routing problem for horticultural products considering road conditions and traffic (modeling and optimization)	
	Soysal et al. [87] (M)	Inventory routing problem considering perishability and shelf life in food supply chains (modeling and simulation)	
Demand Fulfillment	Widodo et al. [89] (M)	Harvest planning considering plant growth and loss of fresh agricultural products to maximize demand satisfaction (modeling and optimization)	Investigating multiple possibilities for demand control in an ornamental horticultural context
	Ludwig [90] (M)	Production planning considering the effects of risks on demand fulfillment (modeling and optimization)	

Table 4. *Cont.*

Planning Area	Literature (Q:Qualitative; M:Model-Based)	Focus and Methodology	Gaps
	Matos et al. [91] (M)	Discounts of fresh horticultural products based on quality decay (modeling and optimization)	
	Pina et al. [92] (M)	Dynamic pricing of horticultural food products considering remaining shelf life (modeling and optimization)	

All in all, a huge number of planning problems have to be resolved in practice and claims for structured decision support. However, up to now, many decisions have been taken based on experience and rules of thumb in practice without data-driven support and recognition of the actual position in hierarchical planning. Examples of this missing professionalization and systematization of supply chain planning can be found throughout all of our case companies. Com 9 for example solves a classic vehicle routing problem for their basic tour plan by hand and on an aggregated level once a year. Every change to the resulting standard tours is then resolved ad-hoc on the respective delivery day. Equal situations are found for example for physical distribution planning at company Com 10 and production scheduling at Com 8. This shows a large gap between current state-of-the-art research and problem-solving in practice. Many companies are even not aware of which decisions they make unconsciously. We find that managers in the sector are not aware of many planning problems in physical distribution planning or demand fulfillment, especially demand control. This again outlines the practical relevance and need for a comprehensive and structured planning problem overview with current research and solution approaches, which we provide.

One approach to handle the large number of planning problems identified in practice could be the introduction of capable ERP systems, which are widely used in many other industries such as manufacturing. Standard ERP systems, however, do not fit the circumstances in horticulture because dealing with "living products" requires sector-specific solutions. The existing sector-specific systems, however, can be characterized as island automation [93]. Our study confirms this observation of Verloop et al. [93]. We find that information systems are mostly used for the communication or control of specific procedures such as ordering or operational distribution planning for certain customers. However, they do not really serve as a basis for comprehensive data-driven decision support and lack interfaces to respective standard software tools. So the systems can be characterized as closed ones. In line with the key limitations for effective usage of ERP systems identified by Akkermans et al. [94] and the obstacles in horticulture identified by Verdouw et al. [45], we find that there is a lack of functionality beyond managing transactions. Uncertainties and sudden changes facing high dynamics in planning limit the development of ERP systems for companies in the horticultural supply chains [93]. These conditions lead to a situation in which enhanced, sector-tailored ERP systems are not only necessary for small and medium-sized companies, but also larger enterprises, that typically implement more comprehensive ERP systems [45], which could benefit from such a development. So similar to agricultural supply chains there is a need for further action to improve the ERP implementation in horticultural practice [95]. Our systematization of planning problems can thereby help to define suitable modules for such systems.

6. Conclusions

Within this study, we have explored and systematized the planning problems in ornamental horticultural supply chains. Our exploratory investigation emphasizes the need for structured decision support in this sector. Up to now many decisions have been taken based on experience and rules of thumb, and necessary data is not recorded appropriately. The interviewees who are people in charge of companies in the sector, however, have already recognized this obstacle and highlighted their willingness to advance digitization and professionalization. Research on decision support systems for the various planning problems along the supply chain that reflects the practically relevant aspects and specifics of ornamental horticultural supply chains is thus highly appreciated. Our reviews of pertinent literature in the various planning areas, however, emphasize that there are still numerous gaps in the literature that need to be filled. Besides that, our systematization and characterization of planning problems can also support the development of data-driven decision-support systems that are tailored to the requirements of the horticultural sector. So far, existing systems lack suitable data-driven decision support modules, interfaces to standard software and data analytic tools, as well as comprehensive planning support.

Limitations

Not all planning problems identified are equally relevant for every supply chain role in the sector, i.e., producers, wholesalers, and retailers. Our systematization thus presents more of a comprehensive overview rather than a dedicated systematization for specific companies. However, as vertical integration is common in the sector, the companies will typically face large parts of the planning tasks outlined within our study. Nevertheless, our contribution is not without limitations. As we have relied on expert interviews with selected managers of companies in the sector, our results cannot be seen as representative in a quantitative sense. However, by following the guidelines of Mayring [19] and reaching theoretical saturation, we believe that our presentation of planning problems represents the typical challenges that small and medium-sized companies face in ornamental horticulture. Furthermore, our findings are derived from interviewees with companies operating in the German market. This was necessary to create a unified context and to limit our data collection to a reasonable scale. As Germany is one of the largest producing horticultural markets in Europe [1,2] and has similarities with several other non-auction-driven markets, we believe that our planning matrix is also applicable to companies in other regions worldwide. Future research could, however, compare our systematization derived from the German markets to other markets. A special interest could be a comparison of auction-driven with non-auction-driven markets, or a comparison of the situation on different continents, as horticulture is very dependent on the ecological environment.

Author Contributions: Conceptualization, A.H.; methodology, M.D. and A.H.; validation, M.D. and A.H.; formal analysis, M.D.; investigation, M.D.; data curation, M.D.; writing—original draft preparation, M.D. and A.H.; writing—review and editing, M.D. and A.H.; visualization, M.D.; supervision, A.H.; project administration, A.H.; funding acquisition, A.H. All authors have read and agreed to the published version of the manuscript.

Funding: This research is part of a project called PlantGrid. The project is supported by funds from the Federal Ministry of Food and Agriculture (BMEL) based on a decision of the Parliament of the Federal Republic of Germany via the Federal Office for Agriculture and Food (BLE) within the aegis of the Innovation Support Program. The APC was funded as part of this project as well.

Institutional Review Board Statement: Not applicable.

Informed Consent Statement: Informed consent was obtained from all subjects involved in the study.

Data Availability Statement: Not applicable.

Acknowledgments: We thank all participating experts for sharing their information and providing valuable sector-specific insights. We also thank the anonymous reviewers for their constructive suggestions for improving the manuscript.

Conflicts of Interest: The authors declare no conflict of interest. The funders had no role in the design of the study, in the collection, analyses, or interpretation of data, in the writing of the manuscript, or in the decision to publish the results.

References

1. Dirksmeyer, W. (Ed.) *Status quo und Perspektiven des Deutschen Produktionsgartenbaus*; vTI: Braunschweig, Germany, 2009; Volume 330.
2. Verdouw, C.N.; Beulens, A.J.; van der Vorst, J.G.A.J. Virtualisation of floricultural supply chains: A review from an Internet of Things perspective. *Comput. Electron. Agric.* **2013**, *99*, 160–175. [CrossRef]

3. Zentralverband Gartenbau e.V. *Jahresbericht 2020*; Mertz, J., Ed.; 2020; pp. 1–64. Available online: https://www.hortigate.de/publikation/86769/ePaper%3A-ZVG-Jahresbericht-2020/ (accessed on 20 October 2022).
4. Havardi-Burger, N.; Mempel, H.; Bitsch, V. Sustainability Challenges and Innovations in the Value Chain of Flowering Potted Plants for the German Market. *Sustainability* **2020**, *12*, 1905. [CrossRef]
5. De Keizer, M.; van der Vorst, J.G.A.J.; Bloemhof, J.M.; Haijema, R. Floricultural supply chain network design and control: Industry needs and modelling challenges. *J. Chain. Netw. Sci.* **2015**, *15*, 61–81. [CrossRef]
6. Bokelmann, W. Wertschöpfungsketten im Gartenbau. *Landbauforsch. Sonderh.* **2009**, *2009*, 115–129.
7. Mir, S.; Padma, T. Decision support systems for horticulture. *Hortic. Compend.* **2019**, 24010452. [CrossRef]
8. Havardi-Burger, N.; Mempel, H.; Bitsch, V.; Havardi-Burger, N.; Bitsch, V. Driving forces and characteristics of the value chain of flowering potted plants for the German market. *Eur. J. Hortic. Sci.* **2020**, *85*, 267–278. [CrossRef]
9. Van der Vorst, J.G.A.J.; Bloemhof, J.M.; de Keizer, M. Innovative logistics concepts in the floriculture sector. In *System Dynamics and Innovation in Food Networks 2012*; International Center for Food Chain and Network: 2012; pp. 241–251. Available online: https://www.researchgate.net/publication/236867355_Innovative_Logistics_Concepts_in_the_Floriculture_Sector (accessed on 20 October 2022).
10. Ossevoort, R.; van der Vorst, J.G.A.J.; Verdouw, C.N.; Wenink, E. *DAVINC3I: Virtualisation Scenarios for floricultural Trade Network*; Weijers, S., Ed.; Vervoerslogistieke Werkdagen: 2012; pp. 81–92. Available online: https://research.wur.nl/en/publications/davinc3i-virtualisation-scenarios-for-floricultural-trade-network (accessed on 20 October 2022).
11. Tromp, S.O.; van der Sman, R.G.; Vollebregt, H.M.; Woltering, E.J. On the prediction of the remaining vase life of cut roses. *Postharvest Biol. Technol.* **2012**, *70*, 42–50. [CrossRef]
12. Flynn, B.B.; Sakakibara, S.; Schröder, R.G.; Bates, K.A.; Flynn, E.J. Empirical Research Methods in Operations Management. *J. Oper. Manag.* **1990**, *9*, 250–284. [CrossRef]
13. Trautrims, A.; Grant, D.B.; Cunliffe, A.L.; Wong, C. Using the "documentary method" to analyse qualitative data in logistics research. *Int. J. Phys. Distrib. Logist. Manag.* **2012**, *42*, 828–842. [CrossRef]
14. Manuj, I.; Pohlen, T.L. A reviewer's guide to the grounded theory methodology in logistics and supply chain management research. *Int. J. Phys. Distrib. Logist. Manag.* **2012**, *42*, 784–803. [CrossRef]
15. Glaser, B.G.; Strauss, A.L. *The Discovery of Grounded Theory: Strategies for Qualitative Research*; Routledge: London, UK; New York, NY, USA, 2017. [CrossRef]
16. Ellram, L.M. The use of the case study method in logistics research. *J. Bus. Logist.* **1996**, *17*, 93–138.
17. Commission of the EC. *Commission Recommendation of 6 May 2003 Concerning the Definition of Micro, Small and Medium-Sized Enterprises: 2003/361/EC*; 2003. Available online: https://op.europa.eu/en/publication-detail/-/publication/6ca8d655-126b-4a42-ada4-e9058fa45155l7 (accessed on 20 October 2022).
18. Bell, E.; Bryman, A.; Harley, B. *Business Research Methods*, 5th ed.; Oxford University Press: Oxford, UK, 2019.
19. Mayring, P. *Qualitative Inhaltsanalyse: Grundlagen und Techniken*, 12th ed.; Beltz Pädagogik Beltz: Weinheim, Germany, 2015.
20. Gioia, D.A.; Corley, K.G.; Hamilton, A.L. Seeking Qualitative Rigor in Inductive Research. *Organ. Res. Methods* **2013**, *16*, 15–31. [CrossRef]
21. Fleischmann, B.; Meyr, H.; Wagner, M. Advanced planning: Structure of advanced planning Systems. In *Supply Chain Management and Advanced Planning*; Kilger, C., Stadtler, H., Meyr, H., Eds.; Springer Texts in Business and Economics; Springer: Berlin, Germany, 2015; pp. 71–106.
22. Hübner, A.; Kuhn, H.; Sternbeck, M.G. Demand and supply chain planning in grocery retail: An operations planning framework. *Int. J. Retail. Distrib. Manag.* **2013**, *41*, 512–530. [CrossRef]
23. Nam, S.; Shen, H.; Ryu, C.; Shin, J.G. SCP-Matrix based shipyard APS design: Application to long-term production plan. *Int. J. Nav. Archit. Ocean. Eng.* **2018**, *10*, 741–761. [CrossRef]
24. Seuring, S.; Müller, M.; Westhaus, M.; Morana, R. Conducting a literature review: The example of sustainability in supply chains. In *Research Methodologies in Supply Chain Management*; Kotzab, H., Seuring, S., Müller, M., Reiner, G., Eds.; Physica-Verlag: Heidelberg, Germany, 2005; pp. 91–106. [CrossRef]
25. Van der Broek, J.; Smulders, H. Institutional gaps in cross-border regional innovation systems: The horticultural industry in venlo–lower rhine. In *The Social Dynamics of Innovation Networks*; Rutten, R., Benneworth, P., Irawati, D., Boekema, F., Eds.; Routledge: Abingdon-on-Thames, UK, 2014; pp. 183–202. [CrossRef]
26. Geerling-Eff, F.; Hoes, A.C.; Dijkshoorn-Dekker, M. Triple helix networks matching knowledge demand and supply in seven Dutch horticulture Greenport regions. *Stud. Agric. Econ.* **2017**, *119*, 34–40. [CrossRef]
27. Matopoulos, A.; Vlachopoulou, M.; Manthou, V.; Manos, B. A conceptual framework for supply chain collaboration: Empirical evidence from the agri–food industry. *Supply Chain. Manag. Int. J.* **2007**, *12*, 177–186. [CrossRef]
28. De Keizer, M.; Haijema, R.; Bloemhof, J.M.; van der Vorst, J.G.A.J. acHybrid optimization and simulation to design a logistics network for distributing perishable products. *Comput. Ind. Eng.* **2015**, *88*, 26–38. [CrossRef]
29. Rijpkema, W.A.; Rossi, R.; van der Vorst, J.G.A.J. Effective sourcing strategies for perishable product supply chains. *Int. J. Phys. Distrib. Logist. Manag.* **2014**, *44*, 494–510. [CrossRef]
30. Yazdani, M.; Pamucar, D.; Chatterjee, P.; Torkayesh, A.E. A multi-tier sustainable food supplier selection model under uncertainty. *Oper. Manag. Res.* **2021**, *15*, 116–145. [CrossRef]

31. Rath, T. Hybride Wissensrepräsentation in einem rechnergestützten Planungs-und Beratungssystem zur Energieversorgung von Gewächshausanlagen. *Ber. Ges. Inform. Land Forstund Ernährungswirtschaft* **1993**, *5*, 217–221.
32. Vanthoor, B.H.; Stigter, J.D.; van Henten, E.J.; Stanghellini, C.; de Visser, P.H.; Hemming, S. A methodology for model-based greenhouse design: Part 5, greenhouse design optimisation for southern-Spanish and Dutch conditions. *Biosyst. Eng.* **2012**, *111*, 350–368. [CrossRef]
33. de Waal, A.; Meingast, A. Applying the high performance organisation framework in the horticulture and greenhouse sector. *Meas. Bus. Excell.* **2017**, *21*, 136–151. [CrossRef]
34. Van der Vorst, J.G.A.J.; Wenink, E.; Bloemhof, J.M.; de Keizer, M. *DAVINC3I: Developing Innovative Logistics Concepts for International Floriculture Trade Networks*; Weijers, S., Ed.; Vervoerslogistieke Werkdagen: 2011; pp. 225–236. Available online: https://agris.fao.org/agris-search/search.do?recordID=NL2020016733 (accessed on 20 October 2022).
35. Cheng, M.; Luo, H.; Zhong, R.Y.; Lan, S.; Huang, G.Q. Cloud-based auction tower for perishable supply chain trading. In *Disruptive Innovation in Manufacturing Engineering towards the 4th Industrial Revolution*; Bauer, W., Constantinescu, C., Sauer, O., Maropoulos, P., Muelaner, J., Eds.; Fraunhofer Verl.: Stuttgart, Germany, 2014; Volume 25, pp. 329–336.
36. De Keizer, M.; Akkerman, R.; Grunow, M.; Bloemhof, J.M.; Haijema, R.; van der Vorst, J.G.A.J. Logistics network design for perishable products with heterogeneous quality decay. *Eur. J. Oper. Res.* **2017**, *262*, 535–549. [CrossRef]
37. White, G.; Uva, W.F. Developing a Strategic Marketing Plan for Horticultural Firms. *Cornell Univ. Rep.* **2000**. Available online: https://www.docin.com/p-1379631549.html (accessed on 20 October 2022).
38. Engelke, C. Service Differentation and Dimensions of Strategic Orientations in German Retail Horticulture. Ph.D. Thesis, University of Worcester, Worcester, MA, USA, 2020.
39. Jiao, J.; Zhang, Y.; Wang, Y. A heuristic genetic algorithm for product portfolio planning. *Comput. Oper. Res.* **2007**, *34*, 1777–1799. [CrossRef]
40. Helferich, A.; Herzwurm, G.; Schockert, S. QFD-PPP: Product line portfolio planning using quality function deployment. In *Software Product Lines*; Hutchison, D., Kanade, T., Kittler, J., Kleinberg, J.M., Mattern, F., Mitchell, J.C., Naor, M., Nierstrasz, O., Pandu Rangan, C., Steffen, B., et al., Eds.; Lecture Notes in Computer Science; Springer: Berlin/Heidelberg, Germany, 2005; Volume 3714, pp. 162–173. [CrossRef]
41. Hübner, A.H.; Kuhn, H. Retail category management: State-of-the-art review of quantitative research and software applications in assortment and shelf space management. *Omega* **2012**, *40*, 199–209. [CrossRef]
42. Verdouw, C.N.; Beulens, A.J.; Trienekens, J.H.; van der Vorst, J.G.A.J. A framework for modelling business processes in demand-driven supply chains. *Prod. Plan. Control.* **2011**, *22*, 365–388. [CrossRef]
43. McIvor, R.T.; Humphreys, P.K. A case–based reasoning approach to the make or buy decision. *Integr. Manuf. Syst.* **2000**, *11*, 295–310. [CrossRef]
44. Van de Water, H.; van Peet, H.P. A decision support model based on the Analytic Hierarchy Process for the Make or Buy decision in manufacturing. *J. Purch. Supply Manag.* **2006**, *12*, 258–271. [CrossRef]
45. Verdouw, C.N.; Robbemond, R.M.; Wolfert, J. ERP in agriculture: Lessons learned from the Dutch horticulture. *Comput. Electron. Agric.* **2015**, *114*, 125–133. [CrossRef]
46. Faraudo Pijuan, C. EOQ: Optimizing Price and Order Quantity for Growing Items with Imperfect Quality and Carbon Restrictions. Ph.D. Thesis, TecnoCampus, Escola Superior de Ciències Socials i de l'Empresa del (ESCSET). 2021. Available online: https://repositori.tecnocampus.cat/handle/20.500.12367/1798 (accessed on 20 October 2022).
47. Van der Vorst, J.G.A.J.; van Kooten, O.; Luning, P.A. Towards a Diagnostic Instrument to Identify Improvement Opportunities for Quality Controlled Logistics in Agrifood Supply Chain Networks. *Int. J. Food Syst. Dyn.* **2011**, *2*, 94–105. [CrossRef]
48. Verdouw, C.N.; Beulens, A.J.; van der Vorst, J.G.A.J. Virtual logistic networks in dutch horticulture. In Proceedings of the 4th Production and Operations Management World Conference, Amsterdam, The Netherlands, 1–4 July 2012; POMS, JOMSA, Eds.; European Operations Management Association: Brussels, Belgium, 2012; pp. 1–10.
49. Rosset, P.; Cornu, M.; Noël, V.; Morelli, E.; Poumeyrol, G. Time-temperature profiles of chilled ready-to-eat foods in school catering and probabilistic analysis of Listeria monocytogenes growth. *Int. J. Food Microbiol.* **2004**, *96*, 49–59. [CrossRef] [PubMed]
50. Rijgersberg, H.; Tromp, S.O.; Jacxsens, L.; Uyttendaele, M. Modeling logistic performance in quantitative microbial risk assessment. *Risk Anal. Off. Publ. Soc. Risk Anal.* **2010**, *30*, 20–31. [CrossRef]
51. Annevelink, I.E. The IMAG production planning system (IPP) for glasshouse floriculture in its introduction phase. *Acta Hortic.* **1989**, *1989*, 37–46. [CrossRef]
52. Annevelink, I.E. Operational planning in horticulture: Optimal space allocation in pot-plant nurseries using heuristic techniques. *J. Agric. Eng. Res.* **1992**, *51*, 167–177. [CrossRef]
53. Darby-Dowman, K.; Barker, S.; Audsley, E.; Parsons, D. A two-stage stochastic programming with recourse model for determining robust planting plans in horticulture. *J. Oper. Res. Soc.* **2000**, *51*, 83–89. [CrossRef]
54. Jans, R.; Degraeve, Z. Modeling industrial lot sizing problems: A review. *Int. J. Prod. Res.* **2008**, *46*, 1619–1643. [CrossRef]
55. Pauls-Worm, K.G.J.; Hendrix, E.M.T.; Haijema, R.; van der Vorst, J.G.A.J. An MILP approximation for ordering perishable products with non-stationary demand and service level constraints. *Int. J. Prod. Econ.* **2014**, *157*, 133–146. [CrossRef]
56. Guzman, E.; Andres, B.; Poler, R. Models and algorithms for production planning, scheduling and sequencing problems: A holistic framework and a systematic review. *J. Ind. Inf. Integr.* **2021**, *27*, 100287. [CrossRef]

57. De Keizer, M.; Groot, J.J.; Bloemhof, J.M.; van der Vorst, J.G.A.J. Logistics orchestration scenarios in a potted plant supply chain network. *Int. J. Logist. Res. Appl.* **2014**, *17*, 156–177. [CrossRef]
58. Ossevoort, R.; de Keizer, M.; van der Vorst, J.G.A.J.; Wenink, E. *DAVINC3I: Moving towards Responsive Hub Network Designs*; Weijers, S., Ed.; Vervoerslogistieke Werkdagen: 2013; pp. 139–151. Available online: https://www.researchgate.net/publication/283418508_DAVINC3I_moving_towards_responsive_hub_network_designs (accessed on 20 October 2022).
59. Jiang, Y.; Chen, L.; Fang, Y. Integrated Harvest and Distribution Scheduling with Time Windows of Perishable Agri-Products in One-Belt and One-Road Context. *Sustainability* **2018**, *10*, 1570. [CrossRef]
60. Gaggero, M.; Tonelli, F. A two-step optimization model for the distribution of perishable products. *Networks* **2021**, *78*, 69–87. [CrossRef]
61. Holzapfel, A.; Hübner, A.; Kuhn, H.; Sternbeck, M.G. Delivery pattern and transportation planning in grocery retailing. *Eur. J. Oper. Res.* **2016**, *252*, 54–68. [CrossRef]
62. Frank, M.; Ostermeier, M.; Holzapfel, A.; Hübner, A.; Kuhn, H. Optimizing routing and delivery patterns with multi-compartment vehicles. *Eur. J. Oper. Res.* **2021**, *293*, 495–510. [CrossRef]
63. Haselbeck, F.; Killinger, J.; Menrad, K.; Hannus, T.; Grimm, D.G. Machine Learning Outperforms Classical Forecasting on Horticultural Sales Predictions. *Mach. Learn. Appl.* **2022**, *7*, 100239. [CrossRef]
64. Dellino, G.; Laudadio, T.; Mari, R.; Mastronardi, N.; Meloni, C. A reliable decision support system for fresh food supply chain management. *Int. J. Prod. Res.* **2018**, *56*, 1458–1485. [CrossRef]
65. Ni, W.; Tan, S.K.; Ng, W.J.; Brown, S.D. Moving-Window GPR for Nonlinear Dynamic System Modeling with Dual Updating and Dual Preprocessing. *Ind. Eng. Chem. Res.* **2012**, *51*, 6416–6428. [CrossRef]
66. Matsuyama, K. The multi-period newsboy problem. *Eur. J. Oper. Res.* **2006**, *171*, 170–188. [CrossRef]
67. Khouja, M. The single-period (news-vendor) problem: Literature review and suggestions for future research. *Omega* **1999**, *27*, 537–553. [CrossRef]
68. Qin, Y.; Wang, R.; Vakharia, A.J.; Chen, Y.; Seref, M.M. The newsvendor problem: Review and directions for future research. *Eur. J. Oper. Res.* **2011**, *213*, 361–374. [CrossRef]
69. Goyal, S.K.; Giri, B.C. Recent trends in modeling of deteriorating inventory. *Eur. J. Oper. Res.* **2001**, *134*, 1–16. [CrossRef]
70. Broekmeulen, R.A.; van Donselaar, K.H. A heuristic to manage perishable inventory with batch ordering, positive lead-times, and time-varying demand. *Comput. Oper. Res.* **2009**, *36*, 3013–3018. [CrossRef]
71. Chalabi, Z.S.; Bailey, B.J.; Wilkinson, D.J. A real-time optimal control algorithm for greenhouse heating. *Comput. Electron. Agric.* **1996**, *15*, 1–13. [CrossRef]
72. Magarey, R.D.; Travis, J.W.; Russo, J.M.; Seem, R.C.; Magarey, P.A. Decision Support Systems: Quenching the Thirst. *Plant Dis.* **2002**, *86*, 4–14. [CrossRef] [PubMed]
73. Damos, P.; Karabatakis, S. Real time pest modeling through the World Wide Web: Decision making from theory to praxis. *Integr. Prot. Fruit Crop. IOBC-WPRS Bull.* **2013**, *91*, 253–258.
74. Bergez, J.E.; Garcia, F.; Lapasse, L. A hierarchical partitioning method for optimizing irrigation strategies. *Agric. Syst.* **2004**, *80*, 235–253. [CrossRef]
75. Gurovich, L.A.; Ton, Y.; Vergara, L.M. Irrigation scheduling of avocado using phytomonitoring techniques. *Cienc. Inverstig. Agrar.* **2006**, *33*, 117–124.
76. Gonzalez-Araya, M.C.; Soto-Silva, W.E.; Acosta Espejo, L.G. (Eds.) *Harvest Planning in Apple Orchards Using an Optimization Model*; International Series in Operations Research & Management Science; Springer: New York, NY, USA, 2015; Volume 224. [CrossRef]
77. Caixeta-Filho, J.V. Orange harvesting scheduling management: A case study. *J. Oper. Res. Soc.* **2006**, *57*, 637–642. [CrossRef]
78. Higgins, A.J.; Laredo, L.A. Improving harvesting and transport planning within a sugar value chain. *J. Oper. Res. Soc.* **2006**, *57*, 367–376. [CrossRef]
79. Ferrer, J.C.; Mac Cawley, A.; Maturana, S.; Toloza, S.; Vera, J. An optimization approach for scheduling wine grape harvest operations. *Int. J. Prod. Econ.* **2008**, *112*, 985–999. [CrossRef]
80. Elia, A.; Conversa, G. A decision support system (GesCoN) for managing fertigation in open field vegetable crops. Part I-methodological approach and description of the software. *Front. Plant Sci.* **2015**, *6*, 319. [CrossRef]
81. Vidal, T.; Crainic, T.G.; Gendreau, M.; Prins, C. Heuristics for multi-attribute vehicle routing problems: A survey and synthesis. *Eur. J. Oper. Res.* **2013**, *231*, 1–21. [CrossRef]
82. Braekers, K.; Ramaekers, K.; van Nieuwenhuyse, I. The vehicle routing problem: State of the art classification and review. *Comput. Ind. Eng.* **2016**, *99*, 300–313. [CrossRef]
83. Gong, W.; Fu, Z. ABC-ACO for perishable food vehicle routing problem with time windows. In Proceedings of the 2010 International Conference on Computational and Information Sciences (ICCIS 2010), Chengdu, China, 17–19 December 2010; IEEE: Piscataway, NJ, USA, 2010; pp. 1261–1264. [CrossRef]
84. Padilla, M.P.B.; Canabal, P.A.N.; Pereira, J.M.L.; Riaño, H.E.H. Vehicle Routing Problem for the Minimization of Perishable Food Damage Considering Road Conditions. *Logist. Res.* **2018**, *2*, 1–18.
85. Wu, D.; Wu, C. Research on the Time-Dependent Split Delivery Green Vehicle Routing Problem for Fresh Agricultural Products with Multiple Time Windows. *Agriculture* **2022**, *12*, 793. [CrossRef]
86. Ginantaka, A. Fuzzy model for distribution route determination of horticultural Products. In Proceedings of the 7th BANGKOK International Conference on "Recent Trends in Engineering and Technology", Bangkok, Thailand, 3–4 May 2017. [CrossRef]

87. Soysal, M.; Bloemhof, J.M.; Haijema, R.; van der Vorst, J.G.A.J. Modeling an Inventory Routing Problem for perishable products with environmental considerations and demand uncertainty. *Int. J. Prod. Econ.* **2015**, *164*, 118–133. [CrossRef]
88. Kuhn, H.; Schubert, D.; Holzapfel, A. Integrated order batching and vehicle routing operations in grocery retail—A General Adaptive Large Neighborhood Search algorithm. *Eur. J. Oper. Res.* **2021**, *294*, 1003–1021. [CrossRef]
89. Widodo, K.H.; Nagasawa, H.; Morizawa, K.; Ota, M. A periodical flowering–harvesting model for delivering agricultural fresh products. *Eur. J. Oper. Res.* **2006**, *170*, 24–43. [CrossRef]
90. Ludwig, S. A stochastic model to evaluate the effect of uncertainty in pot plant production. *Acta Hortic.* **1991**, *295*, 223–230. [CrossRef]
91. Matos, C.; Maciel, V.; Fernandez, C.M.; Lima, T.M.; Gaspar, P.D. Decision support system to assign price rebates of fresh horticultural products based on quality decay. In *Computational Management*; Patnaik, S., Tajeddini, K., Jain, V., Eds.; Springer International Publishing and Imprint Springer: Cham, Switzerland, 2021; Volume 18, pp. 487–497; Springer eBook Collection. [CrossRef]
92. Pina, M.; Gaspar, P.; Lima, T. Decision Support System in Dynamic Pricing of Horticultural Products Based on the Quality Decline Due to Bacterial Growth. *Appl. Syst. Innov.* **2021**, *4*, 80. [CrossRef]
93. Verloop, C.M.; Verdouw, C.N.; van der Hoeven, R. Plantform: Horizontal cooperation in realizing integrated information systems for potted plants production. In Proceedings of the 7th EFITA Conference, Wageningen, The Netherlands, 6–8 July 2009.
94. Akkermans, H.A.; Bogerd, P.; Yücesan, E.; van Wassenhove, L.N. The impact of ERP on supply chain management: Exploratory findings from a European Delphi study. *Eur. J. Oper. Res.* **2003**, *146*, 284–301. [CrossRef]
95. Guzueva, E.R.; Beibalaeva, D.K.; Abubakarov, M.V. Information and management system for farming. In Proceedings of the 1st International Conference ASE-I—2021: Applied Science and Engineering: ASE-I—2021, Melbourne, Australia, 15–19 November 2021; AIP Conference Proceedings; Salamova, A., Dakhaeva, F., Eds.; AIP Publishing: Melville, NY, USA, 2021; Volume 2442. [CrossRef]

agriculture

MDPI

Review

Potential Role of Technology Innovation in Transformation of Sustainable Food Systems: A Review

Nawab Khan [1], Ram L. Ray [2], Hazem S. Kassem [3], Sajjad Hussain [4], Shemei Zhang [1,*], Muhammad Khayyam [5], Muhammad Ihtisham [6,7] and Simplice A. Asongu [8]

1 College of Management, Sichuan Agricultural University Chengdu Campus, Chengdu 611100, China; nawabkhan@stu.sicau.edu.cn
2 College of Agriculture and Human Sciences, Prairie View A&M University, Prairie View, TX 77446, USA; raray@pvamu.edu
3 Department of Agricultural Extension and Rural Society, King Saud University, Riyadh 11451, Saudi Arabia; hskassem@ksu.edu.sa
4 College of Basic Veterinary Medicine, Sichuan Agricultural University Chengdu Campus, Chengdu 611100, China; 2019603001@stu.sicau.edu.cn
5 School of Economics and Management, China University of Geosciences, Wuhan 430074, China; khyousafzai@cug.edu.cn
6 College of Horticulture and Forestry, Huazhong Agricultural University, Wuhan 430070, China; ihtisham@webmail.hzau.edu.cn
7 College of Landscape Architecture, Sichuan Agricultural University, Chengdu 611130, China
8 School of Economics, University of Johannesburg, Johannesburg 2006, South Africa; asongus@afridev.org
* Correspondence: 14036@sicau.edu.cn

Abstract: Advanced technologies and innovation are essential for promoting sustainable food systems (SFSs) because these technologies can be used to answer some of the critical questions needed to transform SFSs and help us better understand global food security and nutrition. The main objective of this study is to address the question of whether technological innovations have an impact on the transformation of SFSs. There are certain innovations including agricultural land utilization, food processing, production systems, improvement in diets according to people's needs, and management of waste products. This study provides an overview of new technologies and innovations being used with potential to transform SFSs. Applications of emerging technologies in digital agriculture, including the Internet of Things (IoT), artificial intelligence and machine learning, drones, use of new physical systems (e.g., advanced robotics, autonomous vehicles, advanced materials), and gene technology (e.g., biofortified crops, genome-wide selection, genome editing), are discussed in this study. Additionally, we suggest eight action initiatives, which are transforming mindsets, enabling social licensing, changing policies and regulations, designing market incentives, safeguarding against undesirable effects, ensuring stable finance, building trust, and developing transition pathways that can hasten the transition to more SFSs. We conclude that appropriate incentives, regulations, and social permits play a critical role in enhancing the adoption of modern technologies to promote SFSs.

Keywords: technology innovation; food processing; transition pathways; sustainable food systems; digital agriculture; transformation

Citation: Khan, N.; Ray, R.L.; Kassem, H.S.; Hussain, S.; Zhang, S.; Khayyam, M.; Ihtisham, M.; Asongu, S.A. Potential Role of Technology Innovation in Transformation of Sustainable Food Systems: A Review. *Agriculture* **2021**, *11*, 984. https://doi.org/10.3390/agriculture11100984

Academic Editor: Dimitre Dimitrov

Received: 16 September 2021
Accepted: 6 October 2021
Published: 9 October 2021

1. Introduction

Food sustainability is directly linked with sustainable agriculture. Sustainable agriculture is an integrated system of plant and animal production practices that can provide sufficient human food and fiber needs, enhance environmental quality and natural resources, use resources efficiently, sustain farm operations, and enhance the quality of human life for the long term [1]. The critical factors for a sustainable food system (SFS) are fertile land, water, fertilizers, favorable climate conditions, and energy [1]. However, currently, the sustainability of food and agricultural systems is under stress due to the positive global demographic change, including rapid global population growth, increases in food demands, climate change effects, limited water supplies, and the transition from conventional energy sources (oil and gas) to new energy sources. Social, economic, and environmental sustainability are closely interlinked and critical for sustainable agriculture. However, an integrated approach that includes proper use of advanced technology is equally important for sustainable agriculture. Therefore, it is important to use an integrated approach that includes ecosystem services, human capital, and new technologies to produce food sustainably [2].

The population of the world is now more than 7.7 billion, and it is increasing at an annual rate of around 1.07%. Hence, by 2050, the world's population is projected to increase by more than 30% of the current population and reach 10 billion [3–8]. Given the projected growth of the population and income, as well as the headwinds of the climate, meeting the total demand for food in the future will place unprecedented pressure on limited water, fertile land, energy, and potential climate change. The risk of huge and potentially irreversible ecological damage caused by unprecedented pressure is subject to serious academic debate. In addition to the long-term pressure from the inevitable increase in the food demand, several other factors have raised concerns about the sustainability of the agricultural food system to adapt to climate change and environmental stresses. The frequency and/or intensity of such stresses appear to be increasing, and they usually respond in a cascade, with one triggering the other [3,9–12]. Stresses to the agri-food system seem to be increasing due to the growing demand for high-quality nutritional food around the world [11,13]. The projected demographic change will negatively impact agricultural productivity and agricultural expansion, which will stress natural resources by increasing deforestation, water consumption, and greenhouse gas emissions, thus contributing to ecological insufficiency and climate change [3,14]. For example, it is projected that the consumption of meat and dairy products will be increased by 173% and 158%, respectively, between 2010 and 2050 globally [6,15]. The continued increase in demand for animal protein and the corresponding expansion of food production are causing serious concerns.

Efficient resources are required to convert vegetable matter into animal-derived proteins (e.g., meat or milk protein). For example, eight kilograms of vegetable food is required to increase one kilogram of weight in beef cattle [3,16]. Since April 2016, the main goal of the United Nations Decade of Nutrition Action has been to "eliminate all forms of malnutrition". However, some important key points (e.g., economic aspects, nutrition and health, environmental, social, and food security) that were agreed at the Second International Nutrition Conference also focus on developing a "sustainable, resilient and healthy diet food system" [3,17]. For this reason, people have discussed food alternatives including all aspects of integrating food safety and sustainability concepts.

Moreover, it is necessary to adopt emerging technologies (Internet, mobile phones, computers, IoT, etc.) for sustainable agriculture, including food productions with high protein. It is also important to optimize the protein contributions from animals and plants. While this will help promote the sustainability of food systems and biodiversity, it will ultimately provide effective distribution of high-quality protein for the global population [18–21]. In the global context, government and non-government policies and consumers' current intentions to include more plant-based protein in their daily diets [3,22] motivate the use of alternative protein sources for better human health. Some examples of emerging and sustainable protein sources include grains (e.g., wheat and zein), seeds (e.g., chia seeds), leaves (e.g., moringa), legumes (e.g., beans, lentils, peas), microalgae, fungi (e.g., bacteria protein), milk (e.g., whey protein), and insects.

Until now, the future sustainability of food systems, changing diet's role, reducing waste, and increasing agricultural productivity have mainly been studied through existing technologies. For example, a common research question concerns what level of yield increase can be achieved through the spread of new crop varieties, livestock species, animal feed, or changes in agricultural practices, as well as the spread of irrigation and improved management techniques [23,24]. However, as research has shown, even if existing agricultural technologies are widely adopted, flexible diets are fully implemented, and food waste is reduced by half, it will still be a challenge to feed the growing world population while ensuring the well-being of the planet [2,23,25,26]. Thus far, few studies have explored whether the world is adopting more destructive, "wild", game-changing options [6,27,28] that can affect the progress of many required dimensions of food systems simultaneously. Some of these game changers are no longer the realm of imagination; they are already being developed fairly rapidly, reshaping the viability of different sectors. Investment data on agriculture show that several companies are focusing on digital agriculture [29–31]. Digital agriculture includes applications of advanced technologies such as the Internet of Things (IoT), artificial intelligence and machine learning, drones and use of new physical systems (e.g., advanced robotics, autonomous vehicles, advanced materials), and gene technology (e.g., biofortified crops, genome-wide selection, genome editing).

Technology itself is not always transformative. However, if it is applied/used appropriately to address critical issues of any field, including agriculture, it can be crucial for developing innovative solutions. The transformative power of technology depends on economic and administrative strategies, social needs, and socioeconomic status [32]. However, few studies have been conducted on the elements that can promote the transformation of food systems through system innovation [21,33–41]. This study contributes to the extant literature by filling the underlying gap, especially discussing and summarizing how to achieve a positive transformation of SFSs. Specifically, this study provides insights on emerging technologies that can be used to achieve sustainability in food systems.

The rest of this review article is organized as follows: Section 2 describes the methodology and framework of the study. Section 3 elaborates on the technology change and innovations in food systems and transformation accelerators. Section 4 provides key factors of sustainable food systems, and Section 5 presents the conclusion of this study.

2. Methodology and Framework of the Study

In recent decades, several studies have been conducted to solve key issues in agriculture to improve agricultural production and achieve SFSs. With recent advancements in technologies to increase the potential of agricultural development and SFSs, some researchers have made great efforts to promote SFSs. The scientific community's interest in advanced technology has grown exponentially. However, it is always challenging for stakeholders and users to select and implement appropriate technologies to increase agricultural production. Recently, some studies have reviewed articles focusing on the implementation, application, challenges, potential, and prospects of the IoT in smart agriculture, and agricultural and food production systems. However, they mainly focused on research based on the IoT. We reviewed the most important research based on advanced technologies essential to SFSs. Our strategy was to review articles based on advanced technologies used in agriculture. We have reviewed articles that focus on the application of sustainable food, precision agriculture, food safety and nutrition, digital technology, the IoT, and smartphone technology in food.

A tentative framework was designed to assess the contribution of recent innovations towards the current technological transformation process and innovations within food systems which help improve food sustainability. It is purely a qualitative process of achieving the Sustainable Development Goals. The current innovations and future technologies are critical to support SFSs. Some innovations including agricultural land use, food processing and production systems, improved diets according to people's needs, and waste management can play a significant role for SFSs. This study discusses how advanced technologies applied to food systems can be transformed into SFSs (Figure 1). Additionally, eight action initiatives, including changing mindsets, enabling social licensing, changing policies and regulations, designing market incentives, preventing adverse effects, ensuring stable finance, building trust, and developing transitional pathways, are discussed. For this review article, we conducted an intensive literature review to collect the information and fill the underlying gaps to better understand the role of innovative technology in transforming SFSs.

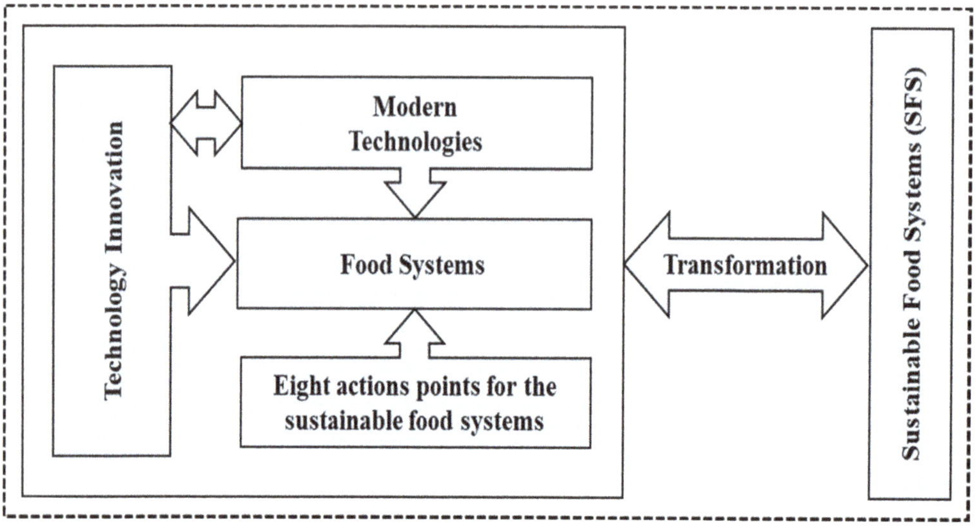

Figure 1. The framework of this study.

3. Technological Transformation and Innovations in Food Systems

Historically, activities dedicated to agri-food production have undergone many technological changes. The changes that have occurred in the past two centuries have been extensively examined [42]. Through the 19th century, the major technological innovations were the bearing, which introduced crucial mechanical modernizations, for instance, lawnmowers, mechanical harvesters, and threshers. With the first canned food entering the market, the increased industrial agriculture, and the gradual demise of household farms in more advanced countries, agricultural marketing had also undergone major changes. At the beginning of the 20th century, animal power was slowly substituted by innovative fuel-based energy sources [43,44].

In 1892, the prototype of the first gasoline tractor was constructed. Simultaneously, there have been numerous technological advances in marketing and food processing, including introduction of innovative packaging forms, expansion of long-distance trade, emergence of new food retail systems, and ongoing urbanization process. The main and lasting technological change of the past century is called the Green Revolution, which was based on the use of new high-yielding wheat varieties and the widespread use of chemical fertilizers and insecticides [43]. The Green Revolution, considered by capital-intensive production processes, led to a significant increase in agricultural productivity. In the United States in 1860, one farm job could feed 4–5 people, and in 1957, this number became 22.8 people [43,45]. At the end of the past century, biotechnology brought major technological variations to the agricultural food sector, followed by nanotechnology and digital technology.

Technological advances in several indicators of human well-being, including hunger, life expectance, and disease prevention, have played a significant role since Neolithic times [46]. Table 1 provides a detailed list of modern technologies such as digital agriculture, cellular agriculture, food processing and safety, gene technology, health, inputs, intensification, replacement food and feed, resource use efficiency, and other technologies and also an explanation of each technology's contribution to the technical advances of food systems. Despite the benefits to humanity of these food and agriculture advances, certain environmental and health indices are continuously declining, particularly in the

21st century. For instance, the conversion of forest land into agricultural land or pastureland has increased air and water pollution and greenhouse gas emissions. In addition, nitrogen and phosphorus usage has been multifarious, and their consumption is continuously increasing [47,48]. Excessive use of nitrogen and phosphorus in agriculture has a significant negative impact on the environment and human health. For example, runoff from agricultural watersheds causes eutrophication of waterbodies. On the other hand, excess nitrogen in the air can impair our ability to breathe and limit visibility [26,49]. The development of inexpensive, fast, or discretionary foods has also contributed to significant malnutrition in many parts of the world [49].

Table 1. Future technology with transformative potential. Modern technologies are divided into (10) categories, covering the entire food system. Table 1 presents a complete explanation of each technology.

Modern Technologies	Explanation of each Technology	Production	Processing	Packaging	Distribution	Consumption	Waste	References
Digital agriculture (DA)	Drones	✓			✓	✓	✓	[50,51]
	Innovative sensors		✓	✓	✓	✓	✓	
	Big data	✓	✓	✓	✓	✓	✓	
	Artificial intelligence	✓	✓	✓	✓	✓	✓	
	Data integration	✓						
	Assistive exoskeletons	✓	✓		✓			
	Disease/pest early warning	✓					✓	
	Robotics	✓	✓	✓	✓		✓	
	On-field robots	✓					✓	
	Sensors for soil		✓	✓	✓		✓	
	Tracking tech for livestock	✓			✓			
	Farm-to-farm virtual market		✓	✓	✓	✓	✓	
	Internet of Things	✓	✓	✓	✓	✓	✓	
	Improved climate forecasts	✓					✓	
	Nano-drones	✓			✓	✓	✓	
	SERS sensors		✓	✓	✓		✓	
	Pest control robotics	✓					✓	
	Nanotechnology		✓	✓	✓		✓	
	Intelligent food packaging		✓	✓	✓		✓	
	Pre-birth sex determination	✓						
	Smartphone food diagnostics		✓		✓	✓		
	Omics data use	✓						
	Traceability technologies		✓	✓	✓	✓		
Cellular agriculture (CA)	Artificial products	✓	✓			✓		[21,52]
	Artificial meat/fish	✓	✓			✓		
	Molecular printing	✓	✓	✓	✓	✓	✓	
Food processing and safety (FPS)	Nanocomposites			✓	✓		✓	[50]
	Food safety tech		✓	✓			✓	
	Whole-genome sequencing	✓	✓				✓	
	Biodegradable coatings			✓	✓		✓	
	Technologies for sustainability		✓	✓	✓		✓	
	Drying/stabilization tech		✓	✓	✓		✓	
	Microorganism coatings			✓	✓		✓	

Table 1. *Cont.*

Modern Technologies	Explanation of each Technology	Production	Processing	Packaging	Distribution	Consumption	Waste	References
Gene technology (GT)	Genome editing	✓						[53]
	GM-assisted domestication	✓						
	Biofortified crops	✓				✓		
	Plant phenomics	✓						
	Synthetic biology	✓						
	Novel perennials	✓						
	Weed-competitive crops	✓					✓	
	RNAi gene silencing	✓						
	Genome-wide selection	✓					✓	
	Apomixis	✓						
	Oils crops	✓						
	Reconfiguring photosynthesis	✓						
	Disease/pest resistance	✓					✓	
	Novel nitrogen-fixing crops	✓					✓	
	Genome selection	✓						
Health (H)	Personalized crops		✓	✓	✓	✓	✓	[54]
Inputs (I)	Soil additives	✓					✓	
	Holobiomics	✓					✓	
	Nano-enhancers	✓					✓	
	Enhanced efficiency fertilizers	✓					✓	
	Nano-fertilizer	✓					✓	
	Micro-irrigation	✓					✓	
	Botanicals	✓					✓	
	Nano-pesticides	✓					✓	
	Macrobials	✓					✓	
	Microbials	✓					✓	
Intensification (In)	Vertical agriculture	✓			✓			[54]
	Electro-culture	✓						
	Irrigation expansion	✓					✓	
Other (O)	Ecological biocontrol	✓					✓	
	3D printing	✓	✓	✓	✓	✓	✓	
	Resurrection plants	✓						
	Battery technologies	✓	✓		✓		✓	
Replacement food and feed (RFF)	Microalgae and cyanobacteria for food	✓	✓			✓	✓	[50,55]
	Seaweed for food	✓	✓			✓	✓	
	Insects for food	✓	✓			✓	✓	
	Omega-3 products for aquaculture	✓						
	Innovation aquaculture feed	✓	✓				✓	
	Microbial protein	✓	✓			✓	✓	
	Dietary additives for livestock	✓					✓	
	Livestock/sea substitutes	✓	✓			✓	✓	
Resource use efficiency (RE)	Circular economy	✓	✓	✓	✓	✓	✓	[50]

Food management technology (e.g., food production, handling supply, and delivery) should be based on hazard analysis, and a critical control point (HACCP) is emerging at an incredible pace, which can be applied for SFSs in the future. According to comprehensive literature reviews, we offer a stock of nearby and potential innovations that can contribute towards the development of SFSs. Each technology is graded corresponding to its role in the value chain (such as manufacturing, production, storage, delivery, usage, and waste) and ready-to-mind ranking [56–58].

The nine technology readiness levels (TRL) include (i) basic principles observed and reported, (ii) technology concept and/or application formulated, (iii) characteristic proof of concept, (iv) component and/or breadboard validation in a laboratory environment, (v) breadboard validation in a relevant environment, (vi) system/subsystem model or prototype demonstration in a relevant environment, (vii) system prototype demonstration in a space environment, (viii) actual system completed and "flight qualified" through test and demonstration, and (ix) actual system "flight proven" through successful mission operations. These technology readiness levels contain the established application of technology on a real-world basis, including fundamental study, concepts discovered, and technology experiments applied [56,57].

This work exercise leads to a few insights. Firstly, as apparent in Table 1, technology covers the whole food chain (production, processing, packaging, distribution, consumption, and waste management). Therefore, various technical solutions can be adapted to tackle particular food system challenges in multiple structural and administrative settings. This complex conduit involving artificial meat, 3D printing, consumers' readiness, nano-drones, smart packaging, etc., provides a genuine chance for structural change. New technology mixes might differ widely based on a country's or region's level of socioeconomic growth and other governmental and institutional constraints.

Secondly, many concerns with digital and smart farming and the substitution of feed and food for fish and livestock are comparatively similar to the vast number of near-ready and advanced technologies, considering the massive size of the industry classes. This is not surprising given the pace of innovation and cost savings due to emerging technology, accompanied by the universal acceptance of these innovations across countries with medium, middle, and high incomes [54]. Additionally, efforts are underway to reduce the demand for livestock goods by offering alternate protein sources and disconnecting animal production with substituted circular feed from the ground, reducing its environmental effects. Increasing the demand for fish relies on decreasing the share of the complete amount of fish captured for livestock feed, which presently is about twelve percent [54].

Thirdly, some near-mature technologies have great capability to be adopted, thereby promoting strategic investments in their dissemination and implementation. There is an urgent need to study how to provide options with minimal disturbance in the current food systems and better understand the factors that may affect their absorption of the scale of transformation. This also highlights the potential contribution of the private sector in promoting the adoption of these technologies and the need to establish a regulatory framework and market structure to ensure that these advancements are fully aligned with public policy goals. Crucially, at least in the medium term, the affordability of these new options will increase, which is more likely to happen as demand becomes clearer and manufacturing processes and supply chains are better established. Finally, the simultaneous implementation of variations of these technologies can drastically accelerate achieving SFSs. This may simultaneously improve sustainable food production and reduce waste while improving human well-being and creating new local business opportunities, as resources are reassessed as part of the process. In addition, this is consistent with the current efforts to revitalize the bioeconomy in many parts of the world [59,60].

Transformation Accelerators

A mode of innovation that requires significant changes in food systems (infrastructure, technology, expertise, and capabilities) and structural reforms of principles, legislation, policy, economies, and governance around them is essential to this process (Table 2). This vision of transformations as a dynamic and integrated mechanism suggests that modern technology alone is not enough to force changes in the food system; instead, it should be supported through a broad spectrum of societal and structural forces that empower its use [32].

Table 2. Technical preparations for future food system technology. The technology readiness score is a (ten-step stage system) evaluation system that helps to assess the maturity of specific technologies. Detailed information about each step, score estimates, technical groups, and initials is provided for each technology.

Food System Technology	Research Initiated	Experimental Proof	Prototype	Implemented	References
Microalgae and cyanobacteria for food				H	
Innovative aquaculture feed				H	
Microbial protein				H	
Insects for food				H	
Seaweed for food				H	
Disease pest resistance				FPS	
Biofortified crops				FPS	
Vertical agriculture				I	
Drying/stabilizing methods				I	
Drones			O		
Battery technologies			O		
Tracking and confinement techniques for livestock			DA		
3D printing			DA		
Improved climate forecasts			DA		
Traceability technologies			DA		
Farm-to-farm virtual marketplace			DA		
Robotics			DA		[11,33,52,61,62]
Disease/pest early warning			DA		
Microbials			H		
Micro-irrigation/fertigation			H		
Dietary additives for livestock			H		
Soil additives			H		
Microbial			H		
Circular economy			H		
Omega-3 products for aquaculture			H		
Irrigation expansion			I		
Oil crops			GT		
Genomic selection			GT		
Genome editing			FPS		
Sustainable processing technologies			FPS		
Biodegradable coatings			FPS		
Food safety techniques			FPS		
RNAi gene silencing			FPS		

Table 2. *Cont.*

Food System Technology	Research Initiated	Experimental Proof	Prototype	Implemented	References
Plant phenomics			FPS		
Big data			DA		
Smartphone food diagnostics			DA		
Intelligent food packaging			DA		
Internet of Things			DA		
Soil sensors			DA		
Advanced sensors			DA		
Holobiomics		H			
Botanicals		H			
Weed-competitive crops		GT			
GM-assisted domestication		GT			
Nano-enhancers		H			
Enhancing efficiency fertilizers		H			
Personalized food		H			
Omic data usage		DA			
Data integration		DA			
Pre-birth sex determination		DA			
On-field robots		DA			
Artificial phenomics		DA			
SERS sensor devices		DA			
Assistive exoskeletons		DA			
Pest control robotics		DA			
Whole genome sequencing		I			
Microorganism coatings		I			
Nanocomposites		I			
Electro-culture		I			
Artificial meat/fish		CA			
Molecular printing		CA			
Genome-wide selection		FPS			
Resurrection plants	FPS				
Apomixis	FPS				
Nano-drones	DA				
Nanotechnology	DA				
Nano-pesticides	H				
Artificial products	CA				
Nano-fertilizers	RE				
Ecological biocontrol	O				
Reconfiguring photosynthesis	GT				
Novel perennials	GT				
Novel nitrogen-fixing crops	GT				
Synthetic biology	GT				

Note: digital agriculture (DA), cellular agriculture (CA), food processing and safety (FPS), gene technology (GT), health (H), inputs (I), intensification (In), other (O), replacement food and feed (RFF), and resource use efficiency (RE).

Transformation is also a deeply political process with winners and losers, which involves choices, consensus, and compromise regarding new directions and pathways. Powerful food system actors provide the right motivations for the continuation of the current status and market share. In comparison, new consumers are much more likely to behave as device disrupters and use it as a means of generating fresh products and appeal (e.g., replacement of meat). Efforts to drive beneficial structural progress and transition must also be compatible with social and political mechanisms that obstruct or catalyze creativity in the sector [50,50].

In reality, this involves creating relationships, dialogue, and faith in the pathway for improving food systems, maintaining governance and regulatory systems to preserve the expected effects of food systems, all necessary conditions for the implementation of modern technologies (Table 2). Emerging innovations benefiting from such improvements include insects for food, meat-generated animal alternatives, food system circularity, and vertical agriculture [11,43,50].

The new device developments (e.g., molecular printing, biodegradables, and customized nutrition) can catalyze technology by incorporating additional devices and tools (e.g., drones) into system developments resulting from extensive social and political shifts that drive transformation [32,60,63]. Technology can also raise undesired lock-ins (e.g., a grower who has practiced and invested heavily in grain production cannot turn quickly to diversified agriculture) [50,64]. To avoid these lock-ins, it is essential to recognize the mechanisms of transition.

4. Eight Action Initiatives for Sustainable Food Systems (SFSs)

We summarize eight action initiatives (e.g., building trust, emerging transition pathways, transforming mindsets, empowering social licensing, changing policies, designing market incentives, safeguarding against undesirable effects, and ensuring stable finance) closely related to technological and structural progress in food systems (Figure 2, Table 3).

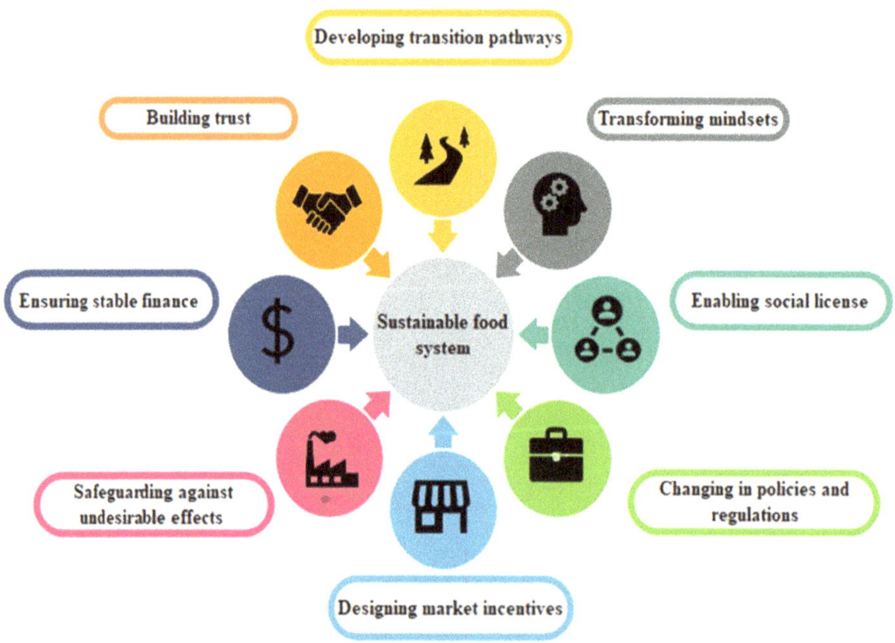

Figure 2. The crucial elements of hastening the transformation of food systems. These elements improve sustainable and healthy diets, productive agri-food systems, and waste management, these three outcomes being essential to achieving SFSs, modified from [11,33].

Table 3. Basic elements of developing and scaling up beneficial effects, and examples from four case study technologies.

Action Initiatives	Examples	References
Developing transition pathways	For all case studies Establish a transition path based on all the above elements. Ensure that everyone, involving those at a disadvantage, can benefit from innovation. Apply adaptive methods to acclimate to changing environments and unintended consequences. Focus on attaining overall goals, not explicit technologies. Local, national, and international pledges and suitable resource apportionment. Case studies dedicated to automation and robotics in agriculture. Endorse health and safety and create employment to attain fair production.	[11,33,50]
Transforming mindsets	For all case studies Boost the acceptance of high-tech products and the handling of nourishment and feed. Case studies specific to microbial proteins in organic waste streams. Treating all types of waste as by-products can be used as valuable inputs for other processes. Accept feed production from organic waste streams, counting human and animal waste.	[11,33,65]
Enabling social licensing and stakeholder dialogue	For all case studies Interact with stakeholders across humanity (comprising consumers, workers, and producers) to ensure transparent development and technology implementation. Case studies specific to grain nitrogen fixation. Focus on food quality to ensure that new crops are as good even if they do not substitute crops. Indications and enhanced environmental footprint, reduction in input usage and waste. Evade vertical integration models that cause industry conspiracy concerns.	[11,50]
Changing policies and regulations expected support	For all case studies Improve and simplify coherent strategies and regulations to ensure proper supervision and enforcement of environmental, social, health, and safety standards throughout food systems. Reduce economic and organizational limitations on technology adoption and dissemination. Case studies specific to personalized nutrition. Apply strong standards on nutrition and health labeling. Develop supervision of the food environment, which will affect personal consumption choices.	[11,33]
Designing market incentives	For all case studies Formulate fiscal and trade policies to cultivate initial markets to achieve economies of scale. Invest in plans to increase awareness of new technologies and their appropriate use. Case studies specific to microbial proteins in organic waste streams. Increase waste costs to encourage alternative uses (for example, enhance waste disposal fees). Provide price help for main inputs to decrease production expenses. Provide support to the traditional feed industry to transition to alternative production.	[11,33]
Safeguarding against undesirable effects Monitor and correct	For all case studies Independent, transparent, and competent regulatory agencies oversee and enforce standards. Establish global eco-friendly, worker, and trade standards to evade offshore external factors. Entail investment to improve the usage of influence valuation and further assurance principles. Case studies specific to grain nitrogen fixation. Monitor land usage to ensure the adoption of technology aids lessen the food system footprint. Monitor the wider adverse effects of extensive adoption of new crops. Monitor the nitrogen concentration in the soil to inform the taxation of excess nitrogen to avoid overuse.	[33,50]

Table 3. *Cont.*

Action Initiatives	Examples	References
Ensuring stable finance Explore and pilot	For all case studies A clear commitment to long-term objectives to support stakeholders in reorienting their assets. Encourage the use of other funding mechanisms to support liable improvement. Persuade long-term funding and approve the extension of the investment timetable to obtain a total return on investments. Case studies dedicated to robotics and automation in agriculture. Promote the application of verified robotics machinery in modern agricultural environments to enhance the visibility and perceivable viability of agricultural food systems.	[11,33]
Build trust vision and values among participants in food systems	For all case studies Establish trust in so-called profits with reason or so-called progressive benefits of the system. Promote transparent production, supply, and management procedures. Develop trust in regulatory agencies that describe and enforce ecological, health, and safety standards. Case studies specific to personalized nutrition. Increase a health-centric machinery platform that equalizes short- and long-term goals.	[11,50]

4.1. Building Trust among Stakeholders of the Food Sector

Every change in the agricultural sector demands the general opinion and support to build new developmental pathways. Technology, cooperation, and a set of collective values regarding the agreement of the results of various food systems are the key factors. These results include the durability, socioeconomic effects, and provenance of the developed food system. Development of trust occupies the central position in this process. Many social and economic networks provide the connecting resources between the food producers or farmers and food consumers or food companies. For the absorbance of technology and a systemic change, many steps are necessary for the actors of food systems, i.e., identification of business opportunity by private companies, identification of systemic change requirement by the government for public welfare, initiation of a dialogue with the mass/public to modify their attitudes, and innovation in policy and market shifts by investments (public or institutional) [32,66]. The Green Revolution in Asia provides an excellent example of these types of systemic variations, which has resulted in increased crop production and utilization and diminishing malnutrition in a little more than a decade [32].

The involvement of the government can remarkably introduce technology to the public. A high-level agreement can be critical in this regard because of ecological and ethical concerns from production to food utilization. For example, suppose the arrangements have a robust scientific base for the desired targets with the participation of public or private sectors for their opinions and discussions. In that case, mechanisms, innovative ideas, different products, incentives, and policies can be developed. The Paris Agreement on greenhouse gas emissions and the Sustainable Development Goals are excellent examples occupying the central position among national and international strategies in the public and private sectors. Managing the prospects of various stakeholders is necessary to gain legitimacy and trust. The best behavior may depend largely on the behavior expected by others. Suppose the benefits of adopting a particular behavior (e.g., using and/or investing in a specific technology) are considered a function of the behavior's popularity, among others. In that case, there may be a vicious or virtuous circle of self-fulfillment expectations, which ultimately accelerates or hinders change [67]. The Green Revolution of the 1960s provides an excellent example in this regard. If the expectation of acceptance of new adoption by other individuals is low, then the target individual would not adopt it; hence, temporary subsidies and incentives can play a crucial role in this regard [68].

4.2. Transforming Mindsets

The actors require an understanding that is ready to accept new information about food systems. There is also a requirement from a similar point of view of decision makers. Humans have a deep relationship with food regarding biological, psychological (especially around naturalness) [11,69], and cultural aspects [50,64], and thus it is still uncertain whether society would accept innovation or not. Hence, innovation's price and security are not the sole factors for innovation to be absorbed in the community. There is a tripartite relationship between people's attitudes to technology, the regulation that can change the market structure, and market actors that play out within a regulatory framework. The need to better understand the technology and transform mindsets arises particularly in the case of technologies whose advantages and disadvantages are still largely unknown (for example, gene editing, reconfiguring photosynthesis, novel nitrogen-fixing crops) [50].

4.3. Empowering Social Licensing and Stakeholder Discussion

There is a strong association between people's expenses on innovative development in social licensing and acceptance of innovation. The development of useful technology and public communication are critical factors in this regard [70,71]. It is possible to get pressure from users, workers, and capitalists to change the technologies being used (e.g., meat alternatives and nano-pesticides) after people become aware of specific issues in the existing technologies. It is compulsory to incorporate these actors. Otherwise, there are fewer chances of adopting innovation even if the invention has enough energy (e.g., genome editing). There may be a constraint to positive change by those who carry out trades routinely. Understanding the utilization of technology is very important for its proper use. Additionally, learning by action or practice is the critical factor in systems based on extensive knowledge [50,72,73], but it may be a loss for the smallholders (e.g., small farmers, suppliers, food processors, and growers) in developing countries.

4.4. Guaranteeing Variations in Strategies and Regulations toward Food System Sustainability

It is important to consider how all investors will respond to a new change in technology by keeping the investors' interests in mind. For example, currently, climate change has inducted a twist in carbon emission policies worldwide. Therefore, investors shall be keen to invest in any new technology possessing low-carbon emission abilities if they feel that this will be monetarily beneficial and rewarding. Once a technological change comes into play, it may become economically accessible, meaning it will have the social effects of being bought and changing policies. Besides this, if people start distrusting, the technology will never become a new product as few will be interested in adopting it. Others will stick towards myths of their own as the benefit of stable reward is their firsthand and selfish courage [67]. Therefore, only little effort is required to make policies favorable, such as subsidizing the projector by using public funds as this cause may attract several investors due to their self-interests [74–76].

4.5. Designing Market Incentives

Any technological change is successful only when all the investments and budgets are well planned and well directed. The hurdles toward a new technology and its application may vary. In a competitive environment, the big companies negate spending on knowledge and research-based technology as they solely intend to invest and make money. However, governments have come forward to play their roles to support such technologies. Governments have to propose a resolution to address this issue by tailoring appetizers such as subsidizing a company to produce on a mass scale, providing them with opportunities in the market, easing the procurement process, and even relaxing tax rates. This act of government is never confined to an old industry only. Instead, governments even try to offer all these helping tools to newly built companies because it is never understood who will eventually introduce better innovations [77]. The government's involvement in incubating innovation and accelerating technological enhancement can offer us new solutions in the market [11,78]. This has been the case with many technologies on our list (Table 1) across all technology groups (drones, algae for feed, plant-based meat substitutes, nano-enhancers, personalized food). Incentives that drive innovation also differ from those that encourage diffusion.

4.6. Safeguarding against Undesirable Effects

There are always hurdles when policymaking for transformation is needed because it is always challenging to correlate investment and technological changes. When the stage of public acceptance comes, it is still a complex situation, and regulating the whole operation is difficult and can even go overlooked [30,79,80]. For example, circular economic approaches in the food sector should always be in accordance with the strict laws which are established across North America and Europe about the re-use of organic waste as animal food (this law came into effect after the bovine spongiform encephalopathy and foot and mouth outbreaks) [81]. A mass-scale or widespread dialogue for a consensus may help more acceptably legitimize or better understand the grounds of disagreement. It may also help us understand how adoption or non-adoption happens, and how a whole struggle of innovation falls under the complexity and lack of social licensing due to not understanding the relevant issues. However, still, great technology cannot be accepted if it conflicts with the myths and traditions of a society deep rooted in the culture [64,69,82].

4.7. Ensuring Stable Finance

Technologies related to food and agricultural products undergo a production that is affected by seasons and regulations. This aspect makes the whole process more challenging. It becomes hard to relate investment and innovation because it favors any failed operation and then starts it again differently. Additionally, the transformational changes become more unpredictable, and their impacts are not easily measurable as accurate environment testing is needed to evaluate the effectiveness. Transformational change requires more creative investment solutions, steady and stable investment plans, and more extended time deployment of persistent investors to encourage a valuable output [50,83]. We need strong support for research and development for a longer time to develop a broader range of technologies for food sustainability (e.g., reconfiguration of photosynthesis, new vaccines for livestock, and genetically modified assisted breeding technologies) [43,65]. In addition, the application of digital knowledge and digital technologies in agriculture may tend towards better solutions, as happened in mobile banking during the mobile phone revolution in the 2000s.

4.8. Emerging Transition Pathways

Most research for prospective food systems is about the effects of substituent models and many other parameters such as food alterations, minimizing waste, and elevated productivity [23,26,54,84]. However, such research has not discussed the mechanisms to convert them to address real-world issues. The term transition pathway is used to demonstrate how to convert these ideas into reality. However, the transition pathway demands a significant amount of information about digital innovation and its effects, goals, and improvement in the framework of public and private institutions, and a systemic path is necessary to obtain the desired results. Hence, accelerators and digital technology can be useful tools for developing these pathways. Digital technologies can provide an innovative solution to enhance the performance and sustainability of agricultural production systems [85,86], described as having economic, social, and ecological aspects [87,88]. Digital technologies can make the food sector more effective, inclusive, and ecologically sustainable, thereby increasing the interests of growers, customers, and society [89]. Moreover, digital technologies can help increase farm productivity, advance resource utilization efficiency, and support environment resilience [86,90]. If digital technologies are implemented/adopted, the improvement in main production, supply chain, and logistics performance and reductions in food loss and waste will be particularly significant. In addition, the COVID-19 pandemic has shown the importance and application of digital technologies in the food industry [91] and has promoted the introduction and adoption of digital technologies for sustainable agriculture and food systems [92]. Despite the numerous benefits promised, such as other major innovation breakthroughs, digital agriculture is not without challenges or risks [33,86,89,93,94]. As reported by the Food and Agriculture Organization (FAO), digital technologies can significantly address the challenges faced by the global agro-food systems at every level of the supply chain [95,96]. The FAO argued that digital technologies at the farm level, such as sensors, robots, and drones, can provide precise information to farmers and help them increase yields in a climate-friendly way. Blockchain technology can enhance traceability and sustainability by monitoring the food chain from the field to the final consumer [97]. The United Nations also explored the opportunities offered by digital technologies in the field of nutrition and concluded that they help provide tailored health advice but warned against their potential threats to the privacy of health information [98]. The FAO argued that "digital technologies can trigger major changes or "disruptions" in the sustainable food system that not only improve efficiency and speed but also redistribute information and power along the value chain" [99]. A similar approach was adopted by the European Commission, which, albeit recognizing the limited spread of digital technologies across the Union, considers them capable of increasing sustainability in the agro-food system [100], thus prioritizing digitalization in the Common Agricultural Policy (CAP) reform 2021–2027 [97,101].

5. Conclusions

Currently, food systems face enormous challenges such as a growing population, competition for resources, global food chain complexity, food consumption, climate change, increased biofuel production, limited food access, unsustainable agricultural practices, lack of farmers' and workers' rights, and food waste [94]. Since society is undergoing transformational progress in the use of telecommunications, including digital agriculture, new physical systems, and renewable energy, technological innovation is bound to play an essential role in the future of SFSs.

The inventory of potential technologies related to food systems has been extended. However, there is an urgent need for a more robust analysis of technological innovation and its potential impact on food security. This research is technically complex, especially in uncertainty and selecting new investment streams identified as funds for research organizations to work on. This research has been invented with a multicultural and socio-governmental perspective to ensure instant innovation where it is required the most, maintain fairness, and adopt diverse ideas.

The technological innovation and advancement in SFSs rely on sufficient investment in rudimentary study and improvement to maintain the research and development process. In the future, several modern techniques will significantly contribute to food systems worldwide. Therefore, there is a dire need to circumvent the bottleneck of the conducive ecosystem, particularly in developing nations, where the prospective influences (positive and negative) of the modernization of technology may be comparatively significant. History demonstrates that technology innovation produces winners and losers. In the short and long term, the considerable agenda in the sustainability of society and food systems is used to deal with several social and agricultural sectors.

Despite the numerous benefits of adopting innovative and advanced agriculture technologies, as with other major innovation breakthroughs, digital agriculture is not without challenges or risks. The major challenges to implement and adopt these technologies to support digital agriculture are the cost and appropriate training to use such technologies.

Finally, and perhaps most significantly, hastening food systems' transition to a positive, ideal state will have to entail social dialogue. Of the eight action initiatives proposed to hasten the systemic revolution of food systems, as a minimum, five revolve around building trust, obtaining a social license, changing mindsets, preventing adverse effects, and developing transitional pathways. Achievements in all these acts will lead to superior health, a better environment, and improved SFSs.

Author Contributions: N.K., R.L.R., H.S.K., S.H., S.Z., M.K., M.I., and S.A.A. developed this concept, including the method and approach to be used; N.K., R.L.R., S.H., and S.Z. outlined the manuscript; N.K., R.L.R., H.S.K., S.Z., M.K., and M.I. developed and outlined the manuscript; N.K., R.L.R., S.Z., S.A.A., and R.L.R. contributed to the methodology and discussion of this manuscript; N.K., S.H., and S.Z. wrote the article. All authors have read and agreed to the published version of the manuscript.

Funding: This research received no external funding.

Institutional Review Board Statement: Not applicable.

Informed Consent Statement: Not applicable.

Data Availability Statement: The data that support our research findings are available from the corresponding author on request.

Conflicts of Interest: The authors declare that they have no conflict of interest.

Abbreviations

Abbreviations

TI	Technology innovation
CA	Cellular agriculture
FPS	Food processing and safety
H	Health
DTP	Developing transition pathways
ICTs	Information communication technologies
GM	Genetically modified
AFS	Agri-food system
FP	Food policy
H	Health
RFF	Replacement food and feed
O	Other
TRLs	Technology readiness levels
ACI	Agri-food consulting
TLU	Tropical livestock unit
GDP	Gross domestic product
IFR	Institute of Food Research
FDA	Food and Drug Administration
MDGs	Millennium Development Goals
SFS	Sustainable food system
DA	Digital agriculture
GT	Gene technology
RUE	Resource use efficiency
TM	Transforming mindsets
PATs	Precision agriculture technologies
R&D	Research and development
FAO	Food and Agricultural Organization
AVC	Agri-food value chain
I	Inputs
In	Intensification
MT	Modern technologies
NASA	National Aeronautics and Space Administration
WHO	World Health Organization
SOFA	State of Food and Agriculture
LEAD	Livestock, Environment and Development
CAC	Codex Alimentarius Commission
FMS	Food management subsystem
MAFF	Ministry of Agriculture, Food and Fisheries

References

1. Morawicki, R.O.; González, D.J.D. Focus: Nutrition and Food Science: Food Sustainability in the Context of Human Behavior. *Yale J. Biol. Med.* **2018**, *91*, 191. [PubMed]
2. Lindgren, E.; Harris, F.; Dangour, A.D.; Gasparatos, A.; Hiramatsu, M.; Javadi, F.; Loken, B.; Murakami, T.; Scheelbeek, P.; Haines, A. Sustainable food systems—a health perspective. *Sustain. Sci.* **2018**, *13*, 1505–1517. [CrossRef] [PubMed]
3. Fasolin, L.H.; Pereira, R.N.; Pinheiro, A.C.; Martins, J.T.; Andrade, C.; Ramos, O.; Vicente, A. Emergent food proteins–Towards sustainability, health and innovation. *Food Res. Int.* **2019**, *125*, 108586. [CrossRef] [PubMed]
4. Khan, N.; Ray, R.L.; Sargani, G.R.; Ihtisham, M.; Khayyam, M.; Ismail, S. Current Progress and Future Prospects of Agriculture Technology: Gateway to Sustainable Agriculture. *Sustainability* **2021**, *13*, 4883. [CrossRef]
5. Steiner, A.; Aguilar, G.; Bomba, K.; Bonilla, J.P.; Campbell, A.; Echeverria, R.; Gandhi, R.; Hedegaard, C.; Holdorf, D.; Ishii, N. *Actions to Transform Food Systems under Climate Change*; CGIAR: Montpellier, France, 2020.
6. Searchinger, T.; Waite, R.; Hanson, C.; Ranganathan, J.; Dumas, P.; Matthews, E.; Klirs, C. *Creating a Sustainable Food Future: A Menu of Solutions to Feed Nearly 10 Billion People by 2050. Final Report*; WRI: Washington, DC, USA, 2019.
7. Roser, M. Future population growth. Our World In Data. 2013. Available online: https://ourworldindata.org/future-population-growth (accessed on 20 September 2020).
8. United Nations. *World Population Projected to Reach 9.8 Billion in 2050, and 11.2 Billion in 2100*; UN DESA: New York, NY, USA, 2017.

9. Maystadt, J.-F.; Ecker, O. Extreme weather and civil war: Does drought fuel conflict in Somalia through livestock price shocks? *Am. J. Agric. Econ.* **2014**, *96*, 1157–1182. [CrossRef]
10. Von Uexkull, N.; Croicu, M.; Fjelde, H.; Buhaug, H. Civil conflict sensitivity to growing-season drought. *Proc. Natl. Acad. Sci. USA* **2016**, *113*, 12391–12396. [CrossRef] [PubMed]
11. Barrett, C.B.; Benton, T.G.; Fanzo, J.; Herrero, M.; Nelson, R.; Bageant, E.; Buckler, E.; Cooper, K.A.; Culotta, I.; Fan, S. *Socio-Technical Innovation Bundles for Agri-Food Systems Transformation*; Cornell Atkinson Center for Sustainability: Ithaca, NY, USA; Springer Nature: London, UK, 2020.
12. Kok, K.P.; Den Boer, A.C.; Cesuroglu, T.; Van Der Meij, M.G.; de Wildt-Liesveld, R.; Regeer, B.J.; Broerse, J.E. Transforming research and innovation for sustainable food systems—a coupled-systems perspective. *Sustainability* **2019**, *11*, 7176. [CrossRef]
13. Barrett, C.B. *Food Security and Sociopolitical Stability*; Oxford University Press: Oxford, UK, 2013.
14. Calicioglu, O.; Flammini, A.; Bracco, S.; Bellù, L.; Sims, R. The future challenges of food and agriculture: An integrated analysis of trends and solutions. *Sustainability* **2019**, *11*, 222. [CrossRef]
15. McLeod, A. *World Livestock 2011–Livestock in Food Security*; FAO: Rome, Italy, 2011.
16. Nadathur, S.; Wanasundara, J.; Scanlin, L. Proteins in the diet: Challenges in feeding the global population. In *Sustainable Protein Sources*; Elsevier: Amsterdam, The Netherlands, 2017; pp. 1–19.
17. FAO, W. Conference outcome document: Rome declaration on nutrition. In Proceedings of the Second International Conference on Nutrition, Rome, Italy, 19–21 November 2014.
18. Aiking, H.; de Boer, J. The next protein transition. *Trends Food Sci. Technol.* **2020**, *105*, 515–522. [CrossRef]
19. Chardigny, J.-M.; Walrand, S. Plant protein for food: Opportunities and bottlenecks. *OCL Oilseeds Fats Crop. Lipids* **2016**, *23*, 6p. [CrossRef]
20. Henchion, M.; Hayes, M.; Mullen, A.M.; Fenelon, M.; Tiwari, B. Future protein supply and demand: Strategies and factors influencing a sustainable equilibrium. *Foods* **2017**, *6*, 53. [CrossRef] [PubMed]
21. Serbulova, N.; Kanurny, S.; Gorodnyanskaya, A.; Persiyanova, A. Sustainable food systems and agriculture: The role of information and communication technologies. *IOP Conf. Ser. Earth Environ. Sci.* **2019**, *403*, 012127. [CrossRef]
22. Niva, M.; Vainio, A.; Jallinoja, P. Barriers to increasing plant protein consumption in Western populations. In *Vegetarian and Plant-Based Diets in Health and Disease Prevention*; Elsevier: Amsterdam, The Netherlands, 2017; pp. 157–171.
23. Springmann, M.; Clark, M.; Mason-D'Croz, D.; Wiebe, K.; Bodirsky, B.L.; Lassaletta, L.; De Vries, W.; Vermeulen, S.J.; Herrero, M.; Carlson, K.M. Options for keeping the food system within environmental limits. *Nature* **2018**, *562*, 519–525. [CrossRef]
24. Foley, J.A.; Ramankutty, N.; Brauman, K.A.; Cassidy, E.S.; Gerber, J.S.; Johnston, M.; Mueller, N.D.; O'Connell, C.; Ray, D.K.; West, P.C. Solutions for a cultivated planet. *Nature* **2011**, *478*, 337–342. [CrossRef]
25. Rockström, J.; Edenhofer, O.; Gaertner, J.; DeClerck, F. Planet-proofing the global food system. *Nat. Food* **2020**, *1*, 3–5. [CrossRef]
26. Willett, W.; Rockström, J.; Loken, B.; Springmann, M.; Lang, T.; Vermeulen, S.; Garnett, T.; Tilman, D.; DeClerck, F.; Wood, A. Food in the Anthropocene: The EAT–Lancet Commission on healthy diets from sustainable food systems. *Lancet* **2019**, *393*, 447–492. [CrossRef]
27. Pikaar, I.; Matassa, S.; Bodirsky, B.L.; Weindl, I.; Humpenöder, F.; Rabaey, K.; Boon, N.; Bruschi, M.; Yuan, Z.; van Zanten, H. Decoupling livestock from land use through industrial feed production pathways. *Environ. Sci. Technol.* **2018**, *52*, 7351–7359. [CrossRef]
28. Walsh, B.; Herrero, M.; Ciais, P.; Obersteiner, M.; van Vuuren, D.; Penuelas, J. New Feed Sources Key to Ambitious Climate Targets. *Carbon Balance Manag.* **2015**, *10*, 26. [CrossRef]
29. Bumpus, A.; Comello, S. Emerging clean energy technology investment trends. *Nat. Clim. Chang.* **2017**, *7*, 382–385. [CrossRef]
30. Froggatt, A.; Wellesley, L. *Meat Analogues: Considerations for the EU*; Chatham House Research Paper; Chatham House: London, UK, 2019.
31. Fears, R.; Canales, C. The Role of Science, Technology and Innovation for Transforming Food Systems Globally. *Cent. Dev. Res. (ZEF) Coop. Sci. Group UN Food Syst. Summit* **2021**, 1–20. [CrossRef]
32. Altshul, H.J.; McMillan, L.; Hall, A. The role of public research agencies in building agri-food bioscience impact and innovation capacity in sub-Saharan Africa: The challenge beyond science capability. *Int. J. Technol. Manag. Sustain. Dev.* **2019**, *18*, 105–125. [CrossRef]
33. Herrero, M.; Thornton, P.K.; Mason-D'Croz, D.; Palmer, J.; Bodirsky, B.L.; Pradhan, P.; Barrett, C.B.; Benton, T.G.; Hall, A.; Pikaar, I. Articulating the effect of food systems innovation on the Sustainable Development Goals. *Lancet Planet. Health* **2021**, *5*, e50–e62. [CrossRef]
34. Klerkx, L.; Begemann, S. Supporting food systems transformation: The what, why, who, where and how of mission-oriented agricultural innovation systems. *Agric. Syst.* **2020**, *184*, 102901. [CrossRef]
35. Gill, M.; Den Boer, A.; Kok, K.; Breda, J.; Cahill, J.; Callenius, C.; Caron, P.; Damianova, Z.; Gurinovic, M.; Lähteenmäki, L. *A Systems Approach to Research and Innovation for Food Systems Transformation*; FIT4FOOD2030: Amsterdam, The Netherlands, 2018.
36. Anderson, C.R.; Bruil, J.; Chappell, M.J.; Kiss, C.; Pimbert, M.P. From transition to domains of transformation: Getting to sustainable and just food systems through agroecology. *Sustainability* **2019**, *11*, 5272. [CrossRef]
37. Voytovych, N.; Smolynets, I.; Hirniak, K. The role of technology innovation in food systems transformation. *Calitatea* **2020**, *21*, 128–134.

38. Spendrup, S.; Fernqvist, F. Innovation in agri-food systems–a systematic mapping of the literature. *Int. J. Food Syst. Dyn.* **2019**, *10*, 402–427.
39. Raheem, D.; Shishaev, M.; Dikovitsky, V. Food system digitalization as a means to promote food and nutrition security in the barents region. *Agriculture* **2019**, *9*, 168. [CrossRef]
40. Deichmann, U.; Goyal, A.; Mishra, D. Will digital technologies transform agriculture in developing countries? *Agric. Econ.* **2016**, *47*, 21–33. [CrossRef]
41. Pingali, P.; Aiyar, A.; Abraham, M.; Rahman, A. *Transforming Food Systems for a Rising India*; Springer Nature: Berlin/Heidelberg, Germany, 2019.
42. Whitfield, K.R. Canning Foods and Selling Modernity: The Canned Food Industry and Consumer Culture, 1898–1945. Ph.D. Thesis, Louisiana State University, Baton Rouge, LA, USA, 2012.
43. Sodano, V. Innovation trajectories and sustainability in the food system. *Sustainability* **2019**, *11*, 1271. [CrossRef]
44. Movilla-Pateiro, L.; Mahou-Lago, X.; Doval, M.; Simal-Gandara, J. Toward a sustainable metric and indicators for the goal of sustainability in agricultural and food production. *Crit. Rev. Food Sci. Nutr.* **2021**, *61*, 1108–1129. [CrossRef]
45. Rasmussen, W.D. The impact of technological change on American agriculture, 1862–1962. *J. Econ. Hist.* **1962**, *22*, 578–591. [CrossRef]
46. Estes, R.J.; Sirgy, M.J. *The Pursuit of Human Well-Being: The Untold Global History*; Springer: Berlin/Heidleberg, Germany, 2017.
47. Tilman, D.; Balzer, C.; Hill, J.; Befort, B.L. Global food demand and the sustainable intensification of agriculture. *Proc. Natl. Acad. Sci. USA* **2011**, *108*, 20260–20264. [CrossRef]
48. Campbell, B.M.; Beare, D.J.; Bennett, E.M.; Hall-Spencer, J.M.; Ingram, J.S.; Jaramillo, F.; Ortiz, R.; Ramankutty, N.; Sayer, J.A.; Shindell, D. Agriculture production as a major driver of the Earth system exceeding planetary boundaries. *Ecol. Soc.* **2017**, *22*, 8. [CrossRef]
49. Swinburn, B.A.; Kraak, V.I.; Allender, S.; Atkins, V.J.; Baker, P.I.; Bogard, J.R.; Brinsden, H.; Calvillo, A.; De Schutter, O.; Devarajan, R. The global syndemic of obesity, undernutrition, and climate change: The Lancet Commission report. *Lancet* **2019**, *393*, 791–846. [CrossRef]
50. Herrero, M.; Thornton, P.K.; Mason-D'Croz, D. Innovation can accelerate the transition towards a sustainable food system. *Nat. Food* **2020**, *1*, 266–272. [CrossRef]
51. Barrett, C.B.; Benton, T.G.; Cooper, K.A.; Fanzo, J.; Gandhi, R.; Herrero, M.; James, S.; Kahn, M.; Mason-D'Croz, D.; Mathys, A. Bundling innovations to transform agri-food systems. *Nat. Sustain.* **2020**, *3*, 974–976. [CrossRef]
52. Graff, G.D.; Silva, F.d.F.; Zilberman, D. Venture Capital and the Transformation of Private R&D for Agriculture. In *Economics of Research and Innovation in Agriculture*; University of Chicago Press: Chicago, IL, USA, 2020.
53. Blay-Palmer, A. *Food Fears: From Industrial to Sustainable Food Systems*; Routledge: Abingdon-on-Thames, UK, 2016.
54. Alexandratos, N.; Bruinsma, J. *World Agriculture towards 2030/2050: The 2012 Revision*; FAO: Rome, Italy, 2012.
55. Canales, C.; Fears, R. The Role of Science, Technology, and Innovation for Transforming Food Systems in Europe. In Proceedings of the UN Food Systems Summit, New York, NY, USA, 23 September 2021.
56. Mankins, J.C. *Technology Readiness Levels*; A White Paper; NASA: Washington, DC, USA, 1995.
57. Hirshorn, S.; Jefferies, S. *Final Report of the NASA Technology Readiness Assessment (TRA) Study Team*; NASA: Washington, DC, USA, 2016.
58. El Bilali, H. Relation between innovation and sustainability in the agro-food system. *Ital. J. Food Sci.* **2018**, *30*. [CrossRef]
59. Martin, M.; Røyne, F.; Ekvall, T.; Moberg, Å. Life cycle sustainability evaluations of bio-based value chains: Reviewing the indicators from a Swedish perspective. *Sustainability* **2018**, *10*, 547. [CrossRef]
60. Boljanovic, J.D.; Dobrijevic, G.; Cerovic, S.; Alcakovic, S.; Djokovic, F. Knowledge-based bioeconomy: The use of intellectual capital in food industry of Serbia. *Amfiteatru Econ.* **2018**, *20*, 717–731. [CrossRef]
61. Pradhan, P.; Costa, L.; Rybski, D.; Lucht, W.; Kropp, J.P. A systematic study of sustainable development goal (SDG) interactions. *Earth's Future* **2017**, *5*, 1169–1179. [CrossRef]
62. Pradhan, P.; Kriewald, S.; Costa, L.; Rybski, D.; Benton, T.G.; Fischer, G.n.; Kropp, J.P. Urban food systems: How regionalization can contribute to climate change mitigation. *Environ. Sci. Technol.* **2020**, *54*, 10551–10560. [CrossRef]
63. Berry, E.M. Sustainable food systems and the Mediterranean diet. *Nutrients* **2019**, *11*, 2229. [CrossRef]
64. Noack, A.-L.; Pouw, N.R. A blind spot in food and nutrition security: Where culture and social change shape the local food plate. *Agric. Hum. Values* **2015**, *32*, 169–182. [CrossRef]
65. Jia, X. Agro-Food Innovation and Sustainability Transition: A Conceptual Synthesis. *Sustainability* **2021**, *13*, 6897. [CrossRef]
66. Wellesley, L.; Happer, C.; Froggatt, A. *Changing Climate, Changing Diets: Pathways to Lower Meat Consumption*; Chatham House Report; Chatham House: London, UK, 2015.
67. Nyborg, K.; Anderies, J.M.; Dannenberg, A.; Lindahl, T.; Schill, C.; Schlüter, M.; Adger, W.N.; Arrow, K.J.; Barrett, S.; Carpenter, S. Social norms as solutions. *Science* **2016**, *354*, 42–43. [CrossRef]
68. Greaker, M.; Midttømme, K. Network effects and environmental externalities: Do clean technologies suffer from excess inertia? *J. Public Econ.* **2016**, *143*, 27–38. [CrossRef]
69. Roman, S.; Sánchez-Siles, L.M.; Siegrist, M. The importance of food naturalness for consumers: Results of a systematic review. *Trends Food Sci. Technol.* **2017**, *67*, 44–57. [CrossRef]

70. Stilgoe, J.; Owen, R.; Macnaghten, P. Developing a framework for responsible innovation. *Res. Policy* **2013**, *42*, 1568–1580. [CrossRef]
71. Ihtisham, M.; Liu, S.; Shahid, M.O.; Khan, N.; Lv, B.; Sarraf, M.; Ali, S.; Chen, L.; Liu, Y.; Chen, Q. The Optimized N, P, and K Fertilization for Bermudagrass Integrated Turf Performance during the Establishment and Its Importance for the Sustainable Management of Urban Green Spaces. *Sustainability* **2020**, *12*, 10294. [CrossRef]
72. Mytelka, L.K. New trends in biotechnology networking. *Int. J. Biotechnol.* **1999**, *1*, 30–41. [CrossRef]
73. Regis, E. *Golden Rice: The Imperiled Birth of a GMO Superfood*; Johns Hopkins University Press: Baltimore, MD, USA, 2019.
74. Asheim, G.B.; Fæhn, T.; Nyborg, K.; Greaker, M.; Hagem, C.; Harstad, B.; Hoel, M.O.; Lund, D.; Rosendahl, K.E. The case for a supply-side climate treaty. *Science* **2019**, *365*, 325–327. [CrossRef]
75. Khayyam, M.; Chuanmin, S.; Haroon Qasim, M.I.; Anjum, R.; Jiaxin, L.; Tikhomirova, A.; Khan, N. Food Consumption Behavior of Pakistani Students Living in China: The Role of Food Safety and Health Consciousness in the Wake of Coronavirus Disease 2019 Pandemic. *Front. Psychol.* **2021**, *12*, 1–16. [CrossRef]
76. Russo, A.; Harstad, B.G.; Lancia, F. Compliance Technology and Self-enforcing Agreements. *J. Eur. Econ. Assoc.* **2019**, *17*, 1–29.
77. Lunn, C.E. The role of green economics in achieving realistic policies and programmes for sustainability. *Int. J. Green Econ.* **2006**, *1*, 37–49. [CrossRef]
78. Bliemel, M.; Flores, R.; De Klerk, S.; Miles, M.P. Accelerators as start-up infrastructure for entrepreneurial clusters. *Entrep. Reg. Dev.* **2019**, *31*, 133–149. [CrossRef]
79. Bryant, C.; Barnett, J. Consumer acceptance of cultured meat: A systematic review. *Meat Sci.* **2018**, *143*, 8–17. [CrossRef]
80. Vanhonacker, F.; Van Loo, E.J.; Gellynck, X.; Verbeke, W. Flemish consumer attitudes towards more sustainable food choices. *Appetite* **2013**, *62*, 7–16. [CrossRef]
81. Preston, F.; Lehne, J. *A Wider Circle? The Circular Economy in Developing Countries*; Chatham House for the Royal Institute of International Affairs: London, UK, 2017.
82. Khan, N.; Ray, R.L.; Kassem, H.S.; Ihtisham, M.; Abdullah; Asongu, S.A.; Ansah, S.; Shemei, Z. Toward Cleaner Production: Can Mobile Phone Technology Help Reduce Inorganic Fertilizer Application? Evidence Using a National Level Dataset. *Land* **2021**, *10*, 1023. [CrossRef]
83. Mazzucato, M. From market fixing to market-creating: A new framework for innovation policy. *Ind. Innov.* **2016**, *23*, 140–156. [CrossRef]
84. Havlík, P.; Valin, H.; Herrero, M.; Obersteiner, M.; Schmid, E.; Rufino, M.C.; Mosnier, A.; Thornton, P.K.; Böttcher, H.; Conant, R.T. Climate change mitigation through livestock system transitions. *Proc. Natl. Acad. Sci. USA* **2014**, *111*, 3709–3714. [CrossRef]
85. Klerkx, L.; Rose, D. Dealing with the game-changing technologies of Agriculture 4.0: How do we manage diversity and responsibility in food system transition pathways? *Glob. Food Secur.* **2020**, *24*, 100347. [CrossRef]
86. Bahn, R.A.; Yehya, A.A.K.; Zurayk, R. Digitalization for Sustainable Agri-Food Systems: Potential, Status, and Risks for the MENA Region. *Sustainability* **2021**, *13*, 3223. [CrossRef]
87. Velten, S.; Leventon, J.; Jager, N.; Newig, J. What Is Sustainable Agriculture? A Systematic Review. *Sustainability* **2015**, *7*, 7833–7865.
88. Béné, C.; Oosterveer, P.; Lamotte, L.; Brouwer, I.D.; de Haan, S.; Prager, S.D.; Talsma, E.F.; Khoury, C.K. When food systems meet sustainability–Current narratives and implications for actions. *World Dev.* **2019**, *113*, 116–130. [CrossRef]
89. Group, W.B. *Future of Food: Harnessing Digital Technologies to Improve Food System Outcomes*; World Bank: Washington, DC, USA, 2019.
90. Trendov, M.; Varas, S.; Zeng, M. *Digital Technologies in Agriculture and Rural Areas: Status Report*; FAO: Rome, Italy, 2019.
91. *COVID-19 and Its Impact on Food Security in the Near East and North Africa: How to Respond*; FAO: Cairo, Egypt, 2020.
92. Mittra, B. *COVID-19 Pandemic Presents Opportunities for Innovation*; TCI Blog—The Tata-Cornell Institute for Agriculture and Nutrition: Ithaca, NY, USA, 2020, 2021.
93. Anib, A.; Gayathri, A.; Shabu, K. Consumer Perception towards Swiggy Digital Food Application Service: A Analytical Study with Special Reference to Ernakulam City. *Int. J. Innov. Technol. Explor. Eng. (Ijitee)* **2019**, *8*, 1–7.
94. FAO, F. *The Future of Food and Agriculture–Trends and Challenges*; Annual Report; FAO: Rome, Italy, 2017.
95. Food and Agriculture Organisation. *Realising the Potential of Digitalisation to Improve the Agri-Food System*; FAO: Rome, Italy, 2020.
96. Prause, L.; Hackfort, S.; Lindgren, M. Digitalization and the third food regime. *Agric. Hum. Values* **2021**, *38*, 641–655. [CrossRef] [PubMed]
97. Samoggia, A.; Monticone, F.; Bertazzoli, A. Innovative Digital Technologies for Purchasing and Consumption in Urban and Regional Agro-Food Systems: A Systematic Review. *Foods* **2021**, *2*, 208. [CrossRef]
98. United Nations System Standing Committee on Nutrition. *Nutrition in a Digital World*; UNSCN: Rome, Italy, 2020.
99. European Commission. A Smart and Sustainable Digital Future for European Agriculture and Rural Areas. 2019. Available online: https://ec.europa.eu/digital-single-market/en/news/eu-member-states-join-forces-digitalisation-european-agricultureand-rural-areas (accessed on 20 September 2020).
100. European Commission. The European Green Deal. 2019. Available online: https://eur-lex.europa.eu/legal-content/EN/TXT/?qid=1588580774040&uri=CELEX:52019DC0640 (accessed on 20 September 2020).
101. United Nations. New Urban Agenda. Draft Resolution Submitted by the President of the General Assembly. 2016. Available online: http://habitat3.org/wp-content/uploads/N1639668-English.pdf (accessed on 24 May 2020).

agriculture

MDPI

Case Report

Leisure Agricultural Park Selection for Traveler Groups Amid the COVID-19 Pandemic

Hsin-Chieh Wu [1], Yu-Cheng Lin [2] and Tin-Chih Toly Chen [3,*]

[1] Department of Industrial Engineering and Management, Chaoyang University of Technology, Taichung City 413310, Taiwan; hcwul@cyut.edu.tw
[2] Department of Computer-Aided Industrial Design, Overseas Chinese University, Taichung City 40721, Taiwan; yclin@ocu.edu.tw
[3] Department of Industrial Engineering and Management, National Yang Ming Chiao Tung University, Hsinchu City 30010, Taiwan
* Correspondence: tcchen@nycu.edu.tw

Abstract: With the widespread vaccination against COVID-19, people began to resume regional tourism. Outdoor attractions, such as leisure agricultural parks, are particularly attractive because they are well ventilated and can prevent the spread of COVID-19. However, during the COVID-19 pandemic, the considerations around choosing a leisure agricultural park are different from usual, and will be affected by uncertainty. Therefore, this research proposes a fuzzy collaborative intelligence (FCI) approach to help select leisure agricultural parks suitable for traveler groups during the COVID-19 pandemic. The proposed FCI approach combines asymmetrically calibrated fuzzy geometric mean (acFGM), fuzzy weighted intersection (FWI), and fuzzy Vise Kriterijumska Optimizacija I Kompromisno Resenje (fuzzy VIKOR), which is a novel attempt in this field. The effectiveness of the proposed FCI approach has been verified by a case study in Taichung City, Taiwan. The results of the case study showed that during the COVID-19 pandemic, travelers (especially traveler groups) were very willing to go to leisure agricultural parks. In addition, the most important criterion for choosing a suitable leisure agricultural park was the ease of maintaining social distance, while the least important criterion was the distance from a leisure agricultural park. Further, the successful recommendation rate using the proposed methodology was as high as 90%.

Keywords: leisure agricultural park; traveler group; COVID-19 pandemic; fuzzy collaborative intelligence

Citation: Wu, H.-C.; Lin, Y.-C.; Chen, T.-C.T. Leisure Agricultural Park Selection for Traveler Groups Amid the COVID-19 Pandemic. *Agriculture* **2022**, *12*, 111. https://doi.org/10.3390/agriculture12010111

Academic Editor: Dimitre Dimitrov

Received: 20 December 2021
Accepted: 12 January 2022
Published: 13 January 2022

1. Introduction

In the late stage of the COVID-19 pandemic, with the popularity of vaccination, domestic tourism gradually recovered. When choosing tourist attractions, some regulations on the prevention of the COVID-19 pandemic are still influential [1]; for example, many tourists tend to choose attractions that are health-oriented, have a relatively mild pandemic, make it easy to maintain social distance, and are well ventilated (to avoid wearing masks for a long time) [2]. These considerations are different to those before the COVID-19 pandemic, when consumers valued the quality of agricultural products [3], the convenience of on-site consumption, the distance from downtown [4], the availability of auxiliary facilities, etc. [5–7]. In addition, many current considerations are quite uncertain [8–10]; for example, whether it is easy to maintain social distance is affected by the number of people entering the park, and it also depends on the control measures of the leisure agricultural park. Furthermore, if all visitors in the park wear masks, good ventilation is not as important. As a result, selecting or recommending a suitable leisure agricultural park has become a challenging task. The motivation of this research is to accomplish this task.

Some relevant references are henceforth reviewed. According to Pan et al. [11], the COVID-19 pandemic has impacted eight major aspects of the agricultural economy, one of which is leisure agriculture. In the view of Ateljevic [12], the normalization of the COVID-19 pandemic on the tourism system may have an impact on regenerative agriculture and transitional tourism. According to statistics supplied by Putian [13], the impact of the COVID-19 pandemic on leisure agriculture includes a substantial reduction in revenues, a reduction in advertising budgets, difficulties in resuming production, unsmooth sales channels, and declining consumer demand. The results of the study by Barrot et al. [14] showed that leisure and agriculture were industries in which employment rates have fallen sharply due to the COVID-19 pandemic. Therefore, Li et al. [4] believed that under the premise that the COVID-19 pandemic can be prevented and controlled, it is the responsibility of the local government to take effective and efficient measures to restore leisure agriculture. Further, in the view of Hsiao and Tuan [10], the dynamic ability of leisure agriculture park operators to change park marketing channels and develop new products or services to respond to the new market can effectively respond to the COVID-19 pandemic. The COVID-19 pandemic has also caused labor shortages in many agricultural activities. This led to a sharp rise in short-term labor wages and increased the burden on farm operators [15]. Leisure agriculture, which allows customers to pick agricultural products by themselves, is a solution. Some studies also concluded that the COVID-19 pandemic has caused suburban agriculture to be replaced by leisure agriculture and other land applications with higher market value [16]. All in all, most past studies have shown that leisure agricultural park operators or employees are facing challenges, but these studies have not explored the difficulties faced by travelers who plan to visit these parks. This study can make up for this deficiency.

To recommend suitable leisure agricultural parks for traveler groups amid the COVID-19 pandemic, a fuzzy collaborative intelligence (FCI) approach is proposed in this study. The reason for adopting a fuzzy approach is to consider the uncertainty brought about by the COVID-19 pandemic. The reason for discussing traveler groups (or group tours) is because the customers of a leisure agricultural park are mostly families. Different family members may have different considerations in choosing a suitable leisure agricultural park, affecting the formation of a consensus among them. FCI is a viable means to solve this problem.

The contribution of this study includes the following:

(1) After the outbreak of the COVID-19 pandemic, the factors affecting travelers visiting a leisure agricultural park have been different to the previous factors. In addition to subjective personal preferences, there is also objective information related to the COVID-19 pandemic. This study is one of the first studies to explore the influence of these factors on travelers' decisions in choosing suitable leisure agricultural parks.
(2) The acFGM method is proposed to enhance the precision of deriving the priorities of factors critical to the selection of a suitable leisure agricultural park.

The remainder of this paper is organized as follows: Section 2 is an introduction of the FCI approach proposed in this study; Section 3 details the application of the FCI approach to a case study in Taichung City, Taiwan, amid the COVID-19 pandemic; Section 4 provides the conclusions of this study, as well as some possible topics for future investigation.

2. Methodology

Without loss of generality, all fuzzy parameters and variables in the proposed methodology are given in or approximated by triangular fuzzy numbers (TFNs). At first, the asymmetrically calibrated fuzzy geometric mean (acFGM) method is proposed for decision makers to derive the fuzzy priorities of criteria that affect their choices.

2.1. *acFGM for Deriving the Fuzzy Priorities of Criteria*

In the beginning, every decision maker is asked to make pairwise comparisons of the relative priorities of criteria. The results by decision maker k are inserted into the following fuzzy judgment matrix: $\widetilde{\mathbf{A}}(k) = \{\widetilde{a}_{ij}(k)\}$ where $\widetilde{a}_{ij}(k)$ are fuzzy sets; $i, j = [1, n]$; $k = 1 \sim K$. The following is true according to Satty [17]:

$$\det(\widetilde{\mathbf{A}}(k)(-)\widetilde{\lambda}(k)\mathbf{I}) = 0 \tag{1}$$

$$(\widetilde{\mathbf{A}}(k)(-)\widetilde{\lambda}(k))(\times)\widetilde{\mathbf{x}}(k) = 0 \tag{2}$$

where $\widetilde{\lambda}(k)$ and $\widetilde{\mathbf{x}}(k)$ are the fuzzy eigenvalue and fuzzy eigenvector of $\widetilde{\mathbf{A}}(k)$, respectively; $(-)$ and $(\times)$ denote fuzzy subtraction and multiplication, respectively. Equations (1) and (2) involve fuzzy multiplication operations, making it difficult to derive the exact values of $\widetilde{\lambda}(k)$ and $\widetilde{\mathbf{x}}(k)$. Fuzzy geometric mean (FGM) is a prevalent method to approximate the solution [18]. However, the accuracy of deriving the fuzzy priorities of criteria using FGM is not always high. To solve this problem, Chen and Wang [19] proposed the calibrated FGM (cFGM) method to improve the accuracy in an efficient manner. The cFGM method has the following steps:

Step 1. Approximate the value of the fuzzy priority of criterion i using FGM as in the following [18]:

$$\widetilde{w}_i(k) \cong (w_{i1}(k),\ w_{i2}(k),\ w_{i3}(k)) \tag{3}$$

where the following applies:

$$w_{i1}(k) = \frac{1}{1 + \sum\limits_{m \neq i} \frac{\sqrt[n]{\prod\limits_{j=1}^{n} a_{mj3}(k)}}{\sqrt[n]{\prod\limits_{j=1}^{n} a_{ij1}(k)}}} \tag{4}$$

$$w_{i2}(k) \cong \frac{1}{1 + \sum\limits_{m \neq i} \frac{\sqrt[n]{\prod\limits_{j=1}^{n} a_{mj2}(k)}}{\sqrt[n]{\prod\limits_{j=1}^{n} a_{ij2}(k)}}} \tag{5}$$

$$w_{i3}(k) \cong \frac{1}{1 + \sum\limits_{m \neq i} \frac{\sqrt[n]{\prod\limits_{j=1}^{n} a_{mj1}(k)}}{\sqrt[n]{\prod\limits_{j=1}^{n} a_{ij3}(k)}}} \tag{6}$$

$\widetilde{w}_i(k)$ is the fuzzy priority of criterion i to decision maker k.

Step 2. Derive the priority of criterion i from the crisp judgment matrix $\mathbf{A}^c(k) = [a_{ij2}(k)]$ using an eigen analysis, as in the following [17]:

$$\det(\mathbf{A}^c(k) - \lambda^c(k)\mathbf{I}) = 0 \tag{7}$$

$$(\mathbf{A}^c(k) - \lambda^c(k)\mathbf{I})\mathbf{x}^c(k) = 0 \tag{8}$$

$$w_i^c(k) = \frac{x_i^c}{\sum\limits_{j=1}^{n} x_j^c} \tag{9}$$

The derived priority is indicated with $w_i^c(k)$.

Step 3. Calibrate the fuzzy priority of criterion i in the following way:

$$w_{i1}(k) \rightarrow w_{i1}(k) + w_i^c(k) - w_{i2}(k) \tag{10}$$

$$w_{i2}(k) \rightarrow w_i^c(k) \tag{11}$$

$$w_{i3}(k) \rightarrow w_{i3}(k) + w_i^c(k) - w_{i2}(k) \tag{12}$$

However, the cFGM method has the following problems:

(1) After calibration, $w_{i1}(k)$ may be negative, which is infeasible.
(2) The range of a fuzzy priority approximated using FGM is usually wider than that of the actual value, which is not considered in the calibration process.

To solve these problems, the acFGM method is proposed, as in the following:

$$w_{i1}(k) \rightarrow \max(w_{i1}(k) + w_i^c(k) - w_{i2}(k),\ w_{i1}(k) \cdot \frac{w_i^c(k)}{w_{i2}(k)}) \tag{13}$$

$$w_{i3}(k) \rightarrow \min(w_{i3}(k) + w_i^c(k) - w_{i2}(k),\ w_{i3}(k) \cdot \frac{w_i^c(k)}{w_{i2}(k)}) \tag{14}$$

The results using various methods are compared in Figure 1.

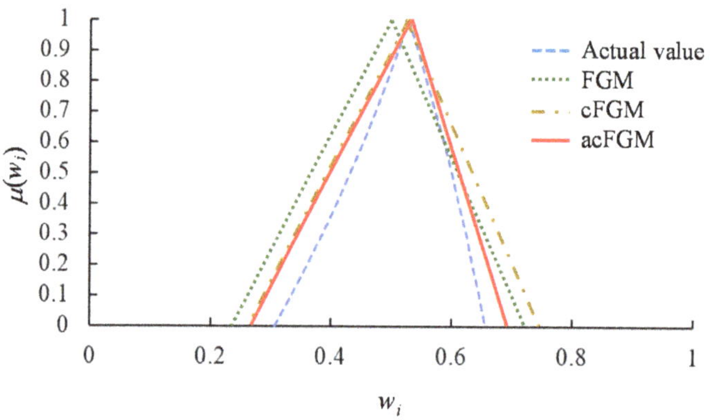

Figure 1. Comparison of the results using various methods.

The fuzzy priorities of a criterion derived by different decision makers are not the same, and need to be aggregated. In addition, some decision makers are more authoritative than others. To consider these, fuzzy weighted intersection (FWI) [20] is applied, as described in the next section.

2.2. FWI for Aggregating the Fuzzy Priorities of Criteria Derived by All Decision Makers

Most of the existing methods aggregate the fuzzy priorities of a criterion derived by all decision makers using fuzzy arithmetic average operators [21–23] and fuzzy geometric mean operators [21–23]. The only difference is the type of fuzzy numbers. However, the aggregation result may be unreasonable [20]. In particular, the aggregation result may be a value with low membership in the fuzzy priority of each decision maker.

In the proposed methodology, FWI [20] is applied to aggregate the fuzzy priorities of a criterion derived by all decision makers, as in the following:

$$\tilde{w}_i(all) = \widetilde{FWI}(\{\tilde{w}_i(k)\}) \tag{15}$$

with the following membership function:

$$\mu_{\tilde{w}_i(all)}(x) = \min_k \mu_{\tilde{w}_i(k)}(x) + \sum_k (\omega_k - \min_l \omega_l)(\mu_{\tilde{w}_i(k)}(x) - \min_l \mu_{\tilde{w}_i(l)}(x)) \tag{16}$$

where ω_k is the authority level of decision maker k; $\sum_{k=1}^{K} \omega_k = 1$. An example is provided in Figure 2, in which the authority levels of the three decision makers are 0.35, 0.15, and 0.5, respectively. The aggregation result is not an empty set, despite the fact that decision makers may lack an overall consensus, as illustrated by Figure 3.

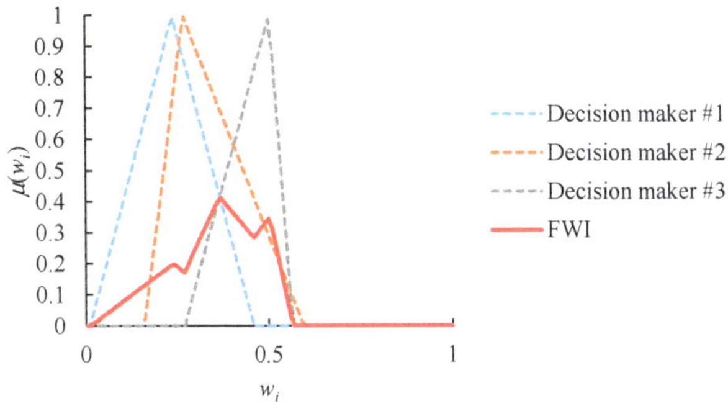

Figure 2. An FWI example.

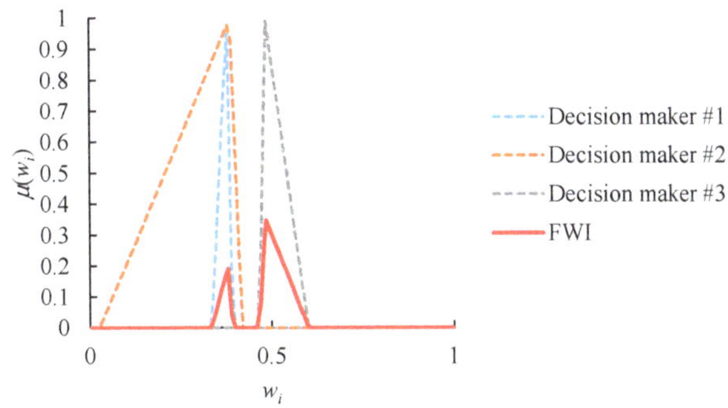

Figure 3. The aggregation result when decision makers lack an overall consensus.

The FWI operator meets the following conditions [20]:

(1) $\widetilde{FWI}(\{\widetilde{w}_i(k)\}) = \widetilde{w}_i(l)$ if $\omega_l = 1$ and $\omega_k = 0\ \forall\ k \neq l$

(2) $\widetilde{FWI}(\{\widetilde{w}_i(k)\}) = \widetilde{FI}(\{\widetilde{w}_i(k)\})$ if $\omega_k = \frac{1}{K}\ \forall\ k$; $\widetilde{FI}$ is the fuzzy intersection operator (i.e., the t-norm).

(3) $\min_l \mu_{\widetilde{w}_i(l)}(x) \leq \mu_{\widetilde{FWI}(\{\widetilde{w}_i(k)\})}(x) \leq \max_l \mu_{\widetilde{w}_i(l)}(x)$

(4) $\frac{\partial \mu_{\widetilde{FWI}(\{\widetilde{w}_i(k)\})}(x)}{\partial \mu_{\widetilde{w}_i(l)}(x)} \propto \omega_l$

Aggregating the fuzzy priorities of a criterion derived by all decision makers using FWI guarantees that values considered highly possible by all decision makers or just the most authoritative decision maker will have high memberships in the aggregation result. In other words, the aggregation result will be more in line with the expectations of all decision makers, and it will be easier for everyone to accept.

One problem with the FWI operator is the polygonal shape of the aggregation result, which increases the computational complexity of subsequent operations. To overcome this difficulty, Wu et al. [24] advised that the aggregation result should be approximated with a TFN, such that their defuzzification results using the center-of-gravity (COG) method [25] are equal, as in the following:

$$\begin{aligned} \widetilde{FWI}(\{\widetilde{w}_i(k)\} \cong\ & (\min(\widetilde{FWI}(\{\widetilde{w}_i(k)\}), \\ & 3COG(\widetilde{FWI}(\{\widetilde{w}_i(k)\}) - \max(\widetilde{FWI}(\{\widetilde{w}_i(k)\}) - \min(\widetilde{FWI}(\{\widetilde{w}_i(k)\}), \\ & \max(\widetilde{FWI}(\{\widetilde{w}_i(k)\}) \end{aligned} \quad (17)$$

where the following applies:

$$COG(\widetilde{FWI}(\{\widetilde{w}_i(k)\})) = \frac{\int_{all\ x} x\mu_{\widetilde{FWI}(\{\widetilde{w}_i(k)\})}(x)dx}{\int_{all\ x} \mu_{\widetilde{FWI}(\{\widetilde{w}_i(k)\})}(x)dx} \tag{18}$$

An example is shown in Figure 4.

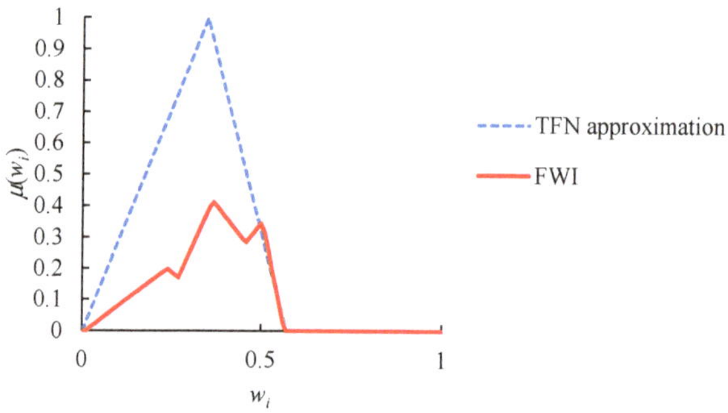

Figure 4. Approximating the aggregation result with a TFN.

2.3. *Fuzzy VIKOR for Evaluating Alternatives*

Subsequently, the fuzzy Vise Kriterijumska Optimizacija I Kompromisno Resenje (fuzzy VIKOR) method [26,27] is applied to evaluate the overall performance of each alternative. The fuzzy VIKOR method comprises the following steps:

Step 1. Determine the best and worst values of each criterion, as in the following:

$$\begin{aligned} \widetilde{p}_i^* &= \max_h \widetilde{p}_{hi} \\ &= (\max_h p_{hi1}, \max_h p_{hi2}, \max_h p_{hi3}) \end{aligned} \tag{19}$$

$$\begin{aligned} \widetilde{p}_i^- &= \min_h \widetilde{p}_{hi} \\ &= (\min_h p_{hi1}, \min_h p_{hi2}, \min_h p_{hi3}) \end{aligned} \tag{20}$$

where $\widetilde{p}_{hi}$ is the performance of alternative h in optimizing criterion i.; $h = 1 \sim H$. $\widetilde{p}_i^*$ and $\widetilde{p}_i^-$ indicate the best and worst performances in optimizing criteria i, respectively.

Step 2. Compute normalized fuzzy distances, as in the following:

$$\begin{aligned} \widetilde{d}_{hi} &= \frac{\widetilde{p}_i^*(-)\widetilde{p}_{hi}}{p_{i3}^* - p_{i1}^-} \\ &= (\frac{p_{i1}^* - p_{hi3}}{p_{i3}^* - p_{i1}^-}, \frac{p_{i2}^* - p_{hi2}}{p_{i3}^* - p_{i1}^-}, \frac{p_{i3}^* - p_{hi1}}{p_{i3}^* - p_{i1}^-}) \end{aligned} \tag{21}$$

Step 3. Compute the values of $\widetilde{S}_h$ and $\widetilde{R}_h$ [28], as in the following:

$$\widetilde{S}_h = \sum_{i=1}^{n} (\widetilde{w}_i(all)(\times)\widetilde{d}_{hi}) \tag{22}$$

$$\widetilde{R}_h = \max_i(\widetilde{w}_i(all)(\times)\widetilde{d}_{hi}) \tag{23}$$

$\widetilde{S}_h$ considers the performances of alternative h in optimizing all criteria, while $\widetilde{R}_h$ highlights the performance of the alternative in optimizing the most important criterion or the worse performance.

Step 4. Compute the value of $\widetilde{Q}_h$ [28], as in the following:

$$\widetilde{Q}_h = \xi \cdot \frac{\widetilde{S}_h(-)\min_r \widetilde{S}_r}{\max(\max_r \widetilde{S}_r) - \min(\min_r \widetilde{S}_r)}(+)(1-\xi) \cdot \frac{\widetilde{R}_h(-)\min_r \widetilde{R}_r}{\max(\max_r \widetilde{R}_r) - \min(\min_r \widetilde{R}_r)} \tag{24}$$

where $\xi \in [0, 1]$.

Step 5. Defuzzify $\widetilde{S}_h$, $\widetilde{R}_h$, $\widetilde{Q}_h$ using the COG method, as in the following:

$$COG(\widetilde{S}_h) = \frac{\int_{all\ x} x\mu_{\widetilde{S}_h}(x)dx}{\int_{all\ x} \mu_{\widetilde{S}_h}(x)dx} \tag{25}$$

$$COG(\widetilde{R}_h) = \frac{\int_{all\ x} x\mu_{\widetilde{R}_h}(x)dx}{\int_{all\ x} \mu_{\widetilde{R}_h}(x)dx} \tag{26}$$

$$COG(\widetilde{Q}_h) = \frac{\int_{all\ x} x\mu_{\widetilde{Q}_h}(x)dx}{\int_{all\ x} \mu_{\widetilde{Q}_h}(x)dx} \tag{27}$$

Step 6. Rank alternatives according to their $D(\widetilde{S}_h)$, $D(\widetilde{R}_h)$, and $D(\widetilde{Q}_h)$ values from the smallest to the largest. The decision maker will have three ranking results, giving him/her a high degree of flexibility, which is an advantage of fuzzy VIKOR over fuzzy technique for order preference by similarity to ideal solution (FTOPSIS) [29,30]; for example, when $D(\widetilde{Q}_h)$ is considered, the top two alternatives are indicated with alternatives $h_{(1)}$ and $h_{(2)}$, respectively. Then, in the view of Opricovic [28], alternative $h_{(1)}$ can be recommended to the decision maker if the following two conditions are met:

$$D(\widetilde{Q}_{h_{(2)}}) - D(\widetilde{Q}_{h_{(1)}}) \geq \frac{1}{H-1} \tag{28}$$

$$D(\widetilde{S}_{h_{(1)}}) = \min_r D(\widetilde{S}_r) \text{ or } D(\widetilde{R}_{h_{(1)}}) = \min_r D(\widetilde{R}_r) \tag{29}$$

3. Case Study

3.1. Background

To validate the effectiveness of the proposed methodology, a standalone leisure agricultural park recommendation system has been developed using Microsoft Access 2019 on a PC with an i7-7700 CPU 272 3.6 GHz and 16 GB RAM, and installed in a travel agency in Taichung City, Taiwan. The following five criteria were considered in building the recommendation mechanism encoded using VBA: the image of the leisure agricultural park, the number of confirmed COVID-19 cases in the city, the easiness of maintaining social distance, the distance to the leisure agricultural park, and the preference for agricultural products or natural facilities in the leisure agricultural park. During August 2021, a total of 10 traveler groups used this system to seek recommendations for suitable leisure agricultural parks. In this case study, the first traveler group is taken as an example to illustrate the application of the proposed methodology.

3.2. Application of the Proposed Methodology

The first few steps of the proposed methodology are a fuzzy analytic hierarchy process (FAHP). FAHP is the incorporation of fuzzy logic into an analytic hierarchy process (AHP), which is a well-known multi-criteria decision-making method based on the pairwise comparison of criteria [17]. The prevalent methods for solving an FAHP problem include FGM [18], fuzzy extent analysis (FEA) [31], and alpha-cut operations (ACO) [32,33].

Fuzzy AHP methods have been widely applied to multi-criteria decision making in agriculture [34–37].

The first traveler group was a family composed of the following three members (i.e., decision makers): father, mother, and daughter. Each decision maker compared the priorities of criteria in pairs. The results are summarized by the following fuzzy judgment matrixes:

$$\tilde{\mathbf{A}}(1) = \begin{bmatrix} 1 & (2,4,6) & 1/(1,1,3) & (2,4,6) & (2,4,6) \\ 1/(2,4,6) & 1 & 1/(3,5,7) & (1,1,3) & 1/(3,5,7) \\ (1,1,3) & (3,5,7) & 1 & (3,5,7) & (2,4,6) \\ 1/(2,4,6) & 1/(1,1,3) & 1/(3,5,7) & 1 & 1/(2,4,6) \\ 1/(2,4,6) & (3,5,7) & 1/(2,4,6) & (2,4,6) & 1 \end{bmatrix}$$

$$\tilde{\mathbf{A}}(2) = \begin{bmatrix} 1 & (3,5,7) & (1,3,5) & (3,5,7) & (1,3,5) \\ 1/(3,5,7) & 1 & 1/(2,4,6) & (2,4,6) & 1/(3,5,7) \\ 1/(1,3,5) & (2,4,6) & 1 & (2,4,6) & (1,3,5) \\ 1/(3,5,7) & 1/(2,4,6) & 1/(2,4,6) & 1 & 1/(3,5,7) \\ 1/(1,3,5) & (3,5,7) & 1/(1,3,5) & (3,5,7) & 1 \end{bmatrix}$$

$$\tilde{\mathbf{A}}(3) = \begin{bmatrix} 1 & (2,4,6) & 1/(1,3,5) & (3,5,7) & (1,3,5) \\ 1/(2,4,6) & 1 & 1/(2,4,6) & (1,3,5) & 1/(1,3,5) \\ (1,3,5) & (2,4,6) & 1 & (2,4,6) & (1,3,5) \\ 1/(3,5,7) & 1/(1,3,5) & 1/(2,4,6) & 1 & 1/(2,4,6) \\ 1/(1,3,5) & (1,3,5) & 1/(1,3,5) & (2,4,6) & 1 \end{bmatrix}$$

The fuzzy priorities of criteria were derived from the fuzzy judgment matrixes using the acFGM method. The results are summarized in Table 1. The fuzzy priorities of criteria are compared in Figure 5. The fuzzy consistency ratios of these fuzzy judgment matrixes were all less than 0.1, showing that they were consistent.

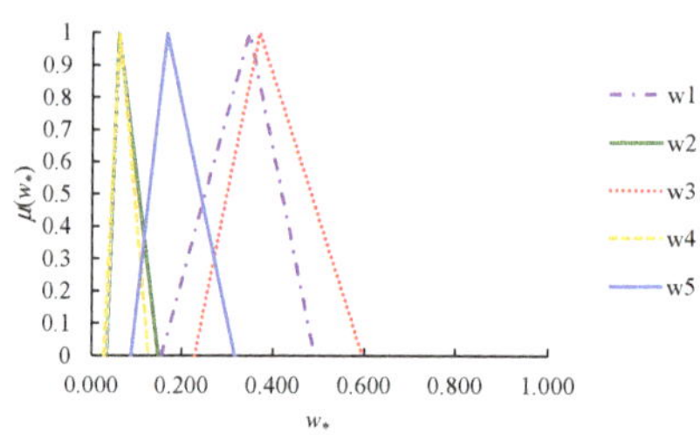

(**a**) Decision maker #1

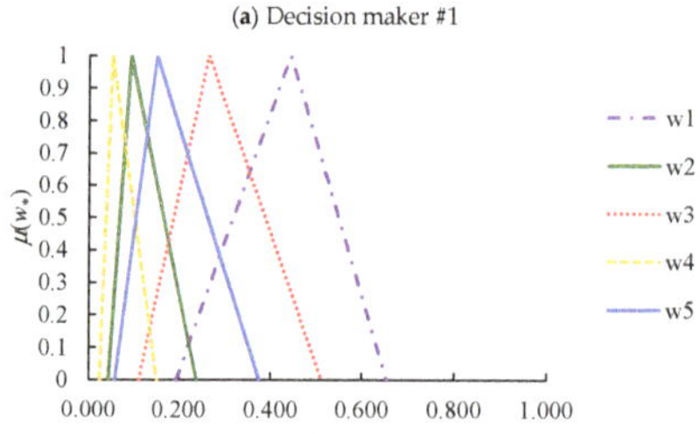

(**b**) Decision maker #2

Figure 5. *Cont.*

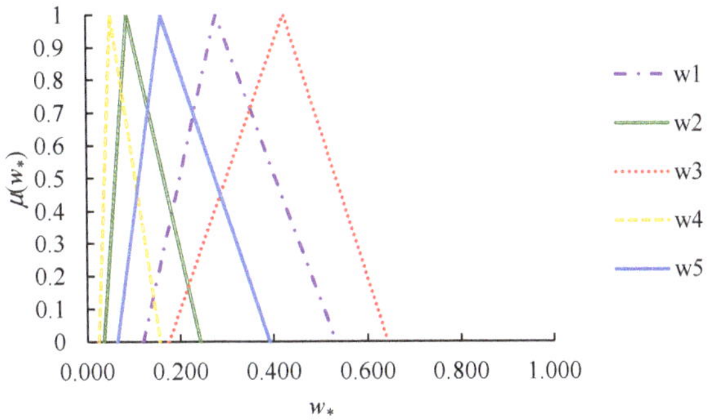

(**c**) Decision maker #3

Figure 5. Comparison of the fuzzy priorities of criteria.

Table 1. Fuzzy priorities of criteria derived by all decision makers.

i	$\widetilde{w}_i(1)$	$\widetilde{w}_i(2)$	$\widetilde{w}_i(3)$
1	(0.155, 0.347, 0.491)	(0.195, 0.442, 0.655)	(0.12, 0.278, 0.532)
2	(0.034, 0.059, 0.147)	(0.044, 0.092, 0.238)	(0.036, 0.087, 0.243)
3	(0.229, 0.369, 0.596)	(0.112, 0.264, 0.511)	(0.175, 0.423, 0.643)
4	(0.027, 0.06, 0.127)	(0.024, 0.051, 0.151)	(0.024, 0.052, 0.156)
5	(0.087, 0.165, 0.316)	(0.057, 0.15, 0.375)	(0.065, 0.159, 0.391)

Subsequently, FWI [38–42] was applied to aggregate the fuzzy priorities derived by all decision makers. The authority levels of the decision makers were subjectively determined by them jointly as 0.5 (father), 0.2 (mother), and 0.3 (daughter), respectively. The aggregation results are shown in Figure 6.

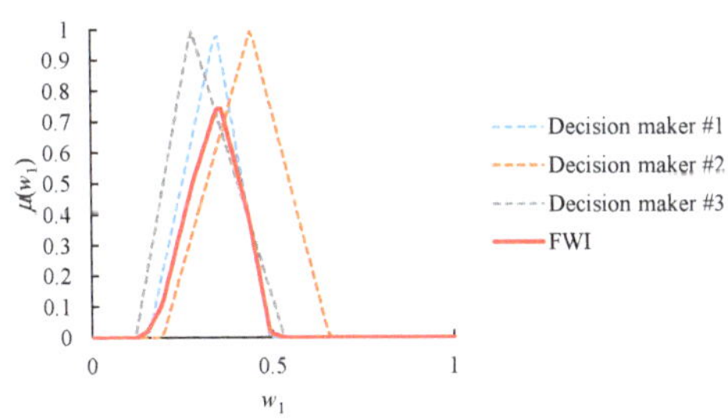

(**a**) The aggregation result of $\widetilde{w}_1$

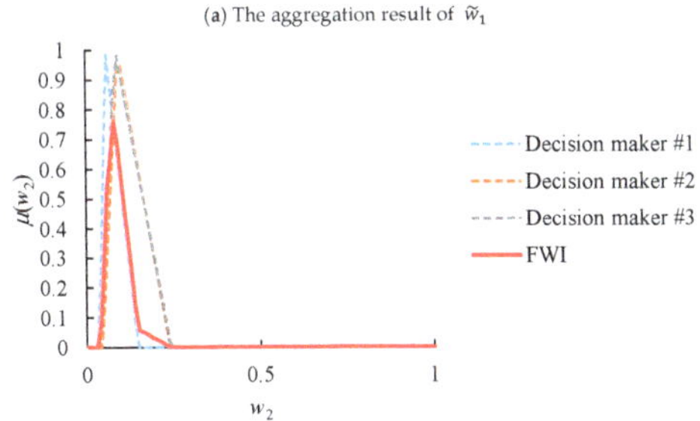

(**b**) The aggregation result of $\widetilde{w}_2$

Figure 6. *Cont.*

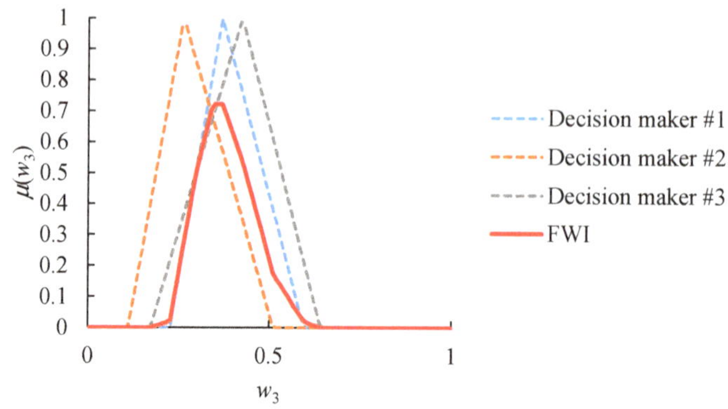

(**c**) The aggregation result of $\widetilde{w}_3$

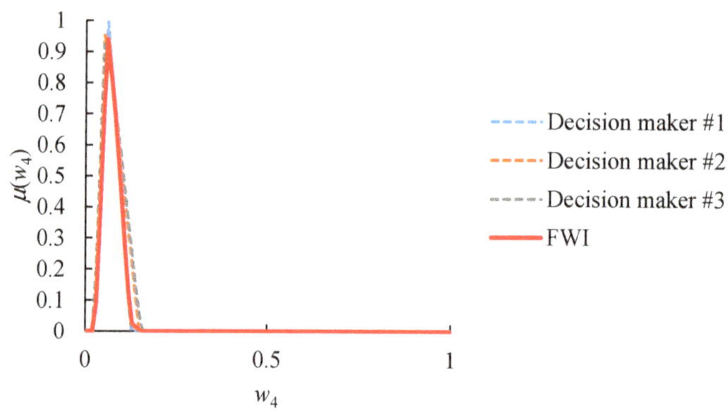

(**d**) The aggregation result of $\widetilde{w}_4$

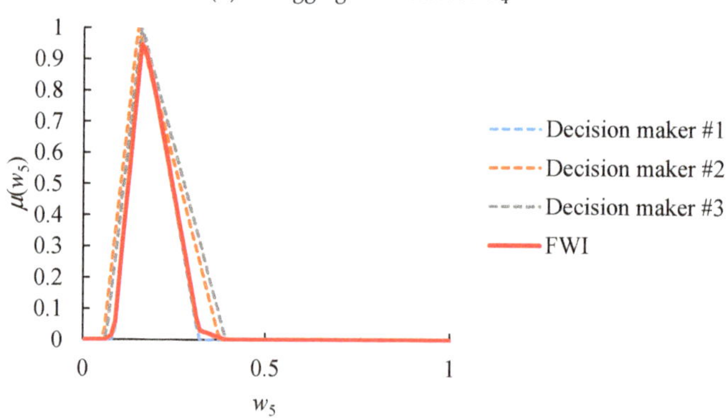

(**e**) The aggregation result of $\widetilde{w}_5$

Figure 6. Aggregation results.

To facilitate subsequent operations, the polygonal aggregation result was approximated with a TFN, as shown in Figure 7.

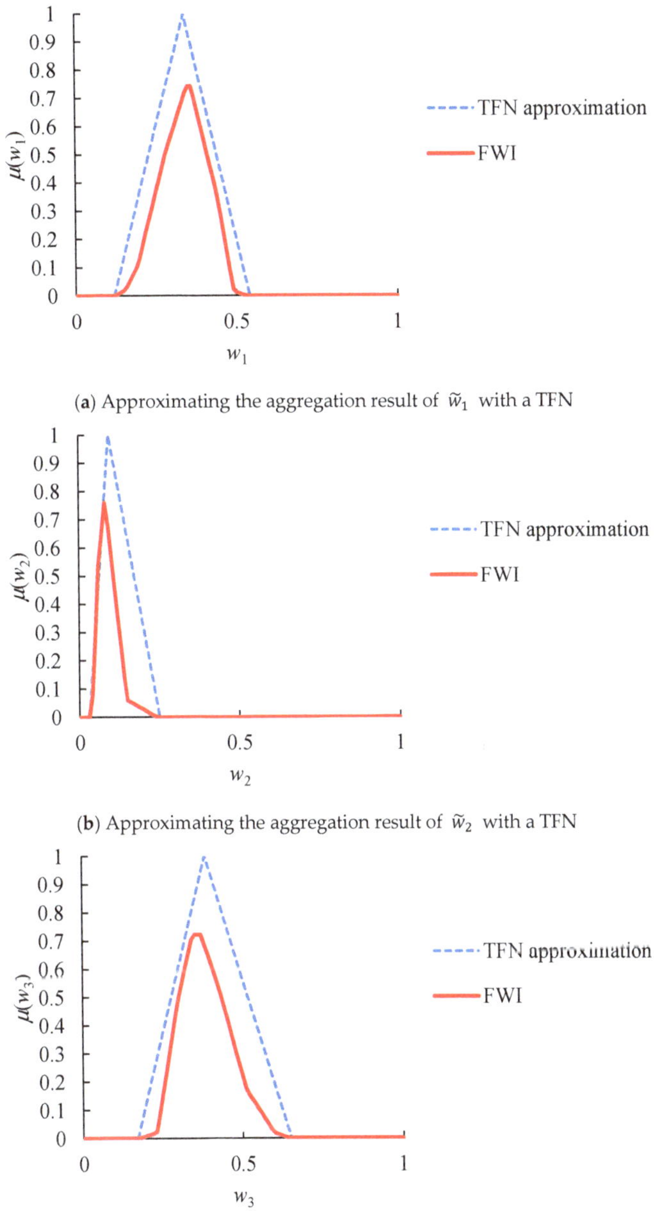

(**a**) Approximating the aggregation result of $\tilde{w}_1$ with a TFN

(**b**) Approximating the aggregation result of $\tilde{w}_2$ with a TFN

(**c**) Approximating the aggregation result of $\tilde{w}_3$ with a TFN

Figure 7. *Cont.*

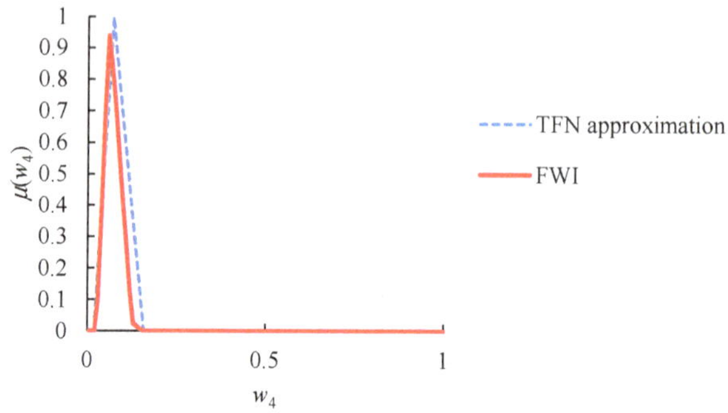

(**d**) Approximating the aggregation result of $\tilde{w}_4$ with a TFN

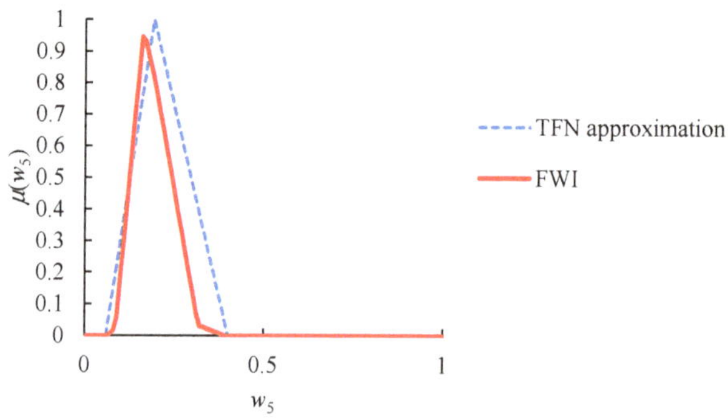

(**e**) Approximating the aggregation result of $\tilde{w}_5$ with a TFN

Figure 7. Approximating the aggregation result with a TFN.

Fuzzy VIKOR was then applied to assess and compare the overall performances of four leisure agricultural parks. The details of these leisure agricultural parks are summarized in Table 2. The performance of a leisure agricultural park was evaluated according to the rules depicted in Table 3. The evaluation results are summarized in Table 4. There was no perfect alternative.

Table 2. Leisure agricultural park details.

h	Area (m^2)	Major Agricultural Products	City	Number of Confirmed COVID-19 Cases *	Distance (min)	Image
1	2,120,000	Shiitake mushrooms, flowers	Taichung	202	57	• Fresh air, rich natural ecology, suitable for hiking
2	23,000	Strawberry	Miaoli	549	64	• Easy fruit picking, only seasonal
3	145,000	Orange, pitaya	Yunlin	22	54	• Cheap, time-consuming
4	500,000	Milk, dairy products, malt	Miaoli	549	57	• Abundant agricultural products and leisure activities, suitable for hiking

*: Since 1 January 2021.

Table 3. Rules for evaluating the performance of a leisure agricultural park.

Criterion	Rule
Image of the leisure agricultural park	$\tilde{p}_{h1}(x_{h1}) = \begin{cases} (0, 0, 1) & \text{if } x_{h1} = \text{"not very interesting (just for killing time)"} \\ (0, 1, 2) & \text{if } x_{h1} = \text{"somewhat interesting"} \\ (1.5, 2.5, 3.5) & \text{if } x_{h1} = \text{"interesting and somewhat healthy"} \\ (3, 4, 5) & \text{if } x_{h1} = \text{"interesting and healthy"} \\ (4, 5, 5) & \text{if } x_{h1} = \text{"very interesting and healthy (enjoyable)"} \end{cases}$ where x_{h1} is the image of the leisure agricultural park.
Number of confirmed COVID-19 cases in the city	$\tilde{p}_{h2}(x_{h2}) = \begin{cases} (0, 0, 1) & \text{if } 0.1 \cdot \min_r x_{r2} + 0.9 \cdot \max_r x_{r2} < x_{h2} \\ (0, 1, 2) & \text{if } 0.35 \cdot \min_r x_{r2} + 0.65 \cdot \max_r x_{r2} \le x_{h2} < 0.1 \cdot \min_r x_{r2} + 0.9 \cdot \max_r x_{r2} \\ (1.5, 2.5, 3.5) & \text{if } 0.65 \cdot \min_r x_{r2} + 0.35 \cdot \max_r x_{r2} \le x_{h2} < 0.35 \cdot \min_r x_{r2} + 0.65 \cdot \max_r x_{r2} \\ (3, 4, 5) & \text{if } 0.9 \cdot \min_r x_{r2} + 0.1 \cdot \max_r x_{r2} \le x_{h2} < 0.65 \cdot \min_r x_{r2} + 0.35 \cdot \max_r x_{r2} \\ (4, 5, 5) & \text{if } x_{h2} \le 0.9 \cdot \min_r x_{r2} + 0.1 \cdot \max_r x_{r2} \end{cases}$ where x_{h2} is the number of confirmed COVID-19 cases in the city.
Easiness to maintain social distance	$\tilde{p}_{h3}(x_{h3}) = \begin{cases} (0, 0, 1) & \text{if } x_{h3} \le 0.9 \cdot \min_r x_{r3} + 0.1 \cdot \max_r x_{r3} \\ (0, 1, 2) & \text{if } 0.9 \cdot \min_r x_{r3} + 0.1 \cdot \max_r x_{r3} \le x_{h3} < 0.65 \cdot \min_r x_{r3} + 0.35 \cdot \max_r x_{r3} \\ (1.5, 2.5, 3.5) & \text{if } 0.65 \cdot \min_r x_{r3} + 0.35 \cdot \max_r x_{r3} \le x_{h3} < 0.35 \cdot \min_r x_{r3} + 0.65 \cdot \max_r x_{r3} \\ (3, 4, 5) & \text{if } 0.35 \cdot \min_r x_{r3} + 0.65 \cdot \max_r x_{r3} \le x_{h3} < 0.1 \cdot \min_r x_{r3} + 0.9 \cdot \max_r x_{r3} \\ (4, 5, 5) & \text{if } 0.1 \cdot \min_r x_{r3} + 0.9 \cdot \max_r x_{r3} < x_{h3} \end{cases}$ where x_{h3} is the area of the leisure agricultural park.
Distance to the leisure agricultural park	$\tilde{p}_{h4}(x_{h4}) = \begin{cases} (0, 0, 1) & \text{if } 0.1 \cdot \min_r x_{r4} + 0.9 \cdot \max_r x_{r4} < x_{h4} \\ (0, 1, 2) & \text{if } 0.35 \cdot \min_r x_{r4} + 0.65 \cdot \max_r x_{r4} \le x_{h4} < 0.1 \cdot \min_r x_{r4} + 0.9 \cdot \max_r x_{r4} \\ (1.5, 2.5, 3.5) & \text{if } 0.65 \cdot \min_r x_{r4} + 0.35 \cdot \max_r x_{r4} \le x_{h4} < 0.35 \cdot \min_r x_{r4} + 0.65 \cdot \max_r x_{r4} \\ (3, 4, 5) & \text{if } 0.9 \cdot \min_r x_{r4} + 0.1 \cdot \max_r x_{r4} \le x_{h4} < 0.65 \cdot \min_r x_{r4} + 0.35 \cdot \max_r x_{r4} \\ (4, 5, 5) & \text{if } x_{h4} \le 0.9 \cdot \min_r x_{r4} + 0.1 \cdot \max_r x_{r4} \end{cases}$ where x_{h4} is the distance to the leisure agricultural park.
Preference for the leisure agricultural park	$\tilde{p}_{h5}(x_{h5}) = \begin{cases} (0, 0, 1) & \text{if } x_{h1} = \text{"very lowly preferred"} \\ (0, 1, 2) & \text{if } x_{h1} = \text{"lowly preferred"} \\ (1.5, 2.5, 3.5) & \text{if } x_{h1} = \text{"moderately preferred"} \\ (3, 4, 5) & \text{if } x_{h1} = \text{"highly preferred"} \\ (4, 5, 5) & \text{if } x_{h1} = \text{"very highly preferred"} \end{cases}$ where x_{h5} is the preference for the leisure agricultural park.

Table 4. Performances of the leisure agricultural parks.

h	$\widetilde{p}_{h1}$	$\widetilde{p}_{h2}$	$\widetilde{p}_{h3}$	$\widetilde{p}_{h4}$	$\widetilde{p}_{h5}$
1	(4, 5, 5)	(3, 4, 5)	(4, 5, 5)	(0, 1, 2)	(1.5, 2.5, 3.5)
2	(3, 4, 5)	(0, 0, 1)	(0, 0, 1)	(4, 5, 5)	(3, 4, 5)
3	(1.5, 2.5, 3.5)	(4, 5, 5)	(0, 0, 1)	(0, 0, 1)	(0, 1, 2)
4	(3, 4, 5)	(0, 0, 1)	(0, 1, 2)	(0, 1, 2)	(4, 5, 5)

Subsequently, the best and worst performances in optimizing each criterion were determined. The results are shown in Table 5.

Table 5. Best and worst performances in optimizing each criterion.

i	1	2	3	4	5
$\widetilde{p}_i^*$	(4, 5, 5)	(4, 5, 5)	(4, 5, 5)	(4, 5, 5)	(4, 5, 5)
$\widetilde{p}_i^-$	(1.5, 2.5, 3.5)	(0, 0, 1)	(0, 1, 2)	(0, 0, 1)	(0, 1, 2)

The normalized fuzzy distance between each leisure agricultural park and the best performance were measured. The measurement results are summarized in Table 6.

Table 6. Normalized fuzzy distance between each leisure agricultural park and the best performance.

h	$\widetilde{d}_{h1}$	$\widetilde{d}_{h2}$	$\widetilde{d}_{h3}$	$\widetilde{d}_{h4}$	$\widetilde{d}_{h5}$
1	(0, 0, 0.29)	(0, 0.2, 0.4)	(0, 0, 0.2)	(0.4, 0.8, 1)	(0.1, 0.5, 0.7)
2	(0, 0.29, 0.57)	(0.6, 1, 1)	(0.6, 1, 1)	(0, 0, 0.2)	(0, 0.2, 0.4)
3	(0.14, 0.71, 1)	(0, 0, 0.2)	(0.6, 1, 1)	(0.6, 1, 1)	(0.4, 0.8, 1)
4	(0, 0.29, 0.57)	(0.6, 1, 1)	(0.4, 0.8, 1)	(0.4, 0.8, 1)	(0, 0, 0.2)

The values of $\widetilde{S}_h$ and $\widetilde{R}_h$ were then computed for each leisure agricultural park. The results are summarized in Table 7. Based on them, the $\widetilde{Q}_h$ of the leisure agricultural park was derived by setting ξ to 0.5.

Table 7. The $\widetilde{S}_h$, $\widetilde{R}_h$ and $\widetilde{Q}_h$ of each leisure agricultural park.

h	$\widetilde{S}_h$	$\widetilde{R}_h$	$\widetilde{Q}_h$
1	(0.01, 0.17, 0.82)	(0.01, 0.1, 0.28)	(0, 0, 0.44)
2	(0.12, 0.61, 1.4)	(0.1, 0.38, 0.65)	(0, 0.35, 0.89)
3	(0.16, 0.85, 1.8)	(0.1, 0.38, 0.65)	(0, 0.41, 1)
4	(0.09, 0.55, 1.45)	(0.07, 0.31, 0.65)	(0, 0.27, 0.9)

The defuzzified values of these performance measures are summarized in Table 8. Based on the defuzzification results, the leisure agricultural parks were ranked, as shown in Table 8. Leisure agricultural park #1 achieved the lowest value of $\widetilde{Q}_h$, followed by leisure agricultural park #4.

Table 8. Defuzzification results.

H	$D(\widetilde{S}_h)$	$D(\widetilde{R}_h)$	$D(\widetilde{Q}_h)$	**Rank**
1	0.296	0.120	0.110	1
2	0.686	0.380	0.395	3
3	0.914	0.380	0.457	4
4	0.663	0.333	0.361	2

3.3. Discussion

According to the results of the case study, the following discussion was presented:

(1) The most suitable leisure agricultural park for the family was leisure agricultural park #1; it had the best image and was the easiest to maintain social distance.
(2) However, the superiority of leisure agricultural park #1 over leisure agricultural park #4 only met the second condition. Therefore, both leisure agricultural parks could be recommended to the family for their consideration.
(3) In contrast, leisure agricultural park #3 ranked last because the family showed the lowest preference for this leisure agricultural park.
(4) A parametric analysis has been conducted to examine the effect of ξ on the ranking result. The results are summarized in Table 9. The superiority of leisure agricultural park #1 over the others was not affected by the value of ξ. In addition, when ξ was set to zero, there was a tie between leisure agricultural parks #2 and #3.

Table 9. Results of the parametric analysis.

ξ	**Ranking Result**
0	1→4→2, 3
0.1	1→4→2→3
0.2	1→4→2→3
0.3	1→4→2→3
0.4	1→4→2→3
0.5	1→4→2→3
0.6	1→4→2→3
0.7	1→4→2→3
0.8	1→4→2→3
0.9	1→4→2→3
1.0	1→4→2→3

(1) The recommendation results to ten traveler groups and their choices are summarized in Table 10. As a result, the successful recommendation rate was 90%, high enough to support the effectiveness of the proposed methodology.
(2) Among the ten traveler groups, seven rated the easiness to maintain social distance as the most important criterion. In contrast, the distance to a leisure agricultural park was considered the least important criterion by most traveler groups.

Table 10. Recommendation results to ten traveler groups.

Group #	Recommendation	Choice
1	Leisure agricultural park #1	Leisure agricultural park #1
2	Leisure agricultural park #5	Leisure agricultural park #5
3	Leisure agricultural park #4	Leisure agricultural park #4
4	Leisure agricultural park #6	Leisure agricultural park #6
5	Leisure agricultural park #1	Leisure agricultural park #1
6	Leisure agricultural park #11	Leisure agricultural park #11
7	Leisure agricultural park #2	Leisure agricultural park #3
8	Leisure agricultural park #9	Leisure agricultural park #9
9	Leisure agricultural park #1	Leisure agricultural park #1
10	Leisure agricultural park #11	Leisure agricultural park #11

(1) Three existing fuzzy group decision-making methods were also applied to this case for comparison. The first was the FGM–FGM–fuzzy weighted average (FWA) method, in which the decision makers' fuzzy judgement matrixes were aggregated using FGM. Then, the fuzzy priorities of criteria were derived using FGM. Finally, the overall performance of each leisure agricultural park was evaluated using FWA. The second method was the FGM–FEA–FWA method, wherein FEA [31] was applied to derive the priorities of criteria in place of the FGM method. The third method was the FGM–FGM–FTOPSIS method, which was similar to the FGM–FGM–FWA method, except that fuzzy TOPSIS was employed to compare the overall performances of leisure agricultural parks. The results obtained using these methods are summarized in Table 11. It can be observed that the ranking results of leisure agricultural parks using existing methods were different from those using the proposed methodology, which is due to the imprecision of these existing methods in deriving the fuzzy priorities of criteria; for example, the fuzzy priorities of criterion $\widetilde{w}_5$ derived by decision maker #1 using various methods are compared in Table 12, showing a significant difference between these results.

Table 11. Ranking results using existing methods.

h	Rank (FGM-FGM-FWA)	Rank (FGM-FEA-FWA)	Rank (FGM-FGM-FTOPSIS)	Rank (Proposed Methodology)
1	1	1	1	1
2	2	2	2	3
3	4	4	4	4
4	3	3	3	2

Table 12. Fuzzy priorities of criterion $\widetilde{w}_5$ derived by decision maker #1 using various methods.

Method	$\widetilde{w}_5$
FGM	(0.079, 0.157, 0.309)
FEA	0.257
acFGM	(0.087, 0.165, 0.316)

4. Conclusions

Visiting leisure agricultural parks has always been an activity for people to relax and pursue health [43–46], especially during the COVID-19 pandemic. At the beginning of the pandemic, people were hesitant to go to leisure agricultural parks for fear of being infected. With the increasing popularity of vaccines, people began to resume this leisure activity. However, during the COVID-19 pandemic, many uncertain factors make choosing a suitable leisure agricultural park a complicated decision, especially for traveler groups. To solve this problem, a fuzzy collaborative intelligence approach is proposed in this study. In the proposed methodology, first, the acFGM method is devised to derive the fuzzy priorities of criteria. Subsequently, FWI is applied to aggregate the fuzzy priorities derived by all decision makers to consider their unequal levels of authority. Based on the aggregation result, the fuzzy VIKOR method is applied to compare the overall performances of leisure agricultural parks.

The proposed methodology has been applied to a case study to examine its effectiveness. The results of the case study are reported as follows:

(1) During the COVID-19 pandemic, the willingness of travelers (especially traveler groups) to go to a leisure agricultural park was quite high.
(2) In choosing a suitable leisure agricultural park, the most important criterion was the easiness to maintain social distance, while the least important criterion was the distance to a leisure agricultural park.
(3) Nine of ten traveler groups followed the recommendations, resulting in a successful recommendation rate of 90%.

The methodology proposed in this research has the following limitations:

(1) The easiness to maintain social distance is directly proportional to the area of a leisure agricultural park. Although such an evaluation method is simple, it may not be practical because in a leisure agricultural park, travelers will only go to part of the area.
(2) Although it is not difficult to write a program to implement the proposed methodology, the proposed methodology is slightly more complicated than some multi-criteria decision-making methods for similar purposes.

After the COVID-19 outbreak, many agricultural activities have encountered difficulties and must change; for example, in response to the shortage of manpower supply, should a farmland owner purchase automated agricultural machinery or change the agricultural products to reduce manpower requirements? The methodology proposed in this study can be applied to make these decisions. In addition, this study applies FWI to aggregate the preferences of decision makers with unequal levels of authority. In future research, different methods can also be proposed to fulfill the same purpose. These constitute some directions for future research.

Author Contributions: All authors equally contributed to the writing of this paper. Data curation, methodology and writing original draft: T.-C.T.C. and H.-C.W.; writing—review and editing: T.-C.T.C., H.-C.W. and Y.-C.L. All authors have read and agreed to the published version of the manuscript.

Funding: This research received no external funding.

Institutional Review Board Statement: Not applicable.

Informed Consent Statement: Informed consent was obtained from all subjects involved in the study.

Acknowledgments: This work was supported by ministry of science of technology of Taiwan.

Conflicts of Interest: The authors declare no conflict of interest.

References

1. Zhong, L.; Sun, S.; Law, R.; Li, X. Tourism crisis management: Evidence from COVID-19. *Curr. Issues Tour.* **2021**, *24*, 2671–2682. [CrossRef]
2. Schmöcker, J.D. Estimation of city tourism flows: Challenges, new data and COVID. *Transp. Rev.* **2021**, *41*, 137–140. [CrossRef]
3. Qiao, L.; Zhou, K.; Zhang, Y. From seed to feed: Organic food leisure park construction. *Nat. Environ. Pollut. Technol.* **2014**, *13*, 429.

4. Li, Z.; Zhang, X.; Yang, K.; Singer, R.; Cui, R. Urban and rural tourism under COVID-19 in China: Research on the recovery measures and tourism development. *Tour. Rev.* **2021**, *76*, 718–736. [CrossRef]
5. Lin, T.C.; Pao, T.P. Leisure activities' selection and motivation. *Int. J. Acad. Res. Bus. Soc. Sci.* **2011**, *1*, 308.
6. Huang, C.M.; Tuan, C.L.; Wongchai, A. Development analysis of leisure agriculture–A case study of Longjing Tea Garden, Hangzhou, China. *APCBEE Procedia* **2014**, *8*, 210–215. [CrossRef]
7. Puvača, N.; Lika, E.; Brkanlić, S.; Breso, E.; Ilić, D.; Kika, T.S.; Brkić, I. The pandemic of SARS-CoV-2 as a worldwide health safety risk. *J. Agron. Technol. Eng. Manag.* **2021**, *4*, 523–532.
8. Sivan, A. Leisure in times of COVID-19: Reflection on Hong Kong and Israel. *World Leis. J.* **2020**, *62*, 322–324. [CrossRef]
9. Du, J.; Floyd, C.; Kim, A.C.; Baker, B.J.; Sato, M.; James, J.D.; Funk, D.C. To be or not to be: Negotiating leisure constraints with technology and data analytics amid the COVID-19 pandemic. *Leis. Stud.* **2021**, *40*, 561–574. [CrossRef]
10. Hsiao, C.Y.; Tuan, C.L. How recreational farm operators use dynamic capabilities to respond outbreak of COVID-19 pandemic. *J. Outdoor Recreat. Tour.* **2021**, 100460. [CrossRef]
11. Pan, D.; Yang, J.; Zhou, G.; Kong, F. The influence of COVID-19 on agricultural economy and emergency mitigation measures in China: A text mining analysis. *PLoS ONE* **2020**, *15*, e0241167. [CrossRef] [PubMed]
12. Ateljevic, I. Transforming the (tourism) world for good and (re)generating the potential 'new normal'. *Tour. Geogr.* **2020**, *22*, 467–475. [CrossRef]
13. Putian, G. Analysis of the impacts of COVID-19 pandemic on leisure agriculture and rural tourism in Zhengzhou and the countermeasures. *Front. Soc. Sci. Technol.* **2020**, *2*, 89–96.
14. Barrot, J.-N.; Basile, G.; Sauvagnat, J. Sectoral effects of social distancing. *Covid Econ. Cent. Econ. Policy Res.* **2020**, *3*, 85–102. [CrossRef]
15. Feuerbacher, A.; McDonald, S.; Thierfelder, K. Peasant farmers and pandemics: The role of seasonality and labor-leisure trade-off decisions in economy-wide models. *Econ. Syst. Res.* **2021**. [CrossRef]
16. Langemeyer, J.; Madrid-Lopez, C.; Beltran, A.M.; Mendez, G.V. Urban agriculture—A necessary pathway towards urban resilience and global sustainability? *Landsc. Urban Plan.* **2021**, *210*, 104055. [CrossRef]
17. Saaty, T.L. Decision making with the analytic hierarchy process. *Int. J. Serv. Sci.* **2008**, *1*, 83–98. [CrossRef]
18. Zheng, G.; Zhu, N.; Tian, Z.; Chen, Y.; Sun, B. Application of a trapezoidal fuzzy AHP method for work safety evaluation and early warning rating of hot and humid environments. *Saf. Sci.* **2012**, *50*, 228–239. [CrossRef]
19. Chen, T.; Wang, Y.C. A calibrated piecewise-linear FGM approach for travel destination recommendation during the COVID-19 pandemic. *Appl. Soft Comput.* **2021**, *109*, 107535. [CrossRef]
20. Chen, T.C.T.; Wang, Y.C.; Lin, C.W. A fuzzy collaborative forecasting approach considering experts' unequal levels of authority. *Appl. Soft Comput.* **2020**, *94*, 106455. [CrossRef]
21. Wei, G. Some induced geometric aggregation operators with intuitionistic fuzzy information and their application to group decision making. *Appl. Soft Comput.* **2010**, *10*, 423–431. [CrossRef]
22. Xia, M.; Xu, Z.; Chen, N. Some hesitant fuzzy aggregation operators with their application in group decision making. *Group Decis. Negot.* **2013**, *22*, 259–279. [CrossRef]
23. Ashraf, S.; Abdullah, S. Spherical aggregation operators and their application in multiattribute group decision-making. *Int. J. Intell. Syst.* **2019**, *34*, 493–523. [CrossRef]
24. Wu, H.C.; Chen, T.C.T.; Huang, C.H.; Shih, Y.C. Comparing built-in power banks for a smart backpack design using an auto-weighting fuzzy-weighted-intersection FAHP approach. *Mathematics* **2020**, *8*, 1759. [CrossRef]
25. Van Broekhoven, E.; De Baets, B. Fast and accurate center of gravity defuzzification of fuzzy system outputs defined on trapezoidal fuzzy partitions. *Fuzzy Sets Syst.* **2006**, *157*, 904–918. [CrossRef]
26. Qin, J.; Liu, X.; Pedrycz, W. An extended VIKOR method based on prospect theory for multiple attribute decision making under interval type-2 fuzzy environment. *Knowl. -Based Syst.* **2015**, *86*, 116–130. [CrossRef]
27. Wang, H.; Pan, X.; He, S. A new interval type-2 fuzzy VIKOR method for multi-attribute decision making. *Int. J. Fuzzy Syst.* **2019**, *21*, 145–156. [CrossRef]
28. Opricovic, S. Fuzzy VIKOR with an application to water resources planning. *Expert Syst. Appl.* **2011**, *38*, 12983–12990. [CrossRef]
29. Chen, T. A FAHP-FTOPSIS approach for choosing mid-term occupational healthcare measures amid the COVID-19 pandemic. *Health Policy Technol.* **2021**, *10*, 100517. [CrossRef]
30. Zandi, P.; Rahmani, M.; Khanian, M.; Mosavi, A. Agricultural risk management using fuzzy TOPSIS analytical hierarchy process (AHP) and failure mode and effects analysis (FMEA). *Agriculture* **2020**, *10*, 504. [CrossRef]
31. Chang, D.Y. Applications of the extent analysis method on fuzzy AHP. *Eur. J. Oper. Res.* **1996**, *95*, 649–655. [CrossRef]
32. Chen, T.C.T. Evaluating the sustainability of a smart technology application to mobile health care: The FGM–ACO–FWA approach. *Complex Intell. Syst.* **2020**, *6*, 109–121. [CrossRef]
33. Chen, T.; Lin, Y.C.; Chiu, M.C. Approximating alpha-cut operations approach for effective and efficient fuzzy analytic hierarchy process analysis. *Appl. Soft Comput.* **2019**, *85*, 105855. [CrossRef]
34. Demirel, N.Ç.; Yücenur, G.N.; Demirel, T.; Muşdal, H. Risk-based evaluation of Turkish agricultural strategies using fuzzy AHP and fuzzy ANP. *Hum. Ecol. Risk Assess. Int. J.* **2012**, *18*, 685–702. [CrossRef]
35. Wang, B.; Song, J.; Ren, J.; Li, K.; Duan, H. Selecting sustainable energy conversion technologies for agricultural residues: A fuzzy AHP-VIKOR based prioritization from life cycle perspective. *Resour. Conserv. Recycl.* **2019**, *142*, 78–87. [CrossRef]

36. Amini, S.; Rohani, A.; Aghkhani, M.H.; Abbaspour-Fard, M.H.; Asgharipour, M.R. Assessment of land suitability and agricultural production sustainability using a combined approach (Fuzzy-AHP-GIS): A case study of Mazandaran province, Iran. *Inf. Process. Agric.* **2020**, *7*, 384–402. [CrossRef]
37. Tashayo, B.; Honarbakhsh, A.; Azma, A.; Akbari, M. Combined fuzzy AHP–GIS for agricultural land suitability modeling for a watershed in southern Iran. *Environ. Manag.* **2020**, *66*, 364–376. [CrossRef]
38. Chen, T.; Wang, Y.C. Recommending suitable smart technology applications to support mobile healthcare after the COVID-19 pandemic using a fuzzy approach. *Healthcare* **2021**, *9*, 1461. [CrossRef] [PubMed]
39. Sharif, M.; Khan, M.A.; Iqbal, Z.; Azam, M.F.; Lali, M.I.U.; Javed, M.Y. Detection and classification of citrus diseases in agriculture based on optimized weighted segmentation and feature selection. *Comput. Electron. Agric.* **2018**, *150*, 220–234. [CrossRef]
40. Chen, T.; Wang, Y.C.; Wu, H.C. Analyzing the impact of vaccine availability on alternative supplier selection amid the COVID-19 pandemic: A cFGM-FTOPSIS-FWI approach. *Healthcare* **2021**, *9*, 71. [CrossRef] [PubMed]
41. Mendas, A.; Delali, A. Support system based on GIS and weighted sum method for drawing up of land suitability map for agriculture. Application to durum wheat cultivation in the area of Mleta (Algeria). *Span. J. Agric. Res.* **2012**, *1*, 34–43.
42. Wu, H.C.; Wang, Y.C.; Chen, T.C.T. Assessing and comparing COVID-19 intervention strategies using a varying partial consensus fuzzy collaborative intelligence approach. *Mathematics* **2020**, *8*, 1725. [CrossRef]
43. Chiu, M.C.; Chen, T. Assessing mobile and smart technology applications for active and healthy aging using a fuzzy collaborative intelligence approach. *Cogn. Comput.* **2021**, *13*, 431–446. [CrossRef]
44. Wojcieszak-Zbierska, M.M.; Jęczmyk, A.; Zawadka, J.; Uglis, J. Agritourism in the era of the coronavirus (COVID-19): A rapid assessment from Poland. *Agriculture* **2020**, *10*, 397. [CrossRef]
45. Chen, T. A fuzzy ubiquitous traveler clustering and hotel recommendation system by differentiating travelers' decision-making behaviors. *Appl. Soft Comput.* **2020**, *96*, 106585. [CrossRef]
46. Deshpande, A. The COVID-19 pandemic and gendered division of paid work, domestic chores and leisure: Evidence from India's first wave. *Econ. Politica* **2021**. [CrossRef]

MDPI
St. Alban-Anlage 66
4052 Basel
Switzerland
Tel. +41 61 683 77 34
Fax +41 61 302 89 18
www.mdpi.com

Agriculture Editorial Office
E-mail: agriculture@mdpi.com
www.mdpi.com/journal/agriculture

www.ingramcontent.com/pod-product-compliance
Lightning Source LLC
Chambersburg PA
CBHW061511220926
51850CB00004B/227
9783036566306